Advances in Plant Cell Biology

Advances in Plant Cell Biology

Edited by **Agatha Wilson**

SYRAWOOD
PUBLISHING HOUSE
New York

Published by Syrawood Publishing House,
750 Third Avenue, 9th Floor,
New York, NY 10017, USA
www.syrawoodpublishinghouse.com

Advances in Plant Cell Biology
Edited by Agatha Wilson

International Standard Book Number: 978-1-68286-357-2 (Hardback)

Printed in China.

Contents

Preface

The main aim of this book is to educate learners and enhance their research focus by presenting diverse topics covering this vast field. This is an advanced book which compiles significant studies by distinguished experts. This book addresses successive solutions to the challenges arising in the area of application, along with it; the book provides scope for future developments.

The study of plant cell biology has enhanced our understanding of the functioning of plants. Diverse applications of this subject, such as in tissue culture, immunology, etc. have been possible today owing to the advances in the study of this discipline. For the students of botany, the understanding of plant cell biology is crucial. This book discusses the essential concepts as well as modern approaches for the study of this field. It also strives to provide in-depth knowledge about the structure and functioning of various types of plant cells. This book will also provide ample amounts of innovative topics for research, which academicians and interested readers can take up and contribute to the field of botany.

It was a great honour to edit this book, though there were challenges, as it involved a lot of communication and networking between me and the editorial team. However, the end result was this all-inclusive book covering diverse themes in the field.

Finally, it is important to acknowledge the efforts of the contributors for their excellent chapters, through which a wide variety of issues have been addressed. I would also like to thank my colleagues for their valuable feedback during the making of this book.

Editor

1

Seslerio uliginosae-Scorzoneretum purpureae (*Festuco-Brometea* class) in the Nida Basin (Małopolska Upland) after 90 years

Krystyna Towpasz, Alina Stachurska-Swakoń*
Institute of Botany, Jagiellonian University, Kopernika 27, 31-501 Kraków, Poland

Abstract

Seslerio uliginosae-Scorzoneretum purpureae is the only xerothermic community of the *Festuco-Brometea* class, where a rare species known as *Sesleria uliginosa* occurs in Poland. A couple of sites hosting this association in the Nida Basin were described in the 1920s and 1950s. All these stands are now situated in the areas designated as xerothermic nature reserves. The study aimed at describing the present-day characteristics of this rare association, the most likely being endemic in Poland, as well as the changes that have occurred in it. The species diversity and the contribution of xerothermic plants have decreased significantly in terms of their number and cover coefficient. Some of the diagnostic species of the association have become extremely rare. *Sesleria uliginosa* is the only one without visible changes in their dominant position in the community. Mosses, that have played a significant role before, were among the vanished species. At the same time, an increase in number and cover of mesophilous meadows species was observed. Mesophilous grasses belong to the group of species growing in significance in the phytocoenoses. The observed changes in species composition were reflected in a statistically significant increase of nitrogen indicator value. The slow process of succession observed in the *Seslerio-Scorzoneretum purpureae* could be attributed to the lack of appropriate land use, e.g. mowing and grazing that have been ceased in the nature reserves.

Keywords: *Sesleria uliginosa*, gypsum grasslands, xerothermic grasslands, species diversity, succession, steppe reserves, ecological indicators, Nida Basin

Introduction

Sesleria uliginosa Opiz is a north-central European grass species occurring mainly in the lowlands and uplands of Europe [1]. The species is recorded in the red lists of vascular plants due to its rare occurrence and the considerable threat to habitats, in which this species occurs: e.g. in Poland, Slovakia, Czech Republic, Latvia, and Romania [2–4]. It could be found in mire vegetation or wet meadows, and rarely in dry calcareous or gypsum grasslands. The rich fen communities with this taxon are classified within the *Caricetalia davallianae* order as *Seslerietum uliginosae* (Hungary [5], Slovakia [6–8]) or *Caricetum davalianae* (Slovakia [9–11], Romania [12]). In Bulgaria the species was found in the *Junco subnodulosi–Schoenetum nigricantis* association [13]. It could occur also in *Molinietum caeruleae* (*Molinietalia* order) as it was found in the Vienna Basin [14] and Scandinavia [15]. Dry calcareous grasslands with *Sesleria uliginosa* are rarely found [11] and are classified as *Deschampsio mediae–Seslerietum uliginosae* [16], or are placed in the *Bromion erecti* alliance of the *Festuco–Brometea* class [17].

Sesleria uliginosa rarely occurs in Poland. It is known from the Małopolska Upland and a few localities occur in the Śląsko–Krakowska Upland [18]. It is encountered in two types of habitats, that are extreme in terms of humidity: wet habitats (which include fens from the *Caricetalia davallianae* order and wet meadows from the *Molinietalia* order) and dry habitats (xerothermic grasslands of the *Festuco–Brometea* class). The presence of wet meadows from the *Molinietalia* order with the contribution of *Sesleria uliginosa* in the area of the Małopolska Upland, was emphasized by Medwecka-Kornaś [19] who, on the basis of representative relevés, described them provisionally as "the community with *Sesleria uliginosa*". Later, communities from the *Caricetalia davallianae* order were studied in detail by Głazek [20,21]. Towpasz and Stachurska-Swakoń [22] studied the extant sites of this species in fen communities, as well as changes that had occurred within them after 25 years. Recently, the occurrence of wet meadows with *Sesleria uliginosa* (*Molinietalia* order) was reported from the Śląsko-Krakowska Upland [23].

The association *Seslerio uliginosae-Scorzoneretum purpureae* was recognized as the only xerothermic community in Poland where *Sesleria uliginosa* occurs (placed in the *Cirsio-Brachypodion pinnati* alliance in the *Festucetalia valesiacae* order of the *Festuco-Brometea* class). The community was described as early as in 1928 by Kozłowska [24] and it verified

* Corresponding author. Email: alina.stachurska-swakon@uj.edu.pl

by Medwecka-Kornaś in 1959 [19]. Kozłowska [24] reported this community from the villages of Chotel Czerwony, Skorocice, and Aleksandrów in the Nida Basin of the Małopolska Upland. Medwecka-Kornaś [19] confirmed the occurrence of this community at the two first sites, and additionally recorded its presence in Winiary. The association only appeared in small patches, on the steep and exclusively northern (north-eastern, north-western) slopes of gypsum hills. The patches have developed on humic rendzinas [19]. The few other localities of this association, however without phytosociological documentation, were reported by Medwecka-Kornaś [25] and Łuszczyński and Łuszczyńska [26] from the Nida Basin. All sites of the *Seslerio uliginosae-Scorzoneretum purpureae* association known up to the 1950s were enclosed in areas designated xerothermic nature reserves. Towpasz and Stachurska-Swakoń found new localities of this rare association during field studies in the Nida Basin in years 2009–2010 [27]. The patches were recorded in the surrounding of the villages of Sielec, Skotniki, Sułkowice, and Wiślica. The exclusive occurrence of *Sesleria uliginosa* (against *S. caerulea*) on the area of the Nida Basin was confirmed on the base of cytological and taxonomical studies [19,27,28].

The aim of this paper is to assess the possibility of maintaining the xerothermic grasslands with *Sesleria uliginosa* in the light of the changes that have occurred since the first description of this community.

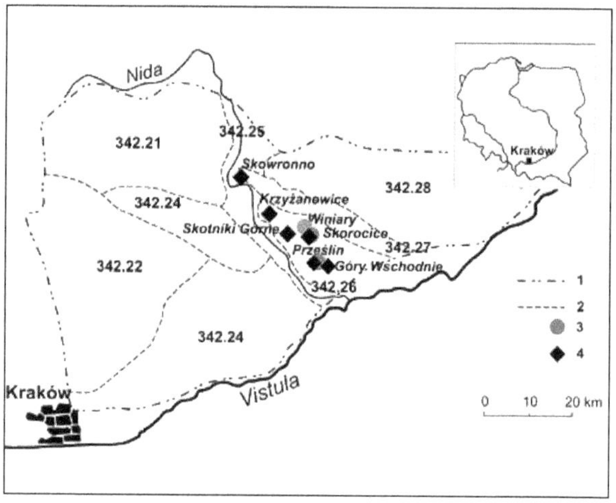

Fig. 1 Location of studied phytocoenoses of *Seslerio-Scorzoneretum* in the Nida Basin, the Małopolska Upland. 1 – border of the Nida Basin; 2 – borders of physiographical units after Kondracki [56]: 342.21– Jędrzejów Plateau, 342.22 – Miechów Upland, 342.23 – Proszowice Basin, 342.24 – Wodzisław Hummock, 342.25 – Nida Valley, 342.26 – Solec Basin, 342.27 – Pińczów Hummock, 342.28 – Połaniec Basin; 3 – relevés made in 1920s, 1959s; 4 – relevés made in 2010.

Material and methods

The phytosociological materials used for the study came from three periods: (***i***) 1920s – the first description of the community by Kozłowska [24] (5 relevés); (***ii***) 1950s – study of Medwecka-Kornaś [19] (7 relevés); (***iii***) 2009–2010 – study of Towpasz and Stachurska-Swakoń [27] (11 relevés).

As the association develops in small patches in very distinctive habitat conditions (northern slopes of gypsum hills in the Nida Basin) and according to available descriptions of the sites, it is highly probable that the relevés were made in the same localities. In the presented study the authors decided to include also a few relevés made by them that originated from other localities; namely nature reserves dedicated to xerothermic vegetation (Fig. 1). The phytocoenoses have a similar character in terms of habitat conditions, management and species composition [27]. Generally, the analysed relevés come from the following nature reserves: "Przęślin", "Skorocice", "Winiary" [19,24,27] and "Góry Wschodnie", "Krzyżanowice", "Skotniki Górne", and "Skowronno" [27].

The similarities between relevés were calculated using the UPGMA classification with cover-abundance values (Ružicka coefficient) and the SYNTAX 2000 software package [29]. Detrended Correspondence Analysis (DCA, MVSP, [30]) was applied with the cover-abundance values, in order to define the main dynamic tendencies along the time gradient. Correlations between DCA axes and selected variables measured for individual relevés were calculated using Kendal's rank correlation coefficient [31].

The synoptic table was made for three periods with constancy degree (frequency of every species occurrence, S) and index of coverage (*D*), that was calculated as total of the mean percentages of coverage of the given species in all relevés, divided by the total number of relevés and multiplied by 100 [32]. The value of that index is independent of the number of relevés and together with the constancy degree and abundance, give a good estimation of the role of species in the communities. The diagnostic value of species follows the work of Medwecka-Kornaś [19], Medwecka-Kornaś et al. [33], and Matuszkiewicz [34].

A total and mean number (per relevé) of diagnostic species from the syntaxonomical units (*Cirsio-Brachypodion pinnati, Festucetalia valesiacae, Festuco-Brometea, Trifolio-Geranietea, Molinio-Arrhenatheretea, Quercetalia pubescentis*) were calculated for each time period. Additionally, the species were divided into the following functional groups: forbs, grasses, and mosses. In order to characterize the environmental preferences of the species composing the phytocoenoses in different time periods, mean ecological indicator values were calculated (light, moisture, reaction, nitrogen; [35]). Diversity indices (the Shannon-Wiener and evenness index) were calculated using the MVSP software [30]. To estimate the statistical significance of differences between the characteristics of three research periods, the non-parametric Kruskal-Wallis test was used.

The nomenclature of vascular plants follows Mirek et al. [36], the nomenclature of bryophytes follows Ochyra et al. [37].

Results

The contemporary field studies approve the distinctiveness and rarity of the *Seslerio-Scorzoneretum purpureae* association in the Nida Basin. The xerothermic grasslands with *Sesleria uliginosa* have always been found on exposed northern, north-western and north-eastern slopes on gypsum hills. The association developed on small areas (8–300 m^2), and was always well bordered with surrounding communities. *Sesleria uliginosa* was one of the dominant species and covered 40–60% of the phytocoenosis. It was accompanied by: *Adonis vernalis, Avenastrum pratense, Brachypodium pinnatum, Campanula sibirica, Thesium linophyllon, Peucedanum oreoselinum* and *Salvia pratensis* (Tab. 1).

The UPGMA dendrogram revealed three clusters of relevés that correspond with the time of relevés origin (Fig. 2). The results indicate a higher similarity between groups of relevés from the 1920s and 1950s, when these areas were managed (probably grazed), as compared to relevés from 2010. Additionally, considering the level of similarity within groups, the similarity coefficients are highest among relevés from the 1920s and lowest among relevés from 2010.

While comparing the relevés made in the 1950s with the present-day, changes in the floristic composition of the *Seslerio-Scorzoneretum purpureae* association can be recorded. The most striking is the smaller number of xerothermic species occurring in particular phytocoenoses (Tab. 1, Tab. 2 and Tab. 3). Out of eight species diagnostic of the association, only *Sesleria uliginosa* remained with a similar cover. Other diagnostic species became rare, especially, *Scorzonera purpurea* which seldom occurred in the patches. The visible changes in floristic composition concern the group of species diagnostic of the *Festucetalia valesiacae* order and the *Festuco-Brometea* class. Many species, once common, no longer occur, such as *Campanula glomerata, Scabiosa canescens, Veronica spicata*. Some taxa, such as *Astragalus danicus* were less frequent. Overall, the total number of species diagnostic of the *Festuco-Brometea* class changed from 56 in the 1950s to 46 in 2010 (Tab. 2). The changes also concerned the *Trifolio-Geranietea* species. We did not find *Anemone sylvestris*, and other taxa were less frequent. The differences in the diagnostic species of xerothermic grasslands were statistically significant over the study period (Tab. 3). It is worth noting an increase of grasses in contemporary phytocoenoses in terms of number of species, degree of cover and index of coverage (Tab. 1, Tab. 2). To the group with a markedly higher coverage index belong: *Avenula pratensis, Brachypodium pinnatum, Briza media* and *Dactylis glomerata*. At the same time, the role of mesophilous species has increased recently as they were characterised by a higher index of coverage. Additionally, single specimens of shrubs (*Crataegus* sp., *Frangula alnus*) were noted in the contemporary patches of phytocoenoses.

A considerable reduction was apparent in the bryophytes layer. The mosses profusely covered the patches (40–60%) and their occurrence was regarded as a characteristic feature of the association. Currently, the coverage of mosses does not exceed 10%. *Abietinella abietina, Homalothecium lutescens* or *Thuidium phillibertii* were among the species now absent.

Diversity indices decreased between 1950 and 2010 (Tab. 3). The number of species recorded in the last period in a single patch was smaller, while at the same time, the cover of the herbaceous layer was higher when compared to the 1950s. The mean indicator value for nitrogen increased with statistical significance, whereas the indicator value for reaction decreased between 1950s and 2010.

The diagram of the DCA showed three groups of relevés as revealed by the 1st Axis. These groups of relevés correspond to the time of their origin (Fig. 3).

Relevés made in the 1920s and in 2010 are located at the opposite ends of the 1st Axis. The diagram also points to the higher homogeneity within the relevés from 1920s and the considerably lower levels present in the relevés from 2010. The indicator of the *Festuco-Brometea* species was most strongly correlated with the 1st DCA Axis (Tab. 4). The only statistically significant correlation with the 2nd Axis was a variable representing the value of the nitrogen indicator.

Tab. 1 The synoptic table of *Seslerio uliginosae-Scorzoneretum purpureae* from 1920 to 2010 in the Nida Basin.

	A		B		C	
	S	*D*	*S*	*D*	*S*	*D*
Ch.(*), D. Ass. *Seslerio uliginosae-Scorzoneretum purpureae*						
Sesleria uliginosa	V	2300	V	3143	V	3909
Galium boreale (reg.)	V	750	V	717	II	136
**Asperula tinctoria*	.	.	V	2536	I	318
** Viola rupestris*	V	1000	IV	326	.	.
** Scorzonera purpurea*	III	202	III	144	I	1
Valeriana angustifolia	.	.	III	74	I	46
** Thalictrum simplex*	I	2	III	4	.	.
Ch. All. *Cirsio-Brachypodion pinnati*						
Veronica austriaca	IV	204	IV	77	I	1
Campanula sibirica	III	202	III	6	II	3
Elymus hispidus	.	.	III	76	.	.
Senecio integrifolius	.	.	III	6	I	46
Seseli annuum	V	206	III	76	.	.
Inula ensifolia	.	.	II	894	I	682
Melampyrum arvense	.	.	I	1	III	94
Carex flacca	.	.	.	.	II	206
Ch. O. *Festucetalia valesiacae*						
Adonis vernalis	.	.	V	79	IV	95
Astragalus danicus	V	402	V	80	I	2
Thesium linophyllon	.	.	III	254	IV	1455
Thymus austriacus	.	.	III	4	.	.
Thymus kosteleckyanus	.	.	.	.	III	455
Erysimum odoratum	.	.	III	74	I	91
Potentilla arenaria	V	402	III	4	.	.
Scabiosa ochroleuca	.	.	III	4	I	1
Thymus glabrescens	.	.	V	79	I	1
Achillea pannonica	.	.	III	6	.	.
Onobrychis arenaria	.	.	III	74	.	.
Ornithogalum collinum	.	.	III	76	.	.
Bromus inermis	.	.	II	3	I	1
Salvia verticillata	.	.	II	3	I	46
Stipa joannis	.	.	II	3	I	1
Galium valdepilosum	.	.	.	.	III	252
Hieracium bauhinii	.	.	.	.	II	48
Ch. Cl. *Festuco-Brometea*						
Euphorbia cyparissias	V	500	V	9	IV	797
Salvia pratensis	.	.	V	149	V	458
Filipendula vulgaris	IV	302	V	716	IV	546
Trifolium montanum	V	750	V	896	IV	639
Centaurea scabiosa	IV	8	V	219	II	47
Polygala comosa	V	402	V	9	II	3
Carex humilis	V	1402	V	2143	I	45
Brachypodium pinnatum	.	.	IV	644	V	1841
Anthyllis vulneraria	V	1500	IV	7	II	48
Plantago media	IV	302	IV	77	II	48
Avenula pratensis	II	200	IV	147	III	683
Dianthus carthusianorum	IV	302	IV	77	III	138
Scabiosa canescens	.	.	IV	7	.	.
Festuca rupicola	V	1900	II	73	IV	1364
Arabis hirsuta	V	10	III	4	.	.
Ranunculus bulbosus	V	402	III	74	II	3
Campanula glomerata	.	.	III	144	.	.

Tab. 1 *(continued)*

Phleum phleoides	II	200	II	3	II	92
Carex caryophyllea	III	300	II	3	.	.
Veronica spicata	.	.	II	3	.	.
Asperula cynanchica	V	206	I	1	III	795
Centaurea stoebe	.	.	.	.	II	48
Onobrychis viciifolia	.	.	I	1	III	5
Abietinella abietina (ml)	V	652	V	826	.	.
Hypnum cupressiforme (ml)	V	10	II	3	.	.
Ch. Cl. *Trifolio-Geranietea sanquinei*						
Medicago falcata	.	.	V	150	III	95
Fragaria viridis	.	.	V	9	III	411
Thalictrum minus	III	202	IV	396	I	160
Galium verum	I	2	IV	256	II	205
Anemone sylvestris	.	.	IV	823	.	.
Peucedanum cervaria	.	.	III	323	II	477
Veronica teucrium	.	.	II	73	I	45
Agrimonia eupatoria	.	.	II	3	II	47
Ch. Cl. *Quercetalia pubescenti-petraeae*						
Primula veris	IV	400	V	576	I	46
Viola hirta	.	.	II	73	.	.
Ch. Cl. *Molinio-Arrhenatheretea*						
Dactylis glomerata	.	.	IV	7	V	365
Galium mollugo	.	.	III	6	I	160
Lotus corniculatus	II	4	II	3	I	46
Leucanthemum vulgare	IV	552	I	1	II	47
Knautia arvensis	V	500	I	1	II	4
Ranunculus acris	.	.	III	4	.	.
Leontodon hispidus	.	.	II	3	.	.
Others						
Briza media	III	104	IV	7	IV	412
Peucedanum oreoselinum	V	402	III	254	II	705
Pimpinella saxifraga	.	.	III	4	II	47
Homalothecium lutescens (ml)	V	500	V	1327	.	.
Thuidium phillibertii (ml)	.	.	V	1217	.	.
Rhytidium rugosum (ml)	V	1250	III	1393	I	.
Fissidens taxifolius (ml)	III	.	II	.	I	.
Plagiomnium affine (ml)	.	.	II	.	I	.

Source of relevés: A – 1920s [24]; B – 1950s [19]; C – 2010 [27]. *D* – index of coverage; ml – moss layer; reg. – regional; *S* – constancy.

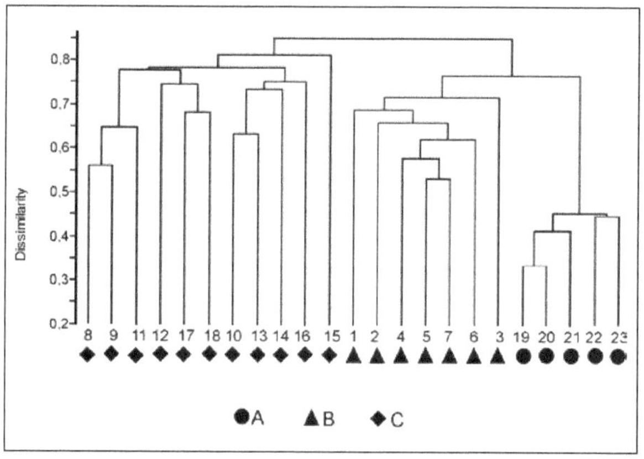

Fig. 2 UPGMA dendrogram based on cover-abundance scale (Ružicka coefficient) for relevés of *Seslerio uliginosae-Scorzoneretum purpureae* made between 1920 and 2010 in the Nida Basin: A – relevés made in 1920s; B – relevés made in 1950s; C – relevés made in 2010.

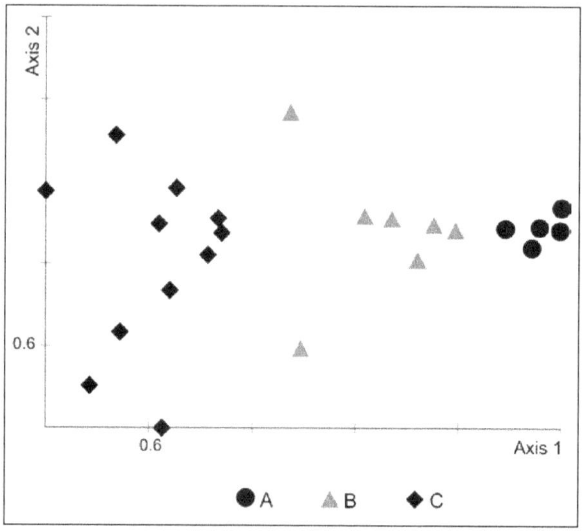

Fig. 3 Ordination of *Seslerio uliginosae-Scorzoneretum purpureae* relevés made between 1920 and 2010 along first two DCA axes (Axis 1 = 15.22%, Axis 2 = 6.98%). A – relevés made in 1920s; B – relevés made in 1950s; C – relevés made in 2010.

Discussion

Analysis of changes in the species composition revealed the process of slow succession, which is currently underway in the phytocoenoses of *Seslerio uliginosae-Scorzoneretum purpureae*. The species richness of the community has decreased. The diagnostic species of the association, *Cirsio-Brachypodion pinnatii* and *Festucetalia valesiacae* have either vanished or become less frequent. This process can be related to the lack of appropriate land use that was stopped or reduced, after the establishment of nature reserves. The majority of the steppe reserves in the area of the Nida Basin (like "Przęślin", "Winiary") was established in 1960. The steppe reserve "Skorocice" was under legal protection from 1947 but the area was used by the local population until the late 1950s [19]. According to localities of the *Seslerio-Scorzoneretum*, it could be assumed that grazing is needed for maintaining the floristic composition of the patches, known from time the community was described. The succession of xerothermic grasslands, caused by lack of appropriate land use, has been observed in different parts of both Poland and Europe (e.g. [38–44]). As the xerothermic species are characterized by a transient seed bank [45–47], the lack of a suitable regeneration niche could lead to a reduced diversity of species. Various experiments have shown that for maintaining calcareous grassland, which has similar habitat conditions to gypsum grasslands, there is a need for different management practices such as grazing, mowing, formation of gaps or even sowing seeds [48–53]. The results of experimental grazing by goats and sheep on the Góra Zborów Nature Reserve (Kraków-Częstochowa Upland), highlighted the fact that hill grazing

Tab. 2 Characteristics of the *Seslerio uliginosae-Scorzoneretum purpureae* from 1920 to 2010 in the Nida Basin.

	A	B	C
Number of relevés	5	7	11
Cover of herb layer (mean value, %)	n.d.	88	99
Cover of moss layer (mean value, %)	n.d.	59	1
Vascular plant species without grass	45	87	74
Grass species	6	13	16
Shrub species	0	0	3
Moss species	7	11	4
Diagnostic species groups			
Ch., D. Ass. *Seslerio-Scorzoneretum*	5	7	5
Ch. All. *Cirsio-Brachypodion pinnati*	3	7	7
Ch. All., O. *Festucetalia valesiaceae*	5	15	14
Ch. Cl. *Festuco-Brometea*	18	26	21
Ch. Cl. *Trifolio-Geranietea*	2	11	12
Ch. O. *Quercetalia pubescentis*	3	3	2
Ch. Cl. *Molinio-Arrhenatheretea*	3	10	9
Others	14	31	25

Values for species groups are given as total numbers. A – 1920s [24]; B – 1950s [19]; C – 2010 [27]. n.d. – lack of data.

Tab. 3 Mean values of characteristics of *Seslerio uliginosae-Scorzoneretum purpureae* in the Nida Basin in last 90 years.

	A	B	C	p_{AB}	p_{AC}	p_{BC}
Ellenberg indicator values						
Light	7.45	7.32	7.36	<0.01	ns	ns
Moisture	3.01	3.08	3.18	ns	ns	ns
Reaction	5.81	6.11	5.62	ns	ns	<0.05
Nitrogen	2.03	2.37	2.60	<0.05	<0.05	<0.05
Shannon index	3.48	3.57	2.99	ns	<0.01	<0.01
Evenness index	0.96	0.96	0.93	ns	<0.01	<0.05
Mean number of vascular plant species	44	50	26	ns	<0.01	<0.01
Mean number of mosses	6	5	2	ns	<0.01	<0.01
Ch., D. Ass. *Seslerio uliginosae-Scorzoneretum purpureae*	6	6	2	<0.01	ns	<0.001
Ch. All. *Cirsio-Brachypodion pinnati*	2	3	2	ns	ns	<0.05
Ch. O. *Festucetalia valesiaceae*	4	8	4	<0.01	ns	<0.05
Ch. Cl. *Festuco-Brometea*	15	14	9	ns	<0.01	<0.01
Ch. Cl. *Molinio-Arrhenatheretea*	2	3	2	ns	ns	ns
Ch. Cl. *Trifolio-Geranietea sanquinei*	1	6	3	<0.01	<0.05	<0.05
Others	12	9	4	ns	<0.01	<0.01

A – 1920s [24]; B – 1950s [19]; C – 2010 [27]. ns – statistically not significant; p_{AB}, p_{AB}, p_{AC} – probabilities on the base of Kruskall-Wallis test.

Tab. 4 Rank correlation (value of Kendall's τ) between 1st and 2nd axis DCA and some variables of the *Seslerio uliginosae-Scorzoneretum purpureae* association.

Variables	Axis 1	Axis 2
Number of species	0.536***	0.075
Cover of herb layer	−0.621***	−0.161
Cover of moss layer	0.584***	0.009
Ellenberg indicator values		
Light	0.091	−0.265
Moisture	−0.259	0.211
Reaction	0.051	0.091
Nitrogen	−0.515***	0.237*
Shannon index	0.581***	0.083
Evenness index	0.549***	0.087
Diagnostic species group		
Ch., D. Ass. *Seslerio uliginosae-Scorzoneretum*	0.651***	0.182
Ch. All. *Cirsio-Brachypodion pinnati*	0.168	0.245
Ch. O. *Festucetalia valesiaceae*	0.303*	0.045
Ch. Cl. *Festuco-Brometea*	0.666***	0.012
Ch. Cl. *Molinio-Arrhenatheretea*	0.141	0.049
Ch. Cl. *Quercetalia pubescenti-petraeae*	0.449**	−0.071
Ch. Cl. *Trifolio-Geranietea sanquinei*	−0.021	0.248
Others	0.657***	−0.033

* $0.01 < p < 0.05$. ** $0.001 < p < 0.01$. *** $p < 0.001$.

creates favourable conditions for plants and animals typical of xerothermic grasslands [48].

As in the presented study, an increase in the cover of high grasses and some perennials during the succession of calcareous grasslands is often reported [38,39,44]. The fast expansion of these species linked with excessive fertilisation of grasslands with nitrogen compounds from the air, is regarded as one of the principal causes of the reduction of the richness of the calcareous grasslands [54]. On the other hand, species with extended vegetative spread have a greater chance to develop and dominate the vegetation. The presence of these plants through the type of growth and development of thick felt, as well as the phenomenon of allelopathy, bring about the reduced possibility of germination for other species. In particular, the calcareous species have difficulty germinating and establishing themselves in thick litter layer and dense vegetation. Very helpful in the formation of gaps are animals such as goats or sheep.

In the course of succession in the old grasslands, the increase in number and coverage of shrubs is usually reported (e.g. [44,49]). Their occurrence often causes a series of changes in species composition. The high density of shrubs result in herb layer being overshadowed, which prevent heliophilous xerothermic species from penetrating and becoming established in such places. However, in our study only single specimens of shrubs were observed. This result could be connected with the conditions of habitats where phytocoenoses of *Seslerio-Scorzoneretum* occur. The shallow soil layer on the northern, rather steep slopes on gypsum hills, and the long distance from source of shrub and tree seeds, could be the reason for the slow establishment of shrubs and trees. Specific chemical and moisture features of gypsum soils are likely to inhibit the growth of tree and shrub seedlings [19]. The significance of soil thickness for maintaining species richness in calcareous grasslands was stressed [38]. Strong influence of climatic conditions, such as temperature or precipitation during the early spring period, was reported in studies where small changes in calcareous grasslands were observed [55].

Our results suggest that without appropriate management, we could expect the change of *Seslerio-Scorzoneretum purpureae* to mesophilous grasslands with a dominance of tall grasses and forbs. However, *Sesleria uliginosa* does not seem to be a threatened species. The occurrence of this grass creating dense tussocks is highly connected with the presence of calcium in the substratum, whereas the moisture conditions are not a limitation for the species. It is believed that *Sesleria uliginosa* belong to the "core species" of the described community, with the tussock form and slow rate of growth.

Acknowledgements

We thank Prof. Anna Medwecka-Kornaś, Dr hab. Józef Mitka and two anonymous reviewers for their insightful comments on the earlier versions of this manuscript.

References

1. Hultén E, Fries M. Atlas of North European vascular plants: north of the Tropic of Cancer. Königstein: Koeltz; 1986. (vol 1).
2. Čeřovský J, Feráková V, Holub J, Maglocký Š, Prochazká F. Cervená kniha ohrozených a vzácnych druhov rastlín a zivocíchov SR a CR. Vyššie rastliny. Bratislava: Príroda; 1991. (vol 5).
3. Witkowski ZJ, Król W, Solarz W, editors. Carpathian list of endangered species. Vienna: WWF and Institute of Nature Conservation, Polish Academy of Sciences; 2003.
4. Zarzycki K, Szeląg Z. Red list of the vascular plants in Poland. In: Mirek Z, Zarzycki K, Wojewoda W, Szeląg Z, editors. Red list of plants and fungi in Poland. Kraków: W. Szafer Institute of Botany, Polish Academy of Sciences; 2006. p. 11–20.
5. Tímár G. Új florisztikai adatok a Soproni-hegységből. Soproni Szemle. 1996;50(4):348–356.
6. Válek P. Další příspěvek k poznaní ekologických poměrů porostů pěchavy bažinné (*Seslerietum uliginosae* Klika) v severovýchodních Čechách. Věst Král Čs Spol Nauk Cl Math-Nath. 1951;5:1–10.
7. Válek P. Půdní vlastnosti subasociací *Seslerietum uliginosae-Caricetosum pulicaris* a *Molinietum coerulae-Caricetosum distantis* u rybníku "Kopičák" u Chlumce n. Cidl. Preslia. 1959;31:14–19.
8. Hájek M, Háberová I. *Scheuchzerio-Caricetea fuscae* R. Tx. 1937. In: Valachovič M, editor. Vegetácia mokradí. Bratislava: Veda; 2001. p. 187–273. (vol 3).
9. Bosáčková E. Ochranársky výskum močiarných biocenóz Turčianskej kotliny (vegetačné pomery význačnejších lokalít). Čs Och Prír. 1974;14:59–102.
10. Škovirová K. Rastlinné spoločenstvá Kláštorských lúk a dolného toku Turca. Kmetianum. 1974;3:205–233.
11. Dítě D, Hájek M, Hájková P. Formal definitions of Slovakian mire plant associations and their application in regional research. Biologia. 2007;62(4):400–408.
12. Coldea GH. Prodrome des associations végétales des Carpates du sud-est (Carpates Roumaines). Documents Phytosociologiques, NS. 1991;13:317–539.
13. Hájek M, Hájková P, Apostolova I. New plant associations from Bulgarian mires. Phytol Balcan. 2008;14(3):377–399.
14. Wagner H. Das *Molinietum caerulae* (Pfeifengraswiese) im Wiener Becken. Vegetatio. 1949;2:128–165.
15. Dierßen K. Vegetation Nordeuropas. Stuttgart: Ulmer; 1996.
16. Trinajstić I. Nomenklaturno-sintaksonomska revizija asocijacije "*Deschampsietum mediae illyricum*" (Zeidler) H-ic 1963 u Hrvatskoj. Agronomski glasnik. 2004;6:401–411.
17. Škovirová K, Očka S. Floristický príspevok k Dielniciam pri Kláštore pod Znievom. Kmetianum. 2005;10:105–130.
18. Zając A, Zając M, editors. Atlas rozmieszczenia roślin naczyniowych w Polsce. Kraków: Laboratory of Computer Chorology, Institute of Botany, Jagiellonian University; 2001.
19. Medwecka-Kornaś A. Roślinność rezerwatu stepowego "Skorocice" koło Buska. Ochr Przyr. 1959;26:1–260.
20. Głazek T. *Ctenidio molluscae-Seslerietum uliginosae* Klika 1943 em. Głazek 1983 – a new association for Poland. Acta Soc Bot Pol. 1984;53(4):575–583.
21. Głazek T. *Lipario-Schoenetum ferruginei* – a new plant association. Fragm Flor Geobot. 1992;37(2):549–562.
22. Towpasz K, Stachurska-Swakoń A. Występowanie *Sesleria uliginosa* (Poaceae) w zbiorowiskach z rzędu *Caricetalia davallianae* na obszarze Niecki Nidziańskiej (Wyżyna Małopolska). Fragm Flor Geobot Polonica. 2009;16(2):305–316.
23. Babczyńska-Sendek B, Barć A. Zbiorowiska łąkowe z *Sesleria uliginosa* (Poaceae) w okolicach Włodowic koło Zawiercia (Wyżyna Śląsko-Krakowska). Fragm Flor Geobot Polonica. 2009;16(2):363–375.
24. Kozłowska A. Naskalne zbiorowiska roślin na Wyżynie Małopolskiej. Rozpr Wydz Mat-Przyr PAU A/B. 1928;67:325–373.
25. Medwecka-Kornaś A. Rezerwaty stepowe nad dolną Nidą. Chrońmy Przyr Ojcz. 1952;8(6):3–20.
26. Łuszczyńska B, Łuszczyński J. Godna ochrony roślinność kserotermiczna Łagiewnik koło Buska-Zdroju w województwie kieleckim. Chrońmy Przyr Ojcz. 1989;45(3):30–35.
27. Towpasz K, Stachurska-Swakoń A. Występowanie *Sesleria uliginosa* (Poaceae) w murawach kserotemicznych (*Festuco-Brometea*) na terenie Niecki Nidziańskiej (Wyżyna Małopolska). Fragm Flor Geobot Polonica. 2011;18(2):321–330.
28. Bielecki E. Badania cytologiczno-systematyczne nad *Oreochloa disticha* Link., *Sesleria uliginosa* Opiz i *S. calcaria* Opiz. Acta Soc Bot Pol. 1955;24(1):145–162.
29. Podani J. SYN-TAX 2000. Computer programs for data analysis in ecology and systematics. User's manual. Budapest: Scientia Publishing; 2001.
30. Kovach WL. MVSP – a multivariate statistical package for Windows, ver. 3.1. Pentraeth: Kovach Computing Services; 2007.
31. Sokal RR, Rohlf FJ. Biometry: the principles and practice of statistics in biological research. New York: W. H. Freeman; 1981.
32. Pawłowski B. Composition and structure of plant

communities and methods of their study. In: Szafer W, editor. The vegetation of Poland. Oxford: Pergamon Press, Polish Scientific Publishers PWN; 1966. p. 241–281.
33. Medwecka-Kornaś A, Kornaś J, Pawłowski B, Zarzycki K. Przegląd ważniejszych zespołów roślinnych Polski. In: Szafer W, Zarzycki K, editors. Szata roślinna Polski. Warszawa: Polish Scientific Publishers PWN; 1977. p. 279–297. (vol 1).
34. Matuszkiewicz W. Przewodnik do oznaczania zbiorowisk roślinnych Polski. Vademecum Geobotanicum. Warszawa: Polish Scientific Publishers PWN; 2005. (vol 3).
35. Zarzycki K, Trzcińska-Tacik H, Różański W, Szeląg Z, Wołek J, Korzeniak U. Ecological indicator values of vascular plants of Poland. Kraków: W. Szafer Institute of Botany, Polish Academy of Sciences; 2002. (Biodiversity of Poland; vol 2).
36. Mirek Z, Piękoś-Mirkowa H, Zając A, Zając M. Flowering plants and pteridophytes of Poland – a checklist. Kraków: W. Szafer Institute of Botany, Polish Academy of Sciences; 2002. (Biodiversity of Poland; vol 1).
37. Ochyra R, Bednarek-Ochyra H, Żarnowiec J. Census catalogue of Polish mosses. Kraków: W. Szafer Institute of Botany, Polish Academy of Sciences; 2002. (Biodiversity of Poland; vol 3).
38. Bąba W. The species composition and dynamics in well-preserved and restored calcareous xerothermic grasslands (South Poland). Biologia, Bratislava. 2004;59(4):447–456.
39. Dzwonko Z, Loster S. Vegetation differentiation and secondary succession on a limestone hill in southern Poland. J Veg Sci. 1990;1(5):615–622.
40. Fischer M, Stöcklin J. Local extinctions of plants in remnants of extensively used calcareous grasslands 1950–1985. Conserv Biol. 1997;11(3):727–737.
41. Michalik S. Changes in xerothermic vegetation during the 20 years' secondary succession on the study area of Grodzisko in Ojców National Park. Prądnik. Prace Muzeum W. Szafera. 1990;2:43–52.
42. Waldon B, Rapacka-Gackowska A. Stan zachowania i problemy ochrony muraw kserotermicznych w dolinach Wisły i Noteci. In: Ratyńska H, Waldon B, editors. Ciepłolubne murawy w Polsce – stan zachowania i perspektywy ochrony. Bydgoszcz: Kazimierz Wielki University in Bydgoszcz Press; 2010. p. 110–127.
43. Ward LK. Management of grassland – scrub mosaics. In: Hillier SH, Walton DWH, Wells DA, editors. Calcareous grasslands: ecology and management. Bluntisham: Bluntisham Books; 1990. p. 134–139.
44. Dzwonko Z, Loster S. Changes in plant species composition in abandoned and restored limestone grasslands – the effects of tree and shrub cutting. Acta Soc Bot Pol. 2008;77(1):67–75.
45. Kalamees R, Zobel M. The role of the seed bank in gap regeneration in a calcareous grassland community. Ecology. 2002;83(4):1017.
46. Grubb PJ. The maintenance of species-richness in plant communities: the importance of the regeneration niche. Biol Rev. 1977;52(1):107–145.
47. Willems JH, Bik LPM. Restoration of high species density in calcareous grassland: the role of seed rain and soil seed bank. Appl Veg Sci. 1998;1(1):91–100.
48. Czylok A, Ślusarczyk M, Tyc A, Waga JM. Wypas zwierząt gospodarskich jako sposób czynnej ochrony krajobrazu i różnorodności biologicznej rezerwatu "Góra Zborów". Prądnik. Prace Muzeum W. Szafera. 2010;20:185–196.
49. Barbaro L, Dutoit T, Cozik P. A six-year experimental restoration of biodiversity by shrub-clearing and grazing in calcareous grasslands of the French Prealps. Biodivers Conserv. 2001;10:119–135.
50. Dolek M, Geyer A. Conserving biodiversity on calcareous grasslands in the Franconian Jura by grazing: a comprehensive approach. Biol Conserv. 2002;104(3):351–360.
51. Krasicka-Korczyńska E, Stosik T. Wpływ oddziaływań zooantropogenicznych na roślinność muraw kserotermicznych. In: Ratyńska H, Waldon B, editors. Ciepłolubne murawy w Polsce – stan zachowania i perspektywy ochrony. Bydgoszcz: Kazimierz Wielki University in Bydgoszcz Press; 2010. p. 80–94.
52. Michalik S, Zarzycki K. Management of xerothermic grasslands in Poland: botanical approach. Coll Phytosociol. 1995;24:881–895.
53. Pärtel M, Kalamees R, Zobel M, Rosén E. Restoration of species-rich limestone grassland communities from overgrown land: the importance of propagule availability. Ecol Eng. 1998;10(3):275–286.
54. Willems JH, Peet RK, Bik L. Changes in chalk-grassland structure and species richness resulting from selective nutrient additions. J Veg Sci. 1993;4(2):203–212.
55. Kammer PM. Developmental responses of sub-dominant grassland species to current weather conditions and their relevance for annual vegetation changes. Folia Geobot. 2002;37:185–204.
56. Kondracki J. Geografia Polski. Mezoregiony fizyczno-geograficzne. Warszawa: Polish Scientific Publishers PWN; 1994.

2

Notes on the self-pollination of *Dendrobium biflorum* (G. Forst.) Sw. (Orchidales, Dendrobiinae)

Agnieszka K. Kowalkowska[1]*, Hanna B. Margońska[2]

[1] Department of Plant Cytology and Embryology, University of Gdańsk, Wita Stwosza 59, 80-308 Gdańsk, Poland

[2] Department of Plant Taxonomy and Nature Conservation, University of Gdańsk, Wita Stwosza 59, 80-308 Gdańsk, Poland

Abstract

The conditions favouring self-pollination are determined for a species of *Dendrobium* Sw. A survey of gynostemium micromorphology in collected flowers of *Dendrobium biflorum* revealed two stages of a unique form of autogamy: on germination of pollen tetrads, pollen tubes reached the stigmatic region without the pollen being displaced from the anther locules. This is the first time for this type of autogamy to be reported in *Dendrobium biflorum*. Pollen grains germinated directly from the locules, bypassing the stigma or falling or sliding down onto the stigmatic surface. The pollen tubes were long and reached the stylar canal. On The Society Islands, differences in exposure of the habitats to weather conditions and altitude gradients influence both composition and form of the vegetation, as well as the occurrence of pollinators. The form of autogamy described here may be the usual method of pollination found on The Society Islands (and even the whole of French Polynesia), where suitable pollinators are absent and/or growth conditions stressful.

Keywords: autogamy, *Dendrobium biflorum*, gynostemium micromorphology, Orchidaceae, The Society Islands

Introduction

Members of the Orchidaceae are well-known for their complex flowers. A particular type of floral morphology, colour pattern, fragrance or reward, all result in the attraction of a specific pollinator, and thus, the tendency for autogamy (automatic self-pollination) is reduced [1–3]. Nevertheless, in populations living in habitats where the occurrence of potential pollinators is low, autogamy is the preferred method of pollination and, in fact, is a rather common phenomenon within the family [3,4]. Autogamy is estimated to occur in approximately 3% of orchid species [2], or even as much as 5–20% [5]. In pollinator-free environments and under stressful growing conditions, the facultative process of self-pollination occurs at the end of anthesis (with the ability for cross-pollination maintained in flower fitness) [6]. Self-pollination mechanisms have been confirmed for two species of *Dendrobium* Sw. [5]. In *D. chryseum* Rolfe and *D. mastersianum* F. Muell. & Kraenzl, entire pollinia, owing to the close proximity of anther and stigma, were observed to fall or slide down onto the stigma, thus resulting in self-pollination.

* Corresponding author. Email: dokakow@ug.edu.pl

Owing to different, recently proposed, taxonomic approaches to the genus *Dendrobium* and the publication of a second, unwarranted, heterotypic description of the species *D. biflorum*, we present our taxonomic treatment of the genus and describe the specimens studied, so as to avoid future errors in identification. *Dendrobium* Sw. [nom cons. Nova Acta Regiae Soc Sci Upsal 1799;6:82. Generitype: *Dendrobium moniliforme* (L.) Sw. typ. cons.] is a large genus containing ca. 600 species, widely distributed in subtropical and tropical regions of Asia and Australia to New Zealand and Oceania. The plants are epiphytic, lithophytic to rarely terrestrial, small to large, forming colonies of various sizes. Their shoots are cane-like or pseudobulbous, few- to multinodal, erect to pendent, whereas leaves are coriaceous or fleshy, conduplicate, sessile or petiolate. The inflorescence may be lateral, axillary to pseudoterminal, racemose to paniculate, with single to many flowers, which are small to medium-sized, often showy, resupinate or not. Lateral sepals are basally adnate to the column foot, forming a distinct mentum. Lip is usually 3-lobed or entire, spurless, mobile or not, attached to the apex of the column foot, embracing the gynostemium, whereas the disc is furnished with calli, ridges, lamellae or other appendages. The column part is prominent, longer than anther, column foot usually very long, sometimes longer than the column part, massive. Stigma is semi-ellipsoid, oblong to semi-ovoid, deeply concave. Rostellum is very short, incumbent, truncate, thickened along the margins. Viscidium is single, narrow, detachable, multilayered or absent. Staminodes have usually massive, wing-like form. Clinandrium is apical, rather voluminous to relatively obscure, entire to 3-lobed. Anther is incumbent, mobile, ellipsoid or ovoid, 2-chambered. Connective is narrow, but very thick, papillate or not, forming a roof in front over the chambers. Pollinia are 4 in 2 pairs, clavate or oblong, apically distinctly attenuate. Caudicles are absent.

Dendrobium biflorum (G. Forst., Fl Ins Austr. 1786;60) Sw., Nova Acta Regiae Soc Sci Upsal. 1799;6:84, non *Dendrobium biflorum* A. Rich, Ess Fl N Zel. 1832;167 (=*Dendrobium cunninghami* Lindl., Bot Reg. 1835;21: sub t. 1756, non *Dendrobium cunninghami* Steud, Nomencl Bot. 2nd ed. 1840;1:490, nom. illeg.) is one of the representatives of the genus [section *Grastidium* (Bl.) J. J. Sm.]. The species is rather widely distributed. It occurs in New Guinea and neighbouring islands, in Oceania from Solomon Islands, Vanuatu, Fiji, Samoa, New Caledonia to The Society Islands. At The Society Islands it was recorded from Maupiti, Bora Bora, Huahine, Tahaa, Raiatea, Mo'orea, Mehetia and Tahiti, where the type-specimens were collected (G. Forster f. 169 LECTO-K!, ISOLECTO-G!). In most habitats *D. biflorum* is medium-sized to large, (20)50–160 cm tall, with rhizome branched, rooted and scaled at nodes. Shoot is 0.2–0.6 cm in diameter, erect to pendent according to size, age and habitat, elongate, cane-like, multi-nodal, lignified, leafy at apical part. The leaves are numerous, distichously arranged, with blades 5–10(17) cm long, 0.4–0.8 cm wide, thin. Inflorescences are axillary, a few at any given time, always 2-flowered with peduncle strongly contracted. Flowers up to ca. 3 cm in diameter when spread, short-lived, white (Fig. 1b,d), becoming creamy-white when older. All tepals are thin and delicate, nearly translucent, lustrous, somewhat similar in general outline. Lip is 3-lobed, spirally recurved at the apex, base tinged with pale pink, lateral lobes with dark and deep brown-red to purple-red margins, lateral fimbriae of the mid-lobe tinged or blotched brown-red to purple-red. Hypochile is more fleshy, base constricted and adnate to the apex of the column foot, its lateral lobes oblique, oblong-triangular to semi-ovate, obtuse to subacute, with anterior margins usually entire, posterior margins erose. Middle lobe is narrowly triangular, attenuate and filiform at the apex, with lateral margins strongly and deeply undulate, erose, crenulate to fimbriate. Its central surface has ca. 3 rows of papillae, whereas the disc is ornamented with oblong, flattened thickening, usually vanishing at the base of the middle lobe. Gynostemium morphology of *Dendrobium biflorum* exhibited features characteristic of cross-pollination. Gynostemium (Fig. 2a,b) is 0.15–0.2 cm

Fig. 1 *Dendrobium biflorum* (G. Forst.) Sw. **a** Epiphytic plants as found in most typical lowland habitats, forming large, dense clusters and colonies on tree trunks and branches. **b** Flowers of lowland plants (Tahiti, Mahape Valley, 2007.09.07). **c** Terrestrial plants (much smaller, erect and more robust) from highland habitats. **d** Flowers of highland plants (Mo'orea, Mount Rotui, 2007.10.22). Phot. H. B. Margońska.

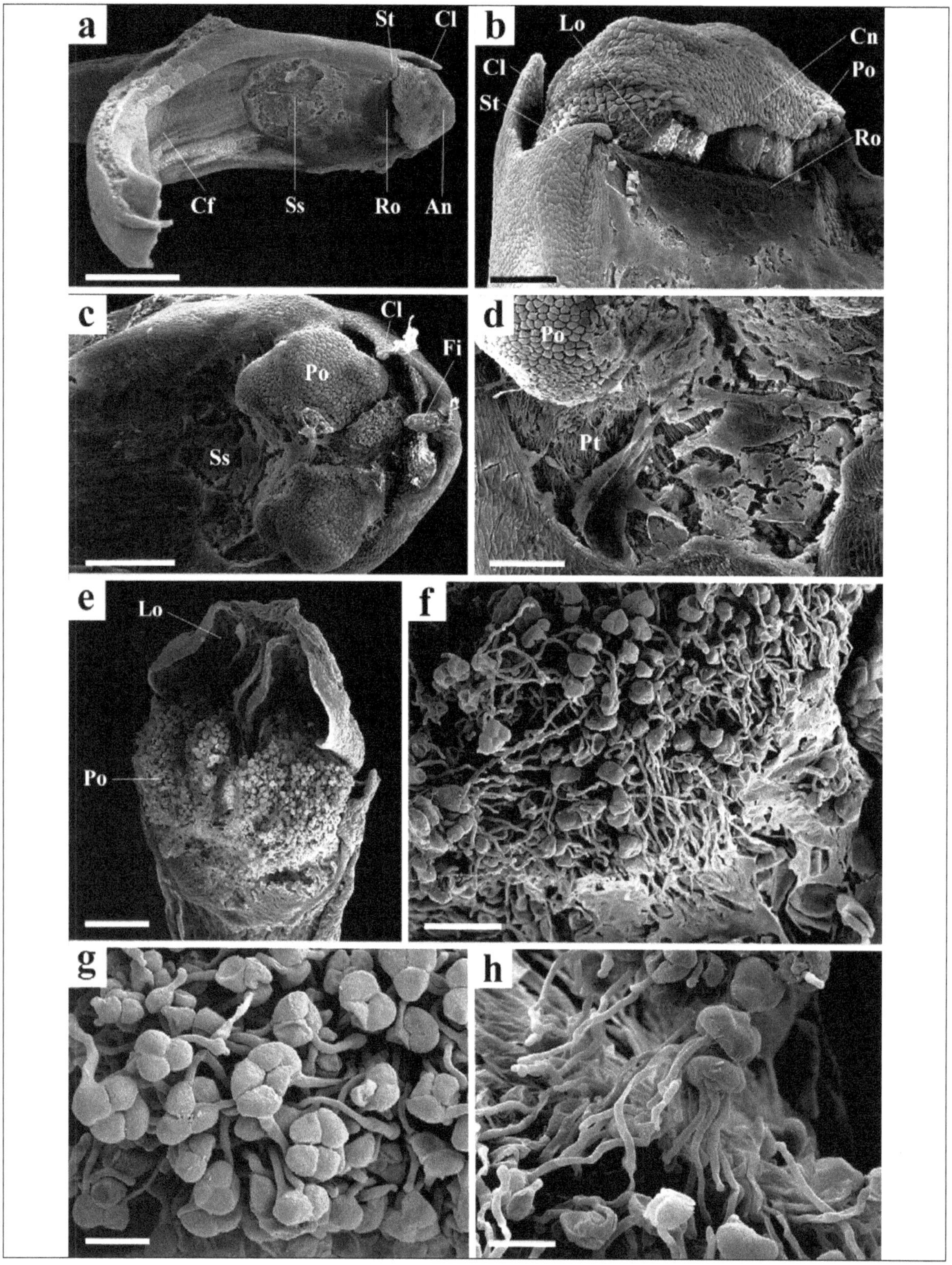

Fig. 2 SEM images showing gynostemium morphology of *Dendrobium biflorum* (G. Forst.) Sw. **a** Anterior view of gynostemium of typical flower (without autogamy) derived from a lowland plant (Fig. 1a,b). **b** Detail of anther with four pollinia. **c** Anterior view of gynostemium of self-pollinated flower from Mo'orea, Belvedere Plateau. **d** Detail of **c** showing long pollen tubes growing directly from locules (without releasing pollen) towards the stigma. **e** Gynostemium of flower of highland plant (Mo'orea, Mount Rotui) in advanced stage of autogamy. **f** Detail of pollinia (from **e**) showing germinating pollen grains growing towards the stigmatic surface. **g,h** Germination of pollen grains in tetrads (detail of **f**). An – anther; Cf – column foot; Cn – connective; Fi – filament; Lo – locules; Po – pollinium; Pt – pollen tubes; Ro – rostellum; St – staminodium; Ss – stigmatic surface. Scale bars: **a** 1 mm; **b,d,e** 200 μm; **c** 500 μm; **f** 50 μm; **g,h** 20 μm. Phot. A. K. Kowalkowska.

long, erect, with column foot longer than the column part. The stigma is semi-ellipsoid, deeply concave, distinctly limbate with erect rim, with secretion. The rostellum is very short, incumbent, truncate, thickened along the margins. The staminodes are rather obscure, subparallel to the column, oblong, wing-like, larger and more massive closer to the stigma, distally attenuate. The clinandrium is apical, rather distinct and 3-lobed. The filament is narrow and incorporated into distal part of clinandrium. The anther is incumbent in natural position, mobile, semi-ovoid, 2-chambered. Both locules are well separated by the connective. The connective is rather narrow, but thick and glabrous, forming a short and wide roof in front over the locules. Pollinia are 4 in 2 pairs, clavate, apically distinctly attenuate. The pollen grains are arranged in tetrads.

Amongst the numerous flower specimens of *D. biflorum* collected, we found several examples of a unique form of autogamy: the pollen tetrads germinated to the stigmatic region without being displaced from the anther locules. In this article, we compare the morphological features of the gynostemium in typical flowers of the species with those of three selected specimens showing different stages of self-pollination. We also discuss reasons for this particular type of autogamy.

Material and methods

A study of the taxonomy of the orchid flora of The Society Islands (French Polynesia) by Margońska, resulted in the analysis of more than 1500 herbarium and spirit-preserved specimens deposited in over 20 collections. The field observations [over 300 specimens studied, records included in Archivum Orchidalium (Arch. Orch.)] and preserved samples (deposited as part of UGDA-HBM) are the result of two scientific expeditions in 2007 and 2009 to The Society Islands (Tahiti, Mo'orea, Rai'atea, Taha'a and Bora Bora). We have chosen three specimens of *D. biflorum* for micromorphological studies and to demonstrate the various stages of autogamy found in this species (Tab. 1). Plant material was fixed in 2.5% glutaraldehyde (GA) and 2.5% paraformaldehyde (PFA) in 0.05 M cacodylate buffer (pH = 7.0) at room temperature. For scanning electron microscopy (SEM), the samples, following dehydration in an ethanol series, were subjected to critical-point drying using liquid CO_2, coated with gold and observed using a Philips XL-30 SEM at an accelerating voltage of 15 kV. For light microscopy (LM), the plant material, following fixation and dehydration in ethanol, was embedded in epoxy resin (Technovit) and stained with 0.05% toluidine blue O (TBO) in 1% aqueous sodium tetraborate solution for 1 min at 60°C on a hot plate [7,8]. Samples were prepared in accordance with procedures described previously [9–11].

Results

Micromorphology

SEM and LM studies revealed floral dissimilarities between plants derived from lowland and highland habitats. Flowers from the lowland part of the Mahape Valley, Tahiti (Fig. 1a,b, Fig. 2a,b), were observed to have developed fully in typical manner. The four pollinia were hidden beneath the anther cap. The stigma was copiously coated with secretion and germination of pollen tetrads was not observed.

Samples gathered from Mo'orea, the Belvedere Plateau (Fig. 2c,d) and the highest crest of Mount Rotui (Fig. 1c,d, Fig. 2c,d) displayed two different stages of self-pollination. In the first set of samples (Belvedere Plateau; Fig. 2c,d), the long pollen tubes germinated directly from the locules, without the release of pollen from the anther, and passed through the copious stigmatic exudate towards the stigma. In the second set of samples (from the crest of Mount Rotui; Fig. 1c,d), more advanced autogamous stages were observed: the anther thecae (locules) were nearly empty and the pollen tetrads had slid down slightly towards the stigmatic surface (Fig. 2e). All pollen grains had germinated (including each component of a tetrad, Fig. 2f–h), their pollen tubes growing towards the stigmatic surface (Fig. 2f, Fig. 3a). Further LM observations of longitudinal sections of the column (Fig. 3c,d) demonstrated that pollen grains germinated after sliding down onto the stigmatic surface and also directly from locules, bypassing the stigma. The pollen tubes grew (Fig. 3a,b) and reached the stylar canal (Fig. 3c,d).

Discussion

A similar autogamous pollination mechanism, where flowers display both autogamous and allogamous traits, occurs in *Epipactis microphylla* (Ehrh.) Sw. [6]. Morphological features of the gynostemium related to the mechanisms of cross-pollination are barely noticeable: a rostellum is present, but very short; it is coated with adhesive substance and is not easily detached from the stigma. Conversely, a few autogamous traits are also displayed. A distinct anther is supported by a short, narrow filament, the clinandrium is poorly developed and pollen grains remain packed in tetrads, even after sliding down to the stigma. We agree with Bonatti et al. [6], who claimed that the

Tab. 1 Samples used for micromorphological studies (deposited in UGDA-HBM) and the results of SEM observations.

Species name	Place and date of collection	Collector and voucher number	Observation of autogamy
Dendrobium biflorum (G. Forst.) Sw.	Tahiti, Mahape Valley, 2007.09.07.	Margońska H. B. 7TAHMahapeVal20092007 07aa	germination of pollen tetrads not observed
Dendrobium biflorum (G. Forst.) Sw.	Mo'orea, Belvedere Plateau 2007.10.12.	Margońska H. B. 21MREABelved 212102007 03a-c	pollen tetrads germinated directly from locules by passing stigma
Dendrobium biflorum (G. Forst.) Sw.	Mo'orea, Mount Rotui, 2007.10.22.	Margońska H. B. 29MREAMtRotui 22102007 04ff	pollen tetrads slightly slid down towards the stigma, mainly germinated from locules bypassing stigma, excluding some pollen tubes emerged from stigmatic surface

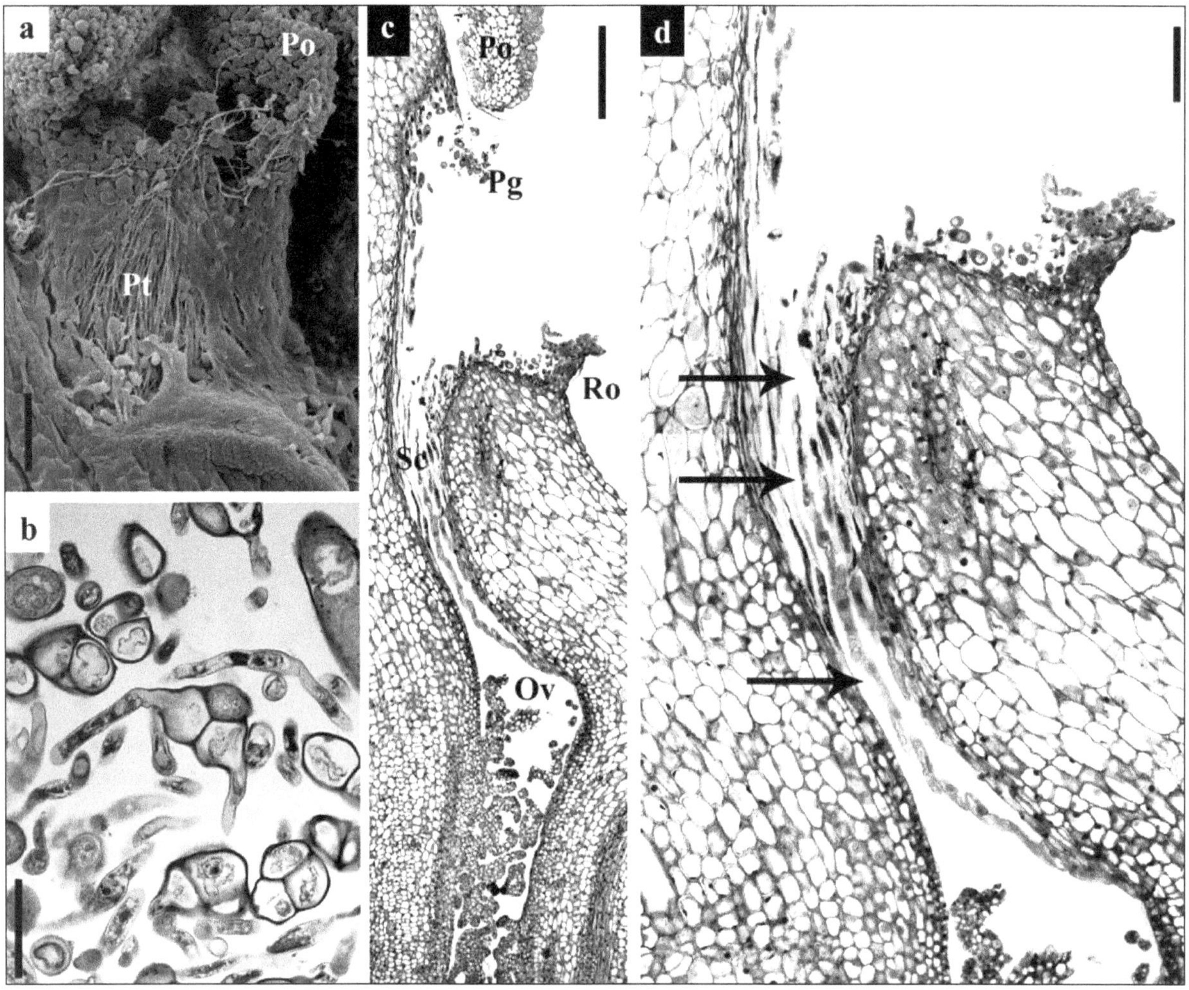

Fig. 3 *Dendrobium biflorum* (Mo'orea, Mount Rotui). **a** Anterior view of gynostemium showing long pollen tubes growing directly from locules (SEM). **b** Germinated pollen grains (LM). **c** Longitudinal section of the column, stylar canal and ovary. The pollen tubes have elongated and reached the stylar canal (LM). **d** Detail of stylar canal (arrowheads) showing pollen tubes growing towards the ovules (LM). Ov – ovary; Po – pollinium; Pg – pollen grains; Pt – pollen tubes; Ro – rostellum; Sc – stylar canal. Scale bars: **a–d** 100 μm; **b** 25 μm; **c** 250 μm. Phot. A. K. Kowalkowska.

adhesive rostellum and the stigmatic exudates, located along the upper border restraining the clinandrium cavity, seem to favour both the retention of released pollen from the locules and growth of the pollen-tube. The extension of pollen tubes in *D. biflorum* was observed both in locules and also after the pollen had slid down slightly onto the stigmatic surface. In flowers of *E. microphylla* (cleistogamous and also chasmogamous species) pollen germinated inside the anther, transiently or totally bypassing contact with the receptive stigma [6,12–16]. Bonatti et al. [6] maintained that in cleistogamous and self-pollinated flowers, pollen tubes emerging from pollen grains retained in the locules were growing beneath the stigmatic surface layer. In *D. biflorum*, we also observed the presence of developing pollen tubes beneath the stigmatic exudate (Fig. 2d, Fig. 3a). Margońska, however, did not observe cleistogamous forms during field work.

A characteristic feature of many Polynesian islands is the enormous diversity of plants to be found there, variation in the floras of individual islands resulting from differences in the degree of exposure to weather conditions. On the humid, windward sides of volcanic islands, especially, tropical phytocenosis occur, whereas on dry, windward slopes, xeromorphic, low forest and scrub prevail and become grass and/or fern vegetation.

The composition and form of vegetation on The Society Islands depends not only on its exposure to seasonal winds and rains, but also on altitude gradients. In most habitats of The Society Islands, *D. biflorum* is recognized as a rather common and abundant plant, growing between 40–1600 m. This epiphytic orchid often forms very large and dense clusters and colonies on tree trunks and branches, usually directly on bark, but always in sunny situations (Fig. 1a,b). In wetter, somewhat colder conditions, at higher elevations, for example cloud forests, they grow in deep moss-covers with many other epiphytic plants, including other orchids. At the highest altitude range for *D. biflorum* , this species grows terrestrially (Fig. 1c,d). It is worth noting that when the habitat changes markedly, even though the flowers remain identical, the plants are much smaller, erect, and robust.

Ridges of Mount Rotui (especially the North side) and its peak, have rather extreme climate conditions (although not unique in Polynesian islands). Each day, until about noon, they are exposed to full sun, resulting in distinctly increased temperatures and aridity. During the afternoon, clouds increasingly gather around the mountain peak, especially from the Western windward side, because winds are forced upward by mountain masses. During the day, the temperature is still high, but humidity increases. Later, this rises as vapour and condenses as it makes contact with high level, colder air. Thus, heavy rains are frequent. Throughout the dry season, as well as an obvious reduction in humidity, a significant increase in temperature is observed. During the rainy season, rainfalls are more frequent, and may even last all day and night [17]. In addition, throughout all this time, the habitats of *D. biflorum* are subject to distinctly windy conditions. Not only do these climatic conditions make it difficult for many plants (lower and more xeromorphic character of the phytocenosis), but they make it especially difficult for pollinators to live in these locations. The unique kinds of autogamy recorded here for *D. biflorum* may be the result of limited opportunity for insect pollination.

Margońska (during the scientific expeditions of 2007 and 2009) and Jean-Yves Meyer, Jean-François Butaud and Jacq Fred (personal communication) did not record observing pollinators at these flowers. Instead, Margońska observed only occasional visits by ants and spiders, but these did not precipitate pollination.

Self-pollination may be the norm in *Dendrobium biflorum*, not only for The Society Islands, but perhaps for the whole of French Polynesia. The specific climatic conditions described above are typical for most of the islands of the Archipelagos, and these result in pollinator-free environments and/or stressful growing conditions. Further research into Pacific Island floras would perhaps help us better explain the occurrence of autogamy in species that are usually cross-pollinated.

Acknowledgements

This work was supported by the National Science Centre grants No. N304 029 32/ 1584 and N303 580439 in Poland. We are thankful to Jean-Yves Meyer, (Délégation a la Recherche, Ministere d'Education, de l'Enseignement Supérieur et de la Recherche, Gouvernement de Polynésie française Papeete, Tahiti), Jean-François Butaud (Consultant en foresterie et botanique) and Jacq Fred (Consultant en Aménagement forestier). The taxonomic and field observations are a part of the database of Hanna B. Margońska - the Arch. Orch. The SEM and LM studies are deposited in the database of Agnieszka K. Kowalkowska - the Database of Orchid Ultrastructure (DOUL).

References

1. Darwin C. The various contrivances by which orchids are fertilized by insects. London: John Murray; 1890.
2. Pijl L, Dodson CH. Orchid flowers: their pollination and evolution. Coral Gables FL: University of Miami Press; 1966.
3. Dressler RL. Phylogeny and classification of the orchid family. Portland OR: Dioscorides Press; 1993.
4. Tremblay RL, Ackerman JD, Zimmerman JK, Calvo RN. Variation in sexual reproduction in orchids and its evolutionary consequences: a spasmodic journey to diversification. Biol J Linn Soc. 2005;84(1):1–54.
5. Catling PM. Auto-pollination in the Orchidaceae. In: Arditti J, editor. Orchid biology: reviews and perspectives. Portland OR: Timber Press; 1990. p. 121–158.
6. Bonatti PM, Sgarbi E, Del Prete C. Gynostemium micromorphology and pollination in *Epipactis microphylla* (Orchidaceae). J Plant Res. 2006;119(5):431–437.
7. Feder N, O'Brien TP. Plant microtechnique: some principles and new methods. Am J Bot. 1968;55(1):123.
8. Ruzin SE. Plant microtechnique and microscopy. New York: Oxford University Press; 1999.
9. Kowalkowska AK, Margońska HB. Diversity of labellar micromorphological structures in selected species of Malaxidinae (Orchidales). Acta Soc Bot Pol. 2009;78(2):141–150.
10. Kowalkowska AK, Margońska HB, Kozieradzka-Kiszkurno M. Comparative anatomy of the lip spur and additional lateral sepal spurs in a three-spurred form (*F. fumeauxiana*) of *Anacamptis pyramidalis*. Acta Biol Cracov Ser Bot. 2010;52(1):13–18.
11. Kowalkowska AK, Margońska HB, Kozieradzka-Kiszkurno M, Bohdanowicz J. Studies on the ultrastructure of a three-spurred fumeauxiana form of *Anacamptis pyramidalis*. Plant Syst Evol. 2012;298:1025–1035.
12. Anderson WR. Cryptic self-fertilization in the Malpighiaceae. Science. 1980;207(4433):892–893.
13. Lord EM. Cleistogamy: a tool for the study of floral morphogenesis, function and evolution. Bot Rev. 1981;47(4):421–449.
14. Pacini E, Franchi GG. Germination of pollen inside anthers in some non-cleistogamous species. Caryologia. 1982;35:205–215.
15. Mayers AM, Lord EM. Comparative flower development in the cleistogamous species Viola odorata. III. A histological study. Bot Gaz. 1984;145(1):83.
16. Caiola MG, Brandizzi F, Canini A. *Hermodactylus tuberosus* L. (Iridaceae) pollen organisation before and after anther dehiscence. Plant Biosyst. 2000;134(3):353–364.
17. Margońska HB, Szlachetko DL. Orchidaceae of Tahiti (Polynesie Française). Gdańsk: Gdańsk University Press; 2010.

3

Macroalgae fouling community as quality element for the evaluation of the ecological status in Vela Luka Bay, Croatia

Gorana Jelic Mrcelic[1]*, Merica Sliskovic[1], Boris Antolic[2]

[1] Maritime Faculty, University of Split, Zrinsko Frankopanska 38, 21 000 Split, Croatia

[2] Institute of Oceanography and Fisheries, Setaliste Ivana Mestrovica 63, 21 000 Split, Croatia

Abstract

One year qualitative and quantitative study of communities of three major taxonomic groups has been carried out at test panles placed in the upper infarlittoral zone of coastal area of Vela Luka Bay, Croatia. A list of 44 taxa was recorded. *Chaetomorpha* sp., *Ulva* sp., *Fosliella farinosa*, *Sphacelaria cirrosa*, *Polysiphonia scopulorum* were the most frequent dominant taxa. Among 27 algal taxa with noticeable presence only three were classified as ESG (Ecological State Groups) I. Low diversity and species richness together with massive presence of the green algae (as *Ulva* sp.) and negligible presence of ESG I taxa, may lead to erroneous conclusion that Vela Luka Bay is eutrophicated area. Low values of biomass and *R/P* (Rhodophyceae by Phaeophyceae ratio) Index together with dominance of Phaeophyta also support conclusion that there is no negative impact of nutrient enrichment on macrophyta fouling community in Vela Luka Bay.

Keywords: artificial habitats, bioindicators, coastal waters, fouling organisms, Vela Luka Bay

Intoduction

The major structuring factors influencing the benthic communities are species recruitment onto a surface, competition between resident organisms and disturbance by predation and/or environmental factors [1], although pollution also influences the development of these communities [2]. Changes in the development of benthic communities caused by organic pollution are often obscured by the interactions between nutrient enrichment and a variety of other ecological factor [3]. In order to distinguish these effects, the use of artificial panels is a good alternative method to field studies [4]. The panels may be easily manipulated and placed under a variety of environmental conditions [1]. Artificial structures are colonized by the most competitive assemblages of floral and fauna species in response to a combination of physical, chemical and biological factors, from the intertidal to the shallow subtidal [5]. Number of factors (age, texture, depth, complexity, inclination and position in the water column) influences colonization of epibiota on artificial panels [6–10]. Vertical surfaces are more densely colonized than horizontal surface [8,11,12].

* Corresponding author. Email: gjelic@pfst.hr

In the last years, the increasing need for stable and comparable criteria of environmental quality in the European aquatic ecosystems, reactivated the use and search of pollution biological indicators [13]. Some ecological indices are focused on the presence/absence of a given indicator species, while others take into account species biomass and abundance, the different ecological strategies adopted by organisms [Feldmans *R/P* (Rhodophyceae by Phaeophyceae ratio) index] [14], the diversity (species richness), or the energy variation in the system through changes in the biomass of individuals [13]. Evidence on the suitability of benthic macrophytes as indicators of effects against different pollution gradients is undoubted [15,16]. Marine benthic macrophytes, as photosynthetic sessile organisms being at the base of food web, are vulnerable and adaptive to human and environmental stress of water and sediment. They respond to aquatic environment representing reliable indicators of its changes [17]. A universal pattern is that highly stressed or disturbed marine environments are inhabited by annual species with high growth rates and reproductive potential, while undisturbed marine environments by perennial species with low growth rates and reproductive potential [18–21]. This was the spark to develop the biotic index EEI (Ecological Evaluation Index), based on the functional-morphological model of Littler and Littler [22] and use it to divide marine benthic macrophytes in two different ecological groups, the late-successional [perennials, ESG (Ecological State Groups) I] and the opportunistic (annuals, ESG II) [23]. Thus, the presence and/or the abundance of some benthic macrophytes could be used for the classification of the Ecological (quality) Status, in the terms of the Water Frame Directive [(WFD) 2000/ 60/ EC) [24]. A cost-effective monitoring system to cover the demands of WFD could include summer destructive samplings, functional-form classification, and use of the EEI. Colonization by local infra-upperlittoral benthic communities can be used as biological indicator of environmental changes because they are exhaustively studied [25] as well as they integrate the environmental changes occurring in marine ecosystems, and they are strongly affected by pollution [26]. Further studies are needed to better understand periphyton response to different substrata types and possible seasonal changes of the community structure, especially studies on pollution impact on the macroalgal communities of artificial substrata [27]. The aim of the present study was to give the development of fouling communities using artificial panels and to demonstrate that several ecological indexes based on the composition and abundance of the test panels phytobenthos in the upper infralittoral zone could give good tools for the rapid assessment of the ecological quality of coastal waters.

Material and methods

The fouling community on polyester test panels was studied in the Vela Luka Bay – Bobovišće, the island of Korcula, the Adriatic Sea (42°50′ N and 16°43′ E). The Vela Luka Bay is closed, shallow and hidden bay, well protected from the north and the south winds. According to Dadić et al. [28], high water oscillations are predominant (in several days periods) during the summer. During the winter, low frequency oscillations are predominant and water exchange is not altered, e.g. water enters in surface layer and exits in bottom layer. Time needed for water exchange ranges from one day to five days depending on sea conditions, wind effects, currents and the stratification in the bay. Water exchange time is longer during the summer when contaminant inflow is the highest. Although the bay is closed and shallow, the oxygen content is high. Water quality analysis show high nitrate, nitrite and phosphorus content, especially in shallow regions of the bay. Phytoplankton structure and biomass shows that Vela Luka Bay is naturally eutrophic area. According to zooplankton biomass this bay can be categorized as the zone D (high zooplankton biomass).

The test-panels were placed in the area of shipyard "Greben" (site Bobovišće) in order to evaluate the environmental impact caused by urbanisation, tourisms and shipbuilding industry.

Four samplings were carried out every three mounths in period between July and July (throughout one year), in order to monitor the characteristics of fouling vegetation on polyester test panels (20 × 20 cm). One metal structure was placed at the quay, and eight test panels were fixed vertically by ropes, at approximately 0.5 m and 2.5 m depth. Panels were collected after 3, 6, 9 and 12 months of immersion. This overall design gave information on succession patterns and potential recruit available at different period of the year. After collection of the panels were kept in the buckets containing 2% formaldehyde for further analysis.

Examination of the panels took place in The Laboratory of Benthos – The Institute of Oceanography and Fishery, Split. The panels were carefully removed from the buckets in the laboratory, placed in an aquarium filled with seawater, and photographed with a digital camera. According to Boduresque [29] and Braun-Blanquet [30] total cover percentage of fouling community was determined for each panel. The structure of the fouling community was estimated using the Constant Area Method. Four subsamples (2 × 2 cm squares) were randomly scratched from each side of the panel, i.e. eight samples per

Tab. 1 The list of algal taxa found on test panels at both depths.

CHLOROPHYTA
Bryopsidophyceae
Blastophysa sp.
Blastophysa rhizopus Reinke
Bryopsis sp.
Dasycladophyceae
Acetabularia acetabulum (Linnaeus) Silva
Ulvophyceae
Cladophora sp.
Chaetomorpha sp.
Chaetomorpha aerea (Dillwyn) Kuetzing
Ulothrix sp.
Enteromorpha sp.
Ulva sp.
Ulvella sp.
Phaeophila dendroides (Crouan) Batters
PHAEOPHYTA
Phaeophyceae
Dictyota dichotoma (Hudson) Lamouroux
Dictyota linnearis (C. Agardh) Gerville
Padina pavonica (Linnaeus) Thivy
Feldmannia irregularis (Kuetzing) Hamel
Colpomenia sinuosa (Mertens) Derbes i Solier
Ectocarpus paradoxus Montagne
Ectocarpus confervoides Kjellmann
Entonema sp.
Myrionema sp.
Myriotrichia sp.
Giraudia sphacelarioides Derbes & Solier
Sphacelaria cirrosa (Roth) C. Agardh
Sphacelaria plumula Zanardini
Sphacelaria tribuloides Meneghini
Stypocaulon scoparium (Linnaeus) Kuetzing
RHODOPHYTA
Bangiophyceae
Goniotrichum alsidii (Zanardini) Howe
Compsopogonophyceae
Erythrotrichia carnea (Dillywyn) J. Agardh
Florideophyceae
Fosliella farinosa (Lamouroux) Howe
Acrochaetium sp.
Acrochaetium davesii (Dillwyn) Naegeli
Aglaothamnion furcellarie (J. Agardh) G. Feldmann
Antithamnion cruciatum (C. Agardh) Naegeli var. profundum G. Feldmann
Ceramium codii (Richards) G. Mazoyer
Heterosiphonia wurdemannii (Bailey) Falkemberg
Dasya ocellata (Grateloup) Harvey
Dasya arbuscula (Dillwyn) C. Agardh
Lophosiphonia cristata Falkemberg
Polysiphonia scopulorum Harvey
Laurencia sp.
Gelidium sp.

Tab. 1 *(continued)*

Falkenbergia rufolanosa (Harvey) Schmitz

panel. Samples were examined by light microscopy (Carl Zeiss – Jena, ocular 10×, lens 8 and 40×). Each sample was carefully sorted and identification at species and functional group-level was attempted. Fouling species were identified using authoritative keys and texts [31–36].

Phytobenthos fouling community structure was analyzed in terms of total coverage, species number, total species coverage, species frequency, Sörensen similarity coefficient [37] qualitative dominance (*DN*%) of main algal systematic groups, *R/P* ratio and classification in ESG groups, in order to get an indication of the state of health of the study area. The total percentage covers of species (*Ri*) and mean cover of species (*RM*) were analyzed according to Boudouresque [29] and Braun-Blanquet et al. [30]. The frequency and classification in ESG groups were estimated only for algal taxa with noticeable presence (>1% cover). Shifts in marine ecosystem structure and function are evaluated by classifying marine benthic macrophytes in two ESGs (I, II), representing alternative ecological states, e.g. pristine and degraded [17]. The total wet weight of fouling organisms (*TW*, g) was weighted on Tecnica – type EXACTA 1200 EB precision electronic scale (within 0.0001g), after the panel were completely scratched.

Results

Fouling biomass ranged from 55 g × m^{-2} (after 3 month of immersion at 2.5 m depth) to 965.25 g × m^{-2} (after 12 month of immersion at 2.5 m depth). Number of algal taxa ranged from 4 to 18 and the number of animal taxa ranged from 2 to 11. Number of algal taxa was considerably higher than the number of animal taxa (except after 6 month of immersion at 0.5 m depth). Total number of algal taxa was 44 (37 taxa at 0.5 m depth and 41 taxa at 2.5 m depth) – Tab. 1. That gave 34 algal species common for both depths, i.e. Sörensen similarity coefficient was 87.18%.

The *R/P* ranged from 0.33 to 1.20 at 0.5 m depth, and from 0.33 to 1.67 at 2.5 m depth. At 0.5 m depth Phaeophyta (except at the inner side of the panel after 6 month and after 12 month) were dominant over Rhodophyta and Chlorophyta with minimum 12.50% and maximum 50%. Qualitative dominance of Chlorophyta ranged between 20% and 42%, while qualitative dominance of Rhodophyta ranged between 12.50% and 40%. At 2.5 m depth there were no dominant algal groups. Qualitative dominance of Chlorophyta ranged between 20% and 40%, while qualitative dominance of Phaeophyta ranged with minimum 23% and maximum 60%, and qualitative dominance of Rhodophyta ranged between 20% and 44%. Fig. 1 and Fig. 2 show the qualitative dominance (*DN*%) of taxa of the main algal groups found on test panels at two different depths.

Chaetomorpha sp. and *Polysiphonia scopulorum* (100%); *Ulva* sp. and *Fosliella farinosa* (87.50%) and *Sphacelaria cirrosa* (81.25%) were the most frequent taxa. Among 27 algal taxa with noticeable presence only three were classified as ESG I (*Acetabularia acetabulum*, *Padina pavonica* and *Fosliella farinosa*) – Tab. 2.

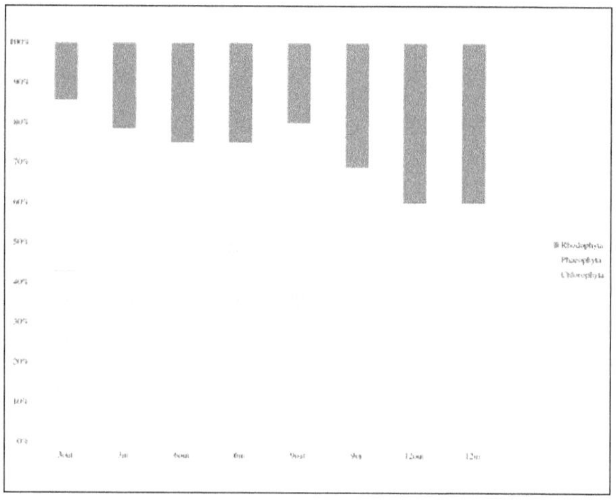

Fig. 1 The qualitative dominance (*DN*%) of taxa of the main algal groups found on test panels (in – inner side; out – outer side) at 0.5 m.

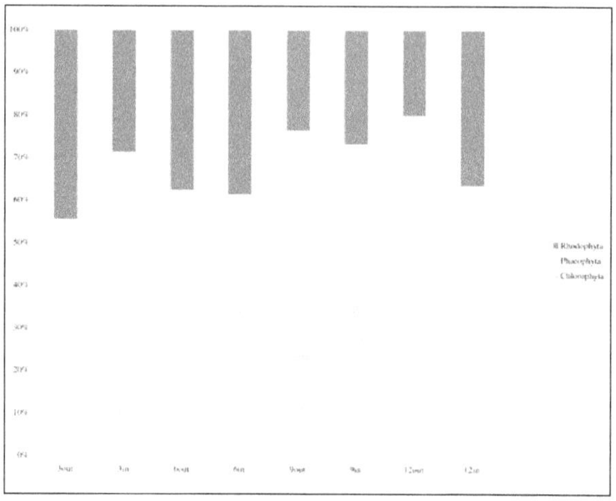

Fig. 2 The qualitative dominance (*DN*%) of taxa of the main algal groups found on test panels (in – inner side; out – outer side) at 2.5 m.

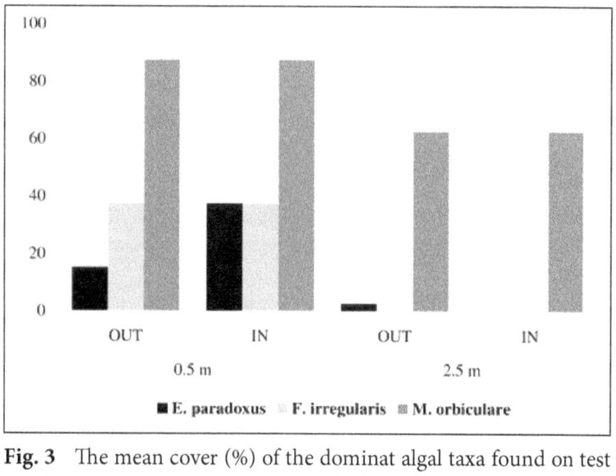

Fig. 3 The mean cover (%) of the dominat algal taxa found on test panels (in – inner side; out – outer side) at 0.5 m and 2.5 m after 3 months of immersion.

Tab. 2 Ecological State Groups classification and the frequency of algal taxa covering at least 1% of the sampling area.

Taxa	ESG	Frquency (*f*/16)
Acetabularia acetabulum (Linnaeus) Silva	I	1
Cladophora sp.	II	2
Chaetomorpha sp.	II	16
Pheophila dendroides (Crouan) Batters	II	7
Rhizoclonium sp.	II	12
Ulotrix sp.	II	4
Ulva sp.	II	14
Dictyota dichotoma (Hudson) Lamouroux	II	7
Dictyota linnearis (C. Agardh) Gerville	II	1
Ectocarpus paradoxus Monatgne	II	7
Entonema sp.	II	5
Feldmannia irregularis (Kuetzing) Hamel	II	9
Giraudya sphacelarioides Derbes i Solier		12
Myrionema orbiculare J. Ag.	II	12
Padina pavonica (Linnaeus) Thivy	I	5
Sphacelaria cirrosa (Roth) C. Agardh	II	13
Sphacelaria plumula Zanardini	II	2
Sphacelaria tribuloides Meneghini	II	3
Aglaothamnion furcellarie (J. Agardh) G. Feldmann	II	3
Antithamnion cruciatum (C. Agardh) Naegeli var. *profundum* G. Feldmann	II	7
Dasya arbuscula (Dillwyn) C. Agardh	II	2
Fosliella farinosa (Lamouroux) Howe	I	14
Heterosiphonia wurdemannii (Bailey) Falkemberg	II	1
Lophosiphonia cristata Falkemberg	II	9
Polysiphonia scopulorum Harv.	II	16

ESG – Ecological State Groups.

Acetabularia acetabulum had frequency 6.25% and *RM* of 3% (*r*), Padina pavonica had frequency 31.25% and *RM* raged between 0% and 15% (form + to 2), while *Fosliella farinosa* had frequency 87.50% and mean cover (*RM*) raged between 0% and 15% (form + to 2).

After 3 months of immersion, total percentage cover was 90% (5 in Braun-Blanquet [30] scale) at 0.5 m at both sides of the panels; while total percentage cover was 90% (5) at the outer side and 70% (4) at the inner side of the panel at 2.5 m. *Myrionema orbiculare*, *Ectocarpus paradoxus* and *Feldmannia irregularis* were dominat algal taxa after 3 months of immersion (Fig. 3).

After 6 months of immersion, total percentage cover was 100% (5) at both depths and both sides of the panels. *Sphacelaria cirrosa* and *Polysiphonia scopulorum* were dominat algal taxa after 6 months of immersion (Fig. 4).

After 9 months of immersion, total percentage cover was 90% (5) at both depths and both sides of the panels. *Cladophora* sp., *Giraudya sphacelarioides*, *Myrionema orbiculare* and *Polysiphonia scopulorum* were dominat algal taxa after 9 months of immersion (Fig. 5).

After 12 months of immersion, total percentage cover was 100% (5) at both depths and both sides of the panels. *Cladophora* sp., *Ulva* sp. and *Polysiphonia scopulorum* were dominat algal taxa after 12 months of immersion (Fig. 6).

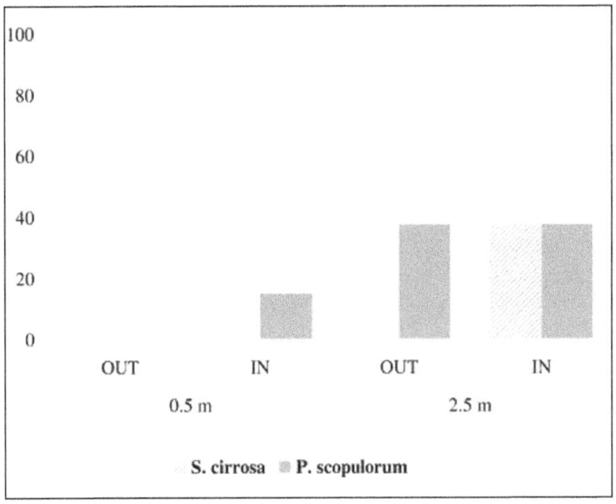

Fig. 4 The mean cover (%) of the dominat algal taxa found on test panels (in – inner side; out – outer side) at 0.5 m and 2.5 m after 6 months of immersion.

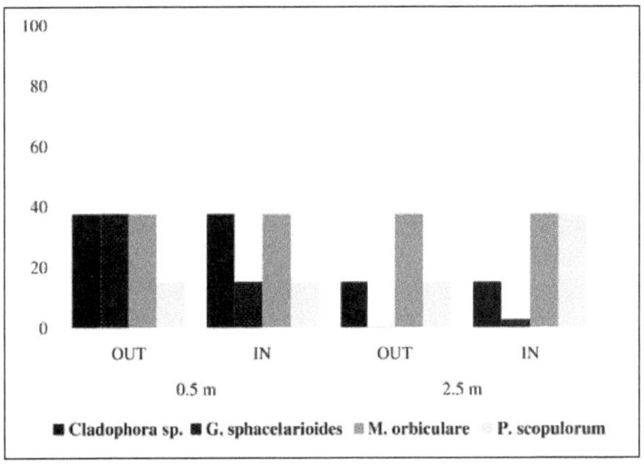

Fig. 5 The mean cover (%) of the dominat algal taxa found on test panels (in – inner side; out – outer side) at 0.5 m and 2.5 m after 9 months of immersion.

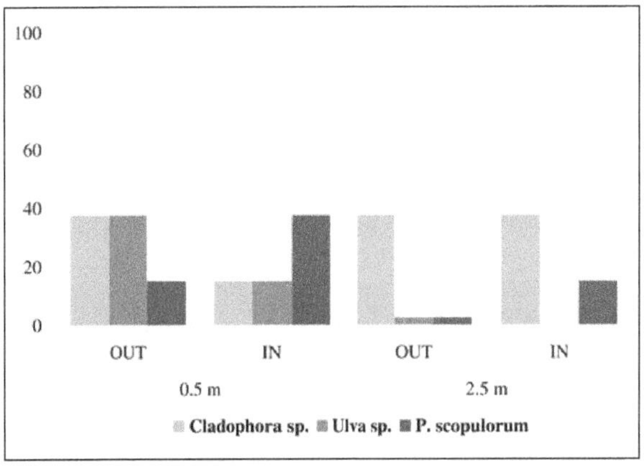

Fig. 6 The mean cover (%) of the dominat algal taxa found on test panels (in – inner side; out – outer side) at 0.5 m and 2.5 m after 12 months of immersion.

Discussion

Organic pollution may increase and speed up the development of fouling communities in originally oligotrophic area [3], or at higher levels, may cause a decrease of biomass and diversity and favor opportunistic species [19,38]. Benthic macrophyta community directly responds to the changes in abiotic variables, reducing its richness, diversity levels and/or enhancing its biomass per surface unit under eutrophication [39]. Therefore it can be used as sensitive bioindicator of ecosystem changes at smaller spatial scales [27]. They provide readable responses with analysis based on a functional group level providing powerful support to traditional species-level analysis. Benthic macrophyte communities have been successfully used as indicators of eutrophication in coastal waters of The Adriatic Sea [40–42].

Fouling biomass was relatively low compare to a similar 12 month study of fouling communities on concrete, plastic and glass panels in Kastela Bay (The Adriatic Sea; from 221 g $\times$ m^{-2} to 61.578 g $\times$ m^{-2}) [43].

Number of found algal taxa were considerably lower compared to 62 benthic algae taxa found on concrete, plastic and glass panels at the entrance of Kastela Bay [43]. Low diversity and species richness may be the result of organic enrichment [27], althought some authors suggest that low diversity is caused by competition between tolerant and non-tolerant species [3].

In polluted areas, the number of algae species decreases, especially Rhodophytes and Phaeophytes, with an increase in the abundance of Chlorophytes [19,44]. Phaeophyceae are extremly sensitive to environmental disturbance [27]. Feldman's *R/P* Index [14], based on marine vegetation, is highly used in the Mediterranean Sea. It was established as a biogeographical index and it is based on the fact that the number of species of Rodophyceae decreases from the Tropics to the Poles. Its application as indicator holds on the higher or lower sensitivity to disturbances of Phaeophyceae and Rhodophyceae [13]. Low *R/P* values and dominance of Phaeophyta over Rhodophyta and Chlorophyta at 0.5 m depth indicated that there was no negative impact of nutrient enrichment on macrophyta. In some cases functional groups was preferable than taxonomic grouping of organisms to reduce spatial and temporal community variability and to discover patterns without loosing important information [23]. Among 27 algal taxa with noticeable presence only three were classified as ESG I (*Acetabularia acetabulum*, *Padina pavonica* and *Fosliella farinosa*). The reason for low presence of late successional ESG I taxa could be the short period of immersion (less than one year).

Although found dominant taxa are very adaptive and represent typical fouler in the Adriatic [45], numerous authors pointed out that macrolagae of genus *Ulva* [46–48] and Cladophora [49] appears in conditions with high nutrient loading [26,46]. The massive presence of the green algae *Ulva* sp. and *Cladophora* sp. along the European coastline is considered to be a reliable indicator for nutrient enriched seawater [50]. Chlorophytes are favoured by an increase of nutrients, and genera such as *Cladophora* and *Ulva* are usually abundant in eutrophicated areas due to their great reproductive capacity and their rapid growth rate [44].

Low diversity and species richness together with massive presence of the green algae (as *Ulva* sp.) and negligible presence of ESG I taxa, may lead to erroneous conclusion that Vela Luka Bay is eutrophicated area. Nevertheless, similar situation may occur in non-polluted environments [13,51,52], e.g. due to short period of immersion. Low values of biomass and *R/P* Index together with dominance of Phaeophyta also support conclusion that there is no negative impact of nutrient enrichment on macrophyta fouling community in Vela Luka Bay.

Conclusions

(*i*) Low biomass, diversity, species richness, *R/P* Index, dominance of Phaeophyta, together with massive presence of green algae (as *Ulva* sp.) and negligible presence of ESG I taxa can result from nutrient enrichment in Vela Luka Bay, but also from short immersion period.

(*ii*) A possibility of wider application of the presence and the abundance of benthic macrophytes for the classification of the Ecological Status is shown.

(*iii*) The analysis of macrophyta fouling communities (functional-form classification, EEI) on artificial test panels obtains simple, cost-effective, non-destructive monitoring system that covers the demands of Water Frame Directive (2000/ 60/ EC).

References

1. Richmond MD, Seed R. A review of marine macro-fouling communities with special reference to animal fouling. Biofouling. 1991;3(2):151–168.
2. Meyer-Reil LA, Köster M. Eutrophication of marine waters: effects on benthic microbial communities. Mar Pollut Bull. 2000;41(1–6):255–263.
3. Mayer-Pinto M, Junqueira AOR. Effects of organic pollution on the initial development of fouling communities in a tropical bay, Brazil. Mar Pollut Bull. 2003;46(11):1495–1503.
4. Rastetter EB, Cooke WJ. Responses of marine fouling communities to sewage abatement in Kaneohe Bay, Oahu, Hawaii. Mar Biol. 1979;53(3):271–280.
5. Juanes JA, Guinda X, Puente A, Revilla JA. Macroalgae, a suitable indicator of the ecological status of coastal rocky communities in the NE Atlantic. Ecol Indic. 2008;8(4):351–359.
6. Connell SD, Glasby TM. Do urban structures influence local abundance and diversity of subtidal epibiota? A case study from Sydney Harbour, Australia. Mar Environ Res. 1999;47(4):373–387.
7. Connell SD. Urban structures as marine habitats: an experimental comparison of the composition and abundance of subtidal epibiota among pilings, pontoons and rocky reefs. Mar Environ Res. 2001;52(2):115–125.
8. Glasby TM, Connell SD. Orientation and position of substrata have large effects on epibiotic assemblages. Mar Ecol Prog Ser. 2001;214:127–135.
9. Svane I, Petersen JK. On the problems of epibioses, fouling and artificial reefs, a review. Mar Ecol. 2001;22(3):169–188.
10. Perkol-Finkel S, Shashar N, Benayahu Y. Can artificial reefs

mimic natural reef communities? The roles of structural features and age. Mar Environ Res. 2006;61(2):121–135.

11. Langhamer O, Wilhelmsson D, Engström J. Artificial reef effect and fouling impacts on offshore wave power foundations and buoys – a pilot study. Estuar Coast Shelf Sci. 2009;82(3):426–432.

12. Knott NA, Underwood AJ, Chapman MG, Glasby TM. Epibiota on vertical and on horizontal surfaces on natural reefs and on artificial structures. J Mar Biol Assoc UK. 2004;84(6):1117–1130.

13. Salas F, Marcos C, Neto JM, Patrício J, Pérez-Ruzafa A, Marques JC. User-friendly guide for using benthic ecological indicators in coastal and marine quality assessment. Ocean Coast Manage. 2006;49(5–6):308–331.

14. Pérez Ruzafa I. Efecto de la contaminación sobre la vegetación submarina y su valor indicador. In: Pérez Ruzafa Á, Marcos C, Salas F, Zamora S, editors. Perspectivas y herramientas en el estudio de la contaminación marina. Murcia: Servicio de Publicaciones de la Universidad de Murcia; 2003. p. 133–147.
15. Chryssovergis F, Panayotidis P. Evolution des peuplements macrophytobenthiques le long d'un gradient d'eutrophisation. Oceanol Acta. 1995;18:649–658.
16. Schiel DR, Wood SA, Dunmore RA, Taylor DI. Sediment on rocky intertidal reefs: Effects on early post-settlement stages of habitat-forming seaweeds. J Exp Mar Biol Ecol. 2006;331(2):158–172.

17. Orfanidis S, Papathanasiou V, Sabetta L, Pinna M, Gigi V, Gounaris S, et al. Benthic macrophyte communities as bioindicators of transitional and coastal waters: relevant approaches and tools. Transit Water Bull. 2007;1(3):45–49.

18. Regier HA, Cowell EB. Applications of ecosystem theory, succession, diversity, stability, stress and conservation. Biol Conserv. 1972;4(2):83–88.

19. Murray SN, Littler MM. Patterns of algal succession in a perturbated marine intertidal community. J Phycol. 1978;14(4):506–512.

20. Sousa WP. The responses of a community to disturbance: The importance of successional age and species' life histories. Oecologia. 1980;45(1):72–81.

21. Schramm W. Factors influencing seaweed responses to eutrophication: some results from EU-project EUMAC. J Appl Phycol. 1999;11(1):69–78.

22. Littler MM, Littler DS. The evolution of thallus form and survival strategies in benthic marine macroalgae: field and laboratory tests of a functional form model. Am Nat. 1980;116(1):25.
23. Panayotidis P, Montesanto B, Orfanidis S. Phytobenthos as a quality element for the ecological status evaluation: a case study of the implementation of the Water Frame Directive (2000/ 60/ EC) in the mediterranean ecoregion. In: Proceedings of the Second Mediterranean Symposium on Marine Vegetation. Athens: UNEP/MAP/RAC/SPA; 2003. p. 211–216.
24. UNEP/MAP. Fact sheets on marine pollution indicators [UNEP(DEC)/MED WG.264/Inf.14]. Athens: UNEP; 2005.
25. Angel D. An application of artificial reefs to reduce organic enrichment caused by net-cage fish farming: preliminary results. ICES J Mar Sci. 2002;59 suppl:324–329.

26. Torras X, Pinedo S, Garcia M, Mangialajo L, Ballesteros E. Assessment of coastal environmental qualita based on littoral community cartography: methodological approach. In: Proceedings of the Second Mediterranean Symposium on Marine Vegetation. Athens: UNEP/MAP/RAC/SPA; 2003. p. 145–153.
27. Mannino AM, Sara G. Effects of fish-farm biodeposition on periphyton assemblages on artificial substrates in the southern Tyrrhenian Sea (Gulf of Castellammare, Sicily). Aquat Ecol. 2007;42(4):575–581.

28. Dadić V, Gačić M, Galasso I, Glavurtić M, Krstulović N, Marasović I, et al. Istrazivanje fizikalnih karakteristika i prijemnog kapaciteta zaljeva Vela Luka s posebnim osvrtom na izgradnju marine. Split: Institute of Oceanography and Fisheries; 1991.
29. Boudouresque CF. Methodea detude qualitative et quantitative du benthos (en paticulier du phytobenthos). Tethys. 1971;3(1):79–104.
30. Braun-Blanquet J, Fuller GD, Conard HS. Plant sociology; the study of plant communities. London: Hafner Pub. Co.; 1965.
31. Coppejans E. Iconographie d'Algues Mediterraneene. Vaduz: Bibliotheca Phycologica; 1983.
32. Costello MJ, Emblow CS, White R. European register of marine species. Aix: Patrimoines Naturels; 2001.
33. Feldmann J. Les algues marines de la côte des Albères. I-III. Cyanophycées, Chlorophycées, Phaéophycées. Rev Algol. 1937;9:149–335.
34. Occhipinti-Ambrogi A. Contributo alla conoscenza dei Briozoi nella Laguna Veneta settentrionale. Boll Mus civ St Nat Venezia. 1981;31:95–109.
35. Parenzan P. Carta d'identita delle conchigllie del Mediterraneo. Bivalvii parte seconda. Tarnto: Bios-Taras; 1976.
36. Riedl R. Fauna e flora del Mediterraneo. Padova: Franco Muzzio Editore; 1991.
37. Sørensen TJ. A method of establishing groups of equal amplitude in plant sociology based on similarity of species content and its application to analyses of the vegetation on danish commons. Kgl Dan Vid Selsk Biol Skr. 1948;4:1–34.
38. Hargrave BT, Thiel H. Assessment of pollution-induced changes in benthic community structure. Mar Pollut Bull. 1983;14(2):41–46.

39. Mirto S, La Rosa T, Gambi C, Danovaro R, Mazzola A. Nematode community response to fish-farm impact in the western Mediterranean. Environ Pollut. 2002;116(2):203–214.
40. Iveša L, Lyons DM, Devescovi M. Assessment of the ecological status of north-eastern Adriatic coastal waters (Istria, Croatia) using macroalgal assemblages for the European Union Water Framework Directive. Aquat Conserv. 2009;19(1):14–23.
41. Orlando-Bonaca M, Lipej L. Benthic macroalgae is bioindicators of the ecological status in the Gulf of Trieste. Varstvo Narave. 2009;22:63–72.

42. Sliskovic M, Jelic-Mrcelic G, Antolic B, Anicic I. The fouling of fish farm cage nets as bioindicator of aquaculture pollution in the Adriatic Sea (Croatia). Environ Monit Assess. 2010;173(1–4):519–532.

43. Span A, Antolic B, Hell Z. Istrazivanje bioloskog aspekta obrastanja na uronjenim staklenim i plasticnim plocama i betonskim stupicima. Split: Institute of Oceanography and Fisheries; 1990.

44. Borowitzka MA. Intertidal algal species diversity and the effect of pollution. Mar Freshwater Res. 1972;23(2):73–84.

45. Igic LJ. Prikaz obrastaja u Jadranskom i drugim svjetskim morima - biolosko znacenje. Pomorski Zbornik. 1995;33:329–356.

46. Golubic S. Effect of organic pollution on benthic communities. Mar Pollut Bull. 1970;1(4):56–57.

47. Ballesteros E, Pérez M, Zabala M. Aproximación al conocimiento de las comunidades algales de la zona infralitoral superior en la costa catalana. Collect Bot. 1984;15:69–100.

48. Rodríguez-Prieto C, Polo L. Effects of sewage pollution in the structure and dynamics of the community of *Cystoseira mediterranea* (Fucals, Phaephyceae). Sci Mar. 1996;60:253–263.

49. Belsher T. Analyse des répercussions des pollutions urbaines sur le macrophytobenthos de la Méditerranée (Marseille, Port-Ventres, Port-Cros). Aix: Paul Cézanne University; 1977.

50. Council Directive for a legislative frame and actions for the water policy, 2000/ 60/ EC. Official Journal of the European Communities. 2000.12.22.

51. Ibanez F, Dauvin JC. Long-term changes (1977 to 1987) in a muddy fine sand *Abra alba-Melinna palmata* community from the Western English Channel: multivariate time-series analysis. Mar Ecol Prog Ser. 1988;49:65–81.

52. Craeymeersch JA. Applicability of the abundance/biomass comparison method to detect pollution effects on intertidal macrobenthic communities. Hydrobiol Bull. 1991;24(2):133–140.

4

Identification of apomixis in the Kentucky bluegrass (*Poa pratensis* L.) using auxin test

Janetta Niemann*, Andrzej Wojciechowski, Jowita Janowicz

Department of Genetics and Plant Breeding, Poznań University of Life Sciences, Dojazd 11, 60-632 Poznań, Poland

Abstract

The effect of auxin treatment on the formation of apomictic – endospermless grains in *Poa pratensis* was investigated in the study. Four polish cultivars: Skiz, Eska 46, Alicja, Ani as well as six breeding lines: POB 13, SK-W-33, SKW-15, SKW-35A, SK-W-35A and SK-W-35B were tested by embryological methods to reveal the reproduction pathway in greenhouse and field conditions. The obtained results showed that four genotypes, i.e. Alicja, Ani, Eska 46 and SK-W-35B are facultative apomictic and the other have formed seeds only on the apomictic way. In facultative apomicts it was found that in many cases both endospermless ovules containing embryo sacs with embryos or embryo sacs with well developed endosperm were formed. In the case of apomictic genotypes only endospermless seeds were observed.

Keywords: apomixis, embryo sac, *Poa pratensis*, auxin test, endosperm, ovules, parthenogenesis

Introduction

Kentucky bluegrass *Poa pratensis* L. is a valuable crop, especially under temperate climatic conditions. It could be used for many purposes including livestock feeding and lawn planting. *P. pratensis* reproduces primarily by apomictic way [1] and is characterized by facultative apomixis, i.e. the same genotype can yield both apomictic and amphimictic seeds [2]. There are two main processes of apomictic reproduction: unreduced embryo sac formation from a somatic cell of the nucellus and parthenogenetic development of apomictic grains in such aposporous embryo sac [3]. Because seeds do not develop without pollination, the type of apomixis is classify as pseudogamous, i.e. combining the development of an unfertilized ovule and the hybrid endosperm [4]. The most widespread form of apomixis in Kentucky bluegrass is apospory, which ensures preservation of the maternal genotype in seedlings [5]. This is the reason of intensive work concerning apomixis research aimed at producing monogenotypic bluegrass cultivars [6–8]. These experiments cannot be performed without studying different types or apomixis mechanisms and developing reliable methods of apomixis identification in wild and cultivated forms of *Poa*. There are different techniques of apomixis screening and estimating their frequency [9,10]. But almost all of them are based on morphological stability of progenies or cytological observations that are difficult and time consuming. The simple method of apomixis identification in Gramineae is an auxin test, developed by Matzk [11]. It is based on the finding that in Poaceae a simple application of synthetic auxin at anthesis induces grain formation without fertilization. This is a procedure [11–13] which estimated the asexual potential as the frequency of auxin induced, endospermless grains containing one or more parthenogenetic embryos. Because parthenogenesis is one of the two main steps in asexual seed production, it gives an estimation of the apomictic capacity of selected plants. Auxin test could be useful for the grass breeders, which should be testing new breeding lines, before the registration of a new variety of *P. pratensis* for DUS (distinctness, uniformity and stability) tests in the Research Centre for Cultivar Testing (COBORU) at Słupia Wielka when it is required to write the percentage of apomixis [14].

The aim of this study was to investigate the parthenogenetic response of different *Poa* genotypes on auxin treatment in the field and greenhouse conditions.

* Corresponding author. Email: niemann@up.poznan.pl

Material and methods

Four polish cultivars: Skiz, Eska 46, Alicja, Ani and six breeding lines: POB 13, SK-W-33, SKW-15, SKW-35A, SK-W-35A and SK-W-35B of *P. pratensis* L. were investigated. The seeds of these genotypes originated from the Małopolska Plant Raising – HBP Ltd., Poland. All genotypes were grown both in a greenhouse and in the field conditions in nursery of Department of Genetics and Plant Breeding, Poznań University of Life Sciences. First, the individual genotypes of *Poa* were cloned in earlier mentioned two environments and further studies on these clones were conducted. It was ten clones for each genotypes. For identification of plants being either incapable or capable for parthenogenesis, modified auxin test was used [11]. Plants were treated with 2,4-dichlorophenoxyacetic acid (2,4-D) in may 2009 and 2010. In 2009 – the first year of our research panicles were sprayed directly before anthesis with 2,4-D in concentrations 40, 80 and 120 mg × L^{-1}. However, the results obtained were comparable, so we decided that panicles will be sprayed only with an 75 mg × L^{-1} solution of 2,4-D. Each year five inflorescences from three chosen clones of each genotype were soaked for one minute in these solution of synthetic 2,4-D. The paper bags were used for isolation. For further observation, one treated and one untreated inflorescence from three clones of each genotype were taken in the same time. Spikelets were sampled from the experimental field and from the greenhouse 25 days after spraying. Six spikelets from the panicle were collected and fixed in the solution of ethyl alcohol:chloroform:acetic acid (6:3:1) for 24 h and stored in 70% ethanol in the refrigerator until use. Additional, spikelets from untreated inflorescences were also sampled to check the grain induction in the absence of the auxin treatment. Embryological analysis of embryo and endosperm formation were made with the help of paraffin section methods [15,16]. There were 720 ovules analyzed altogether. The observation of endosperm and embryos development in embryo sacs (ES) were made under the Nicon microscope.

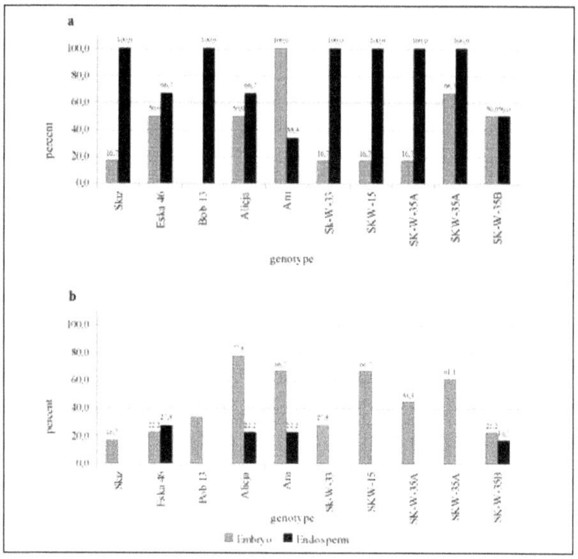

Fig. 1 Percentage of embryos and endosperm observed in: **a** auxin untreated; **b** auxin treated florets of 10 genotypes of *Poa pratensis* in the field condition.

Results

Ovules from the untreated spikelets collected from both analyzed environments contain ES with or nuclear or cellular endosperm almost in all tested genotypes (Fig. 1, Fig. 2). Moreover, it was observed that in untreated ovules collected from the field there were the high percent of developed endosperm (100% in six genotypes) and low percent of embryos (from 0.0% – Pob13 to 100% only in Ani). In contrast, untreated ovules collected from the greenhouse contained higher percent of embryos (100% in 5 genotypes) and lower percent of endosperm than from the fields.

Auxin test analysis showed that ten studied *Poa* genotypes differs considerably in the mode of reproduction. In the field condition in four analyzed genotypes, i.e. Eska 46, Alicja, Ani and Sk-W-35B only 72.2–83.3% of seeds were formed in the apomictic way, because in addition to ES with embryos and without endosperm (Fig. 1, Fig. 3a), could be observed ES with normally developed embryos and endosperm (Fig. 3b). This data indicate, that these genotypes are facultatively apomictic. However, six remaining genotypes: Skiz, Pob13, SK-W-33, SKW-15, SKW-35A and SK-W-35A showed mostly asexual mode of reproduction and only ovules containing one or two embryo sacs with differentiated embryos and visible polar nuclei (Fig. 3c) and without endosperm were observed (Fig. 3c,d). In the greenhouse only in ovules of one genotype – Skiz, the high percent of ES with embryo and without endosperm (94.4%) was observed (Fig. 3e). But the data obtained in our experiment pointed that in the greenhouse the effect of the auxin treatment was a bit better than in the field.

General percentage of ovules without endosperm and with developed endosperm was depending on genotypes and on environment (Tab. 1). After auxin test in the field the highest percent (100%) of ES without endosperm – apomictic seeds was observed in four genotypes, i.e. Skiz, Pob13, Sk-W-33 and Sk-W-35A. However, in the greenhouse the highest number

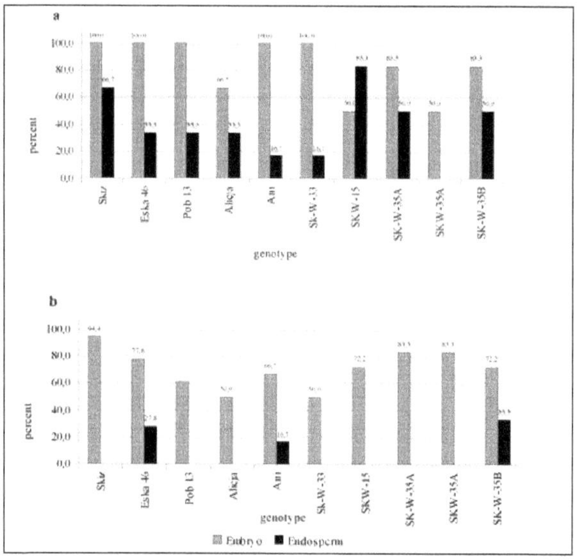

Fig. 2 Percentage of embryos and endosperm observed in: **a** auxin untreated; **b** auxin treated florets of 10 genotypes of *Poa pratensis* in the greenhouse.

of ES without endosperm was in the same genotypes as in the field conditions – except the breeding line SK-W-33 (only 44.4%).

Generally, in auxin treated spikelets taken from the field in all analyzed genotypes ES with differentiated embryos (16.7–77.8%) were observed (Fig. 1). Cellular endosperm was found only in four genotypes, i.e. Eska 46 (27.8%), Alicja (22.2%), Ani (22.2%) and Sk-W-35B (16.7%). In contrast to the field condition, in spikelets after auxin treatment collected from the greenhouse, there was higher percent of ES with embryos (50–94.4%; Fig. 2). Cellular endosperm was found only in ES of three genotypes, i.e. Ani, Eska 46 and SK-W-35B (16.7–33.3%).

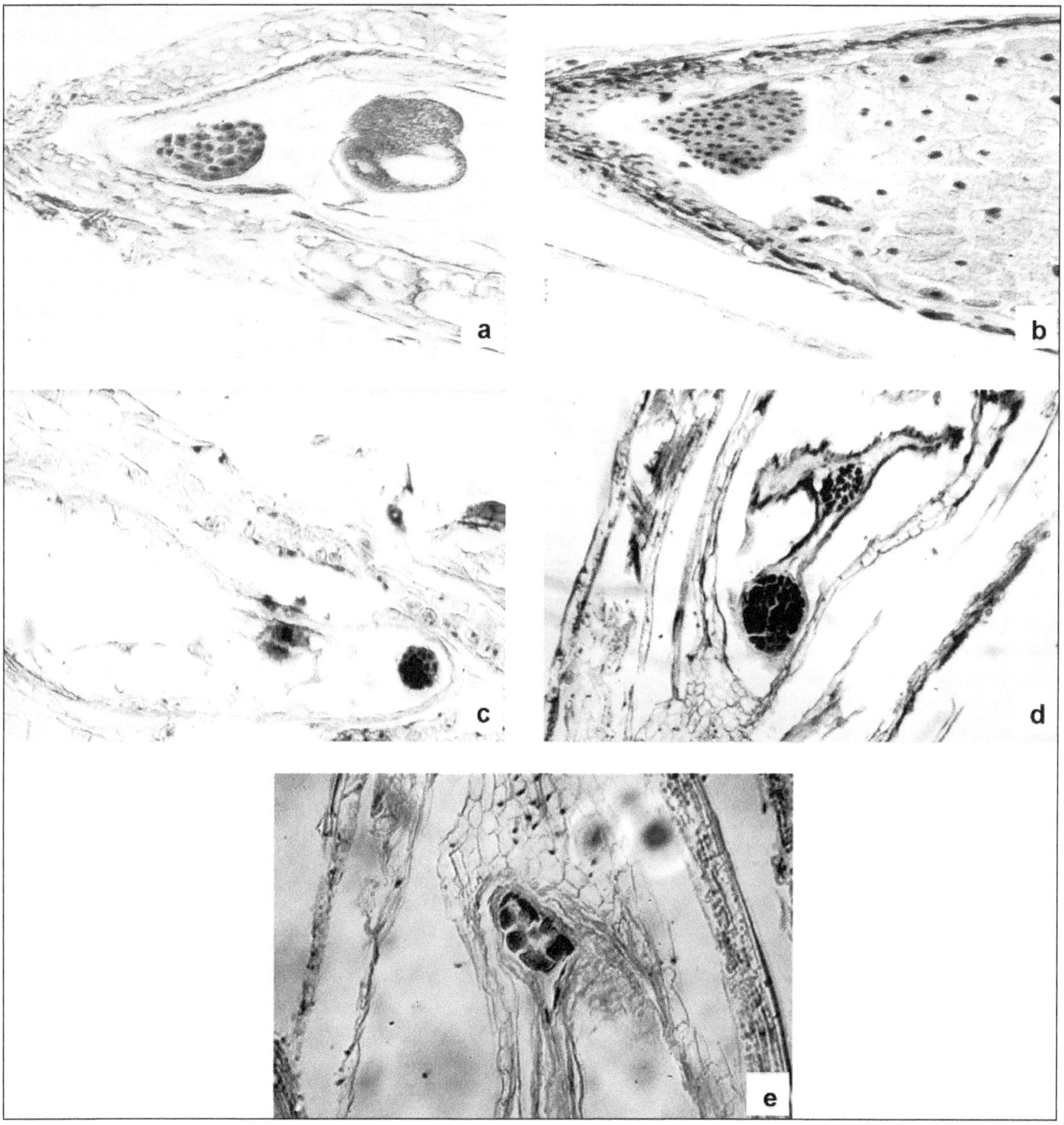

Fig. 3 Embryological analysis of auxin-induced ovules and grains in *Poa pratensis* genotypes. **a–d** Ovules collected from the field condition. **a** Ovule with globular shape embryo of Pob 13 line. To the right of embryo, antipodal cells are visible. **b** Embryo and developed, cellular endosperm in Eska 46 spiklet. **c** Globular embryo and polar nuclei formed in spikelets of Ani cultivar. **d** Two embryos formed in two ES of Alicja cultivar. **e** Ovule with embryo formed in one of two embryo sacs of Skiz cultivar, collected from the greenhouse.

Discussion

The study and other publications clearly demonstrated the importance of determining the characteristics of seeds reproduction in the Kentucky bluegrass (*P. pratensis* L.). In our research the effect of 2,4-D, genotypes and environments on asexual way of seeds formation was investigated. It was shown that auxin test could be a good method for estimation the frequency of apomixis in *Poa*. It could be very useful especially because of the fact that in Kentucky bluegrass the most common mode of reproduction is a facultative aposporous parthenogenesis [17]. Moreover, in addition to the apomictic pathway, sexual reproduction involving fusion of reduced gametes is possible. Our analysis confirmed the thesis presented above, because four of ten studied *Poa* genotypes proved to be a facultative apomicts (Eska 46, Alicja, Ani and Sk-W-35B). This also confirms the assumption adopted in our earlier work [18], where we indicate on two possibilities of seeds formation in Alicja and Ani cultivar – apomictic and sexual reproduction pathway. Expression of parthenogenesis was fairly stable within the analyzed clones, suggesting a strong genetic control. Although differences between environments and genotypes were evident. For example in the greenhouse after 2,4-D application in dose 75 mg × L^{-1}, more apomictic grains were formed comparing to the field conditions. Although, the dose of synthetic auxin applied in our experiments is comparable with another authors, i.e. Mazzucato [13] – 80 mg

Tab. 1 Average percent of ovules without endosperm for ten *P. pratensis* genotypes analyzed by the auxin test in two environments.

	Embryo sacs without endosperm (%)			
	Field conditions		Greenhouse	
	2,4-D		2,4-D	
Genotypes	treatment	untreatment	treatment	untreatment
1. SKIZ	100.0	0.0	100.0	33.3
2. ESKA 46	72.2	33.3	72.2	66.7
3. POB 13	100.0	0.0	100.0	66.7
4. ALICJA	77.8	33.3	100.0	66.7
5. ANI	77.8	66.7	83.3	83.3
6. SK-W-33	100.0	0.0	100.0	83.3
7. SKW-15	100.0	0.0	100.0	16.7
8. SK-W-35A	100.0	0.0	100.0	50.0
9. SKW-35A	100.0	0.0	100.0	100.0
10. SK-W-35B	83.3	50.0	66.7	50.0

× L^{-1}, Matzk [11] – 50–100 mg × L^{-1} or Szabo and Papp [19] – 100 mg × L^{-1}. Effects of the environment on the expression of apomixis were reported by several authors [15,20,21], but differences were always minor comparing to results obtained in presented work. More frequent cases of apomictic ovules observed in the greenhouse might be explained presuming that parthenocarpy can also be induced by natural stress such as heat. Grains with mature embryos without endosperms are found only in plants with genetically established parthenogenesis [11]. Our data indicate that in fully apomictic genotypes the development of ES was independent of environment. It was observed ES with embryos and without endosperm only. In contrast to the earlier researcher [11,19], for observation the embryo and endosperm development we used the paraffin section methods. Probably that is why our data are not the same as for example in Matzk experiment [11], where after auxin treatment sexual plants formed seeds without endosperm and embryos at all. In our study developed endosperm (cellular endosperm) was observed in the ovules containing ES formed on a sexual way after auxin treatment. We confirm earlier observations which have been done by Szabo and Papp [19]. In their work, both the embryo and the weekly developed endosperm in sexual plants were observed.

The percent of endospermless seeds induction was satisfactory in our experiment. Values ranging between 60 and 100% were slightly higher than reported by Matzk, 50–100% [11], who used different synthetic auxins in combinations with other phytohormones, too. The frequency of induction shows a tendency to a genotype specificity, it was low in some genotypes (i.e. Alicja, Ani, SK-W-35B and Eska 46) and high in others (Skiz, SkW-15).

It is concluded that by using the auxin test four of ten studied *Poa pratensis* genotypes formed both apomictic and sexual seeds (Eska 46, Alicja, Ani and Sk-W-35B). Moreover, in greenhouse auxin treatment is slightly more effective than in the field condition, so this method could be especially recommended for greenhouses, where there is no rain and there is less diseases or insects.

It is therefore concluded that auxin test is a workable technique for determining the percentage of parthenogenesis in breeding programs. However, for the future research, the clearing-squash technique [22], could be recommended, which seems to be a good alternative, especially due to their timesaving.

Acknowledgements

The authors would like to thank Małopolska Plant Raising – HBP Ltd., Poland for providing *Poa* genotypes used in this research.

This work is part of the project "Flowering biology in chosen cultivars and strains of *Poa pratensis* with particular emphasis on apomictic way of seeds obtaining" founded by the Polish Ministry of Agriculture and Rural Development.

References

1. Müntzing A. Apomictic and sexual seed formation in *Poa*. Hereditas. 1933;17(2):131–154.
2. Agafonov AV, Sukhareva NB, Baturin SO, Struzhkina OA. Identification of apomixis in the Kentucky bluegrass (*Poa pratensis* L.) using SDS-PAGE of endosperm storage proteins. Biol Bull. 2004;31(1):36–41.
3. Wojciechowski A, Rokicki M. Apomixis and its importance in practical agriculture. Prace Komisji Nauk Rolniczych I Komisji Nauk Leśnych. 2006;100:212–218.
4. Nijs APM. Experimenting with apomixis and sexuality in *Poa pratensis* L. Apomixis Newsl. 1990;2:33–38.
5. Matzk F. Studies towards novel method in breeding the apomictic species *Poa pratensis* L. In: den Nijs APM, Elgersma A, editors. Fodder crops breeding: achievements, novel strategies and biotechnology: Proc 16th Meeting Fodder Crops. Eucarpia. Wageningen: Pudoc; 1991. p. 191–192.
6. Matzk F. Past and new project of apomixis research at Gatersleben. Apomixis Newsl. 1992;4:26–30.
7. Barcaccia G, Mazzucato A, Albertini E, Zethof J, Gerats A, Pezzotti M, et al. Inheritance of parthenogenesis in *Poa pratensis* L.: auxin test and AFLP linkage analyses support monogenic control. Theor Appl Genet. 1998;97(1):74–82.
8. Albertini E, Barcaccia G, Porceddu A, Sorbolini S, Falcinelli M. Mode of reproduction is detected by Parth1 and Sex1 SCAR markers in a wide range of facultative apomictic Kentucky bluegrass varieties. Mol Breeding. 2001;7:293–300.
9. Czapik R. How to detect apomixis in Angiospermae. Polish Bot Stud. 1994;8:13–21.
10. Leblanc O, Mazzucato A. Screening procedures to identify and quantify apomixis. In: Savidan Y, Carman JG, Dresselhaus T, editors. The flowering of apomixis: from mechanisms to genetic engineering. Mexico: CIMMYT, IRD, European Commission DG VI; 2001. p. 121–134.
11. Matzk F. New efforts to overcome apomixis in *Poa pratensis* L. Euphytica. 1991;55(1):65–72.
12. Matzk F. A novel approach to differentiated embryos in the absence of endosperm. Sex Plant Reprod. 1991;4(2):88–94.
13. Mazzucato A, Falcinelli M, den Nijs APM. Estimation of parthenogenesis frequency in Kentucky bluegrass with auxin-induced parthenocarpic seeds. Crop Sci. 1996;36(1):9.
14. Kłyś A, Jankun A. The biology of reproduction in *Poa* L. (Poaceae). Wiad Bot. 2002;46(3-4):19–28.

15. Grazi F, Umaerus M, Åkerberg E. Observations on the mode of reproduction and the embryology of poa pratensis. Hereditas. 1961;47(3–4):489–541.

16. Grober K. Zytologische und embryologische Untursuchungen am Nucellus apomiktischer Sippen der Wiesenrispe (*Poa pratensis* L.). Kulturpflanze. 1986;34:147–158.

17. Bashaw EC, Funk RC. Apomictic grasses. In: Fehr WR, Fehr EL, Jessen HJ, editors. Principles of cultivar development. New York: Macmillan; 1987. p. 40–82. (vol 2).

18. Niemann J, Wojciechowski A, Świtoniak C. Studies on apomixis in chosen cultivars and breeding strains of *Poa pratensis*. Łąkarstwo w Polsce (Grassland Science in Poland). 2009;12:121–129.

19. Szabo Z, Papp M. Parthenogenetic capability of three species in *Poa pratensis* L. aggregation. Acta Biol Szeged. 2005;49:147–148.

20. Hovin AW, Berg CC, Bashaw EC, Buckner RC, Dewey DR, Dunn GM, et al. Effects of geographic origin and seed production environments on apomixis in Kentucky bluegrass. Crop Sci. 1976;16(5):635.

21. Williamson CJ. The influence of light regimes during floral development on apomictic seed production and on variability in resulting seedling progenies of *Poa ampla* and *P. pratensis*. New Phytol. 1981;87(4):769–783.

22. Herr JM. A new clearing-squash technique for the study of ovule development in Angiosperms. Am J Bot. 1971;58(8):785.

5

The detection of thermophilous forest hotspots in Poland using geostatistical interpolation of plant richness

Marcin Kiedrzyński*, Janina Jakubowska-Gabara

Department of Geobotany and Plant Ecology, Faculty of Biology and Environmental Protection, University of Łódź, Banacha 12/16, 90-237 Łódź, Poland

Abstract

Attempts to study biodiversity hotspots on a regional scale should combine compositional and functionalist criteria. The detection of hotspots in this study uses one ecologically similar group of high conservation value species as hotspot indicators, as well as focal habitat indicators, to detect the distribution of suitable environmental conditions. The method is assessed with reference to thermophilous forests in Poland – key habitats for many rare and relict species. Twenty-six high conservation priority species were used as hotspot indicators, and ten plant taxa characteristic of the *Quercetalia pubescenti-petraeae* phytosociological order were used as focal habitat indicators. Species distribution data was based on a 10 × 10 km grid. The number of species per grid square was interpolated by the ordinary kriging geostatistical method. Our analysis largely determined the distribution of areas with concentration of thermophilous forest flora, but also regional disjunctions and geographical barriers. Indicator species richness can be interpreted as a reflection of the actual state of habitat conditions. It can also be used to determine the location of potential species refugia and possible past and future migration routes.

Keywords: biodiversity hotspots; geostatistical interpolation; focal habitats; indicator species; ordinary kriging; Poland; thermophilous forests

Introduction

Attempts to analyze biodiversity hotspots on a regional scale should combine compositional and functionalist criteria [1]. Compositional criteria have typically relied on determining the taxonomical richness or the occurrence of rare, threatened and endemic species [2–4]. The functionalist approach focuses on dynamic processes and indication of possible changes in the distribution of hotspots and their characteristics in the future [1,5].

On a regional scale, it seems to be difficult to produce a single map of hotspots for different groups of species that differ ecologically [6], which is based on different diversity measures [4]. The detection of hotspots in this study uses one ecologically similar group of high conservation value species as hotspot indicators, as well as focal habitat indicators, to detect the distribution of suitable environmental conditions.

Species representative of key habitats of regional biodiversity may be used as indicators of potential habitats for rare species. The study assumes that patterns of focal habitat distribution can be included in the functionalist approaches to hotspot detection. This approach is similar to that of Hughes et al. [7] who propose that emphasis should be placed on preserving the connectivity of widely-dispersed species, combined with intensive protection of quite localized hotspots.

The study evaluates this method with regard to thermophilous forests on Polish territory. They play an important role as habitats for many rare and threatened plants, and are one of the most species-rich forest ecosystems in Central Europe [8–10]. In the European Union, most are protected in the Natura 2000 network as natural habitats [10,11]. Concentrations of thermophilous plants in Central Europe are typical of the complexes of forest, scrub, fringe and grassland vegetation that remained in the relict “dry islands” [12–14]. The long-lasting existence of the Central European forests with helio- and thermophilous flora was probably possible in combination of specific habitat conditions and the regime of anthropo-zoogenic, fire disturbance and large herbivore pressure or special forestry practices, such as coppicing [9,15–19]. During recent decades, a rapid extinction of open forests has been observed as a result of the expansion of mesophilic trees and shrubs after cessation of such traditional forms of forest use as grazing and litter raking, and the overall anthropogenic eutrophication of soils [9,20,21]. As the Polish territory to the north of the Carpathians and the Sudetes Mts. offers various habitat conditions in the glacial, periglacial and highland landscapes, the question arises:

* Corresponding author. Email: kiedmar@biol.uni.lodz.pl

Handling Editor: Jacek Herbich

how large are the areas with a concentration of the rarest thermophilous forest plants in these different landscapes?

The aim of this study was twofold: (*i*) to indicate the hotspots of thermophilous forest flora in Poland using high conservation priority species as indicators, and (*ii*) to detect areas with suitable conditions for thermophilous forest flora using the focal habitat indicators. Generalization of indicator species richness was made by using geostatistical interpolation methods [22,23].

Material and methods

Study area

The study was conducted in the Polish territory, mostly lowland and upland areas between the Carpathians, the Sudetes and the Baltic Sea in Central Europe, stretching between the north latitudes 49°00′–54°50′ and the east longitudes 14°07′–24°08′. The study area is located in the transition zone between the temperate oceanic (Atlantic) climate in the west and the moderate continental climate in the east. The potential natural vegetation is mainly mesophilic oak-hornbeam forests, beech forests and mixed pine-oak forests [24]. The patches of thermophilous forests are of extrazonal character [9,25].

Hotspot distribution was analyzed in the following landscapes: young glacial, periglacial (old glacial) and highlands/mountainand in relation to the maximum extents of the Vistulian (Würm) and Wartanian (Riss) Glaciations, according to Marks et al. [26].

Hotspot indicators

The species from the "Red list of vascular plants in Poland" [27] and from the "Polish red data book of plants" [28] that occur most frequently in thermophilous forests and thermophilous fringe communities was chosen. Twenty-six taxa were identified: *Adenophora liliifolia, Carduus collinus, Cephalanthera damasonium, C. longifolia, C. rubra, Cypripedium calceolus, Dictamnus albus, Dorycnium herbaceum, Epipactis microphylla, Euphorbia epithymoides, Festuca amethystina* subsp. *ritschlii, Inula germanica, Lathyrus pisiformis, Buglossoides purpurocaerulea* (=*Lithospermum purpurocaeruleum*), *Melampyrum cristatum, Orchis pallens, Orchis purpurea, Orchis tridentata, Peucedanum alsaticum, Potentilla rupestris, Prunus fruticosa, Quercus pubescens, Rosa gallica, Thesium ebracteatum, Veratrum nigrum* and *Veronica paniculata*. All plant nomenclature is given according to Tutin et al. [29]. The species richness analysis is based on plant distribution data from the Atlas of vascular plant distribution in Poland, ATPOL [30], collected on the basis of a 10 × 10 km grid.

Focal habitat indicators

Indicators of thermophilous forests north of the Carpathians and Sudetes Mts. belong to a well-defined phytosociological group of species characteristic of subcontinental and submediterranean oak forests *Quercetalia pubescenti-petraeae* Klika 1933 corr. Moravec in Beguin et Theurillat [8,25,31,32]. Ten taxa characteristic of the *Quercetalia pubescenti-petraeae* phytosociological order, after Matuszkiewicz [25], were chosen: *Campanula persicifolia, Hypericum montanum, Lathyrus niger, Melittis melissophyllum, Potentilla alba, Pulmonaria angustifolia, Pulmonaria mollis, Ranunculus polyanthemos, Sorbus torminalis*, and *Tanacetum corymbosum* ssp. *corymbosum*. Again, the richness of the above-mentioned indicators (per square 10 × 10 km) was calculated with the use of data on the distribution of species from the Atlas of vascular plant distribution in Poland – ATPOL [30].

Spatial interpolation of species richness

The interpolation of species richness was performed by ordinary kriging (OK) to detect its general spatial pattern. OK is one of the most commonly used methods for geostatistical interpolation of data in the environmental sciences [22,23,33].

Maps of the distribution of indicator species from the ATPOL atlas [30] were digitized with ArcGIS Desktop™ 9.2 software [34]. Midpoints were generated for the squares (10 × 10 km) and species richness was calculated for each. Because the data being arranged on a regular grid at a fairly high density, hence the values on the map of interpolation uncertainties were evenly distributed [22,35].

To determine the spatial structure of species richness, a semivariogram was calculated [22,36]. Basic spatial parameters of semivariogram such as nugget variance (*Co*), structural variance (*C*), sill (*Co*+*C*) and range (*A*) were calculated [22,37]. The best fit model, spherical, exponential or Gaussian, was selected. Models were assessed using two methods. The first was the residual sum of squares (residual SS), which provides an exact measure of how well the model fits the variogram data: the lower the reduced sum of squares, the better the fit of the model. The second was the regression coefficient r^2, which indicates how well the model fits to the variogram data, although it is not as sensitive or robust as the residual SS for best-fit calculations [37]. The parameters of the model that best fit the empirical semivariogram were further used for spatial interpolation of data. For search methods, a radius equal to the range (*A*) in the model that best fit the semivariogram, and a cell size of 10 × 10 km was used. Geostatistical analysis was performed using GS+9™ Gamma Design Software [37].

The interpolation produced raster data assigned with values for indicator species richness. Visualizations of the interpolation results, statistics and correlation analysis were performed in ArcGIS Desktop™ 9.2: spatial analysis module, multivariate analysis module and band collection statistic tool [34].

Results

Modeling the spatial structure of indicator species richness

The exponential model demonstrated the best fit to the empirical semivariogram demonstrating the spatial structure of richness of hotspot and habitat indicator species in the analyzed area (Tab. 1). In comparison with the other models, it had the smallest residual SS (0.0067 for hotspot indicator richness and 0.076 for habitat indicator richness) and a correlation coefficient r^2 closest to unity (0.901 and 0.940, respectively). Its nugget variance was moderate (0.106 and

0.46) and the range was the greatest (40 800 m and 63 000 m), indicating the distance to which the richness of the studied species is spatially dependent. In line with the results of variogram modeling, the OK interpolation was based on the exponential model. The radius of the search used in OK was the determined range, i.e. 40 800 and 63 000 m.

Tab. 1 Parameters of variogram model types examined to choose a model that best fitted to the experimental data of richness of hotspot and habitat indicators of thermophilous forests in Poland.

Variogram model type	*Co*	*Co+C*	*A* (m)	Residual SS	r^2
Richness of hotspot indicators					
Spherical	0.028	1.074	30 900	0.0117	0.826
Expotential	**0.106**	**1.078**	**40 800**	**0.0067**	**0.901**
Gaussian	0.148	1.074	26 673	0.0116	0.828
Richness of habitat indicators					
Spherical	0.16	3.98	43 000	1.191	0.836
Expotential	**0.46**	**4.01**	**63 000**	**0.076**	**0.940**
Gaussian	0.61	3.96	38 105	0.181	0.840

Values in bold type – parameters of the best fitted model.

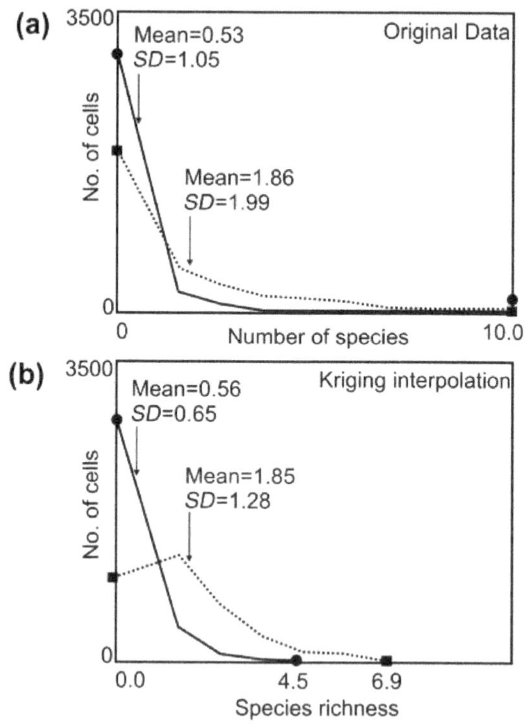

Fig. 1 The schedule of cells in raster layers which presents the richness of the hotspot indicators (solid line) and habitat indicators (dashed line) of thermophilous forests in Poland. **a** Original data. **b** Data after ordinary kriging interpolation; species richness calculated per square 10 × 10 km. *SD* – standard deviation.

Species richness, its interpolation and correlations

The original species number of both hotspot and habitat indicators range from 0 to 10 species per 10 km square (Fig. 1a). However, although the species pool of habitat indicators (10) is less than half of hotspots indicators (26), mean richness of habitat indicators is 3 times higher (Fig. 1a). The number of cells (i.e. area occupied on the maps) with close to mean values of habitat indicator richness was emphasized after the spatial interpolation of data (Fig. 1a,b). In both cases, the kriging interpolation reduced the species richness ranges, from 10 per square to 4.5 for hotspot and 6.9 for habitat indicator species. The correlation coefficient between raster layers, which represent the richness of hotspot and habitat indicators, is 0.56 for original data, and 0.64 after ordinary kriging interpolation.

Hotspot distribution

The upland areas in the south of the Wartanian Glaciation (the Polish Uplands and the Sudetic Foreland) are the richest in hotspots. In this part of the country occur 10 squares with original number of indicator species equal 7 and above (Fig. 2a, , squares No. 2–11) and five hotspots, where the interpolated richness ranged from 3.1 to 4.0 species and in three of them to 4.5 (Fig. 2b). The hotspots with richness more than 2.0 occupy 10% of this part of the country (Tab. 2).

The second center of hotspots is located in the Wielkopolska Lowland and Masurian Lakeland, in the young glacial landscape (Fig. 2b). However, only one hotspot with original number of species equal 9 (Fig. 2a, square No. 1) and with interpolated richness up to 4.0 is located in this part of the country: the Bielinek nature reserve (Fig. 2a,b). The rest of hotspots with interpolated richness up to 3.0 are dispersed and isolated (Fig. 2b); the areas with richness more than 2.0 occupy 1.8% of the area.

A large disjunction between areas with hotspots was identified in central Poland (Fig. 2b), where the areas with richness of hotspot indicators more than 2.0 occupy only 0.1% (Tab. 2). This part of Poland is situated between the Vistulian and the Wartanian Glaciations in the old glacial landscape (Fig. 2b).

The distribution of suitable habitats i.e. areas with richness of indicator species is higher than 2.0 (close to mean value 1.85 – for the whole study area) is similar to general pattern of hotspots distribution (Fig. 2b,c). Suitable habitats are widely distributed (areas with richness >2.0 occupy 36.0%) than hotspots (areas with richness >2.0 occupy 4.5%) and therefore can indicate possible landscape connections between hotspots. Five areas with high richness of habitat indicators but without hotspots occur: one in the eastern part of the Polish Uplands, with richness of habitat indicators up to 6.9 and four in the north-eastern and central parts of the country, with richness up to 6.0 (Fig. 2b,c).

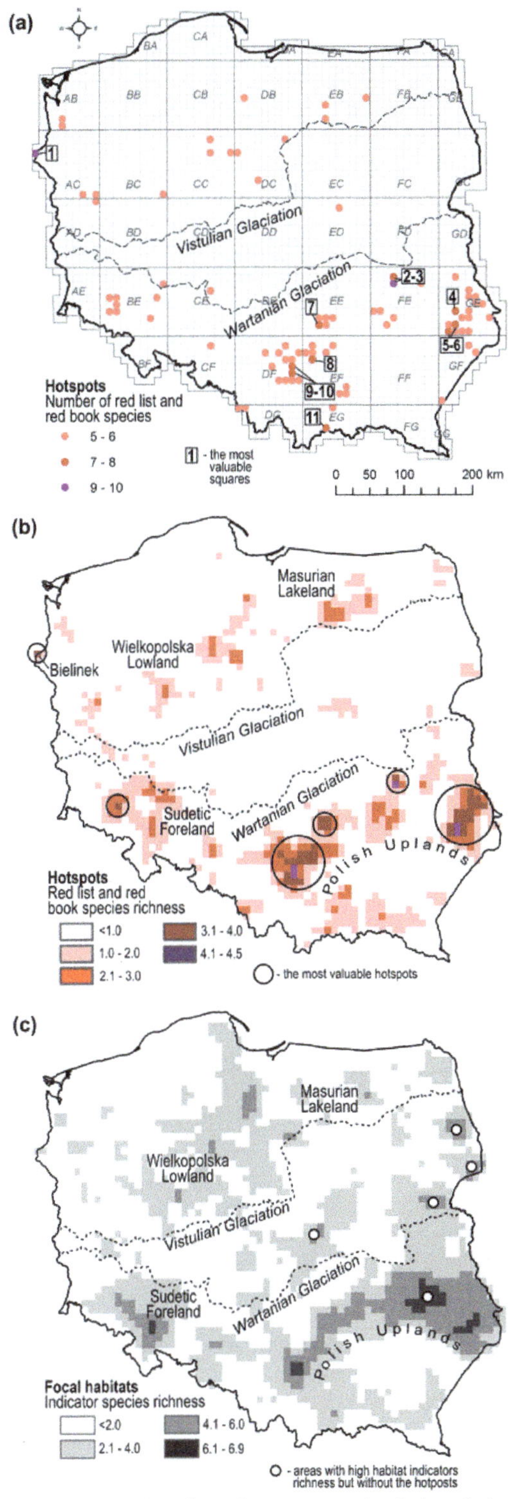

Fig. 2 Thermophilous forest hotspots in Poland identified according to original data (**a**), and geostatistical interpolation (**b**), compared with the distribution of their focal habitats (**c**). The most valuable squares in (**a**): 1 – Bielinek, 2–3 – Puławy-Kazimierz Dolny, 4 – Izbica, 5–6 – Zamość-Zwierzyniec, 7 – Bolmin, 8 – Racławice, 9–10 – Ojców-Bolechowice, 11 – Pieniny. Species richness in **b** and is**c**expressed by ordinary kriging interpolation. Maximum glaciation extents after Marks et al. [26]: Vistulian Glaciation (115–12 × 10^3 years BP), Wartanian Glaciation (210–130 × 10^3 years BP).

Tab. 2 The area (km^2 × 100) and percentage cover (%) of increasing interpolated richness of hotspot indicators in the three parts of the study area distinguished according to the glaciation ranges.

Parts of the study area according to the glaciation ranges	Interpolated richness of hotspot indicators							
	>1.0		>2.0		>3.0		>4.0	
	A	%	*A*	%	*A*	%	*A*	%
North of the Vistulian	126	10.8	21	1.8	1	0.1	0	0.0
Between Vistulian and Wartanian	32	4.1	1	0.1	0	0.0	0	0.0
South of the Wartanian	390	33.4	118	10.1	41	3.5	5	0.4

Discussion

The geostatistical analysis indicates that the maximal distance for significant spatial correlation is 40.8 km for hotspot species richness and 63.0 km for habitat indicator richness. Moreover, the mean value for hotspot indicator richness is less than one third of that of the habitat indicators. These results imply that the hotspot indicators, and by extension, the most threatened thermophilous species, occupy rather isolated environmental islands in comparison to the habitat indicators, which are rather well dispersed. Moderate positive correlations were seen between the hotspot and habitat indicators: the correlation coefficient was found to be 0.56 for the original data and 0.64 after interpolation. Some areas with high habitat indicator richness but with a low incidence of hotspot indicators were detected. They could act as sites where the threatened thermophilous plants may be found in the future, or possibly as environmental corridors.

The ordinary kriging method produced generalized map of indicator species richness. The result can be used as an background layer in phylogeographical studies, spatial modeling, metapopulation demography and landscape interaction modeling. Spatial patterns in actual species distribution are easier to identify after the data is subjected to interpolation. Interpolation of floristic data eliminates also the possible lack of precision or random errors in the location of species, or uneven precision in the examination of the study area. It is important to note that the data was analyzed on a large spatial scale, and was collected for many years by many researchers. The results of this study confirm that ordinary kriging is a good method for the identification of spatial patterns in species richness, as previously noted by other studies [4,38–40].

The distribution of the most important areas for rare plants occurring in thermophilous forests in Polish territory was identified. They are located in the Polish Uplands of the southern and southeastern part of the country, and in the Sudetic Foreland. In both cases, these are areas with a high concentration of calcium in the soil, a favorable microclimate on the southern hill slopes and the nearest location to the glacial refugia in southern and southeastern Europe [41].

The lowland center of concentration of the studied species is located in the young glacial landscape of the Wielkopolska and Masuria regions, within the maximum extent of the

Vistulian Glaciation [26]. The occurrence of thermophilous plants is here connected with the presence of young soils rich in calcium and a pronounced topography [42]. Of special importance is the Bielinek nature reserve on the steep slope of the Odra river valley, which represents the only location of *Quercetum pubescenti-petreae* submediterrianian oak forests in the country [43,44]. Analysis of the occurrence of the rarest plants from the Polish red list and red book confirms the high concentration of endangered forest thermophilous flora at this site, in comparison with other lowland hotspots.

The area of central Poland was found to be a regional disjunction in hotspots of rare thermophilous forest flora. The richness of hotpots and focal habitat indicators was found to be low to the south of the extents of Pleistocene glaciers (Vistulian and Wartanian), where sandy deposits and humid habitats are most common [26]. These areas constitute geographic barriers for thermophilous plants.

An analysis of species richness distribution could also be useful for defining the possible entry areas of thermophilous plants into the Polish territory during their Holocene migration (Fig. 3), which have been discussed in the literature for the last few decades [45–47]. The present study suggests that great migration possibilities existed from the Ukrainian Uplands and along the northern peripheries of the Carpathians to the Polish Uplands, and from the Pannonian Basin through the Moravian Gate and the Beskid Niski Mts (Fig. 3). Our analysis also emphasizes the role of the Poprad River and the Dunajec river valleys as corridors for the thermophilous flora to pass through the Carpathians, which was previously suggested by Cyunel [48] and Pawłowski [49]. Taking into consideration indicated habitat barriers in the central part of the country, the migration to the Wielkopolska and the Masuria regions probably took place from northwestern and northeastern directions, respectively (Fig. 3), as suggested by Paul [47]. The current rate of loss of open forests is so great that, in the case of thermophilous forest flora, the anthropogenic isolation of forest complexes and limited dispersion abilities of plants has resulted in the formation of more isolated in situ refugia, and only a minor migration of species can be observed.

In order to manage biodiversity, it is important to know where it is and how it is arranged at different spatial scales [50]. There is an increasing awareness that conservation on a small scale is not sufficient to protect the processes at a landscape level. The identification of priority areas, or hotspots, is a key component of conservation biogeography [51,52]. As the results of hotspot detection depend on criteria and produce alternative maps of potential conservation prioritization [53], other, more structural and functional criteria should be incorporated in hotspot detection [1], especially in regions which are not rich in endemic species, like the study area presented in this paper.

Modeling the distribution of potential habitats for species is usually based on a variety of environmental features [54–56]. As was shown our study, one of such feature could be the presence of habitat indicator species, since areas with similar environmental characteristics may serve as refugial for groups of species living in a similar habitat. The richness of habitat indicators is an expression of bioclimatic and soil conditions and biological interactions. The richness of habitat indicators can also be understood as a diagnosis of the actual state of habitats on a large spatial scale [57]. Using a species representative of a habitat as an indicator is most justified in areas with well-defined plant or animal ecological assemblages. The use of a combination of hotspot and habitat indicators could give very informative results on regional spatial patterns in flora and fauna.

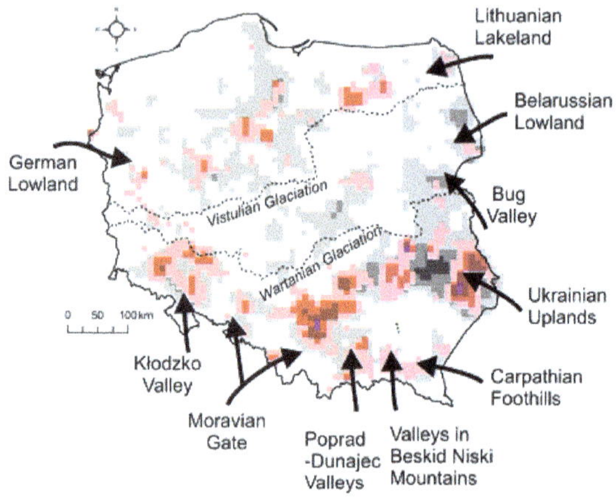

Fig. 3 Potential post-glacial migration routes of thermophilous forest flora into Polish territory on the background of distribution of their current hotspots and focal habitats according to Fig. 2. Maximum extents of Vistulian and Wartanian Glaciations after Marks et al. [26].

Acknowledgments

We wish to thank Professor Ladislav Mucina for the very constructive comments on the previous version of the manuscript. We thank the editors and anonymous reviewers for their remarks and comments, which helped us to improve the manuscript. This study was financially supported by Department of Geobotany and Plant Ecology, University of Łódź.

Authors' contribution

The following declarations about authors' contributions to the research have been made: concept of the study: MK; GIS-geostatistical analysis: MK; writing: MK, JJG.

References

1. Whittaker RJ, Araújo MB, Jepson P, Ladle RJ, Watson JEM, Willis KJ. Conservation biogeography: assessment and prospect. Divers Distrib. 2005;11(1):3–23.
2. Whittaker RJ, Willis KJ, Field R. Scale and species richness: towards a general, hierarchical theory of species diversity: towards a general theory of diversity. J Biogeogr. 2001;28(4):453–470.
3. Mittermeier RA, Turner WR, Larsen FW, Brooks TM, Gascon C. Global biodiversity conservation: the critical role of hotspots. In: Zachos FE, Habel JC, editors. Biodiversity hotspots. Berlin: Springer Berlin Heidelberg; 2011. p. 3–22.
4. Feng JM, Zhang Z, Nan RY. Non-congruence among hotspots based on three common diversity measures in Yunnan, south-west China. Plant Ecol Divers. 2011;4(4):353–361.
5. Williams PH, Araújo MB. Integrating species and ecosystems monitoring for identifying conservation priorities. Eur Conserv. 2000;4:17–18.
6. Cardoso da Silva JM, Cardoso de Sousa M, Castelletti CH. Areas of endemism for passerine birds in the Atlantic forest, South America:

bird areas of endemism in the Atlantic Forest. Glob Ecol Biogeogr. 2004;13(1):85–92.

7. Hughes TP, Bellwood DR, Connolly SR. Biodiversity hotspots, centres of endemicity, and the conservation of coral reefs: biodiversity of coral reefs. Ecol Let. 2002;5(6):775–784.
8. Chytrý M. Thermophilous oak forests in the Czech Republic: syntaxonomical revision of the *Quercetalia pubescenti-petraeae*. Folia Geobot. 1997;32(3):221–258.
9. Jakubowska-Gabara J. Decline of *Potentillo albae-Quercetum* Libb. 1933 phytocoenoses in Poland. Vegetatio. 1996;124(1):45–59.
10. Bergmeier E, Petermann J, Schröder E. Geobotanical survey of wood-pasture habitats in Europe: diversity, threats and conservation. Biodivers Conserv. 2010;19(11):2995–3014.
11. European Commission. Directive 92/43/EEC of 21 May 1992 on the conservation of natural habitats and of wild fauna and flora. J Eur Communities. 1992;206:7–50.
12. Ellenberg HH. Vegetation ecology of Central Europe. 4th ed. Cambridge: Cambridge University Press; 1988.
13. Mucina L, Kolbek J. *Festuco-Brometea*. In: *Mucina* L, Grabherr G, Ellmauer T, editors. Plant communities of Austria. Jena: Gustav Fischer Press; 1993. p. 420–492. (pt 1).
14. Hájková P, Roleček J, Hájek M, Horsák M, Fajmon K, Polák M, et al. Prehistoric origin of extremely species-rich semi-dry grasslands in the Bílé Karpaty Mts. Preslia. 2011;83(2):185–204.
15. Vera FWM. Grazing ecology and forest history. Oxfordshire: CABI; 2000.
16. Ralska-Jasiewiczowa M, Nalepka D, Goslar T. Some problems of forest transformation at the transition to the oligocratic/*Homo sapiens* phase of the Holocene interglacial in northern lowlands of Central Europe. Veg Hist Archaeobot. 2003;12(4):233–247.
17. Roleček J. Formalized classification of thermophilous oak forests in the Czech Republic: what brings the Cocktail method? Preslia. 2007;79:1–21.
18. Jamrichova E, Szabo P, Hedl R, Kunes P, Bobek P, Pelankova B. Continuity and change in the vegetation of a Central European oakwood. Holocene. 2013;23(1):46–56.
19. Szymura TH, Szymura M. Spatial variability more influential than soil pH and land relief on thermophilous vegetation in overgrown coppice oak forests. Acta Soc Bot Pol. 2013;82(1):5–11.
20. Kwiatkowska AJ, Wyszomirski T. Decline of *Potentillo albae-Quercetum* phytocoenoses associated with the invasion of Carpinus betulus. Vegetatio. 1988;75(1–2):49–55.
21. Hédl R, Kopecký M, Komárek J. Half a century of succession in a temperate oakwood: from species-rich community to mesic forest. Divers Distrib. 2010;16(2):267–276.
22. Li J. A review of spatial interpolation methods for environmental scientists. Canberra: Geoscience Australia; 2008.
23. Li J, Heap AD. A review of comparative studies of spatial interpolation methods in environmental sciences: performance and impact factors. Ecol Inf. 2011;6(3–4):228–241.
24. Matuszkiewicz W. Potential natural vegetation of Poland, scale 1:300000. Warsaw: Polish Academy of Sciences; 1995.
25. Matuszkiewicz JM. Zespoły leśne Polski. Warsaw: Polish Scientific Publishers PWN; 2001.
26. Marks L, Ber A, Gogołek W, Piotrowska K, editors. Geological map of Poland, scale 1:500000. Warsaw: Polish Geological Institute; 2006.
27. Zarzycki K, Szeląg Z. Red list of the vascular plants in Poland. In: Mirek Z, Zarzycki K, Wojewoda W, Szeląg Z, editors. Red list of plants and fungi in Poland. Cracow: W. Szafer Institute of Botany, Polish Academy of Sciences; 2006. p. 9–20.
28. Zarzycki K, Kaźmierczakowa R, editors. Polish red data book of plants. Pteridophytes and flowering plants. Cracow: W. Szafer Institute of Botany, Polish Academy of Sciences; 2001.
29. Tutin TG, Heywood VH, Burges NA, Valentine DH, Walters SM, Webb DA, editors. Flora Europaea. Cambridge: Cambridge University Press; 1968. (vol 2–5).
30. Zając A, Zając M, editors. Distribution atlas of vascular plants in Poland. Cracow: Laboratory of Computer Chorology, Institute of Botany, Jagiellonian University; 2001.
31. Müller T. *Quercetalia pubescenti-petraeae* Klika 33 cor. Moravec in Béguin et Theurillat 84. In: Oberdorfer E, editor. South German plant communities. 4. Forests and scrubs. Jena: Gustav Fischer Press; 1992. p. 119–137.
32. Roleček J. Vegetation types of dry-mesic oak forests in Slovakia. Preslia. 2005;77(3):241–261.
33. Bello-Pineda J, Hernandez-Stefanoni JL. Comparing the performance of two spatial interpolation methods for creating a digital bathymetric model of the Yucatan submerged platform. Pan Am J Aquat Sci. 2007;2(3):247–254.
34. ArcGIS 9.2. Redlands, CA: ESRI; 1999.
35. Burrough PA, Mcdonnell RA. Principles of geographical information systems. Oxford: Oxford University Press; 1998.
36. Isaaks EH, Srivastava RM. An introduction to applied geostatistics. Oxford: Oxford University Press; 1989.
37. Robertson GP. GS+: geostatistics for the environmental sciences. Plainwell, MI: Gamma Design Software; 2008.
38. Valley RD, Drake MT, Anderson CS. Evaluation of alternative interpolation techniques for the mapping of remotely-sensed submersed vegetation abundance. Aquat Bot. 2005;81(1):13–25.
39. Hernandez-Stefanoni JL, Ponce-Hernandez R. Mapping the spatial variability of plant diversity in a tropical forest: comparison of spatial interpolation methods. Env Monit Assess. 2006;117(1–3):307–334.
40. Yasrebi J, Saffari M, Fathi H, Karimian N, Moazallahi M, Gazni R. Evaluation and comparison of ordinary kriging and inverse distance weighting methods for prediction of spatial variability of some soil chemical parameters. Res J Biol Sci. 2009;4(1):93–102.
41. Szafer W. Szata roślinna Polski niżowej. In: Szafer W, Zarzycki K, editors. Szata roślinna Polski. 2nd ed. Warsaw: Polish Scientific Publishers PWN; 1972. p. 17–188.
42. Czubiński Z. Zagadnienia geobotaniczne Pomorza. Poznań: Poznańskie Towarzystwo Przyjaciół Nauk; 1950. [Badania fizjograficzne nad Polską zachodnią; vol 2(4)].
43. Celiński F, Filipek M. Flora i zespoły roślinne leśno-stepowego rezerwatu w Bielinku nad Odrą. Poznań: Poznańskie Towarzystwo Przyjaciół Nauk; 1958. (Badania fizjograficzne nad Polską zachodnią; vol 4).
44. Matuszkiewicz JM, Kozłowska AB. Przegląd fitosocjologiczny zbiorowisk leśnych Polski – ciepłolubne dąbrowy. Fragm Flor Geobot. 1991;36(1):203–256.
45. Pawłowska S. Charakterystyka statystyczna i elementy flory polskiej. In: Szafer W, Zarzycki K, editors. Szata roślinna Polski. Warsaw: Polish Scientific Publishers PWN; 1972. p. 129–206. (vol 1).
46. Babczyńska-Sendek B. Problemy fitogeograficzne i syntaksonomiczne kserotermów Wyżyny Śląskiej. Katowice: Wydawnictwo Uniwersytetu Śląskiego; 2005.
47. Paul W. Xerothermic species of the genus *Campanula* in Poland – a model for the phylogeographical assessment of reconstruction of post-glacial migration routes. Ann UMCS Biol. 2012;67(1):27–36.
48. Cyunel E. Studia nad rozmieszczeniem gatunków kserotermicznych w polskich Karpatach Zachodnich. Fragm Flor Geobot. 1959;5(3):409–441.
49. Pawłowski B. Szata roślinna gór polskich. In: Szafer W, Zarzycki K, editors. Szata roślinna Polski. Warsaw: Polish Scientific Publishers PWN; 1972. p. 189–252. (vol 2).
50. Richardson DM, Whittaker RJ. Conservation biogeography – foundations, concepts and challenges: conservation biogeography:

foundations, concepts and challenges. Divers Distrib. 2010;16(3):313–320.

51. Dunstan PK, Bax NJ, Foster SD, Williams A, Althaus F. Identifying hotspots for biodiversity management using rank abundance distributions: identifying hotspots using rank abundance distributions. Divers Distrib. 2012;18(1):22–32.

52. Keppel G, van Niel KP, Wardell-Johnson GW, Yates CJ, Byrne M, Mucina L, et al. Refugia: identifying and understanding safe havens for biodiversity under climate change. Glob Ecol Biogeogr. 2012;21(4):393–404.

53. Dimitrakopoulos P, Memtsas D, Troumbis AY. Questioning the effectiveness of the Natura 2000 special areas of conservation strategy: the case of Crete. Glob Ecol Biogeogr. 2004;13(3):199–207.

54. Franklin J. Mapping species distributions: spatial inference and prediction. Cambridge: Cambridge University Press; 2009. (Ecology, biodiversity and conservation).

55. Franklin J. Moving beyond static species distribution models in support of conservation biogeography: moving beyond static species distribution models. Divers Distrib. 2010;16(3):321–330.

56. Kiedrzyński M, Zielińska KM, Kiedrzyńska E, Jakubowska-Gabara J. Regional climate and geology affecting habitat availability for a relict plant in a plain landscape: the case of *Festuca amethystina* L. in Poland. Plant Ecol Divers. 2014.

57. Stewart JR, Lister AM, Barnes I, Dalen L. Refugia revisited: individualistic responses of species in space and time. Proc Biol Sci B. 2010;277(1682):661–671.

6

Pollination and floral biology of *Adonis vernalis* L. (Ranunculaceae) – a case study of threatened species

Bożena Denisow[1], Małgorzata Wrzesień[2], Anna Cwener[2]*

[1] Department of Botany, University of Life Sciences in Lublin, Akademicka 15, 20-950 Lublin, Poland

[2] Institute of Biology and Biochemistry, Maria Curie-Skłodowska University, Akademicka 19, 20-033 Lublin, Poland

Abstract

Although the knowledge of pollination systems of rare and threatened species is one of the principles for development of optimal conservation and management strategies, the data about their pollination requirements are scarce or incomplete. Different problems are listed (xerothermic habitat disappearance, overgrowing of patches, plant biology i.e., slow plant growth, problems with seed germination) among the possible causes of *Adonis vernalis* being threatened, but until now no consideration was given to the flowering biology and pollination.

The observations of flowering biology of *A. vernalis* (Ranunculaceae), a clonal species, were conducted in an out-of-compact-range population, in the Lublin Upland, Poland (51°18′55″ N, 22°38′21″ E), in 2011–2013. The reproductive potential of *A. vernalis* is related to the population age structure, pollination syndrome, and breeding system. The flowers exhibit incomplete protogyny. The dichogamy function is supported by different (biological, morphological) mechanisms. Stigma receptivity occurred about one day before anthers started shedding self-pollen, and pollen viability was increasing gradually during the flower life-span (66.3% in distal anthers vs. 77.3% in proximal). The decrease in pollen production and in pollen viability coincided with the lowest degree of seed set, irrespective of the pollination treatment. Pollen vectors are necessary for efficient pollination, as the proportion of pistils setting fruits after open pollination (41–82.1%) was significantly higher compared to spontaneous self-pollination (only 5.5–12.3%). The pollination requirements together with pollen/ovule ratio (P/O = 501) indicate a facultative xenogamous breeding system in *A. vernalis*. Therefore, in the conditions of the global lack of pollinators, improper pollination may weaken the population by leading to a decrease in the proportion of recombinants, and in addition to other factors, may accelerate extinction of small *A. vernalis* populations.

Keywords: *Adonis vernalis*; self-pollination; open-pollination; pollen; viability; seed set

Introduction

Conservation of biodiversity refers to different levels of ecological organization (species, biocenoses, ecosystems). In conservation efforts, special interest is devoted to the protection of rare and endangered species, many of which may impact the stability of pollination webs, although mutualistic relations are highly asymmetrical [1,2]. The first step to protect various ecological interactions between organisms is to understand the particular species biology and ecology. Surprisingly, our knowledge of species properties, especially of pollination requirements is still incomplete, even in well-researched floras [3]. Meanwhile, the information on the plant-pollinator relationship and plant requirements for the gene flow is particularly important for proper maintenance of "healthy" populations of rare and endangered species within the metapopulation system [4]. In Poland, detailed studies on pollination systems and reproductive strategies are available for less than 20% of the endangered species, but in recent years, attempts have been made to fulfil the gap [5]. One of the species with the conservation priorities at the national and internationals levels is *Adonis vernalis* L. (Ranunculaceae), a species characteristic for xerothermic grasslands of the continental-submediterranean type, belonging to the *Festucetalia valesiaceae* order [6]. Most populations of *A. vernalis* have declined in size considerably in many European countries in recent decades [7–12], and therefore, as a threatened species, it is maintained under the CITES agreement.

In Poland, the populations of *A. vernalis* are located out of the compact range [13] and the causes of the species being threatened are complex. They include overexploitation of populations (gathering for medical purposes), habitat disappearance (i.e. intensive agricultural practices) [14–16] or overgrowing of patches due to secondary succession of grasslands [17,18]. Some causes are within the biology and

* Corresponding author. Email: acwener@wp.pl

Handling Editor: Łukasz Łuczaj

ecology of the plant, i.e., slow growth, or problems with seed germination [19].

In the case of clonal species, including *A. vernalis*, vegetative reproduction seems to be a crucial strategy for maintenance of individuals in the community [20–22]. However, generative propagation is essential from the genetic and evolutionary perspective. It contributes to improvement of the genetic variability of individuals (outcrossing) within the population and between different populations [23].

Adonis vernalis produces protogynous flowers, in which maturation of anthers and stigmas occurs at different times; receptive stigmas are presented before anther dehiscence [24]. Dichogamy was evolved to reduce interference between pollen import and export, and the effective pollination period is closely linked with the duration of stigmatic receptivity and pollen accessibility [25,26]. A high level of outcrossing is expected in dichogamous flowers. In protogynous Ranunculaceae, a high degree of self-fertility is possible, which suggests that protogyny is to some extent incomplete [27,28].

Pollen characteristics, i.e. pollen viability, pollen germination, or pollen-tube survival are essential in the process of generative reproduction [29–31]. The quantity and quality of *A. vernalis* pollen may differ greatly between populations [9,32]. Fruiting and seed set is dependent on fertilization, which is the result of pollination; therefore, failure of pollination may be fundamental for reproduction of entomophilous species [30].

Since no knowledge of pollination biology is available, it seems reasonable to fulfill the gap for *A. vernalis*. The specific aims of the study were (***i***) to describe the details of flower biology, especially the time of stigma receptivity and anther dehiscence, (***ii***) to find out whether pollen characteristics may have an impact on the reproductive success and (***iii***) to determine to what degree insect visitors influence the seed set.

Material and methods

Study site

The observations were conducted in 2011–2013 in a population of the species in the Lublin Upland at Pliszczyn site (51°18′55″ N, 22°38′21″ E), Poland. *Adonis vernalis* was a component of the *Adonido-Brachypodietum pinnati* association (Libb. 1933) Krausch 1960 of the *Festuco-Brometea* Br.-Bl. et R. Tx. 1943 class [33]. The population was situated on a loess-covered slope of the river Bystrzyca valley with south inclination. The *A. vernalis* population was estimated at ca. 300 individuals. The xerothermic patch was dominated by grass and meadow species [*Medicago falcata* L., *Brachypodium pinnatum* (L.) P. Beauv., *Galium verum* L. s.str., *Euphorbia cyparissias* L., *Veronnica spicata* L., *Salvia pratensis* L., *Phleum phleoides* (L.) H. Karst., *Chamaecytisus ratisbonensis* (Schaeff.) Rothm.].

Study species

Adonis vernalis (Ranunculaceae) is a herbaceous early spring perennial, characterized by clonal, clump growth. A single clump consists of 1-several modules that may consist of 1–5 different individuals (ramets). Modules are distinguished, since there are some difficulties in determination of plant individuals and separation thereof from the clump. Development of clumps involves several phases, during which the number of virginile and juvenile individuals in clumps increases with an increase in the clump area and the number of flower shoots developing from one bud increases with the clump age. The flowering of an individual plant does not occur before the third or fourth year of its life history [34,35].

The bisexual flowers of *A. vernalis* have typical entomophilous features. The flowers are cup-shaped, actinomorphic, and bright-colored. The gynoecium consists of many 1-ovuled carpels. Flowers produce pollen as floral reward and due to the early blooming period may be valuable for bee-like hymenoptera [32]. The flowers are considered to be protogynous [36]. Fructification is reported to be highly changeable among separate clumps, populations, and growing seasons [35,37].

Study methods

During the full flowering phase in the population, the clumps ($n = 40$) were randomly chosen and marked in order to establish the proportion of vegetative and generative shoots. The clump was defined according to Jankowska-Błaszczuk [35] as a spatially distinguishing group of modules (i.e. a specially distinguishing group of shoots) that may consists of a different number of individuals (i.e. ramets). Two weeks after the end of flowering, all fruiting shoots were counted and the proportion of fruiting shoots to the total flowering shoots was calculated.

Flower development

To determine the temporal separation of stigma receptivity and anther dehiscence, the flower development was monitored in 12–20 randomly chosen flowers, from the bud stage until the end of pollen presentation. Every day we recorded the degree to which stigmas were exerted, and the number of dehiscing anthers. Based on this observation, floral stages (1–7) were distinguished. The number of anthers per flower ($n = 12–28$ flowers) and the number of pistils per flower ($n = 12–28$ flowers) were established.

Stigma receptivity

Timing of stigma receptivity was determined at each of the seven stages of floral development. 30% hydrogen peroxide was applied for detection of peroxidase activity (SPA) [38]. The entire gynoecium from flowers was placed on a glass slide and coated in a drop of H_2O_2. Stigmas that produced bubbles within 2–3 min were considered receptive. The number of receptive stigmas was counted for each individual flower under a binocular (NIKON SMZ-2B). The observations were made in laboratory conditions on plants transferred from the study site and grown in pots in an unheated, ventilated experimental greenhouse (day and night air temperature analogous to the outdoor conditions). After the experiment, the plants were removed to the natural patch.

Pollen quantity and quality

The anther size (determined as a dry weight of 100 anthers) and the mass of pollen in the anthers were determined

Tab. 1 The percentage proportion of vegetative and generative shoots per clump of *Adonis vernalis* and the flower longevity with duration of stigma and pollen presentation in flowers in the years 2011–2013.

Year	Proportion of shoots (%)			Duration of (days)		
	vegetative	flowering	fruiting	flower life-span	stigma receptivity	pollen presentation
2011	42.6 ^{a} ±12.3	47.4 ^{a} ±7.8	28.5 ^{a} ±7.3	12.2 ^{b} ±3.1	8.3 ^{b} ±2.1	10.0 ^{b} ±2.4
2012	54.6 ^{a} ±7.7	45.4 ^{a} ±32.5	30.6 ^{a} ±12.1	11.5 ^{b} ±2.6	8.1 ^{b} ±1.6	9.4 ^{b} ±2.1
2013	59.1 ^{a} ±21.6	40.9 ^{a} ±17.8	39.4 ^{a} ±9.3	7.3 ^{a} ±4.1	4.7 ^{a} ±1.6	5.6 ^{a} ±1.1
Mean	52.1 ±19.2	44.6 ±12.5	32.8 ±6.7	10.3 ±2.8	7.5 ±2.5	8.4 ±3.9

Means in columns with the same letters are not significantly different at $P = 0.05$, for vegetative, generative and fruiting shoots according to Kruskal–Wallis test, and for floral traits according to HSD Tukey test. Means are presented as mean ±*SD*.

in proximal, middle, and distal anthers. Fully-grown and unopened anthers were randomly collected from appropriate parts of the androecium before anther dehiscence. The anthers were separated from filaments and placed in weighed glass containers. Four replications with $n = 100$ anthers for each position was prepared for analyses. The glass containers were placed in a dryer (ELCON CL 65) at ca. 33°C for 3–5 days and weighed. Then the pollen was rinsed from the anthers once with pure ether (1–2 ml) and then 4–6 times with 70% ethanol (2–8 ml). The accuracy of the pollen rinsing was checked using a binocular.

To calculate pollen-ovule ratio (P/O) [39], the number of pollen grains per anther ($n = 10$) and the number of ovules in the flower were counted and the pollen/ovules ratio was estimated ($n = 12$–15 flowers).

The value of pollen was evaluated by its viability in a standard 2% acetocarmine solution and observed on microscopic slides. Pollen from proximal, middle, and distal anthers ($n = 300$ pollen grains from each position) was studied. A Nikon Eclipse E 200 light microscope was used.

Pollination requirements

The apocarpous gynoecium has multiple distinct carpels, each of which contains a single ovule and developed single-seeded achenes, and does not split open when ripe [40]. The degree of pollination was established as the proportion of fruits (i.e. seeds) set in relation to the number of pistils. The number of pistils and seeds was detected separately in each flower; 15–22 flowers were used for each pollination treatment. Two treatments were designed to investigate pollination ecology:

(*i*) spontaneous self-pollination – the experimental flowers from the bud stage were bagged with isolators (mesh size <1 mm) to prevent insect visitations;

(*ii*) open-pollination – natural conditions; the flowers were exposed to pollinator visits.

Insect visitors were observed during the time of the full bloom phase of the population by walking of 2–3 persons, through the population in an area of ca. 200 m^2 (4 × 50 m transects). The total sampling time was roughly 40 hours (all observers and years combined).

Data analysis

Data are presented as means with standard deviation (*SD*). Whenever possible, parametric statistical analysis was applied [41]. One-way ANOVA was used to test the effect of year on the number of anthers per flower, the dry mass of anthers, the mass of pollen in anthers, pollen viability, and for effect of the pollination treatment (open-pollination vs. spontaneous self-pollination) on the seed set. Post hoc comparison of means was tested by the HSD Tukey test. The time for stigma receptivity vs. pollen presentation was compared using the independent samples *t*-test. The Kruskal–Wallis test was used to determine the differences for non-normally distributed data (the proportion of vegetative, generative, and fruiting shoots among years). The Pearson correlation coefficient was used to test relationship between the seed set and the mass of pollen produced in flowers, and between the seed set and pollen viability. The level of statistical significance for all analyses was at $\alpha = 0.05$. All analyses were performed using Statistica ver. 9.0 (StatSoft Poland, Cracow).

Results

Flowering and fructification

The number of flowers produced per plant varied depending on the number of modules evolved per plant and was changeable among the growing seasons. The number of generative shoots was the lowest for a 1-module clump (mean = 6.6 ±4.1 *SD*), and the most abundant flowering was characteristic for a 4-module clump (mean = 17.3 ±2.1 *SD*). The proportion of the collective fruit set was 26.4–38.2% in relation to flowering shoots. Consequently, the highest number of fruiting shoots was found for 3–4 module clusters. No year effect was found for both the number of generative shoots (Kruskal–Wallis test: $H = 1.69$, $df = 2$, $P = 0.428$) and the number of vegetative shoots ($H = 2.21$, $df = 2$, $P = 0.331$). The highest proportion of collective fruits developed was noted in 2013, and the value was 22–30% higher compared to that obtained in the years 2011–2012, but the difference was not statistically significant ($H = 0.24$, $df = 2$, $P = 0.884$; Tab. 1).

Flower stages

The *Adonis vernalis* flower life-span, from corolla opening till wilting, was 10.3 days ±2.8 *SD*, on average. In 2011 and 2012, the flower life-span averaged 12.2 and 11.5 days, respectively, and was longer than in 2013 (mean = 7.3 days, $F_{2.28} = 29.37$, $P = 0.024$). The flower development allows distinguishing 7 relatively distinct stages (Fig. 1). Stage 1 – tight yellowish bud, no stigma receptivity and no pollen release; stage 2 – loose bud and beginning of stigma receptivity; stage 3 – corolla opened, elongation of styles and receptivity of stigmas in lower pistils; stages 4–5 with increasing receptivity of stigmas in >30% of pistils, dehiscence of first anthers; stage 6 – a systematic decrease in the number of receptive stigmas, pollen release and fruit development; stage 7 – no stigma receptivity was noted, pollen presentation disappeared, while the number of set fruits systematically increased. The flowers spent a similar time in stigma (mean = 7.5 ±2.6 *SD*) and pollen presentation (mean = 8.4 ±2.6 *SD*; $t = 8.9$, $P > 0.05$).

Androecium and gynoecium characteristics

The anthers of *Adonis vernalis* are spirally arranged at the flower bottom; their ripening was gradual and pollen shedding started from the distally positioned anthers. The apocarpous gynoecium consists of many simple carpels, ranging from 32 to 86 per single flower (mean = 51.3 ±16.2 *SD*). The number of anthers per flower varied significantly from year-to-year ($F_{2.11} = 19.37$, $P = 0.034$; Tab. 2). The mean number of anthers per flower was 120.2 ±48.2 *SD*; the mean number of pollen grains per anther was 214.3 ±201.6 *SD*; and the calculated pollen/ovule (P/O) ratio amounted to 501.

In general, the year effect was established for the dry mass of anthers ($F_{2.61} = 11.6$, $P = 0.041$), the mass of pollen produced in anthers ($F_{2.11} = 24.42$, $P = 0.022$), the mass of pollen per flower ($F_{2.11} = 3.76$, $P = 0.021$), and pollen viability ($F_{2.11} = 27.60$, $P = 0.016$). For example, the mass of produced pollen was similar in 2011 and 2013, and decreased in 2012 by 60% and 52%, respectively.

Tab. 2 The characteristic of androecium, the dry mass of anthers, the mass of produced pollen and the pollen viability in *Adonis vernalis* in 2011–2013.

Year	Number of anthers per flower	Dry mass of 100 anthers (mg)	Mass of pollen per 100 anthers (mg)	Viability (%)
2011	131.3 ^{c} ±31.2	9.2 ^{b} ±2.6	2.8 ^{b} ±2.1	83.5 ^{b} ±7.4
2012	105.7 ^{a} ±28.6	5.7 ^{a} ±1.9	1.1 ^{a} ±1.6	36.1 ^{a} ±3.9
2013	123.4 ^{b} ±17.3	6.9 ^{a} ±2.8	2.3 ^{b} ±1.5	95.4 ^{c} ±4.0
Mean	120.1 ±2.1	6.3 ±3.6	2.1 ±2.9	71.7 ±12.3

Means in columns with the same letters are not significantly different at $P = 0.05$ (HSD Tukey test). Means are presented as mean ±*SD*

The mass of pollen produced in anthers ($F_{2.24} = 3.28$, $P = 0.047$) and the pollen viability ($F_{2.24} = 5.38$, $P = 0.032$) were related to the position of stamens in the androecium (Fig. 2a,b). In general, the pollen mass was the lowest in the distally positioned anthers (mean = 1.4 mg ±0.7 *SD*), and was lower by 44% and 36% compared to the mass of pollen produced in the mid and proximal anthers, respectively. Also the lowest pollen viability was noted in the distal anthers (mean = 66.3% ±26.1 *SD*), and the value increased gradually reaching 71.6% ±27.6 *SD* in the mid-positioned anthers, and viability reached 77.3% ±27.7 *SD* in the proximal anthers.

There was a positive correlation between the seed set and the mass of pollen produced in flowers (Pearson correlation

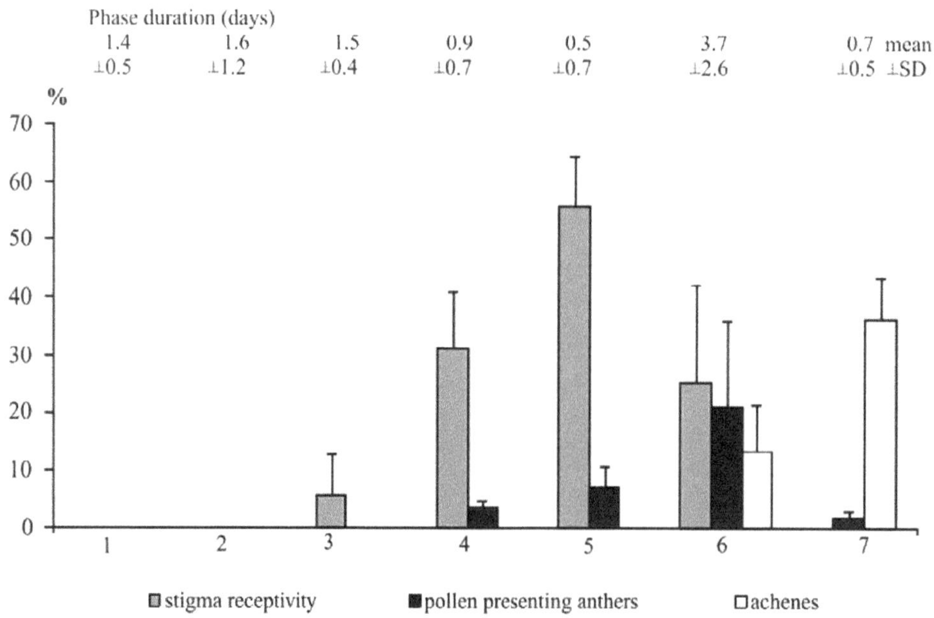

Fig. 1 The duration of floral phases (1 – greenish bud, 2 – loose bud, 3–7 corolla opened) and the proportion of receptive stigmas in gynoecium, the proportion of pollen presenting anthers in androecium and the proportion of achenes per flower of *Adonis vernalis* in consecutive stages of flower development. Means are calculated from 2011–2013. *SD* are given – vertical bars.

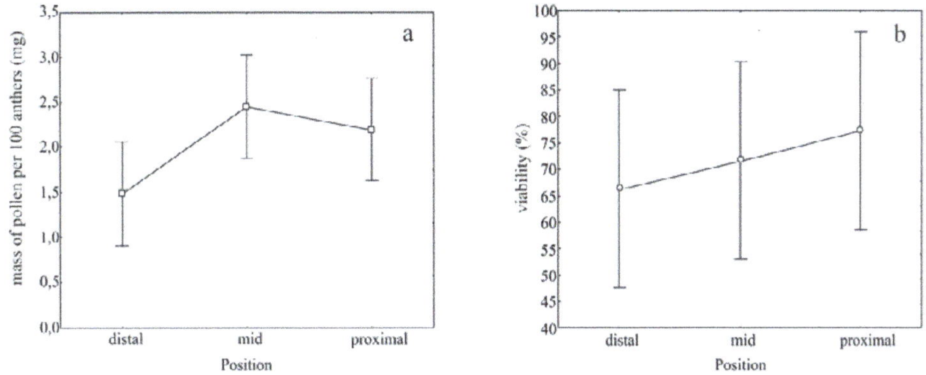

Fig. 2 The effect of the stamens position in the androecium on the mass of pollen in anthers (**a**), and the pollen viability in *Adonis vernalis* (**b**; means calculated from 2011–2013). Vertical bars show 95% confidence intervals.

$r = 0.708$, $P < 0.05$), and between the seed set and pollen viability ($r = 0.630$, $P < 0.05$). The lowest mass of pollen produced in anthers (mean = 1.1 mg per 100 anthers) and the lowest percentage of viable pollen grains with normal morphology (mean = 36.1%) were recorded in 2012, when the number of seeds developed was also the lowest (mean = 41.3% ±12.6 *SD*, after open-pollination and 5.5% ±2.3 *SD* after self-pollination).

Insect visitors and seed set

In the population of *A. vernalis* studied, the *Chlamydatus* sp. (Heteroptera) was found to be the most frequent insect flower visitor. Beetles from the genus *Mordellistena* (Fig. 3a), *Anthonomus* sp., and *Cantharis* sp. were also noted. Among bee-like Hymenoptera, solitary bees (81.2%), *Bombus* species (9.6%) and *Apis mellifera* L. (8.2%) were recorded. Among solitary bees: *Osmia rufa* L., *Andrena albicans* Misident, *Andrena cineraria* L., *Andrena* sp., and *Halictus* sp. were noted, while among bumblebees: *Bombus terrestris* L., *B. pascuorum* Scop., *B. lapidarius* L., were observed. However, the visitation rate of bees was very low (1–2 insect visitors per hour per 50-m transect).

A highly significant effect of the pollination treatment (open-pollination vs. pollinator exclusion $F_{1.28} = 29.37$, $P = 0.004$) and the growing season ($F_{2.33} = 19.43$, $P = 0.027$) on the seed set was revealed for the experimental *Adonis vernalis* plants (Fig. 3b–d, Fig. 4). Irrespective of the study years, a 7–9 fold higher number of achenes were developed after open pollination. The seed set after spontaneous self-pollination was 5.5–12.3%, and it was the highest after open pollination, i.e. 41–82.1%. In the case of both pollination treatments, the lowest number of seed set was noted in 2012.

Discussion

During their life history, the clumps of *A. vernalis* undergo several phases of development. In the studied population, we observed 1-4 module clumps, similarly to two other studies [35,37]. The clump structure had an impact on the number of developed flowers. Although the number of developed fruiting shoots was variable among the years, the relationship between the number of fruiting shoots and the number of flowering shoots correlated positively. This observation

Fig. 3 The flower of *Adonis vernalis* with *Mordellistena* beetle (**a**) and the seed set after open-pollination (**b,c**) and after spontaneous self-pollination (**d**).

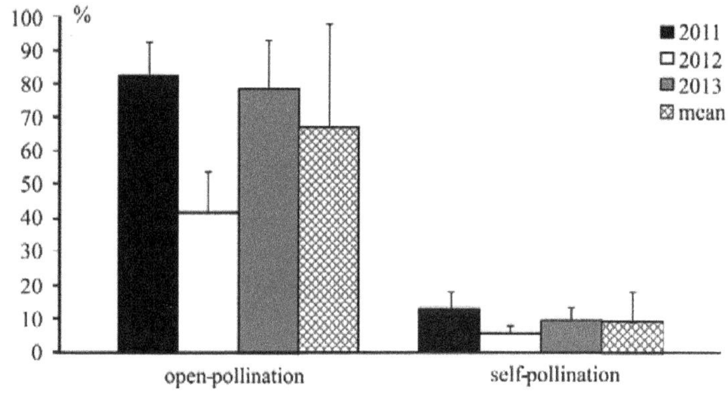

Fig. 4 The proportion of seed set in relation to pistils after open-pollination and after spontaneous self-pollination (insect visitors excluded) in *Adonis vernalis* in the years 2011–2013. Mean values and *SD* (vertical bars) are given.

coincides with the conclusion of two other studies [34,35] that the population age structure influences the reproductive potential of *A. vernalis*.

To our knowledge, this is the first study to reveal the significant impact of the pollination requirements and the breeding system on the seed set in *A. vernalis*. The comparison of the proportion of pistils setting fruits (i.e. seeds) after open-pollination and spontaneous self-pollination (i.e. pollinators exclusion) treatments indicates that pollen vectors are necessary for efficient pollination. In the current study, we noted different groups of bees (Hymenoptera) as insect visitors similar to findings for other *A. vernalis* populations [32,42], but their activity in flowers was very scarce (1–2 insect visitors per hour per 50 m transect). This coincides with the opinion presented in Chmura et al. [37] about the low frequency of entomofauna on *A. vernalis*. The low frequency of flower visitors to *Adonis* flowers may be related to low early spring temperatures, as air temperature during flowering was usually below 15°C. Several studies have demonstrated that lower insect visitation is caused by temperature. In a spring ephemeral *Adonis ramose* Franch, Kudo [43] found that the activity of visiting insects was dependent on temperature, and cool temperatures restricted their frequency. However, bumblebees, as visitors that forage at ambient temperatures of 5°C or less, are reported as foragers to early spring flora of xerothermic swards in Poland [44]. Bumblebees were also noted in our study but their frequency was very low, surprisingly much lower than that of solitary bees. One of the possible causes may be the low population size and density of bumblebee individuals in the study area [45].

For the early spring flora, the low activity of flower visitors can be a decisive factor in reduction of outcrossing frequency. However, it is necessary to note that the interactions between flowers and their pollinators are complex, and many factors (not only environmental) may affect the relation. Among others, the energy budget of pollinators in relation to the food reward offered seems to play a significant role in the evolution of the plant pollination system [36,46]. The flowers of *A. vernalis* offer only pollen as floral reward, therefore they are of low energetic potential. Chittka et al. [47] found that nectarless species receive fewer insect visits than concurrently blooming nectar-producing species. According to the "energy balance" theory, the species that are not able to provide high caloric reward develop alternative strategies to ensure pollen transfer [46]. In the case of *A. vernalis*, the population structure and the clump architecture seems to be an important tool supporting gene flow. Flowers within the clump appear close to each other and therefore may contribute to reduction of energy spent by potential pollinators for the flights among the flowers. Compact blooming flowers can be visited in rapid succession and the pollen transfer between the flowers is more likely. Additionally, for the one-ovuled pistils, theoretically, only one pollen grain is necessary to produce a single seed; hence, there is no need for the pollinator to make repeated visits for efficient pollination.

Mating and sleeping Hemiptera (mainly *Chlamydatus* sp.) and Coleoptera insects were observed in the flowers, which provided evidence for combined flower function (food, shelter), thereby implying a strong impact of *A.vernalis* on insect biodiversity. Their function as pollen agents and pollinators cannot be excluded either. In contrast to Knuth [42], we did not observe dipterans. Several factors operating simultaneously may contribute to these discrepancies, one of which was abundant blooming of co-flowering *Euphorbia cyparissias*, very willingly visited by dipterans.

Other possible factors that may limit the seed set in *A. vernalis* are associated with internal biological parameters. We found a positive relationship between the qualitative and quantitative traits of pollen and the reproductive success of *A. vernalis*. The reduced pollen production together with the decrease in pollen viability coincided with the lowest degree of seed set in 2012, irrespective of the pollination treatment applied. Pollen viability is considered as fundamental for fertilization [29,38]. In our study, the pollen viability varied significantly among the years (36.1–95.4%). Considerable differences of pollen viability were also noted in other *A. vernalis* populations (43–98.5%) [32]. These observations indicate high dependence of microsporogenesis from external factors. The adverse effect of abiotic conditions, e.g. air temperature or drought, on pollen grain formation and reproductive success is widely accepted [29]. Temperature stress or a low precipitation level during flowering reduced seed yield in different species of crop plants e.g. *Brassica napus* L. [48], and wild species, e.g. *Linum usitatissimum* L. [49]. Furthermore, occurrence of the persistent tapetum is specific for *A. vernalis*. The persistent tapetum is known to be able to block nutrient transportation and initiate the process of pollen grain abortion or induce the development of microspores into unviable pollen grains [40]; therefore, this anatomical trait may cause reproductive problems in *A. vernalis*.

Also disorders in pollen productivity may contribute to poor seed set, as an adequate amount of pollen is necessary for complete fertilization of available ovules [25,30]. In our study, the lowest pollen mass produced in anthers was noted in the year of the lowest seed yield. The dependence shows that the male derivative potential of plants may be critical for reproductive function, as pointed out by [31].

The flowers of *A. vernalis* exhibit incomplete protogyny based on Lloyd and Webb's criteria [24]. Medvegy et al. [36] described *A. vernalis* as a protogynous species. In our study, the protogyny turned out to be relatively ineffective in reducing self-pollination, as seed set was noted after spontaneous autogamy (5–12%). Among Ranunculaceae species, incomplete protogyny was also described in *Aquilegia canadensis* L. or *Anemone canadensis* L. [27,28]. Although the protogyny is incomplete, the dichogamy function is supported by different (biological, morphological) mechanisms. Stigma receptivity occurred about one day before anthers started shedding self-pollen, and pollen viability was increasing gradually during the flower life span. The changes in pollen viability to support the reproductive function are regarded as a common trait among different, mainly dichogamous, Angiosperm plants [50].

Although the reproductive efficiency after spontaneous self-pollination is rather low, in comparison to open cross-pollination, and may not guarantee an adequate reproductive output for individuals, it can act as assurance for reproduction, i.e. in the condition of the lack of pollinators or while colonizing new areas. Given the extremely low pollinator visitation rates documented for *A. vernalis* [32,37], selection towards increased self-fertility in order to facilitate reproductive assurance, as shown elsewhere in rare species [51] cannot be excluded. In the flowers of *A. vernalis*, the morphological androecium traits may provide self-pollination. The stamens are inserted in whorls around the style; the long and flexible filaments facilitate anther movement and may allow gravitational transfer of the pollen load onto the stigmas, possibly from proximal anthers, which are the closest to pistils and produce pollen with the highest viability.

Likewise, the functioning of dichogamy depends on how the pollinators act in flowers. The incomplete protogyny may provide pollination if pollinators tend to avoid flowers in the female phase. There is no information about the correlation between floral visits and floral phases for *A. vernalis*; in this study, we did not monitor insect visitation rates in relation to the floral phases either and thus the implications of our findings cannot be fully evaluated. The preferences of flower visitors for the female or male phase have been documented for different plants, e.g. *Geranium sylvaticum* L. [52]. The omission of female phase flowers, as non-rewarding, is particularly likely to occur in *A. vernalis* as non-nectariferous floral phases are frequently avoided by different pollinators [53].

Furthermore, the mating system defined on the basis of the pollen/ovule ratio (P/O = 501) classifies *A. vernalis* as a facultative xenogamous species and may explain the occurrence of spontaneous autogamy as a mechanism for reproductive assurance. According to Cruden [39], the pollen/ovule ratio is a valuable reflection of the breeding system of a plant and indicates the reproductive requirements.

In conclusion, the presence of insect pollinators is necessary for the seed set and reproductive success of *A. vernalis*, although further investigations are necessary to determine the insect visitors effectiveness in pollen transfer. However, in the conditions of the global lack of pollinators, poor pollination may weaken the population by leading to a decrease in the proportion of recombinants, and in addition to other factors, may accelerate extinction of small *A. vernalis* populations.

Acknowledgments

We are immensely grateful to our three anonymous Reviewers for helpful comments, remarks, suggestions and for the amazing support given for the manuscript. Our thanks go also to Rafał Gosik, for the insect visitors identification. The material from protected species was collected in compliance with Polish law under permit from the Regional Nature Conservator in Lublin. Research was supported financially by the Ministry of Science and Higher Education of Poland as a part of statutory activities of the Department of Botany, University of Life Sciences in Lublin (project OKB/DS/2) and the Department of Geobotany, Institute of Biology and Biochemistry, Maria Curie-Skłodowska University.

Authors' contributions

The following declarations about authors' contributions to the research have been made: design of the study: BD, MW; field work: MW, BD, AC; laboratory work: BD; writting the manuscript: BD, MW.

References

1. Bascompte J, Jordano P. Plant-animal mutualistic networks: the architecture of biodiversity. Ann Rev Ecol Syst. 2007;38(1):567–593.
2. Potts SG, Biesmeijer J, Kremen C, Neumann P, Schweiger O, Kunin WE. Global pollinator declines: trends, impacts and drivers. Trends Ecol Evol. 2010;25(6):345–353.
3. Zych M, Jakubiec A. Pollination of Polish red list plants: a preliminary statistical survey. Acta Agrobot. 2008;61(1):85.
4. Schemske DW, Husband BC, Ruckelshaus MH, Goodwillie C, Parker IM, Bishop JG. Evaluating approaches to the conservation of rare and endangered plants. Ecology. 1994;75(3):584.
5. Zych M, Stpiczyńska M. Neither protogynous nor obligatory outcrossed: pollination biology and breeding system of the european red list *Fritillaria meleagris* L. (Liliaceae). Plant Biol Stuttg. 2012;14(2):285–294.
6. Zając A, Zając M. Elementy geograficzne rodzimej flory Polski. Cracow: Jagiellonian University; 2009.
7. Davis SD, Heywood VH, Hamilton AC, editors. Centres of plant diversity. A guide and strategy for their conservation 1. Europe, Africa, Southwest Asia and the Middle East. Cambrige: International Union for Conservation of Nature; 1994.
8. Schnittler M, Günther KF. Central European vascular plants requiring

priority conservation measures – an analysis from national red lists and distribution maps. Biodivers Conserv. 1999;8(7):891–925.

9. Denisow B, Wrzesień M, Cwener A. The estimation of Adonis vernalis populations in chosen patches of Lublin Upland. Acta Agrobot. 2008;61(1):3.
10. Cwener A. Threatened xerothermic vascular plants in the Działy Grabowieckie mesoregion (Wyżyna Lubelska Upland, SE Poland). In: Mirek Z, Nikel A, editors. Rare, relict and endangered plants and fungi in Poland. Cracow: W. Szafer Institute of Botany, Polish Academy of Sciences; 2009. p. 155–162.
11. Bilz M, Kell SP, Maxted N, Lansdown RV. European red List of vascular plants. Luxembourg: Publications Office of the European Union; 2011.
12. Korotchenko I, Peregrym M. Ukrainian steppes in the past, at present and in the future. In: Werger MJ, van Staalduinen MA, editors. Eurasian steppes. Ecological problems and livelihoods in a changing world. Plant and vegetation. Dordrecht: Springer Netherlands; 2012. p. 173–196. (vol 6).
13. Meusel H, Jäger E, Weinert F. Vergleichende Chorologie der Zentraleropäischen Flora. Jena: Gustav Fischer Verlag; 1965. (vol 3).
14. Forycka A, Szczyglewska D, Buchwald W. Stock-talking of *Adonis vernalis* L. in the selected localities in Poland. Bull Bot Gard Mus Collect. 2004;13:55–58.
15. Sudnik-Wójcikowska B, Moysiyenko II, Slim PA. Dynamics of the flora of windbreaks in the agricultural landscape of steppes in southern Ukraine. Biodiv Res Conserv. 2006;1-2:77–81.
16. Parnikoza I, Vasiluk A. Ukrainian steppes: current state and perspectives for protection. Ann UMCS Biol. 2011;66(1).

17. Świerczyńska S. Problemy zachowania zbiorowisk stepowych na podstawie badań prowadzonych na Lubelszczyźnie. Prądnik: Ojcowski Park Narodowy. Muzeum im. prof. Władysława Szafera; 1990. (vol 2).
18. Michalik S, Zarzycki K. Management of xerothermic grasslands in Poland: botanical approach. Colloq Phytosociol 1995;24:881–895.
19. Mihalik E, Gocs K, Kálmán K, Medvegy A. Reproductive success of the individuals with different age in a planted *Adonis vernalis* L. population. Acta Biol Debr Oecol Hung. 2000;11(1):270–275.
20. Harper JL. Biologia populacyjna i ekologia organizmów klonalnych. Moduły i rozgałęzienia a pobieranie składników pokarmowych. Wiad Ekol. 1986;32(4):327–359.
21. Falińska K. Plant population biology and vegetation processes. Cracow: W. Szafer Institute of Botany, Polish Academy of Sciences; 1998.
22. Czarnecka B. Spatiotemporal patterns of genets and ramets in a population of clonal perennial *Senecio rivularis*: plant features and habitat effects. Ann Bot Fenn. 2008;45(1):19–32.

23. Charlesworth D, Charlesworth B. Inbreeding depression and its evolutionary consequences. Ann Rev Ecol Syst. 1987;18(1):237–268.

24. Lloyd DG, Webb CJ. The avoidance of interference between the presentation of pollen and stigmas in angiosperms. I. Dichogamy. N Z J Bot. 1986;24(1):135–162.

25. Lloyd DG, Schoen DJ. Self- and cross-fertilization in plants. I. Functional dimensions. Int J Plant Sci. 1992;153(3):358.

26. Sanzol J, Herrero M. The "effective pollination period" in fruit trees. Sci Hortic. 2001;90(1-2):1–17.

27. Douglas KL, Cruden RW. The reproductive biology of *Anemone canadensis* (Ranunculaceae): breeding system and facilitation of sexual selection. Am J Bot. 1994;81(3):314.
28. Griffin SR, Mavraganis K, Eckert CG. Experimental analysis of protogyny in *Aquilegia canadensis* (Ranunculaceae). Am J Bot. 2000;87(9):1246.
29. Dafni A, Firmage D. Pollen viability and longevity: Practical, ecological and evolutionary implications. Plant Syst Evol. 2000;222(1-4):113–132.
30. Larson BMH, Barrett SCH. A comparative analysis of pollen limitation in flowering plants. Biol J Linn Soc. 2000;69(4):503–520.

31. Aizen MA, Harder LD. Expanding the limits of the pollen limitation concept: effects of pollen quantity and quality. Ecology. 2007;88(2):271–281.
32. Denisow B, Wrzesień M. The study of blooming and pollen efficiency of *Adonis vernalis* L. in xerothermic plant communities. J Apic Sci. 2006;50(1):25–32.
33. Matuszkiewicz W. Przewodnik do oznaczania zbiorowisk roślinnych Polski. Warsaw: Polish Scientific Publishers PWN; 2001.
34. Jankowska-Błaszczuk M. Diagnostyczne właściwości struktury przestrzennej populacji *Adonis vernalis* L. In: Faliński J, editor. Dynamika roślinności i populacji roślinnych. Warsaw: Białowieża Geobotanical Station of Warsaw University; 1991. p. 193–200.
35. Jankowska-Błaszczuk M. Biologia populacji miłka wiosennego *Adonis vernalis* L. w rezerwacie "Skowronno". Ochr Przyr. 1995;52:47–58.
36. Medvegy A, Szőllősi R, Mihalik E. Temporal floral sex allocation in protogynous *Adonis vernalis* L. In: Proceedings of the 2nd European Congress of Conservation Biology. Prague: Czech University of Life Sciences, Faculty of Environmental Sciences; 2009. p. 148.
37. Chmura D, Adamski P, Denisiuk Z. Spatiotemporal aspects of the occurrence of clonal steppe plant *Adonis vernalis* L. in the southern Poland. Cas Slez Semskeho Muzea A. 2012;61(3):245–250.

38. Dafni A, Maués MM. A rapid and simple procedure to determine stigma receptivity. Sex Plant Reprod. 1998;11(3):177–180.

39. Cruden RW. Pollen-ovule ratios: a conservative indicator of breeding systems in flowering plants. Evolution. 1977;31(1):32.

40. Gostin IN. Scanning electron microscopy investigations regarding *Adonis vernalis* L. flower morphology. Analele Univ Din Oradea Fasc Biol. 2009;16(2):80–84.
41. Stanisz A. Przystępny kurs statystyki z zastosowaniem Statistica na przykładach z medycyny. Cracow: Statsoft Polska; 2007.
42. Knuth P. Handbuch der Blütenbiologie. Leipzig: Verlag von Wilhelm Engelman; 1898. (vol 2).
43. Kudo G. Ecological significance of flower heliotropism in the spring ephemeral *Adonis ramosa* (Ranunculaceae). Oikos. 1995;72(1):14.

44. Chmura D, Adamski P, Denisiuk Z. How do plant communities and flower visitors relate? A case study of semi-natural xerothermic grasslands. Acta Soc Bot Pol. 2013;82(2):99.

45. Goulson D. Foraging strategies of insects for gathering nectar and pollen, and implications for plant ecology and evolution. Perspect Plant Ecol Evol Syst. 1999;2(2):185–209.

46. Heinrich B, Raven PH. Energetics and pollination ecology. Science. 1972;176(4035):597–602.

47. Chittka L, Thomson JD, Waser NM. Flower constancy, insect psychology, and plant evolution. Naturwissenschaften. 1999;86(8):361–377.

48. Young LW, Wilen RW, Bonham-Smith PC. High temperature stress of *Brassica napus* during flowering reduces micro- and megagametophyte fertility, induces fruit abortion, and disrupts seed production. J Exp Bot. 2004;55(396):485–495.
49. Gusta LV, O'Connor BJ, Bhatty RS. Flax (*Linum usitatissimum* L.) responses to chilling and heat stress on flowering and seed yield. Can J Plant Sci. 1997;77(1):97–99.

50. Rodkiewicz B, Śnieżko R, Fryk B, Niewęgłowska B, Tchórzewska D. Embriologia "Angiospermae" rozwojowa i eksperymentalna. Lublin: Maria Curie-Skłodowska University Press; 1996.

51. Steffan-Dewenter I, Tscharntke T. Effects of habitat isolation on pollinator communities and seed set. Oecologia. 1999;121(3):432–440.

52. Asikainen E, Mutikainen P. Preferences of pollinators and herbivores in gynodioecious *Geranium sylvaticum*. Ann Bot. 2005;95(5):879–886.

53. Klinkhamer PGL, de Jong TJ, Linnebank LA. Small-scale spatial patterns determine ecological relationships: an experimental example using nectar production rates. Ecol Lett. 2001;4(6):559–567.

7

Calculated characters of leaves are independent on environmental conditions in *Salix herbacea* (Salicaceae) and *Betula nana* (Betulaceae)

Katarzyna Marcysiak*
Department of Botany, Kazimierz Wielki University, Ossolińskich 12, 85-093 Bydgoszcz, Poland

Abstract

The goal of the study was to check if the shape-describing characters, calculated as ratios of the morphological measured traits are more stable, compared to the latter, and can be treated as independent on environmental conditions. The test was based on the example of leaves of *Salix herbacea* and *Betula nana*. The individuals of the two populations of *S. herbacea* from Tatra Mts. were divided into two groups: with bigger and smaller leaves. The two populations of *B. nana* came from different substrata: the first one, collected from the mire on the lower altitude, had bigger leaves, and the second, collected from the granite plateau and higher altitude, had smaller leaves.

For both species, the measured traits were generally more variable than the ratios calculated on their basis, as expressed by the variation coefficients. The results of Students' t-test analyses showed statistically significant differences between the two groups of *S. herbacea* and the two populations of *B. nana* with respect to almost all the measured characters, and no such differences for the calculated traits, reflecting the leaf shape.

As the differentiation of the leaf size was probably bound to the environmental factors, the lack of the dependence of the leaf shape on the leaf size could lead to a conclusion of independence of the leaf shape on the environment conditions.

Keywords: morphologic characters, synthetic characters, *Salix herbacea*, *Betula nana*

Introduction

Morphological traits are basic for the description of a plant. Recently, their role has been decreasing in the face of rapid development of molecular biology and modern methods of DNA analysis, treated as more reliable and more sensitive, especially for reflecting the intra- and inter-populational diversity [1,2]. Still, morphological characters remain important, as they are crucial for the identification of taxa. Various kinds of morphological features, concerning both vegetative and generative organs, are still used in taxonomic studies and analysis of diversity [3–10]. The main criticism of morphological characters is connected with their dependence on changing environmental factors, such as precipitation, temperature, insolation [11–15]. The proposed solution is to use synthetic characters, describing rather the shape of the plant structures, than the size, being regarded as much more stable [5].

The goal of the present study is to test whether the shape-describing characters are more stable and can be treated as independent on environment when compared to measured traits, on the example of leaves. The assumption was that the variability of the measured morphological traits would be greater than in the case of shape describing characters. It was also expected that individuals of various leaf size, sampled in the same geographical region, should not show statistically significant differences with regard to synthetic, shape-describing traits. If the differentiation of the leaf size can be bound with environmental factors, the independence of the leaf shape on the leaf size may lead to a conclusion of independence of the leaf shape of the environmental conditions.

To check the hypothesis, the leaves of two taxonomically unrelated species, interesting as subjects of different biological studies, were examined: *Salix herbacea* L. (Salicaceae) and *Betula nana* L. (Betulaceae).

Material and methods

Species decriptions

Salix herbacea, an arctic-alpine dwarf shrub, is relatively well known [16–23]. Its genetic diversity and phylogeography has been lately characterized by Alsos et al. [24]. The willow has been also interesting as an important fossil material, but recently it has gained more attention as a possibly endangered species in the perspective of the climate warming. Still, a thorough examination of its morphological variability in the whole range is lacking. In the Tatra Mountains, two morphological

* Email: marc@ukw.edu.pl

types, differing with regard to the leaf size, have been reported: besides the typical form, f. *macrophylla* Sér. having leaves 23–27 mm long [25].

Betula nana is a small, circum-boreal shrub. In its southernmost localities the birch forms small, relic populations, while in the northern part of the species range it is important as one of few, if not the only one, woody plants [26–28]. Studies on Betula nana focused on its occurrence in the early Holocene floras [29–31], hybridization with *Betula pubescens* [28,32–36] and importance in forming the mountain birch (*Betula pubescens* var. *tortuosa*) [32–38]. In Scandinavia, *B. nana* builds vast populations and shows low morphological variation. However, the morphological differences between tundra and mire populations, as well as decreasing of the leaf size in severe climatic conditions, were described [28]. The relation of the leaf size to altitude for other birch species was also reported [39].

Measurements and analysis

For each taxon, samples from two populations were analyzed, a sample consisted of approximately 30 specimens, for every specimen 5–10 leaves were collected. Individuals growing no closer than 3 m to each other were selected, to avoid sampling the same genet. The location of sampled populations are given in Tab. 1.

Measurements were taken on the dried material. Leaves were scanned and then the scans were measured with digiShape 1.9.177 (Cortex Nova), teeth were counted from the scans and synthetic traits were calculated (Tab. 2, Tab. 3).

To find possible two morphological types of leaves of *Salix herbacea* in the Tatra Mountains [25], the k-means method of the cluster analysis was used, on the basis of means of individuals of two populations together. For comparisons between two distinguished groups of the willow and between two populations of the dwarf birch, Student' t-test was performed. The deviation of means of traits was checked on the box and whiskers graphs and the variation coefficients were calculated. For all calculations and analysis STATISTICA 9.0 (StatSoft, Inc.) was used.

Results

K-mean clustering allows distinguishing two groups of 60 individuals of *Salix herbacea*: group A with bigger leaves, and group B having smaller leaves, consisting of 23 and 37 individuals, respectively (Fig. 1a, Tab. 2).

For *Salix herbacea*, the values of the variation coefficients, calculated for characters describing the leaf size (Tab. 2) reached more than 15%, while for characters illustrating the leaf shape only up to 10%, with the exception of the character 12 (19.21%). The number of teeth (character 6) and the number of teeth/perimeter (character 13) had variation coefficients about 20%. The coefficient of the petiole length (character 3) was highest (almost 30%; Fig. 2).

Leaves traits of *Betula nana* were generally less variable. For characters 1–5 the coefficient values ranged from 10.53 to 22.27%, for characters 7–11 – from 4.11 to 6.94%, for the character 12, it equaled 14.71% and for character 6 and 13, 14.98% and 14.06%, respectively. Again, the petiole length (character 3) was the most variable trait, with the variation coefficient: 22.27% (Fig. 2).

The analyses of characters' means and standard deviations values gave analogical findings, as they are shown for the example of *Salix herbacea* only (Fig. 1a–c).

Two groups of *Salix herbacea* individuals were created on the basis of measured characters (1–5), so in the t-test these features differed the groups statistically significantly, as well as the number of teeth (6). All synthetic traits, characterizing the shape of leaves (7–12) and the leaf margin (13), showed no statistically significant differences (Tab. 2).

In the case of the dwarf birch, the t-test showed statistically significant differences for traits: 1–2 and 4–6. From among the measured characters, only the petiole length (3) did not differ the populations significantly. The differences found for the synthetic traits were not significant, with the exception of character 13 (Tab. 3).

Discussion

Although within the two studied populations of *Salix herbacea* in the Tatras, the form *macrophylla* was not found, the individuals created two distinct groups. Their values of the leaf length and width were within the limits described in European and local floras [18,26,40]. Still, the differences between the groups with the respect to leaves sizes were great enough to be statistically significant. The morphological variability of *Salix herbacea* was seldom characterized, only the stomatal index was thoroughly studied [19,20,41,42]. Wijk [16,17] reported some correlations between the biomass, the density and the cover of the species in relation to the longevity of the snow cover, as well as the influence of the growing season length on the shoot increment. The material for the present study was not collected in one year (Tab. 1) and the share of the distinguished groups in populations varied. Within the population Sh1, six individuals belonged to the group A (bigger leaves) and 23 to B (smaller leaves), while in the population Sh2, 17 and 14, respectively. This dissimilarity could have been caused by different climatic conditions in particular years. What should be emphasized, the results of Students' t-test analyses for the two groups fulfilled the assumption that the measured traits differed them significantly, while the synthetic ones did not.

Tab. 1 Location of analyzed populations of *Salix herbacea* and *Betula nana*.

Acronym	Taxon	Locality and character of a site	No. of individuals	Alt.	Longitude	Latitude	Year of collection
Sh1	*Salix herbacea*	Poland, Tatra Mts., Świnicka Przełęcz (granite slopes)	29	2000	20°00.38′	49°13.08′	2005
Sh2	*Salix herbacea*	Poland, Tatra Mts., slopes of Rysy (granite slopes)	31	1997	20°04.86′	49°10.45′	2008
Bn1	*Betula nana*	Norway, Buskerud, Geilo (mire)	33	800	8°12.41′	60°32.08′	2006
Bn2	*Betula nana*	Norway, Buskerud, Geilo, Ustetind (granite plateau)	32	1200	8°00.10′	60°27.38′	2006

Tab. 2 Results of t-test for 2 groups of *Salix herbacea*.

No.	Character	Character describes	Group A	Group B	*t*	*df*	*p*
1.	Perimeter of a leaf blade (mm)	leaf size	54.77	42.22	−10.23	58	0.000
2.	Length of a leaf blade (mm)	leaf size	18.56	14.44	−9.66	58	0.000
3.	Length of a petiole (mm)	leaf size	2.82	1.98	−5.69	58	0.000
4.	Width of a leaf blade measured at ½ of its length (mm)	leaf size	14.13	10.58	−9.29	58	0.000
5.	Width of a leaf blade measured at ¼ of its length (mm)	leaf size	12.94	9.71	−8.740	58	0.000
6.	Number of teeth on a leaf blade margin	leaf blade margin	35.06	30.77	−2.68	58	0.009
7.	Apex angle of a leaf blade (°)	leaf shape	169.89	165.11	−1.22	58	0.228
8.	Base angle of o leaf blade (°)	leaf shape	155.17	150.75	−1.10	58	0.277
9.	Ratio of perimeter/length (character 1/character 2)	leaf shape	2.95	2.93	−0.63	58	0.530
10.	Ratio of length/width at ½ (character 2/character 4)	leaf shape	1.33	1.38	1.58	58	0.119
11.	Ratio of width at ¼ and ½ (character 4/character 5)	leaf shape	0.92	0.92	0.08	58	0.933
12.	Ratio of length of a petiole/sum of blade and petiole length [character 3/(character 2 + character 3)]	leaf shape	0.13	0.12	−1.94	58	0.057
13.	Ratio of number of teeth/perimeter (character 6/character 1)	leaf blade margin	0.65	0.74	2.52	58	0.014

Nos. 1–6 represent differences statistically significant at $p < 0.01$.

Tab. 3 Results of t-test for 2 populations of *Betula nana*.

No.	Character	Character describes	Bn1	Bn2	*t*	*df*	*p*
1.	Perimeter of a leaf blade (mm)	leaf size	31.28	28.28	3.69	63	0,000
2.	Length of a leaf blade (mm)	leaf size	8.22	7.41	3.64	63	0.000
3.	Length of a petiole (mm)	leaf size	1.78	1.55	2.49	63	0.015
4.	Width of a leaf blade measured at ½ of its length (mm)	leaf size	8.47	7.76	3.67	63	0.000
5.	Width of a leaf blade measured at ¼ of its length (mm)	leaf size	7.35	6.63	3.83	63	0.000
6.	Number of teeth on a leaf blade margin	leaf blade margin	15.69	12.92	6.80	63	0.000
7.	Apex angle of a leaf blade (°)	leaf shape	165.14	161.22	2.44	63	0.017
8.	Base angle of o leaf blade (°)	leaf shape	165.06	163.74	0.54	63	0.592
9.	Ratio of perimeter/length (character 1/character 2)	leaf shape	3.82	3.83	−0.13	63	0.900
10.	Ratio of length/width at ½ (character 2/character 4)	leaf shape	0.97	0.96	0.78	63	0.440
11.	Ratio of width at ¼ and ½ (character 4/character 5)	leaf shape	0.87	0.86	1.29	63	0.203
12.	Ratio of length of a petiole/sum of blade and petiole length [character 3/(character 2 + character 3)]	leaf shape	0.18	0.17	0.55	63	0.584
13.	Ratio of number of teeth/perimeter (character 6/character 1)	leaf blade margin	0.51	0.46	3.07	63	0.003

Nos. 1, 2, 4–6, 13 represent differences statistically significant at $p < 0.01$.

In the case of *Betula nana* the results were not so clear, although similar. It should be underlined that the two compared populations of the birch grew on different substrata: population Bn1 on the mire, while population Bn2 on the granite mountain plateau (Tab. 1). The populations were not much distant but their altitude also differed: the mire (population Bn1) lied 800 m a.s.l., and the plateau (Bn2) – 1200 m a.s.l. The differences between the leaf sizes of these two samples were statistically significant, with the exception of the petiole length. The smaller leaf size on the plateau could be explained by the influence of the substratum, or of the more severe climatic conditions, as reported by de Groot et al. [28]. Białobrzeska and Truchanowicz [39] also found a tendency for the leaf size to decrease with increasing altitude in their studies on the *Betula pubescens/carpatica* in the Stołowe Mountains in Poland. They also indicated the importance of the insolation for the leaf size, but in the present study both populations growth on the plain ground, under similar exposure conditions. On the other hand, Kovačić and Nikolić [43] described poor correlations between ecological factors and *Betula pendula* leaf shape and size in Croatia.

It should be emphasized that the two populations of the dwarf birch, growing in various environmental conditions, but geographically close, differed statistically significantly in respects of the leaf size, but not of the synthetic characters describing leaf shape. This can be regarded as a proof of independence of the leaf shape of the environmental conditions, at least for the species studied.

Other interesting findings of the present study are similar proportions of variability of the same characters between

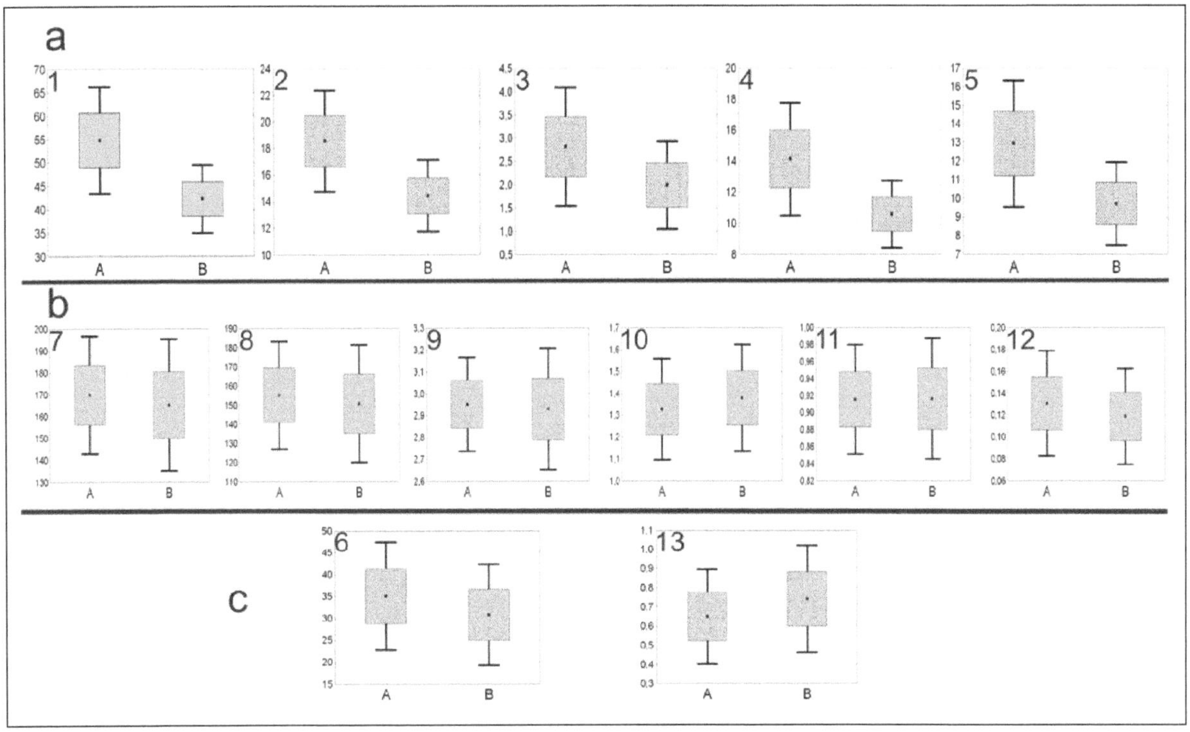

Fig. 1 Analyses of *S. herbacea* characters means for two groups: A – bigger leaves; B – smaller leaves. **a** Characters describing the leaf size. **b** Characters describing the leaf shape. **c** Characters describing the leaf blade margin. According to Tab. 2, numbers of characters, are given in left upper corners. Point – arithmetic mean; box – standard deviation; whiskers – 1.96 standard deviation.

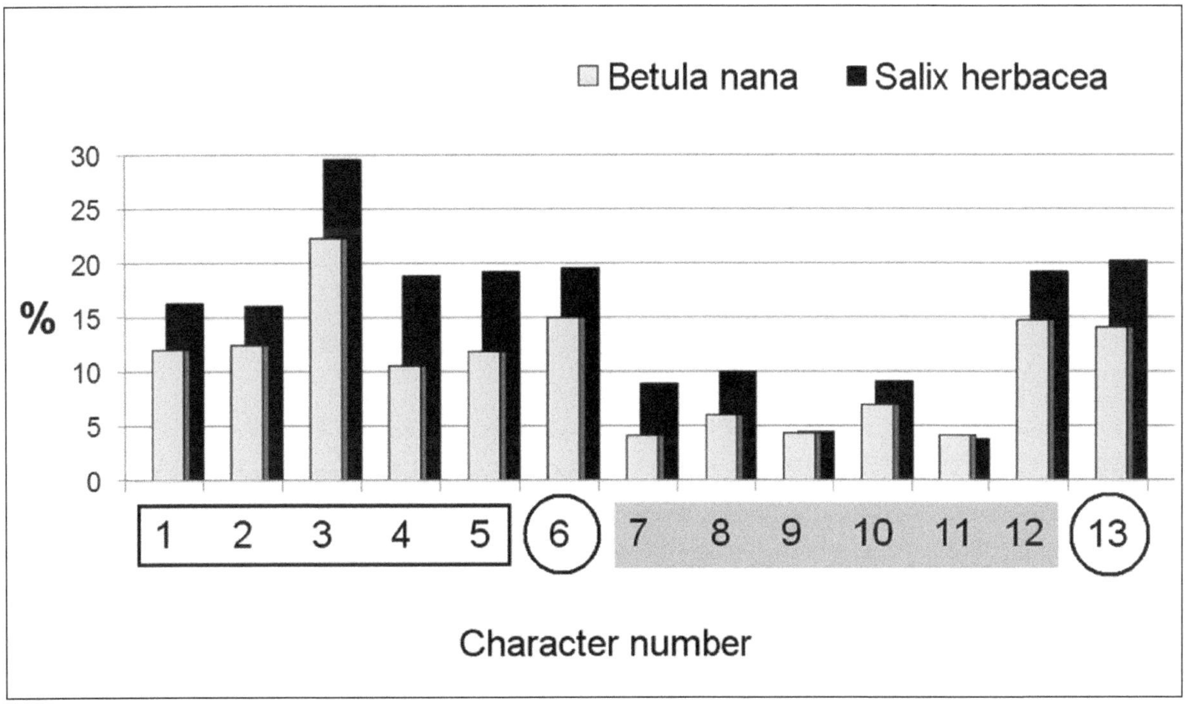

Fig. 2 Coefficients of variation for *S. herbacea* and *B. nana*. Character numbers according to Tab. 2: frame – characters describing the leaf size; shade – characters describing the leaf shape; circles – characters describing the leaf blade margin.

the two species (Fig. 2). The most variable character in both shrubs is the petiole length, and this value divided by the sum of the blade and petiole lengths (i.e. character 12) is one of the two most variable synthetic characters. For both species, the measured features are more variable than the calculated ones, with the exceptions of the previously mentioned character 12, and character 13. The latter describes the margin of the leaf blade and is calculated as the number of teeth (6) divided by the perimeter (1), so the resulting character shows the number of teeth per 1 mm. Both source characters are moderately variable, and the variation coefficients of character 13 calculated on their basis is very close to that of character 6. Thus the variability of the number of teeth and of the number of teeth per 1 mm is almost the same for both species. What is more, the variation coefficients of abovementioned ratios (characters 12 and 13) have almost the same values for both species but this might be accidental.

The higher variability of the petiole length may be just characteristic for this trait, but it also may be a result a measurement technique: some petioles are slightly curved, while measurements are taken along the straight line. Petioles are too short and their curves too weak to be detected. At the same time, the higher variability of the calculated trait 13 (number of teeth per 1 mm) could not be explained by the methods used and so seem to be a typical characteristic of the leaves of both species.

It can be concluded that the study allowed to confirm the hypothesis that calculated ratios are more stable than measured traits and they do not differ significantly leaves of various size. Thus probably they can be regarded as independent of changing environmental conditions, at least for the two dwarf shrubs: *Salix herbacea* and *Betula nana*.

Acknowledgements

I would like to thank Amelia Lewandowska for her great help in the measurement procedures. The collection of the material was possible thanks to the financial support of Kazimierz Wielki University in Bydgoszcz, Poland.

References

1. Levin DA. 50 years of plant speciation. In: Stuessy TF, Hörandl E, Mayer V, editors. Plant systematics: a half-century of progress (1950–2000) and future challenges. Vienna: International Association for Plant Taxonomy; 2001. p. 257–280.
2. Culham A. Molecular systematics: measuring and monitoring diversity. In: Leadlay E, Jury S, editors. Taxonomy and plant conservation. Cambridge: Cambridge University Press; 2006. p. 236–254.
3. Staszkiewicz J. Variability of selected shrub and tree species. Fragm Flor Geobot Polonica. 1997;2 suppl:313.
4. Max KN, Mouchaty SK, Schwaegerle KE. Allozyme and morphological variation in two subspecies of *Dryas octopetala* (Rosaceae) in Alaska. Am J Bot. 1999;86(11):1637–1644.
5. Kremer A, Dupouey JL, Deans JD, Cottrell J, Csaikl U, Finkeldey R, et al. Leaf morphological differentiation between *Quercus robur* and *Quercus petraea* is stable across western European mixed oak stands. Ann For Sci. 2002;59(7):777–787.
6. Philipp M, Siegismund HR. What can morphology and isozymes tell us about the history of the *Dryas integrifolia-octopetala* complex? Mol Ecol. 2003;12(8):2231–2242.
7. Greimler J, Hermanowski B, Jang CG. A re-evaluation of morphological characters in European *Gentianella* section *Gentianella* (Gentianaceae). Plant Syst Evol. 2004;248(1–4).
8. Klimko M, Boratyńska K, Montserrat JM, Didukh Y, Romo A, Gómez D, et al. Morphological variation of *Juniperus oxycedrus* subsp. *oxycedrus* (Cupressaceae) in the Mediterranean region. Flora. 2007;202(2):133–147.
9. Marcysiak K, Boratyński A. Contribution to the taxonomy of *Pinus uncinata* (Pinaceae) based on cone characters. Plant Syst Evol. 2007;264(1–2):57–73.
10. Boratynski A, Marcysiak K, Lewandowska A, Jasinska A, Iszkulo G, Burczyk J. Differences in leaf morphology between *Quercus petraea* and *Q. robur* adult and young individuals. Silva Fenn. 2008;42(1):115–124.
11. Huber H, Wiggerman L. Shade avoidance in the clonal herb *Trifolium fragiferum*: a field study with experimentally manipulated vegetation height. Plant Ecol. 1997;130(1):53–62.
12. Noda H, Muraoka H, Washitani I. Morphological and physiological acclimation responses to contrasting light and water regimes in *Primula sieboldii*. Ecol Res. 2004;19(3):331–340.
13. Marchand FL, Kockelbergh F, van de Vijver B, Beyens L, Nijs I. Are heat and cold resistance of arctic species affected by successive extreme temperature events? New Phytol. 2006;170(2):291–300.
14. Baquedano F, Valladares F, Castillo F. Phenotypic plasticity blurs ecotypic divergence in the response of *Quercus coccifera* and *Pinus halepensis* to water stress. Eur J Forest Res. 2008;127(6):495–506.
15. Fletcher BJ, Press MC, Baxter R, Phoenix GK. Transition zones between vegetation patches in a heterogeneous Arctic landscape: how plant growth and photosynthesis change with abundance at small scales. Oecologia. 2010;163(1):47–56.
16. Wijk S. Performance of *Salix Herbacea* in an Alpine Snow-Bed Gradient. J Ecol. 1986;74(3):675.
17. Wijk S. Influence of climate and age on annual ahoot ancrement in *Salix herbacea*. J Ecol. 1986;74(3):685.
18. Beerling DJ. *Salix herbacea* L. J Ecol. 1998;86(5):872–895.
19. Rundgren M, Beerling D. A Holocene CO_2 record from the stomatal index of subfossil *Salix herbacea* L. leaves from northern Sweden. Holocene. 1999;9(5):509–513.
20. Beerling DJ, Rundgren M. Leaf metabolic and morphological responses of dwarf willow (*Salix herbacea*) in the sub-arctic to the past 9000 years of global environmental change. New Phytol. 2000;145(2):257–269.
21. Stamati K, Blackie S, Brown JWS, Russell J. A set of

polymorphic SSR loci for subarctic willow (*Salix lanata, S. lapponum* and *S. herbacea*). Mol Ecol Notes. 2003;3(2):280–282.

22. Stamati K, Hollingsworth PM, Russell J. Patterns of clonal diversity in three species of sub-arctic willow (*Salix lanata, Salix lapponum* and *Salix herbacea*). Plant Syst Evol. 2007;269(1–2):75–88.
23. Reisch C, Schurm S, Poschlod P. Spatial genetic structure and clonal diversity in an alpine population of *Salix herbacea* (Salicaceae). Ann Bot. 2007;99(4):647–651.
24. Alsos IG, Alm T, Normand S, Brochmann C. Past and future range shifts and loss of diversity in dwarf willow (*Salix herbacea* L.) inferred from genetics, fossils and modelling. Global Ecol Biogeogr. 2009;18(2):223–239.
25. Pawłowski B. Flora Tatr. Rośliny naczyniowe. Warszawa: Polish Scientific Publishers PWN; 1956. (vol 1).
26. Tutin TG. Salicaceae. In: Tutin TG, Heywood VH, Burges NA, Valentine DH, Walters SM, Webb DA, editors. Flora Europaea. Cambridge: Cambridge University Press; 1964. p. 43–55. (vol 1).
27. Boratynska K. Chorologia i rejonizacja leśna brzozowatych (Betulaceae). Cz. II: Rola brzozowatych w poszczególnych regionach geograficznych. Arboretum Kórnickie. 1983;28:69–99.
28. de Groot WJ, Thomas PA, Wein RW. *Betula nana* L. and *Betula glandulosa* Michx. J Ecol. 1997;85(2):241–264.
29. Iversen J. The development of Denmark's nature since the last glacial. København: C.A. Reitzels Forlag; 1973.
30. Ingrouille M. Historical ecology of the British flora. London: Chapman & Hall; 1995.
31. Ralska-Jasiewiczowa M, Goślar T, Madeyska T, Starkel L, editors. Lake Gościąż, Central Poland. A monographic study. Kraków: W. Szafer Institute of Botany, Polish Academy of Sciences; 1998.
32. Elkington TT. Introgressive hybridization between *Betula nana* L. and *B. pubescens* Ehrh. in north-west Iceland. New Phytol. 1968;67(1):109–118.
33. Wagner F, Neuvonen S, Kürschner WM, Visscher H. The influence of hybridization on epidermal properties of birch species and the consequences for palaeoclimatic interpretations. Plant Ecol. 2000;148(1):61–69.
34. Anamthawat-Jónsson K, Thórsson AT. Natural hybridization in birch: triploid hybrids between *Betula nana* and *B. pubescens*. Plant Cell Tiss Organ Cult. 2003;75(2):99–107.
35. Palme AE, Su Q, Palsson S, Lascoux M. Extensive sharing of chloroplast haplotypes among European birches indicates hybridization among *Betula pendula, B. pubescens* and *B. nana*. Mol Ecol. 2004;13(1):167–178.
36. Thórsson AT, Pálsson S, Sigurgeirsson A, Anamthawat-Jónsson K. Morphological variation among *Betula nana* (diploid), *B. pubescens* (tetraploid) and their triploid hybrids in Iceland. Ann Bot. 2007;99(6):1183–1193.
37. Caseldine CH. Changes of *Betula* in the Holocene record from Iceland – a paleoclimatic record or evidence for early Holocene hybridization. Rev Palaeobot Palynol. 2001;117:139–152.
38. Jónsson TH. Stature of sub-arctic birch in relation to growth rate, lifespan and tree form. Ann Bot. 2004;94(5):753–762.
39. Białobrzeska M, Truchanowicz J. Badania systematyczno-biometryczne nad rodzajem *Betula* L. ze Szczelińca w górach Stołowych. Rocznik Dendrologiczny. 1990;39:51–78.
40. Elven R, Karlsson T. Salix L. In: Jonsell B, editor. Flora Nordica. Stockholm: The Bergius Foundation; 2000. p. 117–188. (vol 1).
41. Beerling DJ, Terry AC, Hopwood C, Osborne CP. Feeling the cold: atmospheric CO_2 enrichment and the frost sensitivity of terrestrial plant foliage. Palaeogeogr Palaeoclim Palaeoecol. 2002;182(1–2):3–13.
42. Rundgren M, Björck S. Late-glacial and early Holocene variations in atmospheric CO_2 concentration indicated by high-resolution stomatal index data. Earth Planet Sci Lett. 2003;213(3–4):191–204.
43. Kovačić S, Nikolić T. Relations between *Betula pendula* Roth. (Betulaceae) leaf morphology and environmental factors in five regions of Croatia. Acta Biol Cracov. 2005;47(2):7–13.

8

Uses of tree saps in northern and eastern parts of Europe

Ingvar Svanberg[1], Renata Sõukand[2]*, Łukasz Łuczaj[3], Raivo Kalle[4], Olga Zyryanova[5], Andrea Dénes[6], Nóra Papp[7], Aneli Nedelcheva[8], Daiva Šeškauskaitė[9], Iwona Kołodziejska-Degórska[10,11], Valeria Kolosova[12]

[1] Uppsala Centre for Russian and Eurasian Studies, Uppsala University, Box 514, 751 20 Uppsala, Sweden

[2] Estonian Literary Museum, Vanemuise 42, 51003 Tartu, Estonia

[3] Department of Botany and Biotechnology of Economic Plants, University of Rzeszów, Werynia 502, 36-100 Kolbuszowa, Poland

[4] Institute of Veterinary Medicine and Animal Sciences, Estonian University of Life Sciences, Kreutzwaldi 62, 51014 Tartu, Estonia

[5] V. N. Sukachev Institute of Forest, Russian Academy of Sciences, Academgorodok 50, 660036 Krasnoyarsk, Russia

[6] Natural History Department, Janus Pannonius Museum, Box 158, 7601 Pécs, Hungary

[7] Department of Pharmacognosy, University of Pécs, Rókus 2, 7624 Pécs, Hungary

[8] Department of Botany, Sofia University "St. Kliment Ohridski", Dragan Tzankov 8, 1164 Sofia, Bulgaria

[9] Kaunas Forestry and Environmental Engineering University of Applied Sciences, Liepu 1, Girionys, LT 53101 Kaunas, Lithuania

[10] Warsaw University Botanic Garden, Aleje Ujazdowskie 4, 00-478 Warsaw, Poland

[11] Institute for Interdisciplinary Research "Artes Liberales", University of Warsaw, Nowy Świat 69, 00-046 Warsaw, Poland

[12] Institute for Linguistic Studies, Russian Academy of Sciences, Tuchkov pereulok 9, 199053 Saint-Petersburg, Russia

Abstract

In this article we review the use of tree saps in northern and eastern Europe. Published accounts by travellers, ethnologists and ethnobotanists were searched for historical and contemporary details. Field observations made by the authors have also been used. The presented data shows that the use of tree sap has occurred in most north and eastern European countries. It can be assumed that tree saps were most used where there were extensive stands of birch or maple trees, as these two genera generally produce the largest amount of sap. The taxa most commonly used have been *Betula pendula*, *B. pubescens*, and *Acer platanoides*, but scattered data on the use of several other taxa are presented.

Tree sap was used as a fresh drink, but also as an ingredient in food and beverages. It was also fermented to make light alcoholic products like ale and wine. Other folk uses of tree saps vary from supplementary nutrition in the form of sugar, minerals and vitamins, to cosmetic applications for skin and hair and folk medicinal use.

Russia, Ukraine, Belarus, Estonia, Latvia and Lithuania are the only countries where the gathering and use of sap (mainly birch sap) has remained an important activity until recently, due to the existence of large birch forests, low population density and the incorporation of sap into the former Soviet economic system.

It is evident that gathering sap from birch and other trees was more widespread in earlier times. There are records indicating extensive use of tree saps from Scandinavia, Poland, Slovakia and Romania, but it is primarily of a historical character. The extraction of tree sap in these countries is nowadays viewed as a curiosity carried out only by a few individuals. However, tree saps have been regaining popularity in urban settings through niche trading.

Keywords: sap collection, fermented beverages, non-timber forest products, ethnobotany, forestry, food culture

Introduction

Tree sap has traditionally been a source of nutrients in the spring in the boreal and hemiboreal regions of the northern hemisphere. The famous Arab traveller Ahmad ibn Fadlān observed as early as 921 that the Turkish-speaking Bolgars along the Volga River used fermented birch sap. The German scholar Conrad of Megenberg (14th century) mentions its use as a refreshing drink. The Persian geographer Rashīd al-Dīn also describes in his "Jāmiʿ al-tawārīkh" from the early 14th century that the Uriankhai people in Siberia cut birch trunks and gathered the sap, which was drunk instead of water. Italian botanist Pietro Andrea Mattioli praised its medicinal properties in 1561. According to him it could be used for treating stones in the kidney and the bladder, but also to cure ulcers [1–5]. In his thesis "Disputatio physico-medica votiva" written in 1631 in Tartu, the physician Johannes Raicus praised birch sap and suggested its use in medicine, comparing its healing properties with those of mineral water [6].

* Corresponding author. Email: renata@folklore.ee

It was common practice to tap the liquid from various kinds of trees to obtain a beverage in times of want. The sweet taste of the sap was a welcome change after the long winter in social contexts where sugar was rare. A deficiency of food in early spring was customary in northern Europe until the end of the 19th century, and in eastern Europe until the 1960s. Tree sap has been used for several purposes other than just a source of sugar. Actually, many different products are made out of it, such as fermented beverages, beer, wine, and syrup. Swedish ethnologist Gösta Berg even suggested that tree sap was also probably the most used and one of the most effective medicines against scurvy in earlier times [1].

Internationally, the best known product made of tree sap is the Canadian maple syrup that led to the maple leaf becoming the symbol of the country. Sap from the sugar maple tree (*Acer saccharum* Marshall) and some related species was being processed among indigenous peoples of North America long before Europeans arrived on the continent [7–9]. In northern and eastern Europe maple sap is less frequently gathered, although it is utilised in some regions. Nevertheless, maple sap is valued for its sweet taste, as it contains more sugar than birch sap. Other tree taxa have also been utilised in eastern and northern Europe as sources of sap [1,10].

Birch sap has been considered one of the most profitable non-timber forest products [11,12]. Experiments conducted in Estonia in the 1970s showed that the profit gained from the sap was six times the profit gained from timber [13]. A birch tree can produce 36 l of sap in nine days [14]. In Alaska over 1000 gallons (3785 l) of birch syrup are made every year. The quality of the product is regulated by Alaska Birch Syrup-makers Association and although the production of birch syrup is five times more costly than maple syrup, it has its own market due to its specific taste [15].

There has been little scientific investigation and current research into the technology of production of tree sap in Europe, and only a few of these studies have been published in English, mainly concerning Finland [16–27]. The content of sugar in the European species of birches can reach 0.8% of their weight [25]. The only part of Eurasia where birch sap was industrially gathered was in the Soviet Union. Hence, most of the research on technology, productivity and resources has been published in Russian. In the former Soviet Union, research on birch sap indicates that it can be used against anaemia, cancer, tuberculosis, kidney and liver stones, gout, arthritis, rheumatism, cold and skin diseases. It also has diuretic properties, can be used as worm powder and prevents tooth troubles [28–33]. "Biomos", a medicine, based on birch sap, has been found to heal wounds and burns and work as an antiphlogistic and antisclerotic substance [34]. In veterinary medicine birch sap cures some cattle diseases and increases milk production [35], while in bee farming it is used as extra feed for the bees [36].

There is a rather limited number of modern ethnological or ethnobotanical reports on the use of birch sap for food outside the former Soviet Union [2,10]. Although tree saps have been mostly used as a simple beverage, they also have medicinal and cosmetic functions [14,37,38] as well as folk veterinary applications [39]. Recent research on the use of birch sap for cosmetic purposes has shown promising results [40].

The purpose of this paper is to give an overview of the use of tree saps in northern and eastern Europe (including northern Eurasia). Relying on ethnographic data, travellers' accounts and contemporary ethnobotanical fieldworks, we discuss the means of collection, preservation, food and non-food use of saps of different species, covering especially the genera *Betula* and *Acer*.

Northern Europe

Denmark

Tapping birch sap has a long tradition in Denmark. It is mentioned by botanist Simon Paulli in 1648. Later authors write that it was used not only as a beverage, but also, by adding yeast and malt, made a refreshing ale.

Birch sap has also been used for medicinal purposes [41]. It was included in the "Danish pharmacopeia" in 1772. Birch sap was recommended against hepatitis, rash, intestinal worms and scurvy [42]. Birch sap was also added when making cheese in order to protect it from vermin [43].

A perfume factory in Denmark bought 600 l of birch sap annually in the mid-1950s and used it for making hair-water. It was also used as emergency food in times of food shortages. The sap was also used when making bread [42]. A wine made of birch sap was made by adding yeast to it in earlier days [44].

Norway

Birch sap has been tapped and used in some parts of Norway. In the medieval "Flatey book" ("Flateyjarbók"), completed in 1394, there is a description of how King Sverre and his men "spent two nights in the wilderness and had no food but sap they could suck from the trees" [2]. Norwegian writer Peder Claussøn Friis described in the 1590s from Finnmark that sap was tapped and used by herdsmen and woodcutters [45]. There are many descriptions in the topographical literature from 18th and 19th century about the use of birch sap for food in various parts of Norway. It was also used by the higher classes [2].

Children, especially, have gathered and appreciated birch sap, but adults have drunk it too. Most of the sap produced has been used as a fresh drink, but in some areas they added yeast, and in more recent times also sugar, in order to produce a kind of wine. The sap was also used when making coffee. For children it was a kind of sweet [46].

Sweden

Tapping of sap from the birch tree in the spring used to be very common in Sweden. Cartographer Olaus Magnus made a brief note in 1555 that Scandinavians were tapping birch for sap and using it as a fresh drink [2]. It is also mentioned in a medicinal handbook from 1578. Two species of birch are common all over Sweden: silver birch *Betula pendula* Roth. and downy birch. *Betula pubescens* Ehrh. Particularly the former has been used. In some parts of southern Sweden the Norway maple (*Acer platanoides* L.) has also been utilized for this purpose. Ethnographical records from various provinces give detailed information on the gathering and manufacturing of the sap into drinks or its use as an ingredient in various foodstuffs. A pine-wood pipe was used for the collection [47–49]. Disputes about the right to tap birch sap from trees were taken to court as early as the 17th century [2].

Birch sap has been seen as a good food ingredient and used for gruel (made of birch sap and barley meal), in coffee, and in some areas (such as northern Dalecarlia) made into ale by mixing with malt and yeast [1,47,48]. From Småland there is a description from 1749 saying that the birch sap ale was sometimes flavoured with bog myrtle (*Myrica gale* L.) [50]. At the end of the 19th century and beginning of the 20th century birch sap was mostly gathered and consumed by children [2,51]. Information concerning sap can also be obtained from literature. Poets like Carl Michael Bellman (18th century, "birch sap is my life and liquor is my health") and Erik Axel Karlfeldt (late 19th century, "the sap of the birches is spraying from drilled bark") mention the use of birch sap as a drink [1]. Birch sap was obviously not only used by the peasantry, but also by the aristocracy and bourgeoisie, something that is also proved by recipes in older cook-books. For instance urban taverns provided wine made of birch sap in the late 18th and early 19th century. Many recipes mention additives such as bog cranberry (*Vaccinium oxycoccus* L.) juice, barberry (*Berberis vulgaris* L.) juice, lemon (*Citrus* × *limon*) juice and also, if these were not available, gooseberry (*Ribes uva-crispa* L.) wine (Tab. 1). These recipes probably have a German origin [2].

Besides beverages, vinegar was also made of birch sap [52]. According to a Swedish translation of Wolfgangi Hildebrand's "Magia naturalis" from 1654, some birch juice should be added to cheese to protect it from worms [53].

In some parts of Sweden it was considered as famine food in the springtime. Historian Pehr Arvid Säve [54] writes that poor people on the island of Gotland used to gather birch sap and use it as a beverage. A birch tree could produce around 2 l a day for two weeks on Gotland. During the last real famine in Sweden, in 1867, tree-sap was widely used in the southernmost part of Sweden, for bread, gruel and as a mealtime drink. Birch was therefore referred to as the "poor-man's cow", according to one record from southern Sweden [10].

Birch sap has also been used among the Saami in Sweden, especially among the Forest Saami in Västerbotten. It was used when making gruel. The Saami also used the sap and inner bark from pine (*Pinus sylvestris* L.) and spruce [*Picea abies* (L.) Karst.], particularly the former. It was gathered in early spring. It was dried into flour which was eaten with milk and fat [55].

Maple sap was occasionally gathered as well, for instance in Öland, where it was still collected in the 1930s [2].

As in many other northern and central European countries, there is a renewed interest in using birch sap also in Sweden. Small-scale production of bottled sap exists. Every year courses are run in order to develop an interest for this non-timber wood product.

Birch-wine is also mentioned in several 18th and 19th century texts [56–60]. Wine from birch sap is nowadays made by some home producers, but also by small companies working with local products and gastronomic food. Most famous is the "Grythyttan Björkvin", a rather expensive sparkling birch-wine, developed from an old recipe from 1785. It is produced by a small company in Jämtland and sold by the government owned chain of liquor stores (Systembolaget).

Tab. 1 Additives for the fermentation of tree saps.

Additives	Regions used (time)
Anethum graveolens L. (stems)	BE
Avena sativa L. (sprouting grain)	LT
Berberis vulgaris L. (fruit juice)	SE
Carum carvi L. (dried fruits)	EE (all the time)
Citrus × *limon* (juice)	BE, SE, UA (commonly nowadays)
Citrus × *limon* (peels)	EE (last 80 years), FI (18th century)
Fabaceae (beans)	RU
Hordeum vulgare L. (roasted grains)	UA (now and in the past)
Juniperus communis L. (twigs)	EE (all the time)
Malus sp. (dried fruits)	UA (now and in the past)
Myrica gale L. (herb)	SE (18th century)
Prunus sp. (cherry fruits)	BE
Pyrus sp. (dried fruits)	UA (now and in the past)
Quercus robur L. (bark)	BE
Ribes nigrum L. (twigs)	EE (all the time), LT (new)
Ribes uva-crispa L. (wine)	SE
Syzygium aromaticum (L.) Merrill & Perry (flower buds/clove)	FI (18th century)
Vaccinium myrtillus L. (juice)	BE
Vaccinium oxycoccus L. (fruit juice)	SE
Vitis sp. (rasins)	EE (last 50 years), UA (commonly nowadays)
(Rye) bred	BE, EE, LT, RU, (PL – rarely)
Grains (barley, oats), flour (barley, oats, rye)	EE (until mid-20th cent), FI (modern), LT
Maltose	SE, FI, RU
Sugar	BE, EE (recent), NO, UA
Wax	RU
Yeast	DK, SE, NO

BA – Bosnia and Herzegovina; BE – Belarus; BG – Bulgaria; CZ – Czech Republic; DK – Denmark; EE – Estonia; FI – Finland; GE – Germany; HU – Hungary; LA – Latvia; LT – Lithuania; NO – Norway; PL – Poland; RO – Romania; RU – Russian Federation; SE – Sweden; SK – Slovakia; SI – Slovenia; UA – Ukraine; UK – Scotland and England.

Finland

In Finland and Carelia birch sap was tapped off in the spring and used as a refreshing drink [61,62]. The Saami in Finland also made a refreshing beverage of birch sap [63,64]. The silver birch (*B. pendula*), especially, has been harvested. Tapping birch trunks has a long tradition and was regarded as an important commodity in early modern times. Various techniques were used to collect the sap [65]. In eastern and northern Finland sap was also gathered from the birch tree stumps by making a pit in it. There is a dissertation from Turku in 1759 describing in detail the use of birch sap in Finland. It was obviously widely used both by the peasantry and the nobility. Sap was not only used as a fresh drink but also as beer. It was naturally fermented and could be used during the summer. The dissertation describes how the sap was used for manufacturing syrup, sugar and other goods. It also gives a recipe for birch sap-wine flavoured with cloves and lemon peel [66]. When Swedish historian Nils Reinhold Broocman visited the Governor of Villmanstrand (Lappeenranta) on the shore of the Lake Saimaa in 1754 he was served sparkling wine made of birch sap [2].

Birch trees were regarded as private property and individual trees were even given names (similar to names of cattle) in the area of Vyborg in south-eastern part of the country. The month when the birch gives sap is known as "mahlan aju" or "mahlakuu" in the Finnish folk calendar. This name for the month of April is used in dictionaries by the priest Christfrid Ganander in the late 18th century and by philologist Elias Lönnrot in the mid-19th century, but has also been recorded later [67].

In south-western Finland a kind of ale was made of birch sap mixed with flour and malt [68]. The Dutch diplomat Anthonis Goeteeris noticed as early as 1616 that birch sap was used for brewing beer in Finland [53]. A recipe of traditional birch sap beer from northern Carelia is given by Räsänen [69].

A questionnaire about the use of birch sap was sent out in 1963 and the answers show that it was still tapped in the central and south-western part of the country [65]. A renewed fashion for tapping birch sap has been seen since the 1980s. Currently there is an increasing interest in birch sap as a non-timber forest product, both from researchers and new product developers. Birch water and syrup made of birch sap is available in well-equipped Finnish stores.

Norway maple (*A. platanoides*) grows only in the southernmost part of Finland, and it was therefore more seldom used [68].

Estonia

In Estonia, birch and Norway maple sap has been widely collected for as long as can be remembered. April was known as "mahlakuu" ("sap month") in some parts of Estonia. This name was recorded in 1660 by linguist Heinrich Göseken , who wrote "Aprilis Maala kuu / â fluxu betularum arborum, qve in illo mense fluunt" [70].

In his travelogue from 1615 Adolph Nickolaus von Steinkallenfels describes the gathering of birch sap in Estonia [71]. Linguist August Wilhelm Hupel describes, in 1777, that birch sap was used for making a fresh beverage, but also for making wine and vinegar [72]. Birch sap was not only gathered by ethnic Estonian peasants, but also by Swedish-speaking coast-dwellers [73,74]. The Setu people in south-eastern Estonia also gathered birch sap [2]. In the South Estonian dialect there is a special word for sap-tapping: "tikkama" [75,76].

Norway maple (*A. platanoides*) is less common in Estonia and people did not dare to spoil their yard trees except on the islands of the Baltic Sea, where maple sap is considered "the right sap", while birch sap is not appreciated. However, there are records that boys used to tap maple sap illegally from trunks in manor parks in mainland Estonia [74]. The maple tapping time period was before the snow melted, while birch sap was foraged after the snow was gone. Among common people *B. pendula* has been the preferred sap-tree, although it gives less sap that *B. pubescens*, as it was believed that its sap is sweeter. However, the experiment showed that *B. pendula* gives 158–282 l in 20–27 days and contains less sugar than *B. pubescens*, which gives 67–151 l in 31–38 days and *Acer platanoides* 19–32 l in 32–38 days [77]. It was also believed that a tree growing on a hill had sweeter sap (Fig. 1). Sap was collected using a wooden or metal spout in a drilled hole, but also by striking a scythe in the trunk. A collecting container was placed under the dropping spout and changed as it filled up [78].

Starting from 1972, centralized collection of birch sap was organized through state forest enterprises (Fig. 2). The plans for collecting were fixed, the quality standards for chemical and physical qualities were established, the techniques of collection changed, and opened containers were replaced by relatively hermetic collecting [79]. Two food-factories stocked up and processed sap, adding bog cranberry (*Vaccinium oxycoccos* L.) or quince [*Chaenomeles* spp.] juice, as well as sugar, and sold it in the shops [12,80]. In addition, the birch sap was used to produce shampoo. Over 30 tons of birch sap were collected in 1976 for that purpose [12]. In modern Estonia a few men of enterprise have found use for larger amounts of semi-industrially gathered birch sap in the form of the birch vodka "Kaseviin" and birch wine; also fresh birch sap is sold locally or exported to Italy [81].

Fig. 1 Birch sap is preferably collected on the hills, often near homesteads. Kambja, Tartu County, Estonia, in 2012. Photograph: Raivo Kalle.

With the fall of the Soviet Union, collecting birch (and less frequently Norway maple) sap still remained an important activity for rural people. In spring every year newspapers re-introduce the techniques of collecting and using birch sap. The majority of the population know the taste of birch sap. Historically, most of the maple and a lot of birch sap was used fresh, for drinking and making food (instead of water for tea, soups, and porridges) [82]. When there was a shortage of sugar, particularly during wartime, the tree sap was used to make syrup for sweetening food [12,77]. For some families making syrup is still a kind of ritual entertainment.

The main means of preserving birch sap was through alcoholic fermentation. Besides the plain sour drink that was kept for months in the cellar and used during hay-making, different drinks were produced, like kvass (light ale), beer, and mead. For the fermentation, different additives were used, like twigs of black currant (*Ribes nigrum* L.) and juniper (*Juniperus communis* L.), caraway fruits (*Carum carvi* L.); and later also

Fig. 2 Collecting of birch sap for the food industry in the early 1990s, the last years of Soviet regime, Tiksoja, Tartu County, Estonia. Photograph: Lembit Michelson (EFA.204.0-169211).

raisins and lemon peels. Sometimes, grains or flavour were added after fermentation, to produce specific kvass [12,83]. Although detailed recipes on how to make wine and sparkling wine from birch sap were published as early as the end of the 18th century [72,84], there is no trace of their imitation in ethnographic data. When in the beginning of the 20th century soft drinks became available in shops, the consumption of fresh birch sap dropped considerably [82]. In the 1970s (and even later), birch-sap was processed and was used not only at hay-making time in summer [13,85]. Fermented birch sap is still perceived as a hay-time drink, but many rural households, also without cattle, prepare birch-sap for beverages. Nowadays the fresh birch sap is stored in freezers and is used for instance in the sauna.

The cosmetic use of birch sap was widespread in Estonia. In the 19th century it was believed that one who washes his or her face with the first drops of birch sap will get rid of freckles and their face will stay pale all the summer. Fair skin was valued among the peasantry, as well as among the landowners. It was also said that if one washes the eyes with the first drops of birch or maple sap it will keep them clean and healthy. Sore eyes were treated with birch and maple sap compresses. There are a few reports of using birch sap for treating specific skin diseases [86].

There are also a few reports on the use of other tree species for tapping sap. Folklorist Oskar Loorits [87] has a mid-20th century record of using thicker rowan trees (*Sorbus aucuparia* L.), a few other authors [12,82] mention the tapping of its sap. Ethnographer Aliis Moora [82] also reports the tapping of linden (*Tilia cordata* L.). Recent field work report the modern use of sycamore maple (*A. pseudoplatanus*) on Hiiuma Island. The latter taxon is probably more widely used than was earlier reported, since this maple is widespread in Estonian urban vegetation.

Latvia

As in the other Baltic countries there is a long tradition of drinking birch sap in spring and making beverages from it to be consumed in the months of summer heat. The month of April is known as "sulu mēnesis" in Latvian and "kõļimkū" in Livonian, both meaning "birch sap month", indicating that sap was gathered at that time [88]. Latvian ethnologist August Bielenstein [89] gave a detailed description of the traditional gathering and manufacturing of birch sap among the Latvians (Fig. 3).

Also the Livonian people of northern Latvia had a name for April meaning "birch sap month". They tapped the birch in spring-time and the sap was fermented in a large vessel [90].

Tapping birch sap has been a common by-product in the Latvian forests up until the present day [91]. In the Soviet period Latvia was one of four grand industrial collectors of birch sap in the Union [12]. Birch sap is important in contemporary Latvian culture. Wine and other products made of birch sap are available in stores. Birch sap from Latvia is exported frozen and available in many countries.

Lithuania

Birch sap was gathered in large quantities in the 1830s and 1840s by the Lithuanians. It was used both fresh and fermented [92]. In some parts of Lithuania, birch and Norway maple sap was fermented into a kind of refreshing beer (similarly as in Belarus and Ukraine), at least up until the mid-20th century [93].

Fig. 3 Way of tapping sap from Birch in Latvia in the late 19th century. From Bielenstein [89].

In modern Lithuania, sap is gathered from Norway maple (*A. platanoides*) or birch (*B. pendula*, rarely also *B. pubescens*). Usually, old and thick trees are used for extracting the sap, and every year the same trees are used. The most common procedure is to drill a hole into a tree trunk at a height of about 50 cm. A small wooden gutter (nowadays metal or plastic tube) is inserted into the hole and the sap leaks into a bucket beneath. Maple sap starts flowing in early spring, around March. It is easily checked by breaking off a small twig; if the breakage becomes wet and even starts to trickle it is time to start. The time is described traditionally "after snowmelt but still with frozen ground". The maple gives 3 to 5 l per day. The sap flow depends on weather conditions and lasts from few days up to 3 weeks. While early sap is clear, at the end it becomes dreggy and tends to ferment sooner. Due to the sweet taste, the small amount and the fact that this is the very first fresh food received from plants in the year, maple sap is consumed immediately and is not conserved. When maple sap is finished, one can proceed with birch sap. Again, readiness can be tested by breaking a twig or looking at a fresh tree stump. Birches are much more generous: 10 to15 l out of one hole per day is the revenue for a relatively old (ca. 60 years) birch tree. The bough is cut, or twig broken, and a jar attached. As birches are very common, large amounts of sap were/are extracted and preserved. Additional ingredients may differ: grains of barley or oats, peas, rye flour, a chunk of bread. Especially oats soon sprout shoots and the germinated grains form a solid cover. After a few months – usually during hay-making – this cover is removed and the fermented sap is consumed. It is commonly believed to have revitalizing properties. Ethnologist Vacys Milius [94] mentions the use of twigs of black currant, *Ribes nigrum* L., for sap conservation. He believes this is a rather recent innovation. Birch sap has also been used for washing hair [95].

Scotland and England

The sap of trees has also been collected in Scotland. Several species have been utilized for this, such as birch (*B. pendula*), ash (*Fraxinus excelsior* L.), field maple (*Acer campestre* L.) and sycamore (*A. pseudoplatanus*). The harvesting of birch sap took place in March. The tap was usually made of a hollow elder twig or a grooved peg. A good tree could produce as much as two gallons (approx. 7.5 l) in a single season [96].

The sap was used as a fresh beverage, and sometimes it was made into wine. English gardener Moses Cook gave, in 1675, a quaint recipe for making wine of birch sap [88]. Birch sap was considered to have medicinal qualities. According to Scottish folk medicine it was used in preventing baldness. Queen Victoria is said to have drunk large quantities of birch sap when she was at Balmoral Castle, in order to halt the thinning of her hair [97]. In the Scottish Highlands ash sap was given to newborn children as their first nourishment and in Ireland and England to treat earache [98].

Birch trees were traditionally still tapped in the Highlands as late as the late 1940s, as a rule, the sap was boiled down to make a sweetmeat [8]. However, birch tapping appears, as in other parts of northern Europe, to be making a comeback in Scotland [97].

We also have a few notes from various parts of England where birch sap used to be tapped, for instance from Montgomeryshire, Hampshire and Yorkshire. The trees were tapped in February or early March [2].

In Lincolnshire and the Highlands birch sap wine was made and used as a tonic and for treating rheumatism. The trees were tapped for a few days only in order to not exhaust the tree [99].

Central Europe

Czech Republic

The month of March is called in Czech "březen" – "the month of birches". Tapping sap from birch, maple and beech (*Fagus sylvatica* L.) is described from the Bohemian Forest. The best sap came from the birch, though. In some areas of Bohemia young girls and boys used to gather on 23 March in order to tap birch sap. This was celebrated by eating food and dancing around a birch. The girls consumed the birch sap in order to be healthy and, as grown-up women, fertile [4,100].

Disabled persons went in secret to a birch-tree on the first day of March to cut the bark and to put a piece of linen with a drop of blood into the incision. If the bark inosculated, the impairment was said to be healed. The sap harvested on that day was also assumed to have healing power: it was drunk for good health, against infertility, etc. [101]. Girls washed themselves with birch sap to be beautiful and not to have freckles. Some of them also drank it to be healthy and to have many children in marriage [102].

Germany

Birch sap has been used as medicine among the Germans, especially against lung diseases and gout. Physician Hieronomys Bock describes, in his "Kreuterbuch" (1551), its medicinal use. In the 1880s birch sap was still harvested in the Harz Mountains and in the Thuringian Forest [2]. It was also still used as a cosmetic in the early 20th century [103]. Birch sap was still gathered in East Germany in the 1980s (Fig. 4).

Poland

According to a plethora of ethnographic sources, the use of birch sap (mainly from *B. pendula*, more rarely from *B. pubescens*), was widespread in the 19th century. It was usually drunk fresh, extracted by drilling a hole and making a little trough or pipe from wood, or by breaking a branch. In the mid-20th century this use was (nearly) obsolete in most parts of Poland or practiced mainly as boys' spring entertainment [104–108]. The use of birch sap was most intense and widespread in eastern parts of the country (e.g. Podlasie and Mazovia regions). Processing of birch sap was carried out relatively rarely but is documented in a few regions at the turn of the 19th and 20th century. For example, in the Kozienice Forest (central Poland) it was boiled with rye flour and milk [109]; in the Kurpie region (north-east Poland) birch syrup was produced by boiling off the water from the sap, and used as food sweetener [110]. Birch sap has never been sold in Poland on a large scale; it is only available in health food shops, as a kind of curiosity.

Norway maple (*A. platanoides*) sap was extracted very rarely in Poland, mainly in its eastern part, and, after 1945, in the territories of west and north-eastern Poland, where the settlers from the present Lithuania and Belarus were moved. It was usually drunk fresh, although the inhabitants of the village of Wrzosy (north-east Poland, Polish immigrants from Lithuania) made a fermented drink out of it [93,105,107]. Sycamore (*A. pseudoplatanus*) sap was also occasionally drunk, only fresh, mainly in SE Poland [105,107].

Very rarely a few other species of trees were used to obtain sap, drunk in the fresh state, these were mainly hornbeam (*Carpinus betulus* L.), linden (*Tilia* spp.), sour cherry (*Prunus cerasus* L.) and wild bird cherry (*P. avium* L.) [104,107].

Slovakia

Birch (*B. pendula*) and maple sap were commonly drunk fresh by Slovaks (mainly boys) in 19th century Slovakia [111–113]. Only the Latin name of field maple (*A. campestre*) is mentioned in ethnographic sources [111,112], but sycamore (*A. pseudoplatanus*) was probably commonly used as well.

Fig. 4 Tapping birch sap in Colditz, near Leipzig, in East Germany in April 1985. Photograph: Wolfgang Kluge. Source: http://de.wikipedia.org/w/index.php?title=Datei:Bundesarchiv_Bild_183-1985-0404-003,_Colditz,_Birkensafternte.jpg&filetimestamp=20081204095223

Attempts were made at the beginning of the 19th century to make sugar from maple species (following the North American example) on an industrial scale in the present territory of Slovakia, but later they were abandoned [114].

Eastern Europe

Belarus

In Belarusian, March is called "сакавік" [sakavik], "the month of sap". Some historical data on the use of tree sap in Belarus are present in Rostafiński's ethnobotanical questionnaire of 1883. A few of his respondents reported that the use of both *Betula* spp. and *Acer platanoides* sap was very common there. One of the ways of utilizing the fermented birch and Norway maple sap was boiling it into soup with young shoots of *Aegopodium podagraria* [107]. There is also 19th century information that sap from European aspen (*Populus tremula* L.) was gathered by the Belarusian peasantry [104].

Although it was a popular beverage, it was supposed that one may fall ill with fever having drunk sweet birch sap. In the Brest Region people believed that someone who drank birch sap would have lice [115].

In 1972 Belorussia was the third biggest birch sap producer in the Soviet Union, lagging behind only Ukraine and Russia [12]. There it was suggested that sap can be tapped from mature birch forests for three years before clear cutting, with an estimated yield of 40000 l/ha [116]. Nowadays in the European part of the former Soviet Union birch sap is often sold with lemon juice added. Although this practice is of relatively modern origin it was already reported in Rostafiński's ethnobotanical questionnaire of 1883 in the Belarusian Polesia region (lemon juice and sugar was added and the bottle was closed for 2–4 weeks) [107]. In the times of the Soviet Union a recipe of Belarussian origin was published, it described making kvass with a long storage time: birch sap was closed in a vessel with rye bread, oak bark (*Quercus robur* L.), cherries (*Prunus* sp.) and dill stems (*Anethum graveolens* L.), with no yeast added [117].

In modern Belarus birch sap collecting is regulated by state rules, and all industrially produced sap has been collected by producers authorised by the state. In 2011 over 25000 tons of birch sap was collected, according to a report by the Belarus authorities [118]. Part of it is industrially processed and exported in carton packages with different additives (e.g. *Vaccinium myrtillus* L.) to many European countries (e.g. Germany, France and Baltic States).

Belarusian citizens have the "right to collect birch sap for their own use in the places, allocated for this purposes by the legal entities conducting forestry. It's necessary to apply to the forestry organisation at the place of domicile. [...] Birch sap may be collected in specially protected areas provided it is allowed by the protection and use mode for these territories" [119].

Russian Federation

The English botanist John Tradescant the elder who travelled to Arctic Russia in 1618 observed in the vicinity of Arkhangelsk that the peasantry gathered birch sap and used it as a fresh drink [2]. Early information about sap production and utilization can also be found in a Russian manuscript from 1768. People have traditionally used the sap as a fresh drink as well as for making wine, vinegar, syrup, kvass and confectionery [120,121]. Russian ethnographer Dmitri Zelenin [122] describes the various ways birch sap was utilized by the Russians. It was drunk fresh, but also fermented by adding malt, wax, beans or rye bread. Also Norway maple sap (*A. platanoides*) was gathered. In the 1890s, the peasants in the Vologda area still gathered sap from the Scots pine (*P. sylvestris*) [104].

In the Perm region birch sap was consumed as a beverage and used externally against sores on the shanks. To make ointment, one took two bottles of the sap and 0.06 l of ethyl alcohol and evaporated it on a small fire until the liquid became syrup. The extract was spread on a cloth and put on the sores. In the Transbaykal region the freshly harvested birch sap was given to children with milk during the teething period [123].

Not only the Russians but also many other ethnic groups living within the territory of Russia used sap. Votes, Izhorians and Vepsians, who lived in the Saint Petersburg Governorate at the end of 18th century, used maple sap for making syrup. Birch sap was drunk fresh to improve the health and made into sparkling drinks and vinegar [124]. The Mordvinian used the thickened sap as food. It was also used for medicinal purposes [125]. It was also known as a beverage among Ob-Ugrians [63]. Birch sap for washing faces was still used among Estonian settlers in the Omsk region at the end of 20th century [86]. Yukagirs living in eastern Siberia are known to have utilized the sap of poplars (*Populus* sp.) and willows (*Salix* sp.) [126]. The German botanist and explorer Johannes Gottlieb Georgi [127] writes that the Bashkirs used birch sap water as a refreshing spring beverage. Tofalars and Altays living in the Altay area used thickened birch sap as a sweetener in tea [10]. Ethnographic data on the use of birch sap also exist from many other ethnic groups in Siberia and Russian Far East, such as Evenki, Itelmens and others [1,5,128].

In Soviet times natural and preserved birch sap, birch sap with sugar, "Берёзка" (birch) and "Весенний" (vernal) drinks, birch kvass and maple drinks were available on the retail market while saps with herb and pine needle extracts were widespread [129]. In perfumery and the cosmetics industry birch sap was added to lotions and shampoos [22,35].

Industrial sap collecting in the Soviet Union started in Kazan in the 1920s. In 1937 over 300 tons of birch sap were collected for making wine and kvass in Sverdlovsk [35]. At the end of the 1980s the annual sap production reached more than 70000 tons [129]. Four species of birch (*Betula costata* Trautv., *B. pendula*, *B. platyphylla* Sukaczev, *B. pubescens*) are used for tapping sap in Russia [130]. Daily birch sap extraction ranges from 0.9 up to 13.5 and equates in average 4–5 l per tree of *B. pendula*, *B. platyphylla* and *B. pubescens* [121]. Korean birch (*B. costata*) appears to be of the highest sap productivity (50–78 l per day) as compared with the other birch species [131]. The amount of sap produced depends on the size of the tree, so values given by different authors may vary from a few to a several dozen litres.

After the collapse of the Soviet Union the former birch sap production was completely shut down [132]. However, some regional authorities have recently undertaken efforts to restart birch sap production. For example, the government of the Khabarovsk region has developed such a program [133]. The technology of sap tapping and production specification for "Натуральный березовый сок" (natural birch sap) and "Дальневосточный березовый сок с сахаром" (far east birch sap with sugar) have been developed [132,134].

At the end of the 20th century, Russian researchers proposed the use of coniferous cell sap [129] and phloem sap [135] as prospective substances for medicine, cosmetics, agriculture, household chemicals, etc. Such tree species as *Pinus sylvestris* L., *Picea abies* (L.) Karst., *Abies sibirica* Ledeb., *Tilia* sp. and *Fraxinus* sp. were considered to be the most productive for sap tapping.

Ukraine

In Ukrainian the month of March is known as "березень" [berezen'] – "the month of birches". In Ukraine birch sap is widely used and sold in shops. In 1972, while in the whole of the Soviet Union around 25000 tons of birch sap was industrially collected, the share of Ukraine was 20000 tons of birch sap [12]. It is usually drunk fresh, pasteurized, often with lemon juice. Traditionally it was fermented in wooden barrels with dried fruits (apples and pears) and/or roasted barley added to it. It lasts as long as late summer, and was drunk as a refreshing drink during hay-making and cereal harvest. Nowadays a variety of additives are used, mainly lemon juice, raisins and/or sugar. The use of birch sap decreased after the Chernobyl nuclear catastrophe localized in the main birch sap region of Ukraine, and nowadays some people still tend to make the sap themselves and not buy in the stores out of fear of drinking radioactive sap [136].

Every year the beginning of the collecting season is announced by the media as important news. Short films made during sap collection and recipes for sap preparation and use are quite popular on Ukrainian Internet sites. A large share of commercial sap collection (made by forest enterprises) is exported. The results of a recent study, 2012, on birch sap use in some villages in the Vinnytsia Oblast of Ukraine, conducted by ethnobotanist Iwona Kołodziejska-Degórska, show that both *B. pendula* and *B. pubescens* are used, a variety of sap extraction methods are used (e.g. pipes drilled into the tree, even made from ball pen tubes; one or two cuts with a tube, a grass stalk or a metal sheet bent into a V-shaped rut underneath etc.; Fig. 5, Fig. 6). Most people collect only small amounts (up to 10 l) into plastic bottles (Fig. 7) and use the sap mainly in the fresh state, although the traditional methods of fermentation are still known and used by some individuals and birch sap collection remains an important annual activity. In spite of earlier reports on non-food uses of birch sap in the region, such as medicinal (treating skin diseases internally and externally), a source of vitamins (especially for children), as a diuretic [137] and cosmetic (to remove freckles [138]) none of research participants reported them. For commercial collection of sap plastic bags and glass 3 l jars are used.

Fig. 5 All over Ukraine birch sap is collected into reused plastic bottles, Vinnytsia Oblast, 2012. Photograph: Iwona Kołodziejska-Degórska.

Fig. 6 V-shaped metal spout used in Ukraine, Vinnytsia Oblast, 2012. Photograph: Iwona Kołodziejska-Degórska.

Southeastern Europe

Bosnia, Carinthia and Slovenia

The use of birch, maple and European beech (*F. sylvatica*) sap by herdsmen has been reported from Bosnia [4]. The review of edible plants of Bosnia and Herzegovina mentions only two species of trees whose sap is used: *A. platanoides* and *A. pseudoplatanus* [139].

Nobleman Johann Weikhard von Valvasor describes in his famous "Die Ehre deβ Hertzogthums Crain" [140] from Carniola that a healthy beverage was made of the birch sap. Birch sap was still gathered around Logatec in Inner Carniola

Fig. 7 Bottles covered with birch bark are protected from the sun and are more difficult to spot in the forest, Vinnytsia Oblast, 2012. Photograph: Iwona Kołodziejska-Degórska.

in Slovenia in the mid-20th century [141]. There are records of tapping birch-sap from the Carinthian Slovenes in neighbouring Austria from the end of 19th century [142].

Bulgaria

In the literature there are some data on the use of sap from Tatar maple (*Acer tataricum* L.). The sweet juice is used as food reinforcement, especially for children and young people. Juice flowing from the stems of plants was collected, filtered and compressed. The time of tapping, methods of processing and applications are not specified [143]. Recent field investigations show that tree saps are generally used for medicinal purposes. For instance, rubbing of the roots of the hair with birch (*Betula pendula*) sap is believed to help hair growth.

The reason for the limited use of birch and maple in traditional food and folk medicine among Bulgarians is their fragmented distribution throughout the country. However, grapevines (*Vitis vinifera* L.) are sometimes tapped for their sap. In the Rhodope Mountains on the eve of St. George (6th of May) the grapevine plants are trimmed and sap is collected. In the morning women smear their hair with this sap and throw the rest in a field of barley to grow their hair long as the cereal. It was also believed that grape sap cured sore eyes. Sometimes the sap is collected in glass bottles for medicine when needed [144,145]. The sap is used against dry eyes, and also to remove pigmentation and freckles. Grape vine sap is used against gallstones and bladder stones, boiled in wine. In folk recipes drops from cut branches are called "tears" of vine.

Purulent lichens are treated by brushing them with Cornelian cherry (*Cornus mas*) sap [142]. The juice of the injured stem of ash (*Fraxinus ornus*, *Fraxinus excelsior*) is used as a laxative [143–145].

Hungary

Tapping trees for sap used to be common practice in all forested parts of Hungary. In the north-eastern part of the country, in the so-called "Nyírség" (meaning: "landscape with birch forests"), which was rich in birch trees, collecting birch sap ("birch-water") had been an important way of making money. Hawkers had taken it, in barrels, into the neighbouring cities [146,147]. Birch-sap collection occured in other places too, but no sources mention its collection in commercial quantities [148–158].

Most of the forests of Hungary are turkey oak and hornbeam-oak forests – *B. pendula* is only a rare accompanying species. Where birch occured, it was tapped in the spring. The sap was drunk by herdsmen, forest-workers, only later by children as a delicacy.

Besides *B. pendula*, the sap of Turkey oak (*Quercus cerris* L.) [150,154,156,159,160], sycamore (*A. pseudoplatanus*) [152,153], hornbeam (*Carpinus betulus*) [149], and wych elm (*Ulmus glabra* Huds.) [160] was also collected [*U. glabra*, being a rare species, may be false data, field elm (*Ulmus minor* Mill.) is more common]. In the floodplain of the river Drava an *Ulmus* species – presumably European white elm (*Ulmus laevis* Pall.), which is a common tree in today's floodplain forests – mentioned as "híres szilfák" ("elm celebrities" [161]) was also tapped. Tapping was performed with axes or gimlets; sap was blown or sucked out using reed, elder or nettle stems.

Where water-springs were rare, tree sap was the only drink for herdsmen. The sap of *Q. cerris* was drunk from the early spring until the beginning of frost. Herdsmen in Somogy county built permanent taps into the oaks giving the best sap. The best sap was like "good brown beer", it even frothed. These oaks had been visited for up to 20 years. Taps made of elder stems had grown into the tree-trunk over the years. Turkey oak could also be tapped through the chill cracks; if tapping was performed skilfully, sap spurted out [154,156]. Chill cracks are long vertical cracks formed when superfluous saps freeze inside the tree in the freezing weather. They often occur in *Q. cerris* growing in mesophilous sites too wet for this species.

Birch sap was drunk as a refreshment, and it was also considered an appetizer. It was taken against stomach and lung illnesses, women used it as a cosmetic, especially against freckles. Vinegar and beer was brewed, wine was fermented from sap [147]. Birch sap was also used for coagulating milk

Fig. 8 Collecting birch sap in a Romanian forest. Photograph: Nora Papp.

Tab. 2 The use of tree saps in the countries of northern and eastern Europe.

Species	Preservation method					Purpose of use	
	Fresh	Fermented	Congested into syrup	Pasteurized	Freezing	Food	Medicinal use
Acer platanoides L.	BA, BE, EE, FI, LT, PL, RO, RU, SE	BE, EE, LT, PL	EE,			BA, BE, EE, PL, RU	EE
Acer pseudoplatanus L.	BA, EE, HU, PL, RO, SK					BA, EE, PL,SK, RO	HU
Acer campestre L.	RO, SK					RO, SK	
Acer tataricum L.	BG					BG	BG
Acer sp.	UA	UA					
Betula pendula Roth. and/or *Betula pubescens* Ehrh.	BE, BG, CZ, DK, EE, FI, GE, HU, LA LT, NO, PL, RO, RU, SE, SK, UA, UK	BE, DK, FI, EE, HU, LA, LT, NO, RO, RU, SE, UA	FI, EE, PL, RO	BE, RU, UA	EE, LA	BE, CZ, DK, EE, GE, HU, LA, NO, PL, RO, RU, SE, UA, UK	BG, CZ, DK, EE, GE, HU, RO, BG, SE, UK, SK
B. platyphylla Sukaczev	RU	RU		RU		RU	RU
B. costata Trautv.	RU	RU		RU		RU	RU
Pinus sylvestris L.	SE, RU					SE, RU	RU
Picea abies (L.) Karst.	SE, RU					SE, RU	RU
Sorbus aucuparia L.	EE					EE	
Tilia spp.	EE, PL, RU					EE, PL, RU	RU
Fraxinus ornus L.	BG,						BG
Fraxinus excelsior L.	BG, UK					UK	BG, EE, UK
Fraxinus spp.	RU					RU	RU
Fagus sylvatica L.	BA, CZ, RO					BA, CZ	
Carpinus betulus L.	PL, HU, RO					PL	
Prunus cerasus L.	PL					PL	
Prunus avium L.	PL					PL	
Abies sibirica Ledeb.	RU						
Populus spp.	RU						
Populus tremula L.	BE						
Salix spp.	RU						
Vitis vinifera L.	BG						BG
Cornus mas L.	BG						BG
Ulmus glabra Huds.?	HU						
Ulmus minor Mill.	HU						
Ulmus laevis Pall.	HU						
Quercus cerris L.	HU						
Juglans regia L.	RO						

For abbreviations see Tab. 1.

when making cheese [162]. Herdsmen preferred drinking sap to water, because sap helps with the digestion of heavy food. In veterinary practice it was used against bloating, dysentery and as a diuretic [160]. Sap of *A. pseudoplatanus* was taken against coughs [153].

Tapping of *Q. cerris* was last reported in Hungary in the 1960s [154], although it was forbidden by law then. Today, tree sap collecting has become extinct.

Romania

Both maple and birch sap were occasionally used in Romania, both by Romanians and Transylvanian Hungarians [163–165]. Sap from *A. pseudoplatanus* and *A. platanoides* have been tapped and used fresh by children [163,164], as well as *A. campestre* [164]. In Transylvania the sap from *B. pendula* was known as "virics" ("birch water"). It was extracted by various methods, according to the reports from the 18th to the 20th century. The tree-trunk was slit with an axe (Fig. 8) or a knife at various heights and directions, or drilled to make a hole [146,166,167]. Generally, a funnel made from birch bark, or a stem of reed *Phragmites australis* (Cav.) Steud., elder *Sambucus nigra* L. or hemlock, *Conium maculatum* L. collected in the previous year [168], was applied to the section or a hole, leaking the sap into a piece of pottery [164,169]. Birch sap was given to weak children to strengthen them [163].

When local folk medicine was still practiced in Transylvania, birch sap was sold in barrels in the local town markets, and stored in cellars before use [168]. Virics was used against jaundice and to remove kidney stones, and for coagulating milk when making cheese [162,170]. It has been mentioned as a treatment against cold [164,171], scab, eye disorders and as a diuretic [163,172,173], but also against constipation, goitre, headache and pneumonia. Blended with oil it could be used to heal wounds [168]. Among the external applications, the washing of moles and sunspots, the use as hair colouring with sugar [168], and as a hair conditioner have been documented in the country [174]. Moreover, birch sap has been consumed in spring as a thirst-quenching drink [164,175], and as fermented or boiled for making wine, beer, syrup or vinegar [146,176]. These products were made in special buildings known as "birch-water houses" among the Hungarian-speaking Székely [3]. The utilisation of birch sap is a disappearing practice in contemporary Romania.

Tapping of hornbeam (*C. betulus*), beech (*F. sylvatica*) and walnut (*Juglans regia* L.) is reported from Transsylvania [164]. Also grapevine (*Vitis vinifera*) sap is used among Venetian diaspora and wider to treat eye inflammations [177].

Discussion and conclusions

The presented data shows that tree saps have been harvested in many European countries (Tab. 2). It can be assumed that tree saps were most widely used where there were extensive birch or maple stands, as these two genera produce the largest amount of sap.

Two main uses of tree saps apply to most countries. The first one is nutritional. Their sugar content made tree saps a valuable nutritional resource. This was utilized mainly in the form of fresh and fermented beverages, as the extraction of condensed sugar syrup from European tree species requires more energy than in the case of the North American *Acer saccharum*. The second was medicinal use (Tab. 3). Tree saps were believed to contain some vital substances, hence, for example their use to restore the growth of hair.

Russia, Ukraine, Belarus, Estonia, Latvia and Lithuania are the only countries where the use of tree sap (especially birch sap) has remained an important activity. This is due to the

Tab. 3 Medical/cosmetic applications of tree saps.

Region	*Betula* sap		*Acer* sap
	Medicinal use	**Cosmetic use**	**Medicinal use**
BE	lung diseases, gout		
BG		hair growth	food suplement
CZ	poor health, infertility	for beauty, against freckles	
EE	(prevention of) eye diseases, skin diseases, source for vitamins	washing hair, against freckles and to bleach the skin	(prevention of) eye diseases, tuberculosis, lack of appetite
HU	appetizer, stomach and lung diseases,	against freckles	cough
LT	"revitialization"	washing hair	
PL	"revitalization", kidney stones	washing hair in order to strengthen it	
RO	kidney stones, jaundice, as milk-rennet, scab, diuretic, conjunctivitis cold, pneumonia, constipation, struma, wounds, headache, weakness in children	hair colouring, to remove sunspots and moles, as hair conditioner	
RU	externally against sores, to help children during teething, to improve health	washing face (EE immigrants)	
SE	scurvy, cholera		
UA	treating skin diseases (internal, external), source of vitamins (especially for children), diuretic	to remove freckles	
UK	tonic, rheumatism, first nourishment for new-born children	prevention of baldness	

For abbreviations see Tab. 1.

existence of large birch forests, low population density and the incorporation of sap production into the former Soviet economic system.

There is also a large body of evidence for the extensive use of tree saps from Scandinavia, Hungary, Poland and other central European countries (e.g. Slovakia and Romania) but it has mainly a historical character, as the extraction of tree sap in these countries is nowadays viewed as a curiosity carried out only by some individuals. However, in some regions tree sap is slowly regaining its popularity in urban settings through niche trading (delicatessen, health food shops, etc.).

Remarks

Authors contribution (author initials): initiation and assembling of the article (RS, ŁŁ, IS), "Introduction" (IS, RS, ŁŁ, OZ), tables (RS, AN, NP, ŁŁ, IS), Scandinavia (IS), Estonia (RK, RS, IS), Latvia (IS, RK), Lithuania (DS, IS), British Isles (IS, RS), Poland and Slovakia (ŁŁ), Czech Republic (IS, VK), Belarus (ŁŁ, IS, RS, RK, VK), Russia (OZ, IS, RK, VK), Ukraine (ŁŁ, IK-D), Romania (NP, ŁŁ, AD), Bosnia (IS, ŁŁ), Slovenia (IS), Hungary (AD), Bulgaria (AN).

References

1. Berg G. Nordskandinaviskt – nordeuropeiskt. Rig. 1933;16:118–139.
2. Berg G. Att tappa björklake. Gastronomisk kalender 1968. 1969:35–57.
3. Gunda B. Nyírvíz (szócikk). In: Magyar Néprajzi Lexikon (N–Szé). Budapest: Akadémiai Kiadó; 1987. p. 62. (vol 4).
4. Haberland A. Die volkstümliche Kultur Europas. In: Buschan G, editor. Illustrierte völkerkunde. Stuttgart: Strecker und Schröder; 1926. (vol 2).
5. Manninen I. Überreste der Sammlerstufe und die Notnahrung aus dem Pflanzenreich bei den nordeurasischen, vorzugsweise den finnischen Völkern. Eurasia Septentrionalis Antiqua. 1931;6:30–48.
6. Rein K. Meditsiinitopograafia Tartu linna kohta aastast 1631. Mäetagused. 2010;45:105–144.
7. Eagleson J, Hasner R. The maple syrup book. Boston MA: Boston Mills Press; 2006.
8. Henning DD. Maple sugaring. History of a folk technology. Keystone Folkl Q. 1966;11:239–274.
9. Munson PJ. Still more on the antiquity of maple sugar and syrup in aboriginal Eastern North America. J Ethnobiol. 1989;9:159–170.
10. Eidlitz K. Food and emergency food in the circumpolar area. Uppsala: Almqvist and Wiksells; 1969. (Studia Ethnographica Upsaliensia; vol 32).
11. Ollikainen T. Belarus forestry strategic plan and the non-wood forest products. In: Lund HG, Pajari B, Korhonen M, editors. Sustainable development of non-wood goods and benefits from boreal and cold temperate forests: proceedings of the international workshop. Joensuu: European Forest Institute; 1998. p. 159–162.
12. Rebane H. Voolab kasemahl. Eesti Loodus. 1977;4:223–228.
13. Silm E. Kasemahla varumise kogemustest Tartu Metsamajandis [Master thesis]. Tartu: Eesti Põllumajanduse Akadeemia; 1977.
14. Steinmetz EF. *Betula* species. Pharm Biol. 1963;3:334–351.
15. Helfferich D. Birch: white gold in the boreal forest. Agroborealis. 2003;35:4–12.
16. Kallio H. Towards Nordic co-operation in aroma research: are we on the right track. In: Plant based specialty products and biopolymers: Tema Nord 1997:606. København: Nordic Council of Ministers; 1997. p. 140–146.
17. Kallio H, Lahdenoja ML, Penttinen R. Electrophoretic profiles of birch sap proteins of *Betula pubescens*, *B. pendula* and *B. pendula* forma carelica in Finland with reference to overall composition of sap. In: The 1st International Symposium on Sap Utilization (ISSU) in Bifuka '95, Bifuka, Hokkaido, Japan, April 10–12; 1995. Hokkaido: Hokkaido University Press; 1995. p. 13–21.
18. Kallio H. Aroma of birch syrup. J Agric Food Chem. 1989;37(5):1367–1371.
19. Ahtonen S, Kallio H. Identification and seasonal variations of amino acids in birch sap used for syrup production. Food Chem. 1989;33(2):125–132.
20. Kallio H, Teerinen T, Ahtonen S, Suihko M, Linko RR. Composition and properties of birch syrup (*Betula pubescens*). J Agric Food Chem. 1989;37(1):51–54.
21. Macku C, Kallii H, Takeoka G, Flath R. Aspects on the use of headspace GC on-column injections in flavor research. J Chromatogr Sci. 1998;26(11):557–560.
22. Kallio H. Comparison and characteristics of aroma compounds from maple and birch syrups. In: Charalambous G, editor. Proceedings of the 5th International Flavor Conference, Porto Carras, Greece, July 1–3, 1987. Amsterdam: Elsevier; 1988. p. 241–248.
23. Kallio H, Ahtonen S. Seasonal variations of the acids in birch sap. Food Chem. 1987;25(4):285–292.
24. Kallio H, Rine S, Pangborn RM, Jennings W. Effect of heating on the headspace volatiles of Finnish birch syrup. Food Chem. 1987;24(4):287–299.
25. Kallio H, Ahtonen S. Seasonal variations of the sugars in birch sap. Food Chem. 1987;25(4):293–304.
26. Kallio H, Karppinen T, Holmbom B. Concentration of birch sap by reverse osmosis. J Food Sci. 1985;50(5):1330–1332.
27. Meeuse BJD. Observations on the enzymatic action of maple and birch saps. New Phytol. 1949;48(2):125–145.
28. Balitsky KP, Vorontsova AL. Lekarstvennye rasteniya v terapii rakovykh opukholey. 3rd ed. Rostov-na-Donu: Rostovskoe knizhnoe izdatel'stvo; 1980.
29. Deryabina FI. Materialy po izucheniyu narodnoy meditsiny Komi-Permyatskogo natsional'nogo okruga. 3rd ed. Perm: Perm Farm Inst; 1969.
30. Krylov GV, Stepanov EV. Zelenaya apteka Kuzbassa. 2nd ed. Kemerovo: Kemerovskoe knizhnoe izdatel'stvo; 1979.
31. Kucherov EV, Lazareva DN, Desyatkin VK. Dikorastushchie lekarstvennye rasteniya Bashkirii. Ufa: Bashkirskoe; 1973.
32. Turova AD. Lekarstvennye rasteniya SSSR i ikh primenenie. 2nd ed. Moscow: Meditsyna; 1974.
33. Fedorov A, editor. Rastitel'nye resursy SSSR Tsvetkovye rasteniya ikh khimicheskiy sostav ispol'zovanie. Semeystva Magnoliaceae-Limoniaceae. Leningrad: Nauka; 1984.
34. Beskrovny AM, Khuden'ky YK, Bobylev ET, Baranenko SE. Bioneorganicheskie kompleksy "Biomos" kak perspektivnyy produkt luchevoy obrabotki soka listvennykh porod. In: Proceedings of the All-Union Conference "Podsochka i pererabotka soka listvennykh porod". Lvov: Lvovskoe knizhnoe izdatel'stvo; 1977. p. 56–57.
35. Orlov II. Berezovyy i klenovyy soki dobycha i ispol'zovanie. Moscow: Lesnaya promyshlennost; 1974.
36. Evseeva NV. Bereza tvorit chudesa lechenie dyogtem pochkami list'yami i sokom-. St. Petersburg: Nevskiy prospekt; 2005.
37. Pieroni A. Medicinal plants and food medicines in the folk traditions of the upper Lucca Province, Italy. J Ethnopharmacol. 2000;70(3):235–273.
38. Pieroni A, Gray C. Herbal and food folk medicines of the Russlanddeutschen living in Künzelsau/Taläcker, South-Western Germany. Phytother Res. 2008;22(7):889–901.

39. Blanco E, Macıa MJ, Morales R. Medicinal and veterinary plants of El Caurel (Galicia, northwest Spain). J Ethnopharmacol. 1999;65(2):113–124.

40. Moriyama M, Naru E, Misaki Y, Hayashi A. Effect of birch (*Betula platyphylla* Sukatchev var. *japonica* Hara) sap on cultured human epidermal keratinocyte differentiation. Int J Cosmet Sci. 2009;31(2):159–160.

41. Paulli S. Flora Danica. Det er: Dansk Urtebog. København: M. Martzan; 1648.
42. Brøndegaard VJ. Folk og flora: Dansk etnobotanik. København: Rosenkilde og Bagger; 1978–1979. (vol 1–2).
43. Lange J. Løvtræer. In: Granlund J, editor. Kulturhistoriskt lexikon för nordisk medeltid. Malmö: Allhem; 1966. p. 185–198. (vol 11).
44. Lange A, Seeberg C. Nogle sønderjydske Planter, deres Navne og Stilling i Folkloren. Tønder: H. Chr. Bakkes Boghandel; 1929.
45. Claussøn Friis P. Samlede Skrifter. Kristiania: A. W. Brøgger; 1881.
46. Høeg OA. Planter og tradisjon. Oslo: Universitetsforlaget; 1964.
47. Levander L. Livet i en Älvdalsby före 1870-talet. Stockholm: Norstedt; 1914.
48. Levander L. Övre Dalarnas bondekultur under 1800-talets förra hälft. 3, Hem och hemarbete. Uppsala: Jonsson & Winter; 1947.
49. Svanberg I, Tunón H. Bark, näver och sav. In: Pettersson B, Svanberg I, Tunón H, editors. Människan och naturen: etnobiologi i Sverige. Stockholm: Wahlström & Widstrand; 2001. p. 79–84. (vol 1).
50. Krok S. Urshults pastorats inbyggares seder: tal hållet ... inför Småländska nationen i Uppsala den 11 december 1749. Stockholm: Björck & Börjeson; 1922.
51. Svanberg I. Folklig botanik. Stockholm: Dialogos; 2011.
52. Fischerström J. Nya Swenska economiska dictionnairen. Eller Försök til et allmänt och fullständigt lexicon i Swenska hushållningen och naturläran. Stockholm: Carl Stolpe; 1780. (vol 2).
53. Hildebrand S. En holländsk beskicknings resor i Ryssland, Finland och Sverige 1615–1616: trenne reseberättelser, från de tryckta och handskrivna holländska originalen. Stockholm: Norstedt & Söner; 1917.
54. Säve PA. Skogens sagor eller växtligheten på Gotland, jemte spridda drag ur öns odlingssaga och folklifvet derstädes. Stockholm: I. Haeggström; 1877.
55. Drake SAMC. Västerbottenslapparna under förra hälften av 1800-talet. Uppsala: Almqvist & Wiksells; 1918.
56. Warg C. Hjelpreda i hushållningen för unga fruentimber. Stockholm: Lor. L Grefing; 1755.
57. Synnerberg LN. Svenskt waru-lexicon uti sammandrag ur de mest bekanta författares arbeten, rörande handeln. Göteborg: S Norberg; 1815.
58. Valleria C. Christina Vallerias hushållsbok från tiden omkr. år 1700. Uppsala: Nordblads; 1931.
59. Nylander M. Handbok wid den nu brukliga finare matlagningen. Stockholm: Samel Rumstedt; 1828.
60. Cronquist AW. Illustrerad ordbok öfver näringsämnen och handelsartiklar. Stockholm: Hæggström; 1879.
61. Räsänen M. Charakteristika der finnischen Getreidegetränketradition. In: Ethnologische Nahrungsforschung: Vorträge der zweiten Internationalen Symposiums für ethnologische Nahrungsforschung : Helsinki August 1973. Helsinki: Suomen Muinaismuistoyhdistys; 1975. p. 206–211. (Kansatieteellinen arkisto; vol 6).
62. Vuorela T. Suomalainen kansankulttuuri. Porvoo-Helsinki: Werner Söderström; 1975.
63. Vuorela T. Suomensukuiset kansat. Helsinki: Suomalaisen Kirjallisuuden Seura; 1960.
64. Itkonen TI. Lappalaisten ruokatalous. Helsinki: Suomalais-ugrilainen seura; 1921.
65. Lehtonen JUE. Mahlan käyttö karjan ravintona. Kotiseutu. 1970;2–3:40–42.
66. Kalm P. Oeconomisk Beskrifning öfwer Björkens Egenskaper och Nytta i den Allmänna Hushållningen. (Disp., 33 sid., 4:o). Åbo: J Merckell; 1759.
67. Inkilä AT. Mahlavarvu aika ja "onneton Ollin päiva". Kotiseutu. 1939;1:42–45.
68. Talve I. Finnish folk culture. Helsinki: Finnish Literature Society; 1997.
69. Rasanen M. Savokarjalainen ateria: Savokarjalainen ruoka ja ruokakulttuuri muuttuvassa maailmassa. Helsinki: Otava; 1980.
70. Göseken H. Manuductio ad linguam Oesthonicam, Anführung zur öhstnischen Sprache. Tallinn: Adolph Simon; 1660.
71. Velten C. Adolf Nicklassen vom Steinkallenfels Reisse in Schweden und Liefflandt 1615. Bad Kreuznach: Selbstverlag; 1958.
72. Hupel AW. Topographische Nachrichten von Lief- und Ehstland: Nebst vollständigen Register über alle drey Bände. Riga: J. F. Hartknoch; 1777.
73. Söderbäck P. Rågöborna. Stockholm: Norstedt; 1940. (Nordiska Museets Handlingar; vol 13).
74. Danell G. Ordbok över Nuckömålet. Uppsala: Almqvist & Wiksells; 1951. (Estlandssvenskarnas folkliga kultur; vol 2).
75. Wiedemann FJ. Estnisch-deutsches Wörterbuch. St. Petersburg: Buchdruckerei der Kaiserlichen Akademie der Wissenschaften; 1869.
76. Saareste A. Eesti keele mõisteline sõnaraamat. Stockholm: Vaba esti; 1959. (vol 2).
77. Ostrat A. Andmeid mahla kogumise ja selles leiduva suhkru kasutamise võimalustest. Eesti mets. 1943;5:120–123.
78. Kalle R, Sõukand R. Ajalooline ülevaade eestlaste looduslikest toidu-ja ravimtaimedest. In: Kusmin T, Meikar T, editors. Metsa kõrvalkasutus Eestis. Tartu: Eesti Metsaselts; 2011. p. 29–44. (Akadeemilise Metsaseltsi Toimetised; vol 25).
79. Silm E, Linnus A. Kasemahla varumisest Tartu metsamajandis. Mets Puit Paber. 1980;5:13–14.
80. Raudsepp P. 65 tonni kasemahla. Punalipp. 1976;57:1.
81. Relve H. Metsaomanik müüb kasemahla Itaaliasse. Eesti Mets. 2005;2:15–17.
82. Moora A. Kuidas vanasti kasemahla võeti. Eesti Loodus. 1982;4:241–244.
83. Moora A. Mida vanasti kasemahlast tehti. Eesti Loodus. 1982;5:298–300.
84. Spuhl-Rotalia JG. Kodumaa marjad. Viljandi: A Peet; 1898.
85. Tonka K. Heinaajajook võtab janu ja annab tervist. Järva Teataja. 2010:45.
86. Sõukand R, Kalle R. HERBA: historistlik Eesti rahvameditsiini botaaniline andmebaas [Internet]. 2008 [cited 2012 Jul 15]; Available from:

87. Loorits O. Endis-eesti elu-olu: Lugemispalu metsaelust ja jahindusest. 2nd ed. Tartu: Eesti Kirjandusmuuseum; 2004. (Eesti Rahvaluule Arhiivi Toimetused; vol 21).
88. Ritter von Miklosich F. Die slavischen Monatsnamen. Wien: K. Gerold's Sohn; 1867.
89. Bielenstein AJG. Die Holzbauten u. Holzgeräte der Letten. Ein Beitrag zur Ethnographie, Kulturgeschichte und Archäologie der Völker Rußlands im Westgebiet. Petrograd: Tip. Rossiĭskoĭ Akademii Nauk; 1918. (vol 2).
90. Kettunen L. Livisches Wörterbuch. Mit grammatischer Einleitung. Helsinki: Suomalais-ugrilainen seura; 1938. (Lexica Societatis Fenno-Ugricae; vol 5).
91. Rothbergs U. Forests and forestry in Latvia. In: Paulenka J, Paule L, editors. Conservation of forests in Central Europe: Proceedings of the WWF workshop held in Zvolen, July 7–9, 1994. Zvolen: Arbora; 1994. p. 79–87.
92. Cappeller C, Charpentier J. Hur de gamla litauerna levde. Le Monde Oriental. 1906;1:101–133.
93. Bohdanowicz J. Pożywienie. Zbieractwo – pożywienie głodowe. In: Bohdanowicz J, editor. Komentarze do Polskiego Atlasu Etnograficznego: pożywienie i sprzęty z nim związane. Wrocław: Polish Ethnological Society; 1996. p. 10–21.
94. Milius V. Dubičiai. Vilnius: Mintis; 1989.
95. Kalašaitytė L. Liaudies medicinos priemonės. In: Vėlius N, Jonaitytė A, Merkys V, Milius V, Račkauskas K, Vidugiris A, editors. Gervėčiai. Vilnius: Mintis; 1989. p. 204–219.
96. Edlin HL. Woodland crafts in Britain: an account of the traditional uses of trees and timbers in the British countryside. London: B. T. Batsford; 1949.

97. Milliken W, Bridgewater S. Flora Celtica: plants and people in Scotland. Edinburgh: Birlinn; 2004.
98. Hatfield G. Encyclopedia of folk medicine: old world and new world traditions. Oxford: ABC-CLIO; 2003.
99. Allen DE, Hatfield G. Medicinal plants in folk tradition. An ethnobotany of Britain & Ireland. Portland OR: Timber Press; 2004.
100. von Reinsberg-Düringsfeld O. Fest-Kalender aus Böhmen: ein Beitrag zur Kenntniss des Volkslebens und Volksglaubens in Böhmen. Prague: J. L. Kober; 1862.
101. Vinogradova LN, Usacheva VV. Bereza. In: Tolstoy NI, editor. Slavyanskie drevnosti. Etnolingvisticheskiy slovar'. Moscow: Institut Slavyanovedeniya RAN; 1995. p. 156–160. (vol 1).
102. Grohmann JV. Povery a obyceje v Cechách a na Morave. Prague: Plot; 2010.
103. Marzell H. Volksbotanik. Die Pflanze im deutschen Brauchtum. Berlin: Enckehaus; 1935.
104. Moszyński K. Kultura ludowa Słowian: kultura materialna. Cracow: Polish Academy of Arts and Sciences; 1929.
105. Jankowska B. Ściąganie na wiosnę pitnych soków drzew. In: Gajek J, editor. Polski atlas etnograficzny (map No. 311). Warsaw: Polish Scientific Publishers PWN; 1981. (vol 6).
106. Łuczaj ŁJ. Archival data on wild food plants used in Poland in 1948. J Ethnobiol Ethnomed. 2008;4(1):4.
107. Łuczaj Ł. Wild food plants used in Poland from the mid-19th century to the present. Etnobiologia Polska. 2011;1:57–125.
108. Łuczaj Ł, Szymański WM. Wild vascular plants gathered for consumption in the Polish countryside: a review. J Ethnobiol Ethnomed. 2007;3(1):17.
109. Dekowski JP. Rośliny dziko rosnące w tradycyjnym pożywieniu chłopów kozienickich. In: Kowalska-Lewicka A, editor. Pożywienie ludności wiejskiej. Cracow: Ethnographic Museum; 1973. p. 247–256.
110. Chętnik A. Pożywienie Kurpiów. Jadło i napoje zwykłe, obrzędowe i głodowe. Prace Komisji Etnograficznej. 1936;16:1–134.
111. Holuby JL. Die gewöhnlichsten wildwachsenden Genusspflanzen des Trencsiner Comitates. Verhandlungen des Vereins für Natur-und Heilkunde zu Pressburg. Neue Folge. 1891;7:91–105.
112. Holuby JL. Aus der Botanik slowakischer Kinder des Trentschiner Komitates in Ungarn. Deutsche Botanische Monatsschrift. 1896;19(8–9):126–131.
113. Markus M. Ľudova stravá. In: Mjartan J, editor. Banícka dedina Žakarovce. Bratislava: Vydavatel'stvo Slovenskej Akadémie Vied; 1956. p. 283–324.
114. Stoličná R. Alternative sources of food in Central Europe. In: Lysaght P, editor. Food from nature: attitudes, strategies and culinary practices. Proceedings of the 12th Conference of the International Commission for Ethnological Food Research, Umeå and Frostviken, Sweden, 8–14 June, 1998. Uppsala: The Royal Gustavus Adolphus Academy for Swedish Folk Culture folkkultur; 2000. p. 195–203. (Acta Academiae Regiae Gustavi Adolphi; vol 71).
115. Kruk I, Shved I. Byaroza. In: Vladozina T, Sanjko S, editors. Mifalogya belarusa Entsykl slojn. Minsk: Belarus; 2001. p. 58–59.
116. Rebane H. Kask – meie kodumaa kauneim puu. Eesti Loodus. 1970;10:582–588.
117. Eilart J. Mahlajooks – elusa looduse kevade algus. Eesti Loodus. 1959;2:122–123.
118. Borisovskiy konservnyy zavod ostavil zayavki v leskhozakh na zakupku tystonn berezovogo soka [Internet]. Adzinstva. 2012 [cited 2012 Jun 26]; Available from:
119. Collecting birch sap [Internet]. The State Inspectorate on Wildlife Protection under the President of the Republic of Belarus. 2012 [cited 2012 Jun 26]; Available from:
120. Nikitinsky YY. Surrogaty i neobychnye v Rossii istochniki pishchevykh sredstv rastitel'nogo i zhivotnogo proiskhozhdeniya. Moscow: Gosizdat; 1921.
121. Koldaev VN. Prizhiznennoe ispol'zovanie bereznyakov. Lesn kho-zvo. 1971;4:71–76.
122. Zelenin DK. Russische (Ostslavische) Volkskunde. Berlin: W. de Gruyter; 1927.
123. Toren MD. Russkaya narodnaya meditsina i psikhoterapiya. St. Petersburg: Litera; 1996.
124. Öpik E. Vadjalastest ja isuritest XVIII sajandi lõpul: etnograafilisi ja lingvistilisi materjale F. Tumanski Peterburi kubermangu kirjelduses. Tallinn: Valgus; 1970.
125. Paasonen H. Mordwinisches Wörterbuch. Helsinki: Suomalais-ugrilainen seura; 1992. (Lexica Societatis Fenno-ugricae; vol 23).
126. Jochelson W. The Yukaghir and the Yukaghirized Tungus. Leiden: E. J. Brill; 1910. (vol 1).
127. Georgi JG. Beschreibung aller Nationen des russischen Reichs: ihrer Lebensart, Religion, Gebräuche, Wohnungen, Kleidungen und übrigen Merkwürdigkeiten. St. Petersburg: C. W. Müller; 1776. (vol 2).
128. Steller GW. Beschreibung von dem lande Kamtschatka, dessen einwohnern, deren sitten, nahmen, lebensart und verschiedenen gewohnheiten. Frankfurt-Leipzig: J. G. Fleischer; 1774.
129. Roshchin V, Sultanov V. Complex utilization of tree biomass through harvest of birch sap and coniferous cell sap. In: Proceedings of the 1st International Symposium on Sap Utilization (ISSU) in Bifuka '95, Bifuka, Hokkaido, Japan, April 10-12, 1995. Hokkaido: Hokkaido University Press; 1995. p. 111–115.
130. Zyryanova OA, Terazawa M, Koike T, Zyryanov VI. White birch trees as resource species of Russia: their distribution, ecophysiological features, multiple utilization. Eurasian J For Res. 2010;13:25–40.
131. Izmodenov AG. Siledia. Nachalo teorii. Soki i yagody dikorossov. Khabarovsk: Khabarovskoe knizhnoe izdatel'stvo; 2001.
132. Tolstykh VI, Taghiltsev YG, Kolesnikova RD, Lodygin BS. Soki dal'nevostochnykh berez. In: Proceedings of the 2nd International Conference "Forest bioactive resources", Khabarovsk, September 21-23, 2004. Khabarovsk: Khabarovskoe knizhnoe izdatelstvo; 2004. p. 203–206.
133. Khlynov VI. Ob usloviyakh i merakh po rasshireniyu zagotovki pererabotki i prodazhi prirodnykh produktov v Khabarovskom krae. In: Proceedings of the 2nd International Conference "Forest bioactive resources", Khabarovsk, September 21-23, 2004. Khabarovsk: Khabarovskoe knizhnoe izdatelstvo; 2004. p. 10–17.
134. Taghiltsev YG, Kolesnikova RD. Rasshirenie pishchevykh i tekhnicheskikh produktov i razvitie tekhnologiy proizvodstva bioproduktov. In: Gassinskiy opytnyy les. Problemy mnogotselevogo ispol'zovaniya. Khabarovsk: Khabarovskoe knizhnoe izdatelstvo; 1990. p. 91–98.
135. Khramov AA. Tapping of phloem sap in woody plants. In: Proceedings of the 1st International Symposium on Sap Utilization (ISSU) in Bifuka '95, Bifuka, Hokkaido, Japan, April 10-12, 1995. Hokkaido: Hokkaido University Press; 1995. p. 117–122.
136. Łuczaj Ł. Brzozowy sok, "czeremsza" i zielony barszcz – ankieta etnobotaniczna wśród botaników ukraińskich. Birch sap, ramsons and green borsch – an ethnobotanical survey among Ukrainian botanists. Etnobiologia Polska. 2012;2:15–22.
137. Harchenko MS, Syla VI, Volodars'kyj LJ. Likarski roslyny i i'h zastosuvannja v narodnij medycyni. Kiev: Zdorov'ja; 1971.
138. Shevchuk V. Zelenyj dyvosvit Podillja. Rozpovidi ta porady travoznaja. Vinnycja: Derzhavna kartografichna fabryka; 1997.
139. Redzic SJ. Wild edible plants and their traditional use in the human nutrition in Bosnia-Herzegovina. Ecol Food Nutr. 2006;45(3):189–232.
140. Valvasor JV. Die Ehre dess Hertzogthums Crain: das ist, Wahre, gründliche, und recht eigendliche Belegen- und Beschaffenheit dieses ... Römisch-Keyserlichen herrlichen Erblandes. Laybach: Wolfgang Moritz Endter; 1689.
141. Suhadolnik S. O brezovi vodi na Notranjskem. Slovenski etnograf. 1962;15:106.

142. Grafenauer I. Brezen in brezova voda. Slovenski etnograf. 1962;15:101–106.
143. Stoyanov N. Nashite lekarstveni rasteniya. Sofia: Nauka i Izkustvo; 1972. (vol 1).
144. Georgiev M. Encyclopedia Bulgarian folk medicine. Sofia: Petar Beron; 1999.
145. Nikolov SD. Specilizirana enciklopediia na lechebnite rasteniya v Bulgariya. Sofia: Knigoizdatelska kushta Trud; 2006.
146. Kiss L. A nyírvíz. A Földgömb. 1930;1:1–6.
147. Ecsedi I. A debreceni és tiszántúli magyar ember táplálkozása. Debrecen: A Debreceni Déri Múzeum Évkönyve; 1934.
148. C. Schwalm E. Gyűjtögetés, mint nyersanyag-kiegészítő. In: Bakó F, editor. Palócok III. Eger: Heves Megyei Múzeumok Igazgatósága; 1989. p. 417–423.
149. Bödei J. Adatok Zalabaksa gyűjtögető gazdálkodásához. Néprajzi Értesítő. 1943;2:69–96.
150. Fehér J. Adatok Bernecebaráti gyűjtögető és zsákmányoló gazdálkodásához. Néprajzi Értesítő. 1957;2(3–4):267–292.
151. Kardos L. Az Őrség népi táplálkozása, Gyűjtögetés. Tanulmányok az Őrség monográfiájához. Budapest: Államtudományi Intézet Táj- és Népkutató Osztálya; 1943.
152. Kóczián G. A hagyományos parasztgazdálkodás termesztett, a gyűjtögető gazdálkodás vad növényfajainak etnobotanikai értékelése. Egyetemi doktori disszertáció. (Ethnobotanical evaluation of the cultivated plants of traditional agriculture, and that of wild plant species of gathering livelihood [PhD thesis]. Mosonmagyaróvár: ATE Mezőgazdasági Kar; 1985.
153. Kóczián G. Etnobotanikai vizsgálatok Répáshután. Répáshuta, egy szlovák falu a Bükkben. A Miskolci Hermann Ottó Múzeum Néprajzi kiadványa. 1984;8:229–256.
154. Paládi-Kovács A. A Barkóság és népe. Miskolc: Herman Ottó Múzeum; 2006.
155. Pethő M. Győr-Moson-Sopron megye XX. századi népi táplálkozásának rövid jellemzése. Arrabona. 36(1–2):137–150.
156. Takáts G. Somogyi pásztorvilág. Kaposvár: Somogy Megyei Múzeumok Igazgatósága; 1986.
157. Ujváry Z. A vadontermő növények szerepe a táplálkozásban az abaújzempléni hegyvidéken. Néprajzi Értesítő. 1957;39:232–234.
158. Varga G. A népi táplálkozás Hajdú-Bihar megyében a XX. század elso felében. Debrecen: Hajdú-Bihar Megyei Múzeumi Igazgatóság; 1993.
159. Gunda. A gyűjtögető gazdálkodás emlékei egy gerecse-hegységi tót faluban. Ethnographia. 1938;49(1–2):213–214.
160. Nagy R. Adatok a Baranyamegyei nagyváty növényekkel kapcsolatos szokásaihoz és néphagyományaihoz. Magyar Népnyelv. 1942;4.
161. Kiss GZ. Zsákmányoló foglalkozások: Halászat gyűjtögetés. In: Kiss GZ, editor. A régi Vajszló: 1244–1849. Pécs: Vajszló Nagyközség Önkormányzata; 1994.
162. Gunda B. Tejoltó növények a Kárpátokban. Ethnographia. 1967;78:161–175.
163. Butură V. Enciclopedie de etnobotanică românească. Bucharest: Editura Ştiinţifică şi Enciclopedică; 1979.
164. Péntek J, Szabó A. Ember és növényvilág : Kalotaszeg növényzete és népi növényismerete. Bucharest: Kriterion; 1985.
165. Giurescu CC, Farca E. A history of the Romanian forest. Bucharest: Ed. Academiei Republicii Socialiste România; 1980. (Bibliotheca historica Romaniae; vol 63).
166. Gazda K. Gyermekvilág Esztelneken: néprajzi monográfia. Bucharest: Kriterion; 1980.
167. Bartos T. Magyar szótár: egymást magyarázó szavak és fordulatok tára. Budapest: Corvina; 2002.
168. Györffy I. Viricselés a Székelyföldön. Ethnographia. 1937;48(2):205–220.
169. Szebeni G. A csíki juhászat. Ethnographia. 1962;73(1):54–89.
170. Melius P. Herbárium: az fáknak, füveknek nevekről, természetekről és hasznairól. Bucharest: Kriterion; 1978.
171. Tarisznyás M. A gyűjtögető gazdálkodás hagyományai Gyergyóban. In: Kós K, Faragó J, editors. Népismereti dolgozatok. Bucharest: Kriterion; 1978. p. 25–33.
172. Csedő K, editor. Plantele medicinale şi condimentare din judeţul Harghita. Hargita megye gyógy- és fűszernövényei. Târgu Mureş: Tipografia Tirgu Mures; 1980.
173. Rácz G, Rácz-Kotilla E, Laza A. Gyógynövényismeret. Bucharest: Ceres; 1984.
174. Rácz G, Füzi J, editors. Kovászna megye gyógynövényei. Sfântu Gheorghe: Mezőgazdasági, Élelmiszeripari és Vízügyi Igazgatóság; 1973.
175. Rab J, Tankó P, Tankó M. Népi növényismeret Gyimesbükkön. In: János VA, Károly K, József F, editors. Népismereti dolgozatok. Bucharest: Kriterion; 1981.
176. Dieck A. Die Birke (*Betula* L.) in der Volksmedizin Europas. Curare-Sonderband. 1985;3:85–94.
177. Pieroni A, Quave CL, Giusti ME, Papp N. "We are Italians!": the hybrid ethnobotany of a Venetian diaspora in Eastern Romania. Hum Ecol. 2012;40(3):435–451.

9

Distribution, ecology and conservation status of *Dionysia involucrata* Zaprjag., an endangered endemic of Hissar Mts (Tajikistan, Middle Asia)

Arkadiusz Sebastian Nowak[1]*, Marcin Nobis[2,3], Sylwia Nowak[1], Agnieszka Nobis[2]

[1] Department of Biosystematics, Opole University, Oleska 22, 45-052 Opole, Poland

[2] Institute of Botany, Jagiellonian University, Kopernika 27, 31-501 Kraków, Poland

[3] Laboratory of Biodiversity and Ecology, Tomsk State University, Lenin Prospekt 36, Tomsk, 634050, Russian Federation

Abstract

Dionysia involucrata Zaprjag. (Primulaceae) is known as critically endangered endemic species of Hissar Mountains in Tajikistan. It is reported from few localities mainly in Varzob River valley and its tributaries. The species inhabits steep or overhanging faces of granite rocks in narrow river gorges. During the research all known populations of *D. involucrata* were examined in respect of the habitat conditions and species composition of vegetation plots. We analyzed the population extent of the species in its range in Tajikistan and the main threats in order to assess its conservation status. The detrended correspondence analysis was performed on a matrix of 65 relevés and 49 species (vascular plants and mosses), to classify the phytocoenosis with domination of *D. involucrata* according to their floristic composition in relation to other petrophytic vegetation units. Using our field data regarding present extent of occurrence and area of occupancy we conclude that the threat category of *D. involucrata* should be reassessed from critically endangered to endangered. The species shows decline tendency in extent of occurrence, area of occupancy as well as in number of locations. The vegetation plots with domination of *D. involucrata* have relatively high level of separateness due to different species composition. We define the new association – *Dionysietum involucratae* – representing chasmophytic vegetation of submontane and montane zone in Middle Asia (ca. 1000–1600 m a.s.l.). The plots of *Dionysietum involucratae* were found mainly on granite rocks, on very steep or overhanging faces, on southwestern or southern exposition. The association is rather poor in species with inconsiderable contribution of mosses. Despite the diagnostic species, *Campanula incanescens*, *Carex koshewnikowii* and *Scutellaria hissarica* were the most abundant and frequent taxa within the researched patches of vegetation.

Keywords: endangered species; Pamir Alai Mts; cliff vegetation; chasmophytes; petrophytes; phytosociology; syntaxonomy; Tadzhikistan

Introduction

The loss of biodiversity is of primary concern to the international scientific community, the government agencies responsible for the sustainable use of natural resources at global level and non-governmental organizations alike. The precise evaluation of the conservation status of a particular species is a requirement crucial to the successful prevention its extinction. To this end, the determination of threat degree or, alternatively, the expectation of survival of taxa to which a special significance is attributed is crucial. The biogeographical and ecological features of rare or endemic taxa have been the subject of preferential attention by conservationists [1,2] because they appear to be more exposed to threats. One way of risk assessment is to assign a given taxon to a particular standardized threat category. The IUCN red list categories [3] were defined for this purpose.

Tajikistan is a typical mountainous country located in the central part of the Pamir Alai mountain system. Its vascular plant flora is relatively species-rich, including around 4550 taxa [4]. Results of recent studies have shown that this number is not final, as some taxa new to the Tajik flora have been reported [5–9]. Approximately 30% of vascular plant species known from Tajikistan are generally accepted as endemics [4]. It is worth mentioning that Pamir Alai, belongs to the central Asian mountain system recognized by Conservation International as one of thirty-four biodiversity hotspots [10] and as one of the eleven most important focal point of future plant diversity studies and conservation [11]. At the same time, with very nearly the lowest adaptive capacity to climate instability, Tajikistan is regarded as one of the most sensitive country to climate change in the world [12], and

* Corresponding author. Email: anowak@uni.opole.pl

Handling Editor: Zygmunt Dajdok

the high risk of climate change is one of the most critical factors in the degradation of its vegetation [13]. Although recently several papers concerning Tajik vegetation have been published [7,14–20], comprehensive research on its vegetation cover is necessary.

Alpine rock communities are considered as most unique and interesting plant formations in the mountainous areas of Holarctic and the Mediterranean provinces. Despite not being species-rich, they have drawn attention of botanists because they often consist of numerous specialists adapted to harsh and extreme environments [21–24]. Rocky habitats in central Asia are a refuge for a great many stenochorous plant species, probably on account of their high separateness and marginal position within the Irano-Turanian province. The taxa most frequent in rock vegetation patches in the central Asia include: *Campanula incanescens*, *Poa relaxa*, *Carex koshewnikowii* and *Artemisia rutifolia*. They also consist of other angiosperms such as: *Scutellaria* spp., *Campanula* spp., *Asperula* spp., *Dionysia* spp., *Parietaria* spp., *Penthanema* spp., *Silene* spp. and ferns, including: *Cystopteris fragilis*, *Adiantum capilli-veneris*, *Cryptogramma stelleri*, *Asplenium viride*, *A. ruta-muraria* [16].

The genus *Dionysia* Fenzl is spread throughout the Irano-Turanian and Mediterranean areas of Eastern Europe and southwestern Asia. It includes forty-nine species [25,26]. Most of them are adapted to the harsh, chasmophytic habitat conditions. They create fruticulose, loose or dense tufts and cushions and have reduced leaves, all of which are adjustments to growth under extreme conditions such as vertical niches or overhanging cliff walls at higher altitudes. However, they are not truly xerophytic, but require a certain amount of water and shade, which is provided by these niches [25,27]. Representatives of the genus *Dionysia* grow mainly on granite or dolomite rock faces where the amount of soil, if any, is very scarce and has a markedly low nutrient content. They prefer a Mediterranean-like climate with considerable alpine influences, especially in mean annual and minimum temperatures falling down to −2°C and −23°C respectively. Despite the fact that the members of *Dionysia* are very similar to those from the genus *Primula*, they can be relatively easily recognized by their powdery or woolly farina, long corolla tube and the capsule, which splits into 5 valves towards the base. In Tajikistan, the genus *Dionysia* includes two species. Besides *D. involucrata*, *D. gandzhinae* R. Kam. is also known from southern part of the country. The second species is very narrow endemic as well, known to date only from the locus classicus, which is located in the vicinity of Gandzhina village [28]. In the western part of Hissar Range, another narrow vicariant endemic belonging to the genus, *Dionysia hissarica* Lipsky, was reported. It grows on rock faces in southeastern Uzbekistan [29]. *Dionysia involucrata* is associated with loose rocky communities. However, until now, there has been insufficient investigation into the syntaxonomy of the species. No research on the plant communities and species of saxatile habitats has previously been conducted in the central Asian countries like Tajikistan, Kyrgyzstan, and Uzbekistan. In synthetic works, the rock vegetation has been usually described very briefly as "petriphyton" without the use of any classification methods (e.g. [30]). Only a few studies clarify questions of composition and zonation of rock or scree vegetation [31–33]. However, as they provide no distributional and ecological data, they are of minor importance in the analysis of rocky vegetation and chasmophytic species of Tajikistan.

Although *Dionysia involucrata* appears in several floristic studies [28,30,34], little is known about its ecology and endangerment. Given the extreme rarity of the species and the marked scarcity of data concerning its ecology and present distribution, we aimed to conduct the first-ever assessment of the endangerment and conservation status in accordance with IUCN guidelines and recommendations. In order to facilitate the evaluation, we analyzed the extent of population occurrence (EOO), the area of occupancy (AOO) and the total range. In this paper we also present the coenological condition of the species within the syntaxonomical system of submontane and montane saxatile vegetation of the Pamir-Alai mountain system.

Material and methods

Study species

Dionysia involucrata Zaprjag. (Family: Primulaceae, sect. Dionysiastrum Smoljan., subsect. Involucratae) is a small, 5–10 cm tall, glandular perennial. Its leaves are flat or slightly involute, 4–12 mm long, slightly dentate, with conspicuous, flabellate venations, densely glandular. The leaf veins are raised and farina powdery. Flowers occur in a stalked 3- to 5-flower umbel. The bracts are foliaceous and 2–4 cm long, usually longer than the leaves. Calyx is 7–9 mm long, with teeth three-quarters longer than the tube. Corolla is dark pink to light violet, with tube 3–4 times longer than the calyx, 2–3 cm in diameter, with emarginate lobes, homostylous and glandular-hairy (Fig. 1). Flowering period starts in April, and usually lasts for 3 months. Fruit ripens in June and July. An adult plant produces around 5–15 seeds in each capsule, giving approximately 150 seeds per year.

Dionysia involucrata is endemic to the central section of the Hissar Range located in Tajikistan, where it is restricted to several localities in the rocky gorges of the Varzob River and its tributaries. It was found between 1200 and 2500 meters above sea level: near the mouth of the Mayhura River (Fig. 2, location No. 1), the Sioma Valley (2), the Varzob Valley between Gushark and Kabuty (3), Pugus (4), between Dzhirinodsoy and the mouth of the Takob River (5), near Gazhni (6), in the Khondara Valley (locus classicus; 7), the Zamchurud Valley (8), Shaftimighon (9) and the Horongkong Valley (10), [4,28,29,34,35] (the numbers refer to Fig. 2).

This endemic species is considered to be both rare and critically endangered in Tajikistan [36]. However, it has never been included in the world red data book [37]. The main threats to the population have been attributed to too intensive grazing and direct collection by professional and amateur botanists [36].

Study area

The Hissar Range covers approximately 10 000 km^2 and is situated in north-western part of Tajikistan and eastern part of Uzbekistan, between E 39°06′–39°32′ and N 66°55′–70°48

Fig. 1 *Dionysia involucrata* Zaprjag. in Tajikistan. **a,b** Cushions of *D. involucrata* on pink granite rocks. **c** Flowering individual of *D. involucrata* near Gushark. **d** Young cushion of *D. involucrata* with typical leaves rosette near Pugus. **e** The association of *Dionysietum involucratae* in Takob valley.

(Fig. 2, Fig. 3). This is a typical mountainous area, situated between 850 meters above sea level, in the Varzob River Valley near Dushanbe, and 4886 at the Gaznok peak. The subranges within the Hissar Mountains are: the Osmontal, the Kugizaranga, the Sanginavishta, the Hodzhadaizi and the Odzhuk, along with several smaller ones. The Range is situated in the Mediterranean climate zone. However, the altitude and relief mean that the mountainous character of the climate has a strong influence on the area. According to a recent bioclimatic classification, which primarily considers precipitation and temperature values, the study area should be classified under the Mediterranean macrobioclimate type. This climate type is characterized by a summer drought lasting for at least two consecutive months during which mean precipitation is at least two times smaller than mean temperature value [38]. In the case of Dushanbe, four months

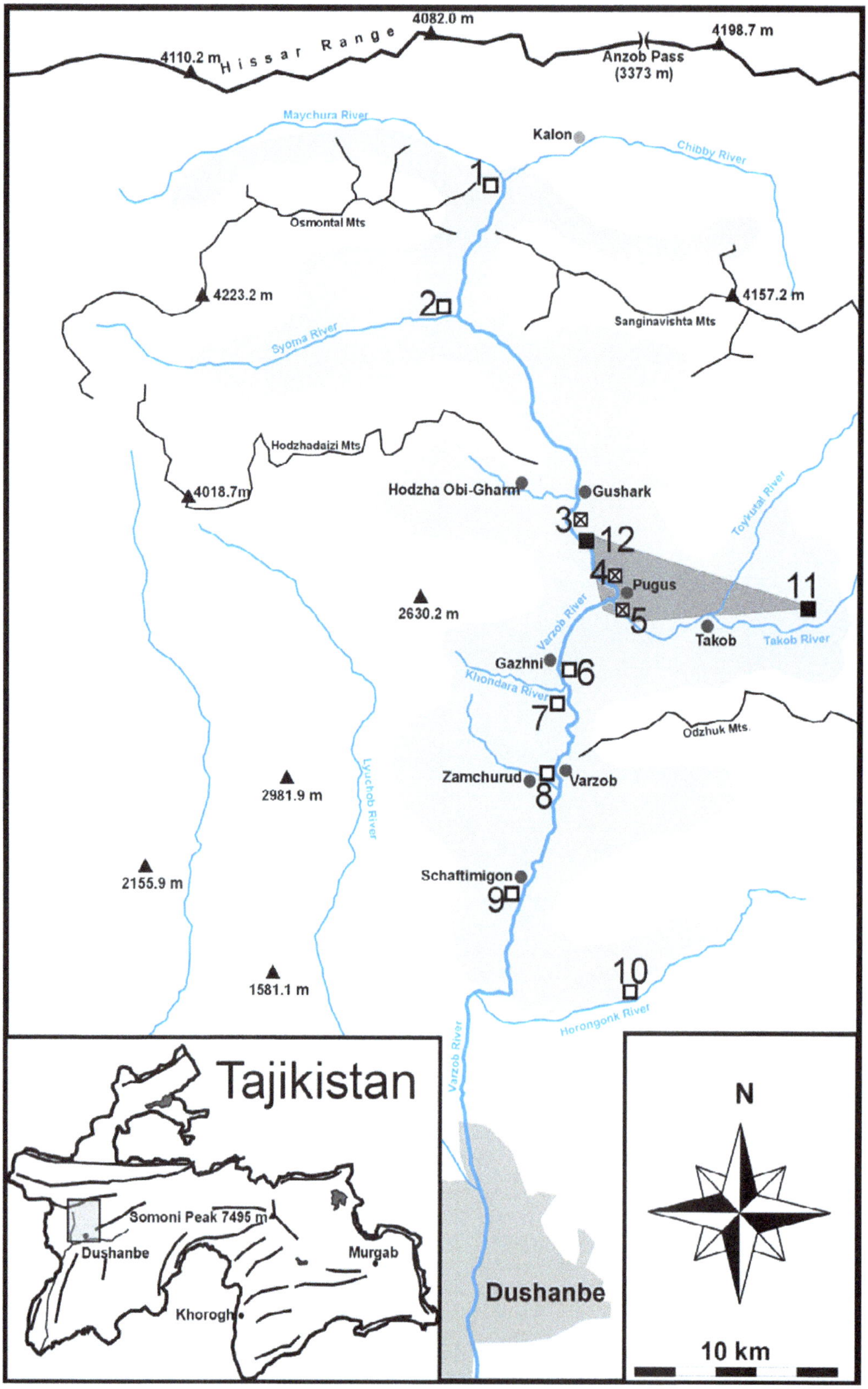

Fig. 2 The location of the study area with the distribution of *Dionysia involucrata* in the Hissar Mts. Location numbers are in accordance with the text and Tab. 1: white square – historical location; crossed square – confirmed historical location; dark square – new location. With hell grey the extent of occurrence, and with medium grey the area of occupancy were highlighted.

Fig. 3 The location of Tajikistan in the Middle Asia.

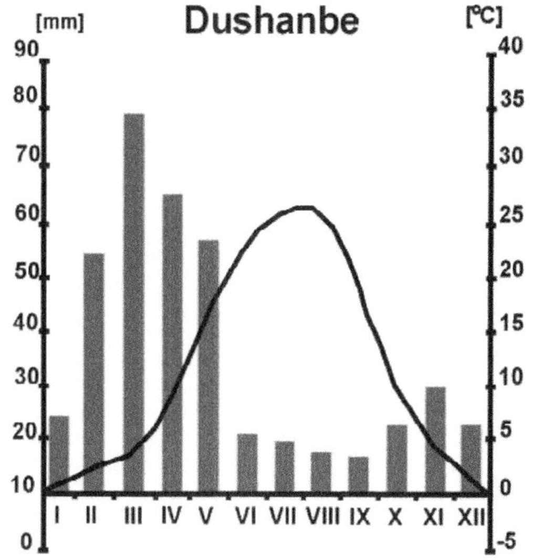

Fig. 4 Climatic characterization of the study area according to the Dushanbe weather station.

in the summer period match this condition (Fig. 4). Other bioclimatic features of the study area also classify it to the Mediterranean macrobioclimate. The annual average temperature is below 25°C (12°C) and the compensated thermicity index is lower than 580, at 542.8). The relatively high index of continentality (*Ic* = 25) demonstrates the considerable influence of the continental climate (subcontinental zone). Generally, this area of the country is characterized by high insolation, a low percentage of cloud cover, high-amplitude annual temperatures, low humidity and low precipitation. In the subtropical regions of Tajikistan, the average temperatures in June are around 30°C. In the temperate zone, which in Tajikstan mainly comprises the high mountains, the climate is much harsher, with average temperatures of between 9.7°C and 13.5°C in July. Annual precipitation in Tajikistan ranges from approximately 70 mm in the Pamirs to around 600 mm in the Hissar Range. The lower perpetual snow limit occurs at an altitude of 3500 to 3600 meters above sea level in the western part of the country, and at one of 5800 meters in its eastern regions [39,40]. These climatic and bioclimatic conditions determine the vegetation types and plant formations in the study areas, where deciduous and juniper forests, xerothermophilous swards and shrubs dominate in the lowlands and the montane belt.

The field studies were conducted in Tajikistan, on a potential area of *Dionysia involucrata* occupancy. The historical range of *Dionysia involucrata* and the adjacent areas in the Varzob River basin were investigated, with a particular emphasis on the valleys of the Takob, Gharm-Chashma, Khondara, Horongkong and Mayhura rivers, as well as several smaller ones. They are all located within the central section of the Hissar range of the Pamir Alai mountain system. Altogether, approximately 100 square km were investigated (Fig. 2, Tab. 1). The sites were phytosociologically examined in respect of the habitat conditions (rock type, elevation, insolation) and species composition of vegetation plots.

Data and analyses

The field studies were conducted in 2008–2013. The phytosociological data were sampled in 2012. A total of 17 relevés were made during the course of the research. The plot size used to sample the vegetation was established in

Tab. 1 A list of localities of *Dionysia involucrata* found during field studies in 2008–2013.

No.	Location	Lat/Long	Area of occupancy (m²)	Population size (cushions)	Altitude
3	To the S from Gushark	38°53′55″ N 68°49′48″ E	50	45	1300
4	Varzob River Valley near Pugus	38°51′23″ N 68°50′29″ E	270	750	1100
5	Lower section of Takob River Valley	38°49′57″ N 68°52′76″ E	185	500	1220
11	Varmanik (middle section of Takob River Valley)	38°50′10″ N 68°56′38″ E	15	20	1600
12	Varzob River Valley between Pugus and Gushark	38°52′13″ N 68°50′0″ E	130	350	1250

such a way as to represent the full floristic composition. It varied from 0.5 to 2 m^2, depending on the plant density and homogeneity of the vegetation cover. All the vascular plants and cryptogams were recorded for each plot. The plant species were recorded in line with the Braun-Blanquet method employing the 7-degree scale (r, +, 1, 2, 3, 4, 5) [41]. The percentage values of the summarized species cover were computed automatically using the Juice program [42] and square rooted during ordination analyses. For the species from different layers (herb and moss), the total cover was calculated on the assumption that species covers can overlap and that they do so independently. Within the moss layer, the covers were assumed to be mutually exclusive. The geographical coordinates, elevation above sea level, substrate type, aspect and slope inclination for each relevé were recorded. The rock type was determined by an analysis of the lithology, pore geometry, mineralogical components, texture, permeability, hardness and pH carried out by a professional geologist (see acknowledgments). Hydrogen ion concentrations were measured in aqueous rock solution, using an Elmetron CP-105 pH meter.

The newly presented syntaxon is proposed in accordance with the "International code of phytosociological nomenclature" [43]. Works by Valachovič et al. [44], Dimopoulos et al. [45] and Nowak et al. [16] were taken into account when distinguishing and ranking the association. The association concept follows Willner [46]. It is based on distinctiveness of floristic composition of the association. The main diagnostic taxon is *Dionysia involucrata*. As differential one in relation to other communities of closely related habitats (community of *Scutellaria hissarica* and of *Tylosperma lignosa*) the *Scutellaria adenostegia* has been indicated.

In order to identify differences in the species composition of the vegetation patches with *Dionysia involucrata* and other phytocoenoses developing on rock faces in submontane and montane zone, the relevés collected during our field studies were compared with 48 relevés sampled by authors in western Pamir Alai Mts. To conduct the ordination analyses, we employed detrended correspondence analysis (DCA) to test the gradient length of the data using and subsequently elected to use unimodal techniques. The 17 relevés we gathered in this study were pooled with the 48 relevés of other communities (Fig. 5). To identify the differences between the vegetation patches, all the relevés were analyzed using the DCA functions of the Canoco software package [47].

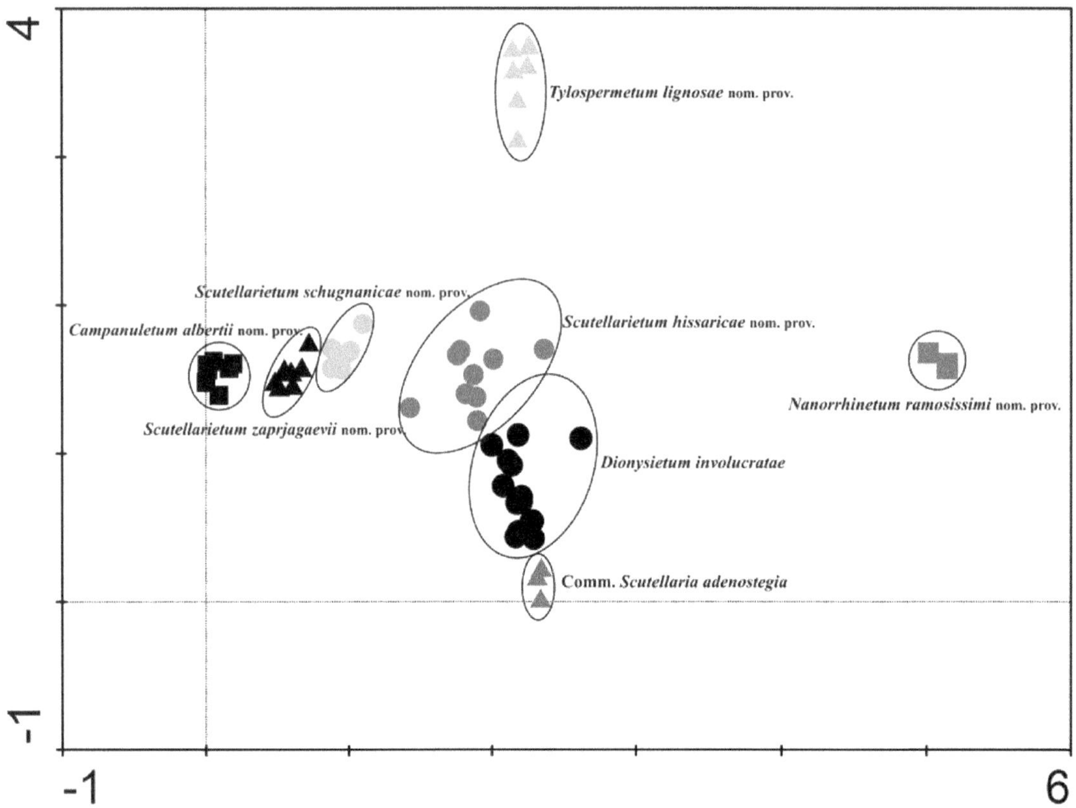

Fig. 5 DCA ordination for all samples of rock faces communities of submontane zone collected by authors in western Pamir Alai Mts in Tajikistan ($N = 65$).

Species nomenclature follows Czerepanov [48] for vascular plants and Ochyra [49] for mosses. The plant material collected during the field studies was deposited in the Middle Asian Mountains Herbarium hosted in OPUN (Opole University, Poland) and KRA (Jagiellonian University, Poland).

Threat evaluation

For the assessment of the current conservation state of *Dionysia involucrata*, we adopted the newest version of the IUCN guidelines for species endangerment evaluation [3]. We analyzed the total potential range, EOO and AOO as the most readily available and important data. To determine the AOO and EOO an exhaustive bibliographical survey was carried out and the herbarium specimens deposited in St. Petersburg (LED: Varzob River valley, leg. Stepanchenko and Stockij, Hissar Mts., 1954.06.25; Varzob River valley, leg. Gubanow I.A., 1963), Tashkent (TASH: Hissar range, Khondara River valley, leg. Krasovskaja, 1982.08.31; Kohondara River alley, leg. Baranova, 1985.05.12; Khondara River valley, leg. Sharapov, 1982.08.31) and Dushanbe (TAD: Khondara valley, Hissar range, leg. Pisjaukova, June 1950).

The extent of occurrence was defined as the area contained within the shortest continuous, imaginary boundary, which can be drawn to encompass all the known, inferred or projected sites of the current occurrence of a taxon. The area of occupancy was defined as the area within the EOO, which is presently occupied by a given taxon.

Results

Historical range and present distribution

Dionysia involucrata was previously known from 10 locations in the Varzob River Valley in the Hissar Mountains, within the following limits: north, Mayhura River Valley (39°01′51.8″N; 68°46′29.7″E); and south, the Horongonk River Valley (38°38′38.5″N; 68°50′52.1″E). All of the locations were checked during the study. A total of 5 populations of *Dionysia involucrata* were found, including two which were previously unknown. The current distribution of the species is generally restricted to the central section of the Varzob Valley, at an altitude of between 1100 and 1600 meters above sea level (Tab. 1). All recorded populations were established on pink granite rock, with the exception of the Takob Valley, where the individuals of *D. involucrata* grow on compact limestone.

Status of endangerment

The total range of *Dionysia involucrata* as known from the literature, herbarium and field data, is no larger than 2285 km^2. Our results show that the present EOO of *D. involucrata* is 11 km^2 and the current AOO is 650 m^2. In accordance with the IUCN categories and criteria, the taxon should be assessed as endangered (EN), because (***i***) the EOO is much lower than 5000 km^2 (criterion B1); (***ii***) the AOO is much lower than 500 km^2 (criterion B2) and, simultaneously, the taxa has only five extant locations [criterion B(a)] while a continuing decline in the EOO, AOO, as well as in the number of locations is observed [criteria B(b)i,ii,iv].

Most of the populations examined were large, with three of the five exceeding 350 individuals. The largest populations consist of more than 700 specimens. A total of over 1700 established plants were recorded (Tab. 1), meeting criterion C, small population size, for the EN category. Seedlings were frequent. Juvenile, mature, flowering and fruiting plants were observed in all the populations.

Ecology and syntaxonomical position

The plots of phytocoenosis with *Dionysia involucrata* were clearly distinguished in the results of the numerical classification of all relevés sampled in the rock habitats of the montane zone of the Pamir Alai Mountains in Tajikistan (Fig. 5). The plots are distinctively separated according to species composition (eigenvalue of first axis – 0.897, of the second – 0.786). Within eight of the distinguished groups, association with the domination of *D. involucrata* takes a central position. Given the structural differences, habitat separateness and floristic composition, this community could be defined as association. Despite insufficient investigation of the whole submontane zone of Tajikistan, the other groups probably should be identified also as associations: *Tylospermetum lignosae* nom. prov., *Scutellarietum hissaricae* nom. prov., *Nanorrhinetum ramosissimi* nom. prov., *Scutellarietum zaprjagaevii* nom. prov., *Scutellarietum schugnanicae* nom. prov., *Campanuletum albertii* nom. prov. and comm. of *Scutellaria adenostegia*. However, further investigations are still needed. The species composition of the plots in question determines the classification of *Dionysietum involucratae* association into the *Campanuletalia incanescentis* M. Nobis, A. Nowak & A. Nobis 2013 within the *Asplenietea trichomanis* class [7,16]. All the communities mentioned, as well as *Dionysietum involucratae* association, should be delimited within a new alliance of rock face vegetation in the subalpine and montane zone of the Pamir Alai Mountains. In view of the most frequent and abundant species, the proposed name for this syntaxon is *Caricion koshewnikovii* nom. prov.

Dionysietum involucratae ass. *nova*

Typus relevé: Tab. 2, relevé 2.

Diagnostic species: *Dionysia involucrata*.

The plots of *Dionysietum involucratae* association were found in all the extant location of the main diagnostic species in the valleys of the Varzob and Takob rivers, within the western Hissaro-Darvasian geobotanical subregion. Obviously, the association has a present distribution range even narrower than that of the diagnostic species and belongs to the endemic vegetation types of Tajikistan. It was found in montane and subalpine zones at the relatively low altitudes of 1085 to 1575 meters above sea level. The association prefers granite rocks of a relatively low cohesion index and acidity (pH 6.0 to 6.5). Only in Varmanik it was also found on compact limestone with fine crevices and fissures. The community develops on rock faces, mainly on southwestern and southern aspects with almost vertical, or even overhanging, inclinations, the approximate mean being 100° (Fig. 6). Owing to the scarce contribution of the species, the association is characterized by a moderate vegetation cover. In general, the total cover of the herb layer ranged between 15 and 55%, with mean value of approximately 25% (Tab. 2). The phytocoenosis is also characterized by a

Tab. 2 *Dionysietum involucratae* in Hissar Mts (Tajikistan).

Relevé number	1	2	3	4	5	6	7	8	9	10	11	12	13	14	15	16	17	CONSTANCY
Day	8	8	8	8	8	8	16	19	8	8	8	8	8	8	8	8	8	
Month	6	6	6	6	6	6	6	8	6	6	6	6	6	6	6	6	6	
Year	2012	2012	2012	2012	2012	2012	2012	2012	2012	2012	2012	2012	2012	2012	2012	2012	2012	
Aspect	SW	S	S	S	SW	SW	W	NW	SE	W	SW	SE	S	S	S	SW	S	
Inclination (degrees)	90	120	120	100	90	85	110	80	85	80	85	75	100	150	150	90	125	
Altitude (m)	1226	1085	1229	1250	1560	1575	1090	1100	1250	1226	1230	1262	1250	1085	1229	1227	1236	
Cover of shrub layer (%)	-	-	-	-	-	-	-	-	-	-	-	5	5	-	-	-	-	
Cover of herb layer (%)	30	30	30	25	20	35	55	35	25	15	25	20	20	30	30	30	15	
Cover of moss layer (%)	0.5	-	-	-	-	-	-	-	0.5	-	1	0.5	0.5	0.5	-	-	-	
Relevé area (m^2)	0.5	1	1	1	1	1	2	1	1	0.5	0.5	2	1	1	1	0.5	1	
pH	-	-	6	-	-	6	-	-	6	-	-	-	-	-	6.5			
Rock type	Gr	Gr	Gr	Gr	Gr	Gr	Gr	Gr	Gr	Gr	Gr	Gr	Gr	Gr	Li	Gr	Gr	
Number of species	6	4	3	5	5	3	3	5	5	3	6	7	5	4	2	3	2	
Diagnostic species																		
Ass. ***Dionysietum involucratae***																		
Dionysia involucrata	3	3	2	2	2	3	4	3	2	2	2	2	2	3	1	3	2	V
All. *Caricion koshewnikowii*																		
Carex koshewnikowii	+	+			+	+		2	1	1	1	+				+	+	IV
Scutellaria hissarica	+			+	+		+	+	+									II
*Scutellaria adenostegia**										+	1			+				I
Tylosperma lignosa							+	+										I
O. *Campanuletalia incanescentis*																		
Campanula incanescens	+	+	2	1				+	+		+	+	+	+	+	+		IV
Cl. *Asplenietea trichomanis*																		
Grimmia pulvinata d											+	+	+	+				II
Parietaria judaica		+	+															I
Others																		
Bryum argenteum d	+								+		+							I
Spiraea baldshuanica b												1	1					I
Bromus tectorum				+		+												I
Spiraea baldshuanica c												+	+					I

Sporadic species: *Allium* sp. 4; *Echinops nanus* 1; *Gagea* sp. 5; *Pseudosedum condensatum* 12; *Steptorhamphus crambifolius* 5.

Locations of samples (according to the numbers of releves): 1, 3, 4, 9, 10, 12, 13, 17 – Takob (385038; 685014); 2, 14 – Varzob (384524; 684856); 5, 6 – Takob (385018; 685505); 7 – Varzob (384525; 684855); 8 – Varzob (384651; 684914); 11, 16 – Takob (385037; 685014); 15 – Varmanik (385010; 685638). Rock type: Gr – granite, Li – limestone. * Differential species.

moderate number of species as far as rupicolous vegetation is concerned, having from two to seven taxa in one relevé (with the average number of species amounting 4). Mosses provide an insignificant contribution to the association and their total cover is no greater than 1%. Despite the diagnostic taxon, *Campanula incanescens*, *Carex koshewnikowii* and *Scutellaria hissarica* have the highest values of constancy and abundance among the vascular plants. In the moss layer, the most important contributors are *Grimmia pulvinata* and *Bryum argenteum*.

Based on this study, we propose the following syntaxonomical classification of the *Dionysietum involucratae* ass. *nova* in Tajikistan:

Class: *Asplenietea trichomanis* (Br.-Bl. in Meier et Br.-Bl. 1934) Oberdorfer 1977

Order: *Campanuletalia incanescentis* M. Nobis, A. Nowak et A. Nobis 2013

Alliance: *Caricion koshewnikowii* nom. prov.

Ass. *Dionysietum involucratae* A. Nowak, M. Nobis, S. Nowak & A. Nobis.

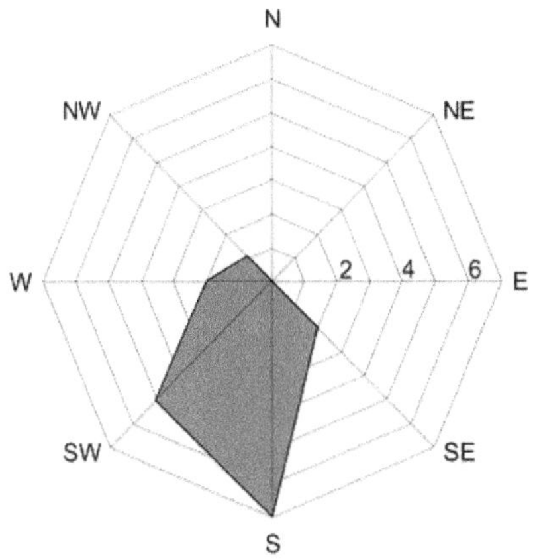

Fig. 6 Exposition of *Dionysietum involucratae* plots.

Discussion

Potential and present distribution of the species

According to Ovchinnikov [34], *Dionysia involucrata* is a long-standing element of what is known as Turanic flora of Irano-Mediterranean origin with highly stenochorous current distribution. The species meets the criteria of endemism [4] and is confined mostly to the granite rocks of the Varzob River Valley in the Hissar Mountains. The area belongs to the Hissaro-Darvasian geobotanical subregion, which is known as the richest as far as vascular plant endemics are concerned [4].

The small population located on limestone near Varmanik is rather ephemeral and untypical in character and could probably be defined as a sink part of the metapopulation [50]. Our field studies and extensive consultations with a geologist and botanists indicated no loose granite outcrops in the subalpine and montane zone in Tajikistan. There is thus little possibility of finding other populations of *D. involucrata* beyond its known historical range. The explanation for this absence can be found in the habitat conditions. This stenochorology is more typical for closely related species from the *Dionysia* genus, such as *D. hissarica*, *D. balsamea*, *D. paradoxa*, *D. lacei*, *D. saponacea*, *D. gandzhinae*, *D. hedgei*, *D. freitagii*, *D. viscidula*, *D. microphylla* and others. They are often confined to just one mountain range [25,35,51] and contribute significantly to the high endemism of isolated mountains within the Irano-Turanian province [4,21–23,52]. This kind of stenochory in chasmophytic plant communities is also observed in many other mountainous areas, especially with Mediterranean-type climates, such as the Bokkoya Mountains in northern Morocco [53,54], as well as on Gibraltar [55], on Crete and mainland Greece [45], in the Caucasus and the mountains of central Asia [22,56], in the Taurus Mountains in Turkey [57,58] and in Galicia in Spain [59]. Both *Dionysia involucrata* and other members of that genus undoubtedly corroborate the importance of rocky environments for the conservation of biodiversity (e.g. [4,60,61]). The specificity and diversity of the rock and scree habitats of Tajikistan is already known to be responsible for the extreme abundance of stenochorous plant species. This results from the extra-zonal nature of these habitats and their very different conditions in terms of humidity, type of substrate rock, insolation, temperature and inclination [4]. Another explanation for the chorological pattern is the concept defining highly distinct rock habitats as isolated, terrestrial "islands" serving as evolutionary traps. The surrounding dispersal barriers, genetic drift and founder effect work effectively on these isolated rock habitats [62–64].

Endangerment status of *Dionysia involucrata*

According to the red data book of Tajikistan, *Dionysia involucrata* was known from three locations and was considered as a critically endangered plant endemic to Tajikistan [36]. However, this assessment neither employs the approach suggested by the IUCN [3] nor takes into account all the previously known populations. For this reason, the re-evaluation of the endangerment status of *D. involucrata* with the use of the available data on spatial distribution, critical habitat and threat processes was vital. There is a global flora published for the region and there are also local checklists that give consideration to all the published and herbarium data on the distribution of the *D. involucrata* [28,34]. Despite this, the available data of use in order for the categorization process to be accomplished, such as fluctuations or the decline of, mature individuals, demographic tendency and so forth are either limited or even non-existent. At present, it is therefore possible neither to conduct an analysis of changes in population sizes, nor to evaluate the probability of the risk of species extinction in a given period of time. The methodology proposed by IUCN [3] requires precise information for the application of some criteria, although it does also enable decisions to be made on the basis of measurable parameters. In this case, only the EOO, the AOO and the present population size could be taken into account.

The detectability of *Dionysia involucrata* appears to be both unhampered by spatial or temporal distribution and unlimited by the biology and morphology of the species. Cushions of the species occupy the rock faces at medium heights and can be seen with the naked eye almost year round, with no need to employ alternative survey techniques. Despite some discrepancies concerning the precise location of the historical populations in relation to the extant ones, we can assess that the species is known from twelve locations, seven of which are probably extinct. *D. involucrata* also has a very narrow EOO of less than 5000 km^2 and an AOO of less than 650 m^2. This distributional feature makes it possible to evaluate the threat status of the species and permits the conclusion that the "endangered" is an appropriate category and should be considered as global. Nevertheless, in order to make this assessment more comprehensive, it seems reasonable to conduct further sampling, not only to ascertain the fluctuations or decline in the numbers of mature individuals more fully, but also to enable ecological and population genetic studies with a view to determining the species' capacity for dispersal among habitat patches and the minimum patch size and connectivity needed to support viable populations.

There is no evidence for the causes of *Dionysia involucrata* decline. Rasulova and Junusov [36] either, apart from suggestions that the withdrawal of the species is related to intensive grazing of goats and sheep and collection by botanists or tourists. The populations growing on the low-lying and accessible ledges of rock walls on flat slopes are, indeed, exposed to be grazed, primarily by goats or sheep. These livestock species are still the main source of livelihood in central Tajikistan and their farming causes considerable environmental degradation. Neither the pressure exerted by tourism on the population of *D. involucrata* in the Hissar mountains nor collection for scientific purposes seem to be relevant in this case, since the numbers and frequency of research studies and tourist expeditions in the Varzob River basin are still quite low. However, with their large attractive flowers, which do embrace values of beauty and ornamentation, *Dionysias* could be threatened by indiscriminate collection.

Another threat is related to road construction and modernization in the Varzob River basin over the few years. The populations of *Dionysia involucrata* occupy walls in closest vicinity of the river and the communication routes running alongside it. The granite quarrying or the fires lit by workers during the modernisation and broadening of these roads could have a negative impact on those populations. This is all the more significant in that *D. involucrata* is not protected by a species conservation law in Tajikistan and the currently existing populations are not included in the aerial protection system. Even though steering the formulation and implementation of public conservation policies in order to encompass the extant location of the species within protected areas is not a full guarantee of effective protection, it might at least increase the chances of halting the withdrawal of the species in Tajikistan [65–68].

Given the high conservation priority and threat status of *Dionysia involucrata*, as well as the difficulties involved in counteracting the threats in situ, we recommend the possibility of ex situ conservation, either by means of seed-banking or by taking measures to cultivate the species in ex-situ collections, due to the difficulties of counteracting the threats.

Position of *Dionysietum involucratae* in the syntaxonomical system of rock vegetation in the Pamir Alai Mountains

Syntaxonomical richness in mountainous areas with the Mediterranean type of macrobioclimate is known to be markedly high in both Europe and Asia (e.g. [45,56,58,59,69–80]). Because of the considerable structural differences between chasmophytic flora and other vegetation types, a separate *Asplenietea trichomanis* (Br.-Bl. in Meier et Br.-Bl. 1934) Oberdorfer 1977 class was designed for rock face, fissure, cleft, crevice and rock ledge communities [68,81]. *Dionysietum involucratae* association undoubtedly belongs to this class. It is also certain that, on account of the considerable frequency and abundance of diagnostic taxa, as well as of Irano-Turanian distributional-types species, the vegetation plots surveyed should be classified in the *Campanuletalia incanescentis* order [7,16]. The order includes phytocoenoses characterized by a high share of petrophytic taxa such as *Campanula incanescens*, *C. lehmanniana*, *Stipa zeravshanica*, *Poa relaxa*, *Scutellaria hissarica*, *Artemisia rutifolia* and others.

However, the ordination of the *Dionysietum involucratae* association at the alliance level still requires further research, particularly on the rocky vegetation of the montane and submontane belt. Recent studies of rupicolous vegetation in the Hissar, Zeravshan, Turkestan and Darvaz Mountains suggests that the new alliance, *Caricion koshewnikowii* nom. prov., needs to be established in order to differentiate the vegetation plots of chasmophytic vegetation developed in the lower altitudinal range of the Pamir Alai Mountains [16]. This type of phytocoenoses is distinctly different from *Asperulo albiflorae-Poion relaxae* [16] on account of both the floristic composition and the habitat conditions, with their mainly lower altitudinal amplitude, higher soil amount and precipitation. For example, in the alpine and nival belts, the most frequently spotted plants are *Asperula albiflora*, *A. czukavinae*, *Lophanthus virescens*, and *Poa relaxa*. In relevés sampled for *Dionysietum involucratae* association and other communities from the subalpine and montane zones, the *Carex koshewnikowii*, *Campanula incanescens*, *Scutellaria hissarica*, *Scutellaria zaprjagaevii*, *Scutellaria adenostegia* and *Syntrichia ruralis* reach the highest levels of abundance and frequency.

Dionysia involucrata has a high diagnostic syntaxonomical value and occurs almost exclusively within its own association. In only a few cases was it also spotted in a community of *Tylosperma lignosa*, the species with a broader distribution in southwest Asia, ranging from western Tajikistan to northern Iran and also occurring on granite rocks. Other sampled phtocoenoses with the optimum of occurrence on limestone and shale outcrops are devoid of any contribution on the part of *Dionysia*. The association is thus clearly related and depends on specific habitat conditions, these primarily consisting of an acidophilous substrate, relatively high precipitation, particularly in the spring period (Fig. 4) and very steep, often overhanging, rock walls. The fact that cushions of *Dionysietum involucratae* have never been observed in large numbers on rock ledges or on mountain tops also seems to be significant. The association prefers rock outcrops in deep gorges where exposure to sunlight is limited by the relatively long daily period in the shade.

Concluding remarks

The chasmophytic vegetation located in central Asia, in one of the world's most important biodiversity hotspots as far as the impact of global change and susceptibility to degradation is concerned, is under considerable threat [10–13]. Many of the species that occur in this environment are in severe danger, as human-caused threats, particularly goat grazing and collecting, are leading to habitat degradation and population damage [4]. Our study has revealed information regarding the geographical distribution and ecological requirements of *D. involucrata*, one of the most threatened and least known species in the region. We hope that this information will be used in conservation measures implemented for this endangered species. Although we shifted the threat category of *D. involucrata* from CR to EN, this species needs further monitoring and conservation on account of the continuing shrinkage of the distribution range. It has to be pointed out that the downgrading of the endangerment status is due neither to effective conservation action nor to the result of any increase in population size, but rather to precise re-evaluation of the species. Still, these isolated populations are particularly important from the ecological, genetic and evolutionary viewpoint and require more attention from conservation biologists [82].

We have also pointed out that *Dionysia involucrata* creates its own phytocoenosis and grows in similar habitats along the rock outcrops of the Varzob River Valley and its tributaries. The population of *D. involucrata* is mainly sensitive to goat grazing. However, the recent increase in the level of activity in the modernization of Tajikistan's road system may well be contributing considerably to species decline as well. We therefore recommend both further population studies on the species and the possibility of ex situ conservation, either by means of seed cryopreservation and/or by cultivating the species in garden collections.

Acknowledgments

We are very grateful to V. Plašek Ph.D., D.Sc. for his help in moss species determination. Special thanks to Prof. A. Bodzioch for determination of rock samples. We thank also two anonymous reviewers who considerably improve the text. The authors wish also to thank F. Abdurahimova from the Nature Protection Team Dushanbe for assistance and help in organizing expeditions. The project was partially funded by the Polish Ministry of Science, grant No. N304 377838/2010.

Authors' contributions

The following declarations about authors' contributions to the research have been made: project idea, field research: ASN, MN; data analysis, preparing the manuscript: ASN, MN, SN, AN.

References

1. Domínguez Lozano F, Carlos Moreno Saiz J, Sainz Ollero H. Biological properties of the endemic and threatened shrub in Iberia *Vella pseudocytisus* subsp. *paui* Gómez Campo (Cruciferae) and implications for its conservation. J Nat Conserv. 2005;13(1):17–30.
2. Rana MS, Samant SS. Threat categorisation and conservation prioritisation of floristic diversity in the Indian Himalayan region: a state of art approach from Manali Wildlife Sanctuary. J Nat Conserv. 2010;18(3):159–168.
3. IUCN Standards and Petitions Subcommittee. Guidelines for using the IUCN red list categories and criteria. Version 11. Prepared by the Standards and Petitions Subcommittee; 2014.
4. Nowak A, Nowak S, Nobis M. Distribution patterns, ecological characteristic and conservation status of endemic plants of Tadzhikistan – a global hotspot of diversity. J Nat Conserv. 2011;19(5):296–305.
5. Nobis M, Nowak A, Zalewska-Gałosz J. *Potamogeton pusillus* agg. in Tajikistan (Middle Asia). Acta Soc Bot Pol. 2010;79(3):235–238.
6. Nobis M. Remarks on the taxonomy and nomenclature of the *Stipa tianschanica* complex (Poaceae), on the base of a new record for the flora of Tajikistan (central Asia). Nord J Bot. 2011;29(2):194–199.
7. Nobis M, Nowak A, Nobis A. *Stipa zeravshanica* sp. nov. (Poaceae), an endemic species from rocky walls of the western Pamir Alai Mountains (Middle Asia). Nord J Bot. 2013;31(6):666–675.
8. Nobis M. Taxonomic revision of the *Stipa lipskyi* group (Poaceae: *Stipa* section *Smirnovia*) in the Pamir Alai and Tian-Shan Mountains. Plant Syst Evol. 2013;299(7):1307–1354.
9. Nobis M, Nowak A, Nobis A, Paszko B, Piwowarczyk R, Nowak S, et al. Contribution to the flora of Asian and European countries: new national and regional vascular plant records. Acta Bot Gall. 2014;161(1):81–89.
10. Mittermeier RA, Gil PR, Hoffman M, Pilgrim J, Brooks T, Goettsch Mittermeier C, et al. Hotspots revisited: Earth's biologically richest and most endangered terrestrial ecoregions. Mexico City: Conservation International, CEMEX; 2005.
11. Giam X, Bradshaw CJA, Tan HTW, Sodhi NS. Future habitat loss and the conservation of plant biodiversity. Biol Conserv. 2010;143(7):1594–1602.
12. Fay M, Patel H. A simple index of vulnerability to climate change. Washington, DC: World Bank; 2008. (Adapting to climate change in Europe and central Asia).
13. Baettig MB, Wild M, Imboden DM. A climate change index: Where climate change may be most prominent in the 21st century. Geophys Res Lett. 2007;34(1):457–469.
14. Nowak AS, Nobis M. Distribution patterns, floristic structure and habitat requirements of the alpine river plant community *Stuckenietum amblyphyllae* ass. nova (Potametea) in the Pamir Alai Mountains (Tajikistan). Acta Soc Bot Pol. 2012;81(2):101–108.
15. Nowak AS, Nobis M. Distribution, floristic structure and habitat requirements of the riparian forest community *Populetum talassicae* ass. nova in the central Pamir-Alai Mts (Tajikistan, Middle Asia). Acta Soc Bot Pol. 2013;82(1):47–55.
16. Nowak A, Nowak S, Nobis M, Nobis A. Vegetation of solid rock faces and fissures of the alpine and subnival zone in the Pamir Alai Mountains (Tajikistan, Middle Asia). Phytocoenologia. 2014;44(1–2):81–104.
17. Nowak S, Nowak A, Nobis M. Weed communities of rice fields in the central Pamir Alai Mountains (Tajikistan, Middle Asia). Phytocoenologia. 2013;43(1–2):101–126.
18. Nowak S, Nowak A, Nobis M, Nobis A. Weed vegetation of cereal crops in Tajikistan (Pamir Alai Mts., Middle Asia). Phytocoenologia. 2013;43(3–4):225–253.
19. Nowak S, Nowak A, Nobis M, Nobis A. *Caucalido platycarpi-Vicietum michauxii* – a new weed association from crop fields of Kyrgyzstan (Middle Asia). Cent Eur J Biol. 2014;9(2):189–199.
20. Nowak S, Nowak AS. Weed communities of root crops in the Pamir Alai Mts, Tajikistan (Middle Asia). Acta Soc Bot Pol. 2013;82(2):135–146.
21. Favarger C. Endemism in the montane floras of Europe. In: Valentine DH, editor. Taxonomy, phytogeography and evolution. London: Academic Press; 1972. p. 191–204.
22. Agakhanjanz O, Breckle SW. Plant diversity and endemism in high mountains of central Asia, the Caucasus and Siberia. In: Körner C, Spehn EM, editors. Mountain biodiversity: a global assessment. New York, NY: Parthenon Publishing Group; 2002.
23. Médail F, Verlaque R. Ecological characteristics and rarity of endemic plants from southeastern France and Corsica: implications for biodiversity conservation. Biol Conserv. 1997;80:269–281.
24. Kazakis G, Ghosn D, Vogiatzakis IN, Papanastasis VP. Vascular plant diversity and climate change in the alpine zone of the Lefka Ori, Crete. Biodivers Conserv. 2007;16(6):1603–1615.
25. Lidén M. The genus *Dionysia* (Primulaceae), a synopsis and five new species. Willdenowia. 2007;37(1):37–61.

26. Valant-Vetschera KM, Bhutia TD, Wollenweber E. Chemodiversity of exudate flavonoids in *Dionysia* (Primulaceae): a comparative study. Phytochemistry. 2010;71(8–9):937–947.

27. Wendelbo P. Studies in Primulaceae: I. A Monograph of the genus *Dionysia*. Bergen: Aarbok University; 1961.

28. Chukavina AP. Flora Tadzhikskoi SSR. Vol. VII. Zontichnye - Verbenovye. Leningrad: Izdatelstvo Nauka; 1984.

29. Smolyaninova LA. Rod *Dionysia* Fenzl. In: Schischkin BK, Bobrov EG, editors. Flora SSSR. Moscow: Izdatelstvo Akademii Nauk SSSR; 1952. p. 208–217. (vol 18).

30. Stanjukovich KW. Rastitelnost. In: Saidmuradov CM, Stanjukovich KW, editors. Tadzhikistan, priroda i prirodnye resursy. Dushanbe: Izdatelstvo Donish; 1982. p. 358–435.

31. Dzhuraev A. Bioloho-ecolohicheskie i zhizneynye formy rastitelnosti pervichnyh osypiey Hissarskoho hrebta. Dokl Akad Nauk Tadzhikskoy SSR. 1970;13:38–50.

32. Dzhuraev A. Rastitelnost pervichnyh osypiey zapoviednika "Ramit". Dokl Akad Nauk Tadzhikskoy SSR. 1972;15(12):85–97.

33. Dzhuraev A. Rastitelnost pervichnyh osypiey ushchelya Takob (Hissarskij hrebet). Dokl Akad Nauk Tadzhikskoy SSR. 1972;15:44–46.

34. Ovchinnikov PN, editor. Flora i rastitelnost ushchelya reki Varzob. Leningrad: Izdatelstvo Nauka; 1971.

35. Zaprjagaev FL. Novyj vid roda *Dionysia* Fenzl. Tr Tadzhikskiej Bazy Akad Nauk SSSR. 1936;2:157–159.

36. Rasulova MR, Junusov SJ. Pokrytosemyannye - Angiospermae. In: Abdusalyamov IA, editor. Krasjaya kniga Tadzhikskoy SSR. Dushanbe: Izdatelstvo Donish; 1988. p. 189–328.

37. IUCN. The IUCN red list of threatened species [Internet]. 2013 [cited 2013 Oct 15].

38. Rivas-Martínez S, Rivas-Sáenz S, Merino AP. Worldwide bioclimatic classification system. Glob Geobot. 2011;1:1–634.

39. Latipova WA. Kolichestvo osadkov. In: Narzikulov IK, Stanjukovich KW, editors. Atlas Tajikskoi SSR. Dushanbe: Akademia Nauk Tadzhikskoi SSR; 1968. p. 54–55.

40. Narzikulov IK, Stanjukovich KW, editors. Atlas Tajikskoi SSR. Dushanbe: Akademia Nauk Tadzhikskoi SSR; 1968.

41. Braun-Blanquet J. Pflanzensoziologie: Grundzüge der Vegetationskunde. 3rd ed. Wien: Springer; 1964.

42. Tichý L. JUICE, software for vegetation classification. J Veg Sci. 2002;13(3):451–453.

43. Weber HE, Moravec J, Theurillat JP. International code of phytosociological nomenclature. 3rd edition. J Veg Sci. 2000;11(5):739–768.

44. Valachovič M, Dierssen K, Dimopoulos P, Hadač E, Loidi J, Mucina L, et al. The vegetation of screes - a synopsis of higher syntaxa in Europe. Folia Geobot Phytotax. 1997;32:173–192.

45. Dimopoulos P, Sýkora KV, Mucina L, Georgiadis T. The high-rank syntaxa of the rock-cliff and scree vegetation of the mainland Greece and Crete. Folia Geobot. 1997;32(3):313–334.

46. Willner W. The association concept revisited. Phytocoenologia. 2006;36(1):67–76.

47. ter Braak CJF, Šmilauer P. CANOCO reference manual and CanoDraw for Windows user's guide: software for canonical community ordination (version 4.5). Ithaca, NY: Microcomputer Power; 2002.

48. Czerepanov SK. Plantae vasculares URSS. Leningrad: Nauka; 1995.

49. Ochyra R, Bednarek-Ochyra H, Żarnowiec J. Census catalogue of Polish mosses. Cracow: W. Szafer Institute of Botany, Polish Academy of Sciences; 2002. (Biodiversity of Poland; vol 3).

50. Gilpin M, Hanski I, editors. Metapopulation dynamics: empirical and theoretical investigation. San Diego, CA: Academic Press; 1991.

51. Trift I, Liden M, Anderberg AA. Phylogeny and biogeography of *Dionysia* (Primulaceae). Int J Plant Sci. 2004;165:845–860.

52. Strid A. Phytogeographical aspects of the Greek mountain flora. Fragm Flor Geobot. 1993;2:411–433.

53. Deil U. Felsgesellschaften beiderseits der Straße von Gibraltar. Hoppea. 1994;55:757–814.

54. Deil U, Hammoumi M. Contribution à l'étude des groupements rupicoles des Bokkoya (Littoral du Rif Central, Maroc). Acta Botánica Malacit. 1997;22:131–146.

55. Sánchez García I, Galán de Mera A, Cortés JE. La vegetación del peñón de Gibraltar. Acta Botánica Malacit. 2000;25:107–130.

56. Ermolaeva OY. The petrophyte plant communities of high mountain limestone area, the West Caucasus. Veg Russ. 2007;10:23–37.

57. Hein P, Kürschner H, Parolly G. Phytosociological studies on high mountain plant communities of the Taurus mountains (Turkey) 2. Rock communities. Phytocoenologia. 1998;28(4):465–563.

58. Parolly G. Phytosociological studies on high mountain plant communities of the South Anatolian Taurus mountains 1. Scree plant communities (Heldreichietea): a synopsis. Phytocoenologia. 1998;28(2):233–284.

59. Ortiz S, Rodriguez-Oubiña J. Synopsis of the rupicolous vegetation of Galicia (north-western Iberian Peninsula). Folia Geobot Phytotax. 1993;28:15–49.

60. Martinelli G. Mountain biodiversity in Brazil. Braz J Bot. 2007;30(4):587–597.

61. Porembski S. Tropical inselbergs: habitat types, adaptive strategies and diversity patterns. Braz J Bot. 2007;30(4):579–586.

62. Snogerup S. Evolutionary and plant geographical aspects on chasmophytic communities. In: Davis PH, Harper PC, Hedge IE, editors. Plant life of south-west Asia. Edinburgh: Botanical Society; 1971. p. 157–170.

63. Deil U. Synvikarianz und Symphylogenie. Zur Evolution von Pflanzengesellschaften. Ber Reinh Tüxen Ges. 1999;11:223–244.

64. Nowak A, Nowak S. The effectiveness of plant conservation: a case study of Opole Province, southwest Poland. Env Manage. 2004;34(3):363–371.

65. Whittaker RJ, Fernandez-Palacios JM. Island biogeography: ecology, evolution, and conservation. Oxford: Oxford University Press; 2007.

66. Hegazy AK, Kabiel HF, Boulos L, Sharashy OS. Conservation approach to the demography and dynamics of protected and unprotected populations of the endemic *Ebenus armitagei* in the western Mediterranean coast of Egypt. J Nat Conserv. 2010;18(3):151–158.

67. del Vecchio S, Giovi E, Izzi CF, Abbate G, Acosta ATR. *Malcolmia littorea*: the isolated Italian population in the European context. J Nat Conserv. 2012;20(6):357–363.

68. Mucina L. Asplenietea trichomanis. In: Grabherr G, Mucina L, editors. Die Pflanzengesellschaften Österreichs. Teil II. Natürliche waldfreie Vegetation. Jena: Gustav Fischer Verlag; 1993. p. 241–275.

69. Poveda JFM, Mercado FG, Tendero FV. Rupicolous vegetation of the betic ranges (south Spain). Vegetatio. 1991;94(2):101–113.

70. Carmona EC, Luque MM, Tendero FV. The plant communities of the *Asplenietea trichomanis* in the SW Iberian Peninsula. 1997;32(4):361–376.

71. Deil U. The class *Adiantetea* in the Mediterranean area - a state of knowledge report. Ann Bot Roma. 1998;56(1):73–78.

72. Reyes-Betancort JA, Wildpret de la Torre W, León Arencibia MC. The vegetation of Lanzarote (Canary Islands). Phytocoenologia. 2001;31(2):185–247.

73. Onipchenko VG. Alpine vegetation of the Teberda Reserve, the Northwestern Caucasus. Zürich: Geobotanisches Institut ETH, Stiftung Rübel; 2002.

74. Eren Ö, Gökçeoğlu M, Parolly G. The flora and vegetation of Bakirli Daği (Western Taurus Mts, Turkey), including annotations on critical taxa of the Taurus range. Willdenowia. 2004;34(2):463.

75. Deil U, de Mera AG, Orellana JAV. Rock and scree plant communities in the Serra de Monchique (SW Portugal). Feddes Repert. 2008;119(5–6):556–585.

76. Terzi M, D'Amico FS. Chasmophytic vegetation of the class *Asplenietea trichomanis* in south-eastern Italy. Acta Bot Croat. 2008;67(2):147–174.

77. Sanda V, Burescu P, Öllerer K. Fitocenozele din România sintaxonomie, structură, dinamică şi evoluţie. Bucharest: Ars Docendi; 2008.

78. Tzonev RT, Dimitrov MA, Roussakova VH. Syntaxa according to the Braun-Blanquet approach in Bulgaria. Phytol Balcan. 2009;15(2):209–233.

79. Golub VB, Grechushkina NA, Sorokin AN, Nikolaychuk LF. Plant communities on rock outcrops in the northwest part of the Black See Caucasian Coast. Veg Russ. 2009;14:3–14.

80. Bergmeier E, Dimopoulos P, Mucina L. Validation of some alliances of the Aegean chasmophytic vegetation of the *Asplenietea trichomanis*. Lazaroa. 2011;32:183–186.

81. Meier H, Braun-Blanquet J. Classe des Asplenietales rupestres, groupements rupicoles. Prodromus Pflanzengesselschaften. 1934;2:1–47.

82. Abeli T, Gentili R, Rossi G, Bedini G, Foggi B. Can the IUCN criteria be effectively applied to peripheral isolated plant populations? Biodivers Conserv. 2009;18(14):3877–3890.

10

The negative impact of intentionally introduced *Quercus rubra* L. on a forest community

Beata Woziwoda[1]*, Dominik Kopeć[1], Janusz Witkowski[2]

[1] Department of Geobotany and Plant Ecology, University of Łódź, Banacha 12/16, 90-237 Łódź, Poland

[2] Regional Directorate of the State Forests in Łódź, Jana Matejki 16, 91-402 Łódź, Poland

Abstract

Some alien woody species used in commercial forestry become invasive and, as invaders, cause major problems in natural and semi-natural ecosystems. However, the deliberate introduction of aliens can bring unintended negative changes also within areas of their cultivation. This paper presents the effects of the intentional introduction of the North-American *Quercus rubra* in European mixed Scots pine-Pedunculate oak forests (POFs): *Querco roboris-Pinetum* (W. Mat. 1981) J. Mat. 1988. Phytosociological data from field research combined with GIS data analysis of the current distribution of Northern Red oak in the studied habitat were used to determine the composition and structure of forest communities in plots with and without *Q. rubra* participation.

The results show that *Q. rubra* significantly reduces native species richness and abundance, both in old-growth and in secondary (post-agricultural) forests. Not one resident vascular plant benefits from the introduction of Northern Red oak and only a few are able to tolerate its co-occurrence. The natural restocking of all native woody species is also strongly limited by this alien tree.

The introduction of Northern Red oak significantly limits the environmental functions of the POF ecosystem and weakens its economic and social aspects. However, its further cultivation is justified from an economic point of view, as the essential function of the studied forests is commercial timber production, and the introduction of this fast growing alien tree supports the provisioning ecosystem services. A clear description of the level of trade-off between the accepted negative and positive effects of the introduction of *Q. rubra* on forest ecosystem services requires further interdisciplinary studies.

Keywords: Northern Red oak; biodiversity; commercial forestry; old-growth and secondary forests; ecosystem services

Introduction

A substantial part of contemporary commercial forestry is based on the cultivation of intentionally introduced non-native woody species [1–3]. This approach is justified by the numerous positive silvicultural properties of the planted trees, such as their fast growth and the wide range of applications of their wood products, as well as the high capability of aliens to adapt to a wide variety of environmental conditions, their easy establishment and their high resistance to negative biotic and abiotic impacts in new areas of introduction. Unfortunately, the same species characteristics, which are valuable for forestry, generate serious problems when the alien becomes invasive [4–10]. Introduced species that spread from plantations to natural and semi-natural areas have a considerable impact on the properties and functions of the ecosystem and cause ecological, economical and social damage [11–13]. However, alien trees planted as monocultures or deliberately introduced as a "biocoenotic admixture" for enrichment of impoverished forest stands can also bring unintended negative changes within the areas of cultivation [14–16]. Invading woody species alter natural regeneration regimes by the utilization of various mechanisms to dominate understory environments and outcompete native species for available resources [17–20]. Self-sown young individuals of planted exotics occupy the niches of autochthonous plants and inhibit both their recruitment and growth [21–24]. An understanding of the positive and negative effects of the introduction of alien species on native ecosystems is fundamental to sustainable management and to proper nature conservation policy [25–28]. It is also an essential part of assessing of the influence of the alien species on ecosystem services [29,30], especially in forests which are altered, disrupted or degraded by commercially-important but invasive plants introduced deliberately [21,31–33].

In this paper, we analyze the effects of Northern Red oak *Quercus rubra* L. introduction on European mixed pine-oak forests. It is one of 582 North American woody species

* Corresponding author. Email: woziwoda@biol.uni.lodz.pl

Handling Editor: Beata Zagórska-Marek

brought to Europe before 1915 [34] and was first intentionally introduced as an ornamental in the XVII century [35,36]. Two hundred years after its successful introduction in forest ecosystems at the end of XVIII century [37], it has become one of the most common and commercially-important broadleaved woody species of alien origin [38,39]. At present, *Q. rubra* is widespread in European forests [40], including Polish ones (the Information System of the National Forest Holding State Forest - SILP), and in the majority of territories it is noted as "established" (Fig. 1).

Quercus rubra is planted in a wide range of forest habitats suitable for deciduous as well as for coniferous trees [36,41]. Foresters describe it as the "wonder plant" (sensu Richardson [4]) due to its numerous positive characteristics [38,42,43], and in the timber industry it is called an "American Beauty" [44]. However, the positive perception of *Q. rubra* is challenged by some studies, which report that it has negative effects on forest biodiversity [45–48]. The inclusion of Northern Red oak to the official list of invasive alien species in the Czech Republic [49], and its description as "invasive" in Poland [50,51] and in Lithuania [52] is noted by the supporters of *Q. rubra* introduction as an unjustified misinterpretation of the facts [53].

In Poland, the misunderstandings between supporters and opponents of further *Q. rubra* cultivation are based on a lack of empirical data concerning the positive and negative ecological, economical and social effects of *Q. rubra* introduction, and by the ambiguity of general rules on non-native species use. Current Polish legislation limits the introduction of alien plants [54], although a number of exceptions exist for species which are used in rational forestry and in agriculture, or are planted in urban, park and garden areas as ornamentals [55]. Similarly, the rules and standards of sustainable forest management from 2003 also banned the introduction of alien trees in Polish forests and even suggested the gradual elimination of some of those already introduced (like *Padus serotina*), but these rules did not concern exotics used in plantations for timber production [56]. Furthermore, the increasing pressure of conservationists who oppose any introduction of alien trees, even in plantations, is regarded as a limitation on forest management and a waste of long-standing and fruitful studies on the acclimatization of non-native woody species useful in commercial forestry. Consequently, as the new regulations concerning forest cultivation [57] do not mention the problem of alien species at all, this issue has reached an impasse.

Even though *Q. rubra* is reported to be invasive [50,51], it was not included in the newest list of alien plants which may be a threat to native species or natural habitats [58], which justifies its further introduction as a timber tree, and the danger of its invasion is not regarded as being serious [53]. The solution to this conflict requires an understanding of both the real and potential, positive and negative, ecological and economical effects of the introduction of *Q. rubra* to forest ecosystems [59].

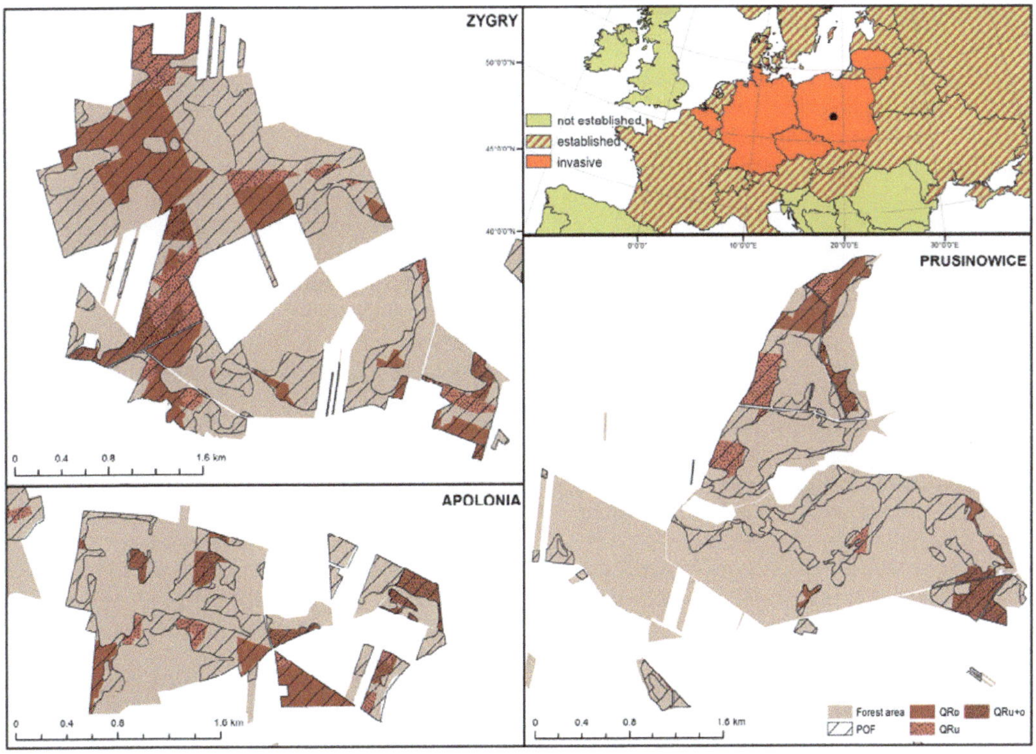

Fig. 1 Location of the study area on the map of *Quercus rubra* distribution in Europe, and the distribution of this alien tree within a mixed Scots pine-Pedunculate oak forest (POF) habitat in three forest complexes in the Poddębice Forest District (central Poland), where Northern Red oak was introduced, and at present it occurs in the overstory (QRo), in the understory (QRu) or in both layers (QRu+o) of the POF (data source: the distribution and statue of *Q. rubra* in European countries is based on NOBANIS data and additionally on Redei et al. [39] Riepšas and Straigytė [52], Verloove [101], and Medvecká et al. [102]. GIS maps were obtained using SILP data from 2011.

This paper addresses five main points: (*i*) how the native plant community is affected by Northern Red oak; (*ii*) how different abundances of *Q. rubra* affect structure and species composition, and whether there are any susceptible/resistant species to this alien species; (*iii*) whether *Q. rubra* influences native sapling growth; and, (*iv*) whether the negative/positive impact of *Q. rubra* on biodiversity depends on site history (old-growth forests versus secondary forests).

Material and methods

Study area

The studies were conducted in forest complexes in the Poddębice Forest District located in the central part of Poland (Fig. 1). The old-growth forests (forests with continuous sequence of tree generations) and secondary forests growing on former agricultural land (ancient and recent forest complexes sensu Peterken [60]) are spread across landscape dominated by agricultural fields [61]. The total woodland area amounts 17.5 thousand ha, constituting 21.1% of the forest district area, in which 4.7 thousand ha of habitat is suitable for Scots pine-Pedunculate oak forest (POF) occurrence. The sites are characterized by well-drained but moist mineral soils, usually sandy loams, which are acidic (pH ~5.0), with low nutrient availability (SILP data).

The POF forests are found between conifer and broad-leaved forests and the habitat is naturally occupied by *Querco roboris-Pinetum* (W. Mat. 1981) J. Mat. 1988 community [62]. The two- or three-layered overstory comprises *Pinus sylvestris* and *Quercus robur* with an admixture of *Betula pendula*, *Populus tremula* and rare examples of *Carpinus betulus*. *Frangula alnus*, *Sorbus aucuparia* and *Juniperus communis* are common subcanopy associates. The herb and moss layers are built from numerous acidophilous, oligo- or mezotrophic species, "spreading" both from conifer and from broad-leaved forest habitats [63]. Species characteristic of coniferous forests – *Vaccinium myrtillus*, *V. vitis-idaea* and *Trientalis europaea*, which prefer semi-light or sunny sites, dominate in the herb layer. *Maianthemum bifolium*, *Melampyrum pratense*, *Luzula pilosa*, *Pteridium aquilinum* and *Calluna vulgaris* are also frequently noted. The species preferring semi-shadow and shadow sites, connected to deciduous forests are numerous, but they usually achieve low abundance. However, the occurrence of *Polygonatum odoratum*, *Peucedanum oreoselinum*, *Viola riviniana*, *Solidago virgaurea*, *Scorzonera humilis*, *Veronica officinalis*, *V. chamaedris*, *Convallaria majalis* and *Fragaria vesca* distinguishes POF communities [62,63].

The studied forests were planted as timber forests, and Scots pine, admittedly suitable to the POF habitat, is much more common here than in natural assemblages. Moreover, in around one fifth of the POF area, the native oak species *Q. robur* has been supplanted by the Northern Red oak [41]. The intentional introduction of this alien tree started at the beginning of the XX century: the oldest *Q. rubra* individuals are 104 years old (SILP data). In the majority of forest patches, *Q. rubra* was introduced as a "biocoenotic admixture", where it was planted in numerous small clusters, 10 × 10 m in area, or as single individuals regularly dispersed within Scots pine stands growing on both in old-growth and in secondary forests, where it has become the second timber tree. At present, forests where *Q. rubra* dominates in the tree stand layer constitute 292.6 ha of the POF (SILP data). This tree occurs both in the overstory (introduced) and in the shrub layer (introduced and/or self-sown) for another 259.9 ha. A further 266.1 ha of the POF contain the plots with *Q. rubra* planted in last 20 years, and where it is noted only in the shrub layer. The 332 isolated patches with introduced *Q. rubra* are widely dispersed within the area of the POF habitat: three examples are presented in Fig. 1. The majority of introduced trees, representing 67% of the introduction area, have already achieved reproductive age.

Data collection

The effects of the introduction of *Q. rubra* to the POF were researched in 2012: this was preceded by an analysis of SILP data on POF habitat distribution, carried out in GIS (ArcGIS 10.0), and on *Q. rubra* occurrence. Field sampling was conducted using phytosociological reléves, sample plots, located in homogenous patches of vegetation, far from the forest edge to avoid border effects, either with introduced *Q. rubra* (QR plots) or without QR (reference plots). The 83 sample plots, including 18 reference ones, were taken within mature timber plantations older than 60 years. In each plot, the cover of all vascular plant species (herbs, shrubs and trees) was estimated using a six-degree cover-abundance scale from "+", representing a few individuals covering less than 1%, to "5", representing plant species covering more than 75% of the plot area [64], which allowed an estimation of both the number of species and the proportion of the area covered by them. Additionally, the number of *Q. rubra* individuals with tree trunk perimeters greater than 10 cm was counted for each sample plot.

Data analyses

For each sample plot, six floristic indices, which characterized the diversity and the condition of forest communities, were counted. The richness of the plant species within the sets of plots with *Q. rubra* and without *Q. rubra* was described by the mean number of species (*Ns*) per sample plot. The species diversity in QR plots and in reference plots was characterized by the Shannon diversity index: $H' = -\Sigma p_i \ln p_i$, where p_i is the proportion of the ith species frequency of the sum of all species frequencies and by the evenness index (J'). The Shannon index and evenness were calculated by MVSP software (version 3.13, Kovach Computing Services). Before this analysis, original species cover values were transformed to mean percentages of species cover (+ = 1.0; 1 = 5.0; 2 = 17.5; 3 = 37.5; 4 = 62.5; 5 = 87.5). For trees and shrubs the cover index was calculated as a sum of percentages cover from each layers in sample plot where the species was noted. We assumed that the cover index for species in sample plot cannot be higher than 100. The differences in abundance of species in QR plots and reference one are described by the sum of percentage cover (*Sci*).

The naturalness of plant community in plots was characterized by the number of species characteristic to a POF phytosociological unit (*NCs*) and by their sum of percentage cover (*SCci*). Species characteristic to POF were categorized according to Matuszkiewicz [63].

Next, the statistical differences between plots with and without *Q. rubra* were analyzed with the Student's *t*-test ($P < 0.05$). The normality was checked with the Shapiro–Wilk test, and homogeneity of variance with Levene's test. Due to the lack of normal distribution of the calculated indexes, a Box–Cox transformation was used. The linear relationship between the number of *Q. rubra* individuals and the value of floristic indices (*H'*, *J'*, *Ns*, *NCs*, *Sci*, *SCci*) was also examined. Pearson's correlation coefficient with its significance (*P* value <0.05) was also calculated, also after Box–Cox transformation. All these statistical analyses were performed using the STATISTICA package (version 10.0, StatSoft Inc., Tulsa). DCA and CCA in the CANOCO program version 4.5 [65] were used to characterize the general pattern of variation of sample composition and species distribution.

In two canonical analyses, the number of *Q. rubra* individuals and the sum of percentages cover of *Q. rubra* from a different cover layer were used as supplemented data. The percentages of species cover were used in canonical analysis: species whose frequencies were lower than 10% were excluded from the analysis.

On the basis of these results, the species resistant (winners) and sensitive (losers) to *Q. rubra* occurrence were indicated. It was assumed that the changes in POF herb layer composition were likely caused by the changes in light condition under the *Q. rubra* canopy [66], which is much closer than in natural POF forests, so the preferences of "winners" and "losers" to the light condition were characterized using Ellenberg's indicator values (*EIV*) [67].

Results

Research carried out in the POFs showed that the occurrence of introduced *Q. rubra* is an important factor leading to changes in the composition of native flora. DCA analysis shows that the sample plots with *Q. rubra* create a separate set of patches, which are displaced relative to the references in the plane of the first ordination axis. The first axis accounts for 70.2% of variance and the second only 4.7% in the data set. The direction of the shifts of the infected area against the reference is consistent with the direction of the vector describing the number of *Q. rubra* individuals and sum *Q. rubra* cover index (Fig. 2).

Impact of *Q. rubra* occurrence on POF biodiversity

The obtained results indicate that *Q. rubra* reduces POF diversity. Its occurrence in the studied forest plots resulted in a statistically significant reduction of biodiversity, the Shannon index (*H'*), total number of species (*Ns*) and number of species characteristic of coniferous forest (*NCs*; Student's *t*-test, $P < 0.005$). QR plots were characterized by a lower sum of all species cover index (*Sci*). No significant differences were found in the evenness index (*J'*), the value of which was not dependent on the presence of *Q. rubra* (Tab. 1).

A strong linear relationship was found between the number of individual *Q. rubra* and the value of the parameters describing biodiversity (Fig. 3). The Pearson correlation index (*r*) reached a value between −0.48 to −0.64, confirming that biodiversity dramatically decreases as the number of

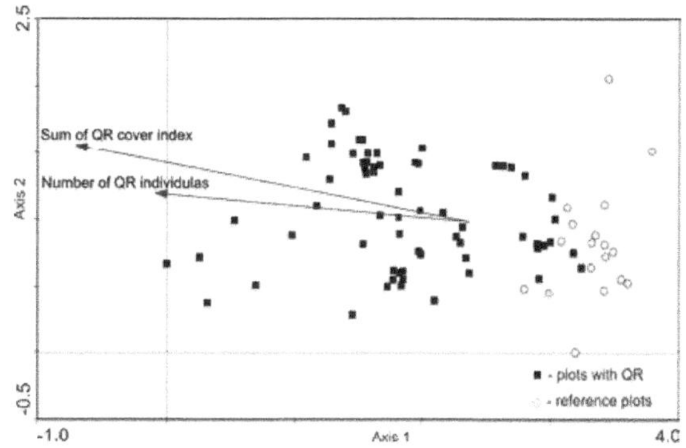

Fig. 2 Distribution of individual samples in the ordination diagram for the first two axes of the DCA with passively projected explanatory variables (number of QR individuals, sum of QR cover index).

Q. rubra individuals increase ($P < 0.005$). Among all analyzed diversity indicators, the Shannon index (*H'*) demonstrated the largest decline.

Indicator values: 'winners' and 'losers'

The total number of herbaceous plants noted in studied POF plots amounted to 38. The QR plots contained 27 species and the reference plots – 29. The results show that no herbaceous species would have a significantly higher average coverage index in the *Q. rubra* plots in comparison to the reference plots. Only one species, *Dryopteris carthusiana* has a slightly higher frequency and mean cover index in communities with the introduced alien. *Polygonatum odoratum Luzula pilosa*, *Vaccinium vitis-idaea*, *Trientalis europaea Festuca ovina* and *Convallaria majalis* are relatively well able to tolerate *Q. rubra* occurrence (Fig. 4). These species grow in a wide range of habitats and they are commonly noted in mixed coniferous as well as in mixed deciduous forests. Six species: *Vaccinium myrtillus*, *Calluna vulgaris Pteridium aquilinum*, *Melampyrum pratense*, *Carex pilulifera* and *Anthoxanthum odoratum* are indicated as very sensitive to *Q. rubra* introduction (Fig. 4). This group includes species more strongly associated with coniferous and mixed coniferous forests. The species sensitive to *Q. rubra* occurrence prefer semi-light or sunny sites and their mean light *EIV* amounts to 5.1, while the species more resistant to *Q. rubra* tolerate shade and semi-shade sites – their mean light *EIV* amounts to 3.8.

The QR plots were rarely settled by *Padus serotina*, another alien North-American species in Polish forests [50], however, its saplings were noted equally often in reference plots.

Impact of QR occurrence on natural restoration of native trees

Plots with *Q. rubra* are distinguished from the reference plots by their significantly lower mean cover index of saplings in the shrub layer and seedlings in the herb layer (Fig. 5) of native trees. The only species which *Q. rubra* introduction has little influence on is *Sorbus aucuparia*. All remaining native woody species fail to compete successfully with *Q. rubra* (Fig. 4).

Tab. 1 Differences in Shannon index, evenness, number of species, number of coniferous forest species and sum of cover index in invaded plots and reference plots.

Biodiversity indices	Average for QR plots	Average for references plots	*t* statistic	*P* value
Shannon index (*H'*)	1.45 ±0.33	1.71 ±0.22	−3.23	<0.005
Evenness (*J*)	0.63 ±0.09	0.61 ±0.07	0.90	ns
Number of species (*Ns*)	7.03 ±3.15	14.72 ±1.58	−8.19	<0.001
Sum of all species cover index (*Sci*)	118.40 ±79.13	261.50 ±59.61	−6.68	<0.001
Number of coniferous forest species (*NCs*)	1.63 ±1.31	3.39 ±1.25	−4.75	<0.001
Sum of coniferous forest species cover index (*SCci*)	23.36 ±36.27	72.39 ±31.45	−5.64	<0.001

The significant of differences was tested by Student's *t*-test ($P < 0.05$) after Box–Cox transformation.

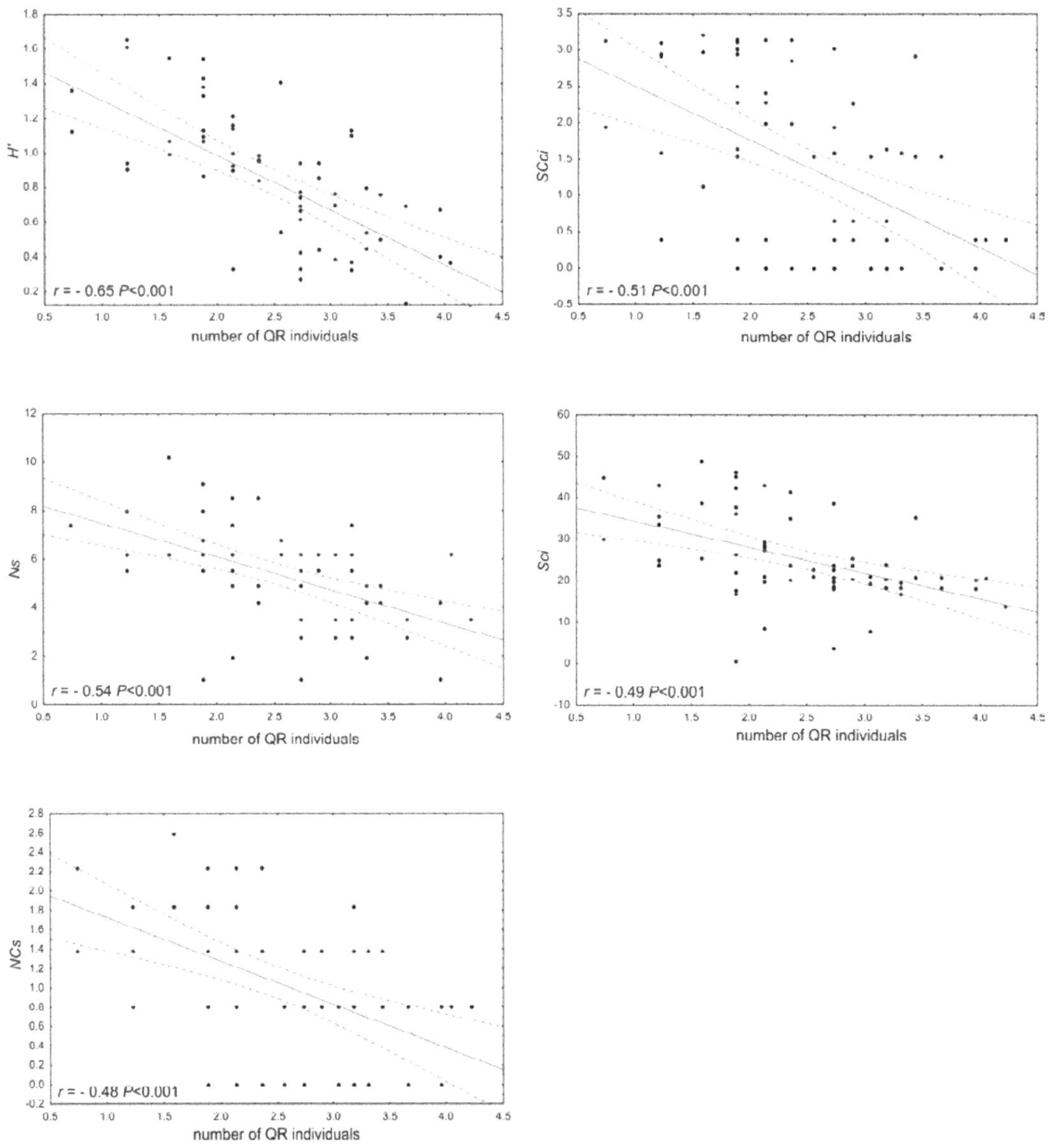

Fig. 3 Linear correlations between number of QR individuals and Shannon index (*H'*), number of species (*Ns*), number of species characteristic to coniferous forests (*NCs*), sum of all species cover index (*Sci*) and sum of cover index of species characteristic to coniferous forests (*SCci*); *r* = Pearson correlation coefficient ($P < 0.001$). All data before analysis were normalized by Box–Cox transformations.

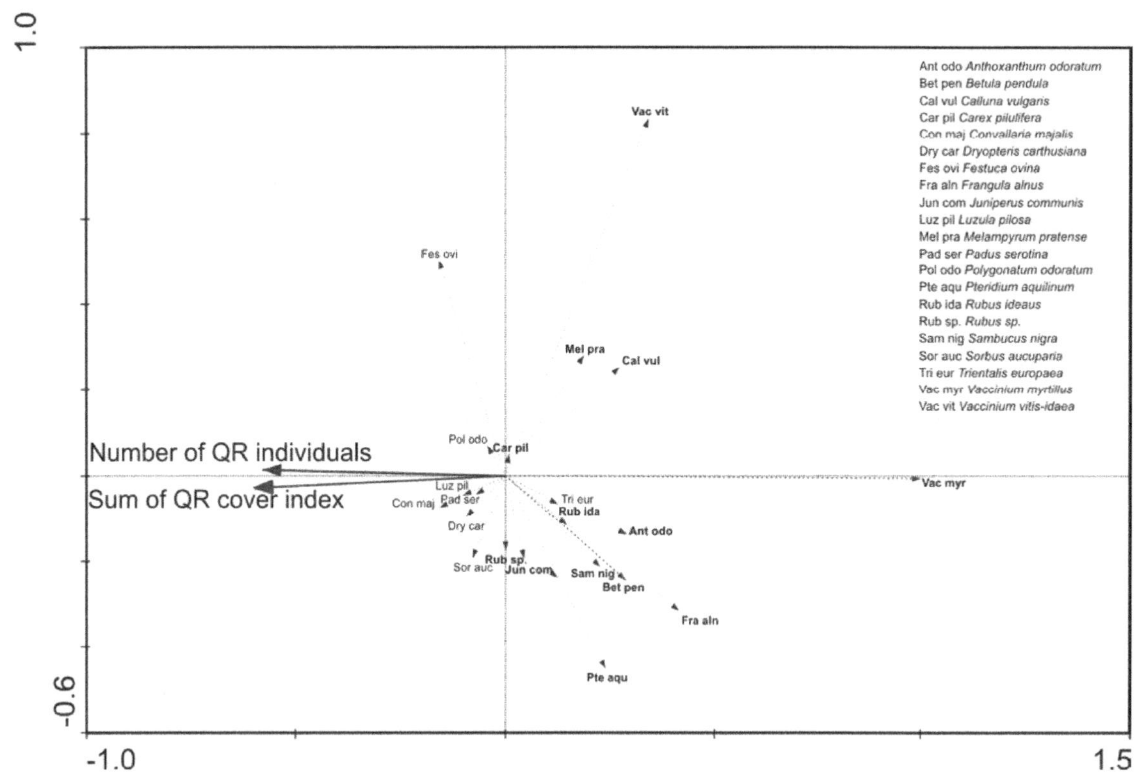

Fig. 4 Distribution of selected plant species in the ordination diagram of CCA for the first two axes with passively projected explanatory variables (number of QR individuals, sum of QR cover index). Eigenvalues were 0.75 for the first axis and 0.07 for the second in the biotic CCA. Species in bold have a significantly lower cover in QR plots (Student's *t*-test, $P < 0.05$).

Effect of *Q. rubra* introduction on forest communities located in old-growth and in secondary forests

Plots with *Q. rubra* growing in areas permanently used as forest land were characterized by a higher average number of species characteristic of POF and by a higher mean cover index than the plots located within forests established on former agricultural lands (Fig. 6).

Discussion

Quercus rubra negatively influences vascular plant species diversity and abundance (Tab. 1 and Fig. 3) and no one native species which naturally occurs in POF benefits from its introduction (Fig. 4). The drastic changes noted in plant cover are probably caused by the decrease of understorey insolation. *Quercus rubra*, like other deciduous tree-species [68–70], gives deep shadow [66], which may lead to the reduction of understorey components, mainly species preferring semi-light and sunny sites. A higher number of *Q. rubra* individuals in a forest patch and the closer *Q. rubra* canopy is associated with a greater reduction of resident plants, regardless of whether it occurs in the shrub layer or in the overstory. A shade which is too deep, created by a close overstorey of broadleaf trees, may also negatively affect shade-tolerant species [70] and complicate the natural restocking of trees [71], including invasive species like *P. serotina* [72,73] (cf: [74,75]) The increase of broadleaf tree density in coniferous forests may also result in the withdrawal of acidophilus vascular plant species as a result of changes in soil conditions [76–78]. However, no studies presently assess the impact of QR on the soil parameters.

The forest patches dominated by intentionally introduced *Q. rubra* tend to be very homogenous and extremely poor in terms of vascular plant occurrence, which may lead to a loss of the distinctiveness of the POF community. Similar negative effects of *Q. rubra* introduction have been reported from German and Lithuanian forests [48,52,79], and, more interestingly, these effects were observed in different types of forest habitat. The introduction of alien species, which outcompete resident ones in a wide range of ecosystems, may cause the homogenization of vegetation [80–85].

Studies show that *Q. rubra* significantly reduces the occurrence and the natural restoration of all native woody species (Fig. 5). Admittedly, the majority of new generation of tree stands in the studied timber forest are artificially planted, spontaneous forest restoration is rarely used, but the abundant self-reproduction of *Q. rubra* and its ability to create numerous stump sprouts after cutting, which is common in its native range [42,86], and which has been observed in European forests [46,52] may make forest regeneration after *Q. rubra* felling significantly more difficult. The vegetative reproduction of *Q. rubra* by re-sprouting may ensure its establishment and long-term maintenance

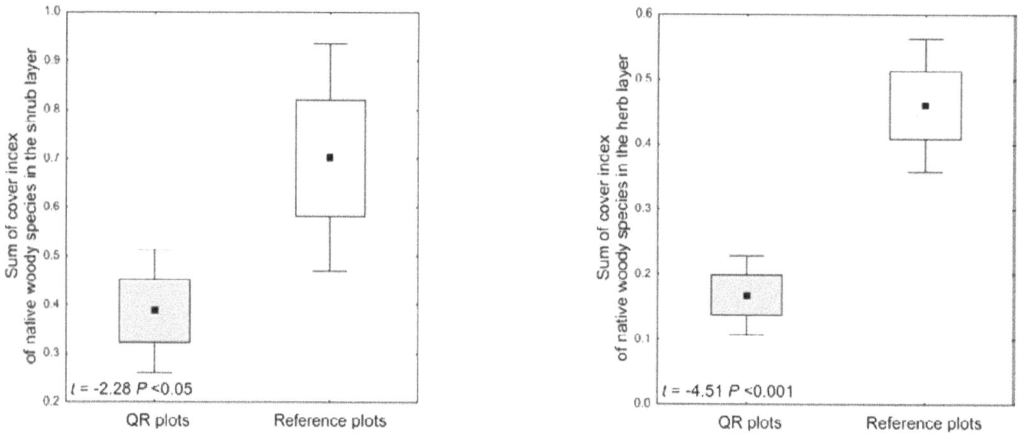

Fig. 5 Sum of native woody species cover index for QR and reference plots. The analysis was conducted separately for the shrub and herb layers. The statistical differences between plots with and without QR were analyzed with a Student's *t*-test ($P < 0.05$). All data before analysis was normalized by Box–Cox transformations.

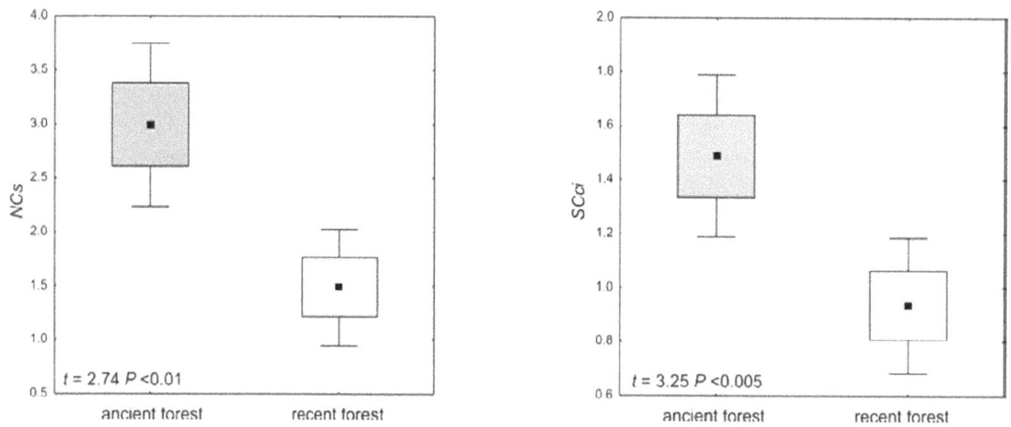

Fig. 6 The average number of coniferous forest species (*NCs*) and sum of coniferous forest species cover index (*SCci*) in ancient and recent QR plots ($df = 63$). The statistical differences between plots with and without QR were analyzed with a Student's *t*-test ($P < 0.05$). All data before analysis was normalized by Box–Cox transformations.

in the POF habitat and the increasing local reproductive pressure of this alien tree may lead to its higher invasiveness [87,88]. In a negative scenario the uncontrolled spread of *Q. rubra* limits or ultimately displaces native species in the POF ecosystem. However, the spontaneous restoration of *Q. rubra* as well as its spread outside areas of introduction, is not yet fully recognized.

Quercus rubra is still recommended as an biocoenotic admixture and is used to rehabilitate abandoned agricultural lands by afforestation [43,89]. However, the forests with introduced *Q. rubra* provide less diversified vascular understories than forests without this alien tree. QR plots located in secondary forests were not found to be richer in species number and abundance than QR plots located within old-growth forests (Fig. 6) and reference plots without the alien tree, implying that *Q. rubra* not only reduces the native forest species diversity (Tab. 2), but also it may inhibit the process of natural succession of post-agricultural plantations into natural communities. The diversification of tree species composition promoted by forest management not always favors biodiversity (for review see [71,90]), and in case of the introduction of *Q. rubra* into the studied forests, it brings a reduction in biodiversity. Although the novel ecosystems constructed by the introduced aliens are described as suitable also for resident species (e.g. Hobbs et al. [91] and Kubiak [92]), no such vascular plant species has yet been found in studied forests with *Q. rubra*.

From an economic point of view, further *Q. rubra* cultivation in the POF habitat is justified only in timber forests and only when the profits obtained from the sale of this exotic wood are higher than profits from domestic products. At present *Q. rubra* does not play an important role in the

Tab. 2 Conflicts and synergies caused by Northern Red oak introduction in mixed Scots pine–Pedunculate oak forests.

Forest ecosystem services	Impact of QR introduction
Provisioning services	
lumber, fibre and fuel wood	strong synergy
food (wild plants, mushrooms)	strong conflict
Supporting services (habitat for species & biodiversity conservation)	
vascular plant diversity	strong conflict
natural restoration of native trees	strong conflict
new habitats for native forest species (in recent forests)	strong conflict (in QR tree stands under 80 years old)
Cultural services	
aesthetic values	synergy (in Autumn)
recreation (mushrooming, hunting) and ecotourism (camping)	conflict

Ecosystem services and production functions are adapted from the Millennium Ecosystem Assessment [29] and Patterson and Coelho [32].

local timber trade. The proportion of wood obtained from *Q. rubra* cultivation in the studied area during the last decade amounted to less than 0.1% of the total acquired wood resources, while the acquisition of Scots pine wood at the same time constituted over 90% and the wood of native oaks constituted 8.4% (SILP data). In total, 92% of *Q. rubra* wood was fuel wood, 5% was used as fiber and only 3% was lumber. However, the current market value of *Q. rubra* fiber and fuel wood is equal to that of native oak wood, but the prices are higher than those of Scots pine wood [93]. Although the current profit from *Q. rubra* timber sales is low (SILP data), it can be expected to significantly increase when the stands reach logging age (100–120 years). The domestic prices of *Q. rubra* lumber are comparable to those of native oaks, and the *Q. rubra* wood is noted as precious and profitable in the timber market [53]. Furthermore, *Q. rubra* grows faster and produces more wood than European oaks [37–39], especially in post-agricultural lands [94,95].

The introduction of commercially important but invasive woody species may be accepted when the anticipated ecological, economical and social benefits exceed the costs associated with its invasiveness [96–98]. Studies show that *Q. rubra* introduction in POF definitely decreases supporting ecosystem services, however, it increases the level of provisioning services (Tab. 2).

The negative impact of *Q. rubra* on native biodiversity may be found not only in commercial forests but also in nature reserves and in nature parks where *Q. rubra* also occurs (e.g. [46,99,100]). However, the observed changes in forest structure and composition may be caused by various factors like litter quality, soil acidity and fertilization, or by different forest management options. Further studies are necessary on the mechanisms of change, on the effects of *Q. rubra* introduction, as well as on its spontaneous spread and management in the future.

Conclusions

(*i*) The introduction of *Q. rubra* causes significant changes in the structure and species composition of POF.
(*ii*) *Q. rubra* negatively influences vascular plant species diversity and abundance. No one native species benefits from *Q. rubra* introduction.
(*iii*) *Q. rubra* inhibits spontaneous restocking of POF trees.
(*iv*) Understanding the mechanisms of the changes seen in habitat conditions caused by *Q. rubra*, as well as its spontaneous spread, requires further studies.

Acknowledgments

The authors would like to thank Anonymous Reviewers for useful comments. This study was financially supported by Department of Geobotany and Plant Ecology, University of Łódź.

Authors' contributions

The following declarations about authors' contributions to the research have been made: study conception: BW, DK; field research: BW, DK; GIS and statistical analysis DK; SILP data analysis JW; analyzed the results and wrote the paper: BW, DK.

References

1. Zobel BJG, van Wyk G, Stahl P. Growing exotic forests. New York, NY: Wiley; 1987.
2. Sedjo RA. The potential of high-yield plantation forestry for meeting timber needs. New For. 1999;17(1–3):339–360.
3. Palmberg-Lerche C, Iversen PA, Sigaud P. Global data on forest plantations resources [Internet]. 2001 [cited 2012 Apr 22]; Available from:
4. Richardson DM. Forestry trees as invasive aliens. Conserv Biol. 1998;12(1):18–26.

5. Rouget M, Richardson DM, Nel JL, van Wilgen BW. Commercially important trees as invasive aliens: towards spatially explicit risk assessment at a national scale. Biol Invasions. 2002;4(4):397–412.

6. Haysom KA, Murphy ST. The status of invasiveness of forest tree species outside their natural habitat: a global review and discussion paper [Internet]. 2003 [cited 2012 Feb 22].

7. Moore BA. Alien invasive species: impacts on forests and forestry–a review [Internet]. 2005 [cited 2012 Feb 20].

8. Kohli RK, Jose S, Pal Singh H, Batish DR, editors. Invasive plants and forest ecosystems. Boca Raton, FL: CRC Press; 2009.

9. Martin PH, Canham CD. Dispersal and recruitment limitation in native versus exotic tree species: life-history strategies and Janzen–Connell effects. Oikos. 2010;119(5):807–824.

10. Richardson DM, Rejmánek M. Trees and shrubs as invasive alien species – a global review. Divers Distrib. 2011;17(5):788–809.

11. Essl F, Moser D, Dullinger S, Mang T, Hulme PE. Selection for commercial forestry determines global patterns of alien conifer invasions. Divers Distrib. 2010;16(6):911–921.

12. Vilà M, Basnou C, Pyšek P, Josefsson M, Genovesi P, Gollasch S, et al. How well do we understand the impacts of alien species on ecosystem services? A pan-European, cross-taxa assessment. Front Ecol Env. 2010;8(3):135–144.

13. Pyšek P, Jarošík V, Hulme PE, Pergl J, Hejda M, Schaffner U, et al. A global assessment of invasive plant impacts on resident species, communities and ecosystems: the interaction of impact measures, invading species' traits and environment. Glob Change Biol. 2012;18(5):1725–1737.

14. Armstrong AJ, van Hensbergen HJ. Impacts of afforestation with pines on assemblages of native biota in South Africa. Afr J. 1996;175(1):35–42.

15. Spellerberg IF, Sawyer JWD. Standards for biodiversity: a proposal based on biodiversity standards for forest plantations. Biodivers Conserv. 1996;5(4):447–459.

16. Vanhellemont M, Verheyen K, Staelens J, Hermy M. Factors affecting radial growth of the invasive *Prunus serotina* in pine plantations in Flanders. Eur J Res. 2009;129(3):367–375.

17. Hutchinson TF, Vankat JL. Invasibility and effects of Amur honeysuckle in southwestern Ohio forests. Conserv Biol. 1997;11(5):1117–1124.

18. Orr SP, Rudgers JA, Clay K. Invasive plants can inhibit native tree seedlings: testing potential allelopathic mechanisms. Plant Ecol. 2005;181(2):153–165.

19. Stinson KA, Campbell SA, Powell JR, Wolfe BE, Callaway RM, Thelen GC, et al. Invasive plant suppresses the growth of native tree seedlings by disrupting belowground mutualisms. PLoS Biol. 2006;4(5):e140.

20. Meiners SJ. Apparent competition: an impact of exotic shrub invasion on tree regeneration. Biol Invasions. 2007;9(7):849–855.

21. Webster CR, Jenkins MA, Jose S. Woody invaders and the challenges they pose to forest ecosystems in the eastern United States. J For. 2006;104(7):366–374.

22. Collier MH, Vankat JL, Hughes MR. Diminished plant richness and abundance below *Lonicera maackii*, an invasive shrub. Am Midl Nat. 2002;147(1):60–71.

23. Huebner CD. Vulnerability of oak-dominated forests in West Virginia to invasive exotic plants: temporal and spatial patterns of nine exotic species using herbarium records and land classification data. Castanea. 2003;68(1):1–14.

24. Lichstein JW, Grau HR, Aragón R. Recruitment limitation in secondary forests dominated by an exotic tree. J Veg Sci. 2004;15(6):721–728.

25. Simberloff D, Parker IM, Windle PN. Introduced species policy, management, and future research needs. Front Ecol Env. 2005;3(1):12–20.

26. van Wilgen BW, Dyer C, Hoffmann JH, Ivey P, Le Maitre DC, Moore JL, et al. National-scale strategic approaches for managing introduced plants: insights from Australian acacias in South Africa: strategic approaches for managing introduced acacias. Divers Distrib. 2011;17(5):1060–1075.

27. Chytrý M, Wild J, Pyšek P, Jarošík V, Dendoncker N, Reginster I, et al. Projecting trends in plant invasions in Europe under different scenarios of future land-use change: projecting future plant invasions in Europe. Glob Ecol Biogeogr. 2012;21(1):75–87.

28. Hulme PE. Weed risk assessment: a way forward or a waste of time? J Appl Ecol. 2012;49(1):10–19.

29. Millennium Ecosystem Assessment. Ecosystems and human well-being: synthesis. Washington, DC: Island Press; 2005.

30. Young SL. What contributions are invasive plant species making to ecosystem services? J Soil Water Conserv. 2010;65(2):31A–32A.

31. de Wit MP, Crookes DJ, van Wilgen BW. Conflicts of interest in environmental management: estimating the costs and benefits of a tree invasion. Biol Invasions. 2001;3(2):167–178.

32. Patterson TM, Coelho DL. Ecosystem services: foundations, opportunities, and challenges for the forest products sector. Ecol Manage. 2009;257(8):1637–1646.

33. Devine K, Fei S. A review of impacts by invasive exotic plants on forest ecosystem services. In: Proceedings of the 17th central hardwood forest conference GTR-NRS-P-78. Newtown Square, PA: USDA Forest Service, Northern Research Station; 2011. p. 425–435.

34. Bucharova A, van Kleunen M. Introduction history and species characteristics partly explain naturalization success of North American woody species in Europe. J Ecol. 2009;97(2):230–238.

35. Hereźniak J. American trees and shrubs in the Polish territories. In: Ławrynowicz M, Warcholińska AU, editors. Łódź: Scientific Society in Łódź; 1992. p. 97–150. (vol 9).

36. Magni Diaz CR. Reconstitution de l'introduction de *Quercus rubra* L. en Europe et conséquences génétiques dans les populations allochtones. Paris: ENGREF; 2004.

37. Król S. Northen red oak – *Quercus rubra* L. in forest environment in western part of Poland. Pr Kom Nauk Rol Kom Nauk Leśnych. 1967;21:419–482.

38. Vansteenkiste D, de Boever L, van Acker J. Alternative processing solutions for red oak (*Quercus rubra*) from converted forests in Flanders, Belgium. In: Proceedings of the COST action E44 conference on broad spectrum utilization of wood at BOKU. Vienna: BOKU; 2005. p. 13–26.

39. Rédei K, Csiha I, Keserű Z, Rásó J, Győri J. Management of red oak (*Quercus rubra* L.) stands in the Nyírség forest region (Eastern Hungary). Hung Agric Res. 2010;3:13–17.

40. Lambdon PW, Pyšek P, Basnou C, Hejda M, Arianoutsou M, Essl F, et al. Alien flora of Europe: species diversity, temporal trends, geographical patterns and research needs. Preslia. 2008;80:101–149.

41. Woziwoda B. Expansion of northern red oak *Quercus rubra* L. supported by anthropogenic factors. In: Mazur S, Tracz H, editors. VIII symposium of protection of forests ecosystem: human threats to forest ecosystems, identification – monitoring – management. Warsaw: Warsaw University of Life Sciences Press; 2008. p. 259–263.

42. Sander IL. *Quercus rubra* L. northern red oak. Washington, DC: U.S. Department of Agriculture; 1990. (vol 2).

43. Murat E. Forest cultivation in details. Warsaw: Wydawnictwo Świat; 2002.

44. AHEC (American Hardwood Export Council) [Internet]. 2005 [cited 2012 Nov 30].

45. Chmura D. Penetration and naturalization of invasive alien plant species (neophytes) in woodlands of the Silesian Upland (Southern Poland). Nat Conserv. 2004;60:3–11.

46. Otręba A, Ferchmin M. Alien tree species as indicators of environmental transformation in Kampinoski National Park. SiM CEPL. 2007;2/3(16):234–244.

47. Jakubowska-Gabara J, Woziwoda B. Decrease of vascular flora diversity in forest communities connected with invasive red oak *Quercus rubra* L. In: Coles S, Dimopoulos P, editors. 52 Symposium of IAVS: vegetation processes and human impact in a changing world. Chania: Wiley-Blackwell; 2009. p. 179.

48. Marozas V, Straigytė L, Šepetienie J. Comparative analysis of alien oak (*Quercus rubra* L.) and native common oak (*Quercus robur* L.) vegetation in Lithuania. Acta Biol Univ Daugavpilensis. 2009;9(1):19–24.

49. Pyšek P, Danihelka J, Sádlo J, Chrtek JJ, Chytrý M, Jarošík V, et al. Catalogue of alien plants of the Czech Republic (2nd edition): checklist update, taxonomic diversity and invasion patterns. Preslia. 2012;84:155–255.

50. Tokarska-Guzik B. The establishment and spread of alien plant species (kenophytes) in the flora of Poland. Katowice: University of Silesia Press; 2005.

51. Institute of Nature Conservation PA of S. Alien species in Poland. Data base [Internet]. 2009 [cited 2012 Feb 18].

52. Riepšas E, Straigytė L. Invasiveness and ecological effects of red oak (*Quercus rubra* L.) in Lithuanian forests. Balt For. 2008;14(2):122–130.

53. Kuc M, Piszczek M, Janusz A. Importance of northern red oak *Quercus rubra* L. in forest ecosystems and economic calculus in Regional Directorate of State Forests in Katowice. SiM CEPL. 2012;33(4):152–159.

54. Obwieszczenie Marszałka Sejmu Rzeczypospolitej Polskiej z dnia 25 sierpnia 2009 r. w sprawie ogłoszenia jednolitego tekstu ustawy o ochronie przyrody. 2009.

55. Dz. U. 2009.151.1220. Art.120. p. 4.

56. Rozwałka Z. Zasady hodowli lasu. Bedoń: Ośrodek Rozwojowo-Wdrożeniowy Lasów Państwowych w Bedoniu; 2003.

57. Haze M, editor. Zasady hodowli lasu. Warsaw: Centrum Informacyjne Lasów Państwowych; 2012.

58. Dz. U. 2011.2010.1260

59. Woziwoda B, Kałucka I, Ruszkiewicz-Michalska M, Sławska M, Sławski M, Tołoczko W, et al. Interdisciplinary research of ecological effects of Northern Red oak *Quercus rubra* L. introduction in forest ecosystems (central Poland) - the principles and aims of study. SiM CEPL. 2012;33(4):181–192.

60. Peterken GF. A method for assessing woodland flora for conservation using indicator species. Biol Cons. 1974;6(4):239–245.

61. Majchrowska A, Woziwoda B. Effects of forest history on the biodiversity of vascular plants flora. In: Holeksa J, Babczyńska-Sendek B, Wika S, editors. The role of geobotany in biodiversity conservation. Katowice: University of Silesia Press; 2009. p. 49–58.

62. Matuszkiewicz JM. Forest communities of Poland. Warsaw: Polish Scientific Publishers PWN; 2001.

63. Matuszkiewicz W. A guide to identification of the plant communities of Poland. Warsaw: Polish Scientific Publishers PWN; 2002.

64. Braun-Blanquet J. Pflanzensoziologie: Grundzüge der Vegetationskunde. 3rd ed. Vienna: Springer-Verlag; 1964.

65. ter Braak CJF, Šmilauer P. CANOCO reference manual and CanoDraw for Windows user's guide: software for canonical community ordination (version 4.5). Ithaca, NY: Microcomputer Power; 2002.

66. Canham CD, Finzi AC, Pacala SW, Burbank DH. Causes and consequences of resource heterogeneity in forests: interspecific variation in light transmission by canopy trees. Can J Res. 1994;24(2):337–349.

67. Ellenberg H, Weber HE, Düll R, Wirth V, Werner W, Paulißen D. Zeigerwerte von Pflanzen in Mitteleuropa. Scr Geobot. 1991;18:1–248.

68. Kwiatkowska AJ. Changes in the species richness, spatial pattern and species frequency associated with the decline of oak forest. Vegetatio. 1994;112(2):171–180.

69. Sonohat G, Balandier P, Ruchaud F. Predicting solar radiation transmittance in the understory of even-aged coniferous stands in temperate forests. Ann Sci. 2004;61(7):629–641.

70. Tinya F, Márialigeti S, Király I, Németh B, Ódor P. The effect of light conditions on herbs, bryophytes and seedlings of temperate mixed forests in Őrség, Western Hungary. Plant Ecol. 2009;204(1):69–81.

71. Barbier S, Gosselin F, Balandier P. Influence of tree species on understory vegetation diversity and mechanisms involveda-a critical review for temperate and boreal forests. Ecol Manage. 2008;254(1):1–15.

72. Knight KS, Oleksyn J, Jagodzinski AM, Reich PB, Kasprowicz M. Overstorey tree species regulate colonization by native and exotic plants: a source of positive relationships between understorey diversity and invasibility: tree species, diversity, and invasibility. Divers Distrib. 2008;14(4):666–675.

73. Robakowski P, Bielinis E. Competition between sessile oak (*Quercus petraea*) and black cherry (*Padus serotina*) : dynamics of seedlings growth. Pol J Ecol. 2011;59(2):297–306.

74. Starfinger U. Die Einbürgerung der Spätblühenden Trubenkirsche (*Padus serotina* Ehrh.) in Mitteleuropa. Landschaftsenwicklung Umweltforchung. 1990;69:119.

75. Godefroid S, Phartyal SS, Weyembergh G, Koedam N. Ecological factors controlling the abundance of non-native invasive black cherry (*Prunus serotina*) in deciduous forest understory in Belgium. Ecol Manage. 2005;210(1–3):91–105.

76. Dzwonko Z. Effect of proximity to ancient deciduous woodland on restoration of the field layer vegetation in a pine plantation. Ecography. 2001;24(2):198–204.

77. Augusto L, Dupouey JL, Ranger J. Effects of tree species on understory vegetation and environmental conditions in temperate forests. Ann Sci. 2003;60(8):823–831.

78. Collier M, Farrell EP. The environmental impact of planting broadleaved trees on acid-sensitive soils. Literature review. Dublin: COFORD; 2007.

79. Straigytė L, Žalkauskas R. Effect of climate variability on *Quercus rubra* phenotype and spread in Lithuanian forests. Dendrobiology. 2012;67:79–85.

80. Richardson DM, Macdonald IAW, Forsyth GG. Reduction in plant species richness under stands of alien trees and shrubs in fynbos biome. Afr J. 1989;149(1):1–8.

81. Winter M, Schweiger O, Klotz S, Nentwig W, Andriopoulos P, Arianoutsou M, et al. Plant extinctions and introductions lead to phylogenetic and taxonomic homogenization of the European flora. Proc Natl Acad Sci USA. 2009;106(51):21721–21725.

82. Peterken GF. Ecological effects of introduced tree species in Britain. Ecol Manage. 2001;141(1–2):31–42.

83. Olden JD, Poff NL. Towards the mechanistic understanding and prediction of biotic homogenization. Am Nat. 2003;162(4):442–460.

84. Rejmánek M, Richardson DM, Pyšek P. Plant invasions and invasibility of plant communities. In: van der Maarel E, Franklin J, editors. Vegetation ecology. Oxford: John Wiley & Sons; 2013. p. 387–424.

85. Hejda M, Pyšek P, Jarošík V. Impact of invasive plants on the species richness, diversity and composition of invaded communities. J Ecol. 2009;97(3):393–403.

86. Hibbs DE. Forty years of forest succession in central New England. Ecology. 1983;64(6):1394–1401.

87. Wangen S, Webster C. Spatial and temporal dynamic of exotic tree invasions: lesson from a shade-tolerant invader, *Acer platanoides*. In: Kumar Kohli R, Jose S, Pal Singh H, Rani Batish D, editors. Invasive plants and forest ecosystems. Boca Raton, FL: CRC Press; 2008. p. 71–85.

88. Dodet M, Collet C. When should exotic forest plantation tree species be considered as an invasive threat and how should we treat them? Biol Invasions. 2012;14(9):1765–1778.

89. Zieliński J, Nowak G. Trees and shrubs used in rehabilitation of industrial waste stockpiles. In: Drozdek ME, editor. Plants for special tasks. Sulechów: Państwowa Wyższa Szkoła Zawodowa; 2011. p. 455–466.

90. Thomaes A, Keersmaeker L, Schrijver A, Vandekerkhove K, Verschelde P, Verheyen K. Can tree species choice influence recruitment of ancient forest species in post-agricultural forest? Plant Ecol. 2010;212(4):573–584.

91. Hobbs RJ, Arico S, Aronson J, Baron JS, Bridgewater P, Cramer VA, et al. Novel ecosystems: theoretical and management aspects of the new ecological world order. Glob Ecol Biogeo. 2006;15(1):1–7.

92. Kubiak D. Lichens of red oak *Quercus rubra* in the forest environment in the Olsztyn Lake District (NE Poland). Acta Mycol. 2006;41(2):319–328.

93. Price list of the wood range. Załącznik nr 1. In: Zarządzenie nr 1/13 Nadleśniczego Nadleśnictwa Poddębice z dnia 2 stycznia 2013 r. Poddębice: Nadleśnictwo Poddębice; 2013. p. 1-4.

94. Bellon S, Tumiłowicz J, Król S, editors. Alien woody species in forest management. Warsaw: Państwowe Wydawnictwo Rolnicze i Leśne; 1977.

95. Danusevičius J, Gabrilavičius R, Danusevičius D. Quality of red oak (*Quercus rubra* L.) stands on abandoned land. Balt For. 2002;8:51–56.

96. Hulme PE, Nentwig W, Pyšek P, Vila M. Common market, shared problems: time for a coordinated response to biological invasions in Europe? Biological invasions: towards a synthesis. Neobiota. 2009;8:3–19.

97. Pejchar L, Mooney HA. Invasive species, ecosystem services and human well-being. Trends Ecol Evol. 2009;24(9):497–504.

98. Yokomizo H, Possingham HP, Hulme PE, Grice AC, Buckley YM. Cost-benefit analysis for intentional plant introductions under uncertainty. Biol Invasions. 2012;14(4):839–849.

99. Danielewicz W, Maliński T. Trees and shrubs of alien origin in Wielkopolski National Park. Rocz Dendrol. 1997;45:65–81.

100. Adamowski W, Dworak L, Ramanjuk I. 102 - Atlas of alien woody species of the Białowieża primeval forest. Phytocoenosis (N.S.). Cartogr Geobot. 2002;supp 14:7–303.

101. Verloove F. Catalogue of the Neophytes in Belgium (1800–2005). Scr Bot Belg. 2006;39:1–89.

102. Medvecká J, Kliment J, Májeková J, Halada Ľ, Zaliberová M, Gojdičová E, et al. Inventory of the alien flora of Slovakia. Preslia. 2012;84:257–309.

11

Effect of osmotic stress on in vitro translational capacity of polysomes and on the composition of polysome-associated proteins in germinating seeds of pea (*Pisum sativum* L.)

Wioletta Brosowska-Arendt*, Stanisław Weidner

Department of Biochemistry, University of Warmia and Mazury in Olsztyn, Oczapowskiego 2, 10-719 Olsztyn, Poland

Abstract

Plant growth throughout the world is often limited by unfavourable environmental conditions. This paper reports results of a study on long- and short-term osmotic stress (−0.5 MPa) followed by a recovery on in vitro translational capacity of polysomes and on the composition of polysome-associated proteins in germinating pea (*Pisum sativum* L.) seeds. Here we show that, under osmotic stress, cytoskeleton-bound polysomes were charaterized by the highest translation activity, which may be indicative of an important role that this population of polysomes plays in the synthesis of the so-called "stress proteins". We also find out that in response to osmotic stress, new proteins (22.01, 96.47 and 105.3 kDa), absent in the unstressed sample, associated with the total pool of polysomes, whereas the protein of 22.95 kDa, which was present in the embryonic tissue of seeds germinating under unstressed conditions, disappeared. These changes may have affected both the stability and the translational capacity of polysomes.

Keywords: germinating, osmotic stress, pea seeds, polyethylene glycol, polysomes, ribonuclease, translation in vitro

Introduction

Changes in the environment, such as drought, salinity, high or low temperature, are an important factor which affects the growth of crops and the volume of crop yields [1–3]. Very important legume grown and consumed extensively worldwide is pea [4]. As a rich source of proteins, carbohydrates, fibre, vitamins and minerals, peas are important in human nutrition [5]. Pea is the fourth leading legume in terms of consumption in the world after soybean, peanuts and bean [6]. One of the major abiotic stress often occurs in Europe is osmotic stresses induced with polyethylene glycol (within −0.5 MPa). The earliest metabolic change caused by water stress is a decreased amount of polysomes [7]. Once the polysomes have been disaggregated, the plant growth is inhibited due to a slower protein synthesis rate [7–9]. Considerable reduction in the amount of polysomes in plant tissues is observable after osmotic stress lasting for just 20–30 minutes [10]. Decrease of the content of polysomes in response to abiotic stresses is connected with the process of "switching" the expression of genes from those participating in the growth and development of plants under unstressed conditions to the ones active in response to stress [11].

In plant tissues, polysomes can occur as free polysomes (FP), endoplasmic reticulum membrane-bound polysomes (MBP) [12], cytoskeleton-bound polysomes (CBP) [13,14] and cytoskeleton-membrane-bound polysomes (CMBP) [15]. Changes in the distribution of polysomes between the particular fractions reflect changes in the complement of proteins, as each population of polysomes is engaged in the synthesis of specific proteins [16]. Thus, the FP population is mainly involved in the synthesis of soluble proteins of the cytoplasm, cellular nucleus, mitochondria and peroxisomes [17]; the MBP population is engaged in the production of secretory proteins, lysosome proteins and the proteins which are an integral component of the plasmatic membrane and intracellular membranes, including the endoplasmic reticulum [16,17]; finally, the CBP and CMBP populations are responsible for the synthesis of cytoskeleton and stress proteins [16,18–21]. The largest share in the total polysome pool consists of the CBP population, which sometimes reaches 70% of the total polysome content [22]. Polysomes contain mRNAs, which may undergo selective translation, result in modification of protein synthesis in response to stress conditions in plants [23,24].

The objective of this study was to indicate differences in the composition of polysome-bound proteins and in the products of in vitro translation in pea (*Pisum sativum* L.) seeds germinating under unstressed conditions and under long- and short-term osmotic stress (−0.5 MPa) followed by post-stress recovery.

* Corresponding author. Email: wioletta.brosowska@uwm.edu.pl

Material and methods

Plant material and germination under normal and osmotic stress conditions

The experiments were conducted on *Pisum sativum* L. cv. six-week TOR seeds, supplied by Torseed S.A. (Toruń, Poland). Seeds were surfaced disinfected in 1% sodium hypochloride for 3 minutes and washed with tap water and sterile water. Disinfected and intact seeds were placed on Petri dishes with two layers of Whatman 2 filter paper No. 1 (Whatman, Maidstone, Kent, the UK) wetted with 40 ml distilled water and germinated in the dark at 20°C for 48 (C48), 72 (C72) and 96 hours (C96). After 48 hours, some seeds germinated under unstressed conditions (distilled water), whose roots were ≥1 mm, were transferred for next 24 h to osmotic stress conditions with polyethylene glycol (PEG), which caused decrease of water potential to −0.5 MPa (S72; 48 h distilled water + 24 h osmotic stress). After short-term osmotic stress, early seedlings of pea were transferred again to unstressed germination conditions for another 24 hours – recovery (SR96; 48 h distilled water + 24 h osmotic stress + 24 h distilled water). At the same time, other pea seeds were germinated under long-term osmotic stress conditions (−0.5 MPa) at 20°C for 96 hours (S96). Having fixed the time of germination, sprouts or early seedlings (according to Gong et al. [25]) were isolated from seeds and used for further tests. All determinations were repeated three times.

Polysome isolation

The tissue (400 mg) was homogenized in 3 ml of cytoskeleton-stabilizing buffer C [22], filtered and centrifuged at 27000 × *g* for 10 min. Buffer C consist of 5 mM HEPES (N-2-hydroxyethylpiperazine-N″-2-ethanesulfonic acid), 10 mM $Mg(OAc)_2$ (magnesium acetate), 2 mM EGTA [ethylene glycol-bis (aminoethyl ether) N,N,N″,N″-tetraacetic acid], 1 mM PMSF (phenylmethylsulfonyl fluoride) adjusted to pH 7.5 with 9.8 mM KOH. This buffer allows sequential isolation of four polysome populations [26]: FP, MBP, CBP and CMBP. The supernatant polysomes included FP and, to prevent their degradation by RNAse, the samples were adjusted to buffer U, consisting of 200 mM Tris-HCl, pH 8.5, 50 mM KOAc (potassium acetate), 25 mM $Mg(OAc)_2$, 2 mM EGTA, 100 mg/ml heparin, 2% PTE (polyoxyethylene-10-tridecyl ether, a non-ionic detergent) and 1% DOC (sodium deoxycholate) [27] by adding ¼ vol. of 4× concentrated of buffer U and held on ice prior to further processing. The remaining polysomes were in the pellet and were sequentially solubilized as followed. The pellet was resuspended in 3 ml of buffer C + 0.5% PTE to disrupt membranes, centrifuged for 10 min at 27000 × *g* to leave MBP in the supernatant (again converted to buffer U to maintain polysome integrity). All the other polysomes were still in the pellet. This pellet was next resuspended in 3 ml of buffer C + 200 mM Tris-HCl, pH 8.5 to disrupt the polysome-cytoskeleton interaction and centrifuged for 10 min at 27000 × *g* to leave CBP in the supernatant (again converted to buffer U). The final pellet was resuspended in 4 ml of 1× concentrated of buffer U and re-centrifuged to release CMBP. All supernatant fractions (FP, MBP, CBP, CMBP) were layered on a 0.5 ml "pad" of 50% (w/v) sucrose in buffer B [50 mM Tris-HCl, pH 7.5, 20 mM KOAc, 10 mM $Mg(OAc)_2$] and centrifuged for 90 min at 300000 × *g* in Beckman 65 Ti rotor. Sprouts or early seedlings of pea were also homogenized in buffer U [27] to solubilize total polysomes.

Polysomes protein separation with the Laemmli method

The content of protein was determined using Bradford's method [28]. The polysome pellets were resuspended in 0.15 ml of lysis buffer, consisting of 0.125 M Tris-HCl pH 6.8, 4% SDS, 10% 2-merkaptoethanol, 20% glycerol and 0.05% bromophenol blue. The resuspended polysomes were boiled at 95°C for 4 min and centrifuged at 27000 × *g* for 3 min, after which they were cooled to 0–4°C.

Proteins connected with the total polysome pool and with particular polysome fractions were separated in 12% SDS-PAGE on an electrophoretic apparatus Mini-PROTEAN GEL II (Bio-Rad). Each time, 40 μg of protein was applied to each lane. Electrophoresis was run for 45 min at the temperature of 0–4°C and 200 V, using electrode buffer containing: 0.025 M Tris-HCl pH 3, 0.192 M glicyne and 1% SDS. The gels were fixed in 5% solution of trichloroacetic acid (TCA) for 30 min, then stained for 90 min with Coomassie Blue according to Laemmli [29], consisting of 0.1% Coomassie Blue R-250, 40% methanole and 10% acetic acid. When the protein had been stained, the gels were destained using a solution containing 25% of methanol and 10% of acetic acid. The gels were scanned with the Labscan 5.0, using an Image Scanner (Amersham). For the analysis of the gels, ScanGel 1.0 (Kucharczyk) was used.

Electrophoresis and Western Blotting of in vitro synthesized proteins

Sediments of the total ribosomal fraction and of particular populations of polysomes, i.e. FP, MBP, CBP and CMBP were dissolved in 0.05 ml of 0.5% PTE solution and centrifuged. For each sample, optical density (*OD*) was determined at the wavelengths of 260 and 330 nm. The difference between the result at 330 nm and the one at 260 nm was the value of the *OD* of a given sample. For an in vitro translation reaction, 2.0 *OD* of polysomes were needed.

Polysomes (2.0 *OD*) were mixed with the components of a translation mixture according to the instruction attached to the "Rabbit Reticulocyte Lysate System" (Promega), incubated at 30°C for 90 min, and then mixtures were placed on ice to stop the reaction. The content of a protein was established with Bradford's method [28]. The reagent mixture was dissolved in a lysing reagent (the contents as above) in a 1:4 ratio, heated at the temperature of 95°C for 4 min, centrifuged at 27000 × *g* for 3 minutes and chilled to 0–4°C. The biotinylated proteins were separated in 12% SDS-PAGE according to Laemmli [29]. 20 μg of the protein was applied to each lane. The control sample, with which it was possible to eliminate the so-called background effect, contained all the components of the translation mixture (except polysomes) as well as the luciferase RNA standard.

The biotinylated proteins separated by electrophoresis transferred to polyvinylidene difluoride membrane (PVDF; Immobilon-P of the pore diameter equal 0.45 μm; Millipore). The electrotransfer was conducted at fixed voltage of 100 V for 60 min at 15–20°C. The buffer for the transfer contained 25 mM Tris, 192 mM glicyne, 20% methanole, 0.1% SDS. Afterwards, biotinylated proteins were visualized by binding Streptavidin-HRP, followed by chemiluminescent detection (Transcend™ Chemiluminescent Translation Detection System; Promega) and scanned using Labscan 5.0 software on an ImageScanner (Amersham). To determine the apparent weight of the translated biotinylated proteins, biotinylated protein standards (Bio-Rad) were used. For analysis, ScanGel 1.0 (Kucharczyk) was run.

Results

SDS-PAGE showed that small-molecule proteins – less than 40 kDa (Fig. 1) – dominated among the ribosomal proteins bound with the total population of polysomes as well as with their particular fractions, such as FP, MBP, CBP and CMBP. In the profile of polysomal proteins isolated from early pea seedlings grown under unstressed conditions and under long-term osmotic stress (−0.5 MPa), obtained after 96 hours of seed germination, qualitative and quantitative changes were observed. It was found out that in response to the osmotic stress the 22.0, 96.5 and 105.3 kDa proteins, absent in the control samples, bound with the total population of polysomes. It is noteworthy that these proteins (22.0 and 105.3 kDa) also bound to all fractions of polysomes (FP, MBP, CBP and CMBP) isolated from seeds germinated under stress conditions. Besides, in response to stress conditions, plants produced such proteins that bound to specific polysomal fractions, e.g. the 50.3 kDa protein which bound to the FP population, 58.9 kDa – to CMBP, and 101.4 kDa – to FP, CBP and CMBP. In turn the 23.0 kDa protein, which was present in pea seeds germinating under unstressed conditions, under osmotic stress was absent, both from the total polysomal population and from any of the polysomal fractions (Fig. 1). Furthermore, it was also observed that under the effect of water stress, the amounts of 23.5, 24.9, 50.3, 58.9 and 110.0 kDa proteins bound with the total population of polysomes increased in comparison with the control (Fig. 1). Some other proteins of the particular polysomal populations formed during the water stress of −0.5 MPa also occurred in either larger or smaller quantities than in the control. The proteins bound with MBP were observed in smaller amounts while the ones bound with FP increased (Tab. 1).

In vitro translation primed by total polysomes and their particular fractions (FP, MBP, CBP and CMBP) indicated that there were differences in the newly synthesised proteins, both during the seed germination under the unstressed conditions and between the control sample versus the samples treated with osmotic stress. These differnces were both quantitative and qualitative. During the germination under the unstressed conditions, the in vitro synthesis of 108.2 and 158.3 kDa proteins decreased, while that of 20.7 and 21.9 kDa proteins increased (Fig. 2). Under the effect of both long- and short-term osmotic stress (S96, S72), the 20.7, 21.9, 26.6, 108.2 and 158.3 kDa proteins were observed to be synthesized more effectively, and therefore appeared in larger quantities than under the unstressed conditions (C96, C72, respectively). However, during the recovery after the short-term osmotic stress (SR96), the synthsis of these proteins decreased in comparison with the sample exposed to the stress (S72).

With respect to in vitro translation which involved particular populations of polysomes isolated from 96-hour pea seedlings grown under the unstressed conditions and under long-term water stress of the intensity of −0.5 MPa, differences were mainly observed between the 108.2 and 158.3 kDa proteins. In pea seedlings grown under the unstressed conditions, the 108.2 kDa protein was synthesized mainly by the MBP as well as CBP and CMBP fractions, while the 158.3 kDa protein – by the FP, MBP, CBP populations and, only weakly, by CMBP. For comparison, in pea seedlings growing under long-term osmotic stress, the 108.2 kDa protein was synthesized in vitro mainly by the MBP and CBP populations and, to a smaller extent, by the FP and CMBP fractions, while the 158.3 kDa protein – by the MBP and CBP populations. It was also demonstrated that the most important differences between the osmotic stress and control samples appeared in the proteins which were synthesized on the CBP population. In response to the stress conditions, these proteins (>66 kDa) were more intensively synthesized than under the unstressed conditions (Fig. 3).

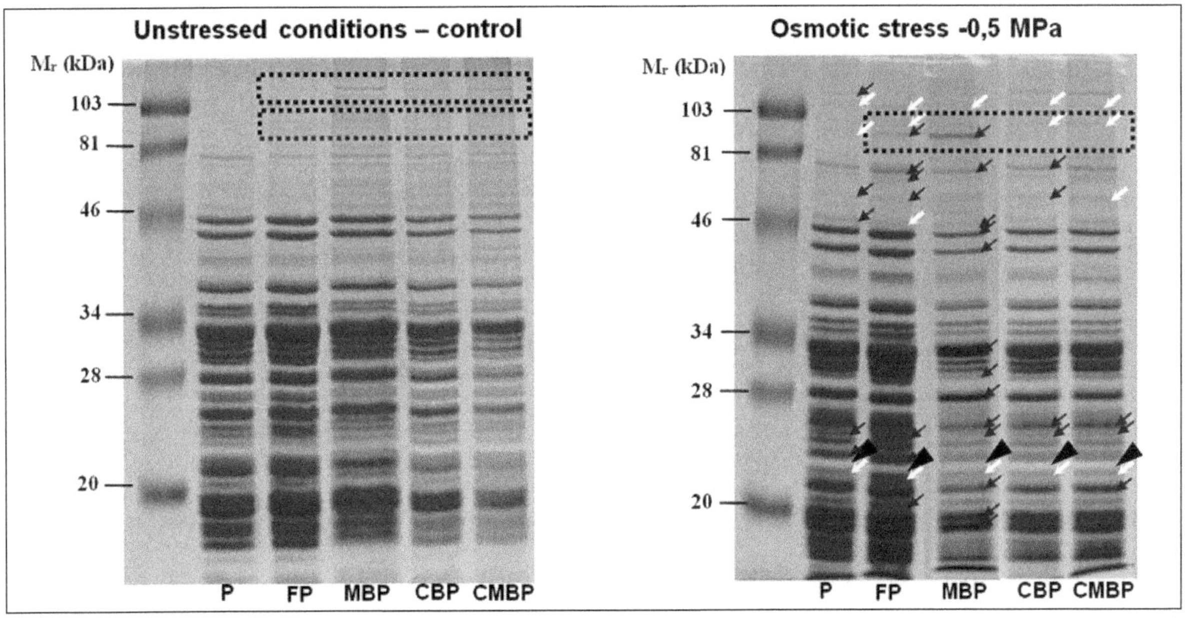

Fig. 1 SDS-PAGE of proteins bound to the total polysomal population and to particular polysomal fractions: FP, MBP, CBP and CMBP, isolated from embryonic tissue after 96 h of pea seed germination under unstressed (control) and long-term osmotic stress (−0.5 MPa) conditions. Molecular weight (kDa) of protein markers are given to the left. Black arrowheads – proteins which were absent under stress conditions; black frames – differences between proteins within a given sample; grey arrows – proteins different in quantities between control and stress samples; white arrowheads – proteins which were joined to polysomes during stress. P – total polysomal population.

Tab. 1 Differences in proteins associated with the total population of polysomes and with their particular fractions: FP, MBP, CBP, CMBP, isolated from embryonic tissue after 96 h of germination under unstressed (control) and long-term osmotic stress (−0.5 MPa) conditions.

	Control					Osmotic stress −0.5 MPa				
Protein (kDa)	**P**	**FP**	**MBP**	**CBP**	**CMBP**	**P**	**FP**	**MBP**	**CBP**	**CMBP**
110.0	+	+	+	+	+	↗	+	+	+	+
105.3	–	–	–	–	–	+	+	+	+	+
101.4	+	–	+	–	–	+	+	+	+	+
96.5	–	+	+	+	+	+	↗	↗	+	+
77.8	+	+	+	+	+	+	↗	↗	↗	+
67.4	+	+	+	+	+	+	↙	+	+	+
58.9	+	+	+	+	–	↗	↗	+	↗	+
50.3	+	–	+	+	+	↗	+	↙	+	+
46.1	+	+	+	+	+	+	+	↙	+	+
32.0	+	+	+	+	+	+	+	↙	+	+
30.5	+	+	+	+	+	+	+	↙	+	+
28.0	+	+	+	+	+	+	+	↙	+	+
25.0	+	+	+	+	+	+	+	↙	↗	↗
24.9	+	+	+	+	+	↗	↗	↙	↙	↙
23.5	+	+	+	+	+	↗	↗	+	+	+
22.9	+	+	+	+	+	–	–	–	–	–
22.0	–	–	–	–	–	+	+	+	+	+
20.2	+	+	+	+	+	+	+	↙	+	↗
19.9	+	+	+	+	+	+	↗	+	+	+
18.3	+	+	+	+	+	+	+	↙	+	+
17.3	+	+	+	+	+	+	+	↙	+	+

"+" – presence; "–" – absence; "↗" – increased; "↙" – decreased of amount of a protein fraction joined with the total polysomal population and/or particular polysomal fractions in an experimental sample.

Discussion

The PEG used in our experiments is a neutral polymer which binds water, thus simulating soil drought stress [30,31]. Scientific experiments which simulate naturally ocurring conditions enable us to expand our knowledge of a given environmental stress factor and plant reactions to this stress, thus making it possible to obtain crops which are more tolerant to a given stress factor.

Osmotic stress inhibits the process of seed germination and subsequent plant growth [32–37] as well as the process of polysome formation [36,38,39]. Changes in the profiles of polysomes, which contain mRNAs undergoing translation, enable us to follow very precisely the rate of in vivo synthesis of proteins in cells [40].

The changes in the composition of polysomal proteins presented in this paper may play an important role in the regulation of translation under stress conditions and can result in the differentiated translational activity of particular fractions of polysomes, which had been demonstrated earlier [19,20,39,41,42]. The end products from in vitro translation of polysomes also reflect the activity of mRNAs that respond to stress [43]. Synthesis of most proteins under environmental stresses is reduced but translation of individual mRNA species is differentially regulated [23,44–46]. Kawaguchi et al. [45] showed that less than 1% of the stress-induced mRNAs increased significantly in polysome association, what can suggest that proteins encoded by this mRNAs were synthesized at higher levels during stress conditions. In turn on Shenton et al. [47] suggested that under low intensity of oxidative stress

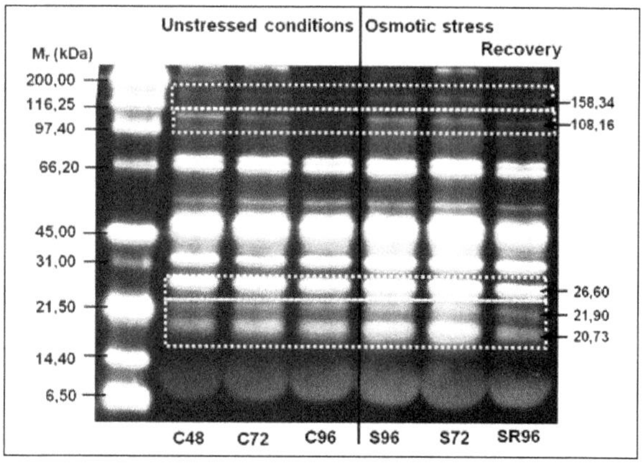

Fig. 2 Chemiluminescent detection of biotinylated proteins which were synthesized in vitro by total polysomal isolated from embryonic tissue of pea seeds germinated under unstressed conditions after 48 h (C48), 72 h (C72), 96 h (C96), under long-term (S96; 96 h osmotic stress) and short-term (S72; 48 h distilled water + 24 h osmotic stress) osmotic stress (−0.5 MPa) and during recovery after short-term osmotic stress (SR96; 48 h distilled water + 24 h osmotic stress + 24 h distilled water) separated by SDS-PAGE. Biotinylated proteins were visualized by binding Streptavidin-HRP, followed by chemiluminescent detection. Differences between the proteins are presented in the boxes and the arrows indicate precisely these protein fractions.

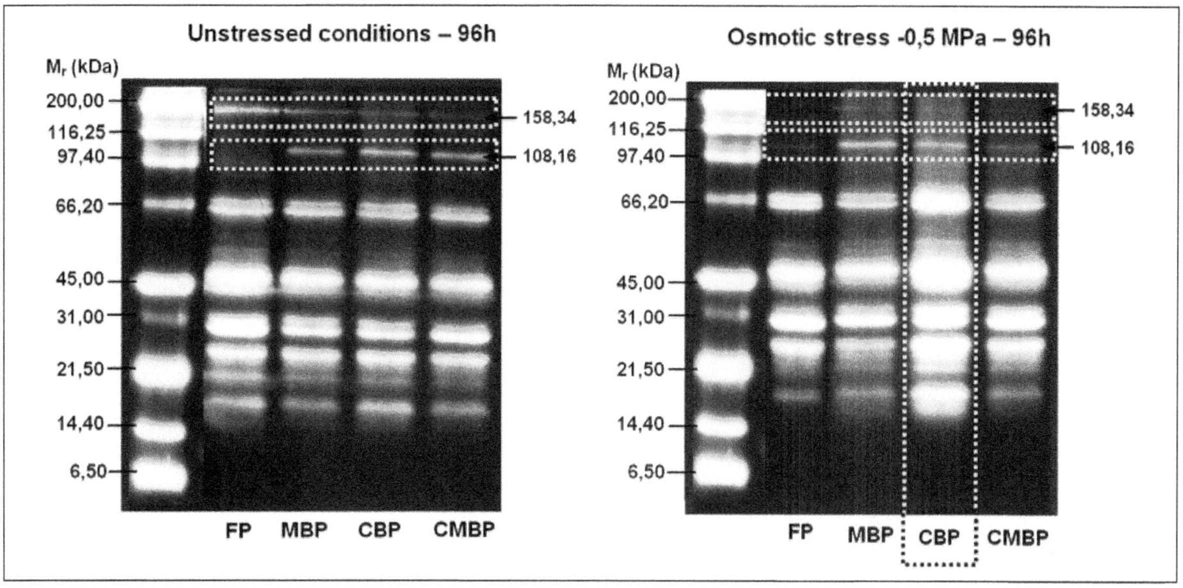

Fig. 3 Chemiluminescent detection of biotinylated proteins which were synthesized in vitro by particular polysomal fractions: FP, MBP, CBP and CMBP, isolated from embryonic tissue of pea seeds germinated under unstressed conditions (C96) and under long-term (S96) osmotic stress (−0.5 MPa) separated by SDS-PAGE. Biotinylated proteins were visualized by binding Streptavidin-HRP, followed by chemiluminescent detection. Differences between the proteins are presented in the boxes and the arrows indicate precisely these protein fractions.

translation of stress protective proteins increased, whereas under high intensity – mRNAs increased in polyribosome association but translation activity does not increased. Under the influence of osmotic stress, in this study, some proteins were synthesized in vitro more intensively by the total population of polysomes than in the optimal conditions too. Higher translational activity of polysomes during cold acclimation of wheat seedlings and wounding of potato tubers, compared to the control conditions, was also observed by Perras and Sarhan [48] and Morelli et al. [49], respectively. The most important change was the more effectively synthesis of five polypeptides under stress, suggesting that their higher expression was associated with tolerance capacity to this stress. The expression of specific proteins during stress conditions may be determined by presence of IRES (internal ribosome entry site) at mRNAs 5' UTR region. Spriggs et al. [50] described that 10–15% of all mRNAs may be translated by IRES and there is coordinated translation of subsets IRESs during stress. This is probably alternative mechanisms of translation initiation selectively recruit mRNAs to polysomes during cell stress.

Davies et al. [19] described important role of the cytoskeleton in the translation process in plants. They suggested that initiation or elongation of the translation process, or else both of these stages, were faster on the cytoskeleton-bound polysomes (CBP and CMBP) than on polysomes not bound with the cytoskeleton (FP, MBP). Changes in translational activity may be determined by dynamics of the actin cytoskeleton. Depolimerization of actin filaments (e.g. under stress conditions) reduce translational activity, because components of the translational machinery (e.g. polysomes, initiation factors) are dispersed. In turn on, association of translational components with cytoskeleton may prevent diffusion of factors and thus increase translational activity [51]. The highest translational activity under water stress, in this study, was also demonstrated by the CBP. Higher translational activity of the CBP population than that of the FP or MBP fractions in bean tissues had been shown previously by Klyachko et al. [41]. Their results confirmed some subsequent findings reported by Weidner et al. [20], who observed that most of exogenous 14C-amino acids were incorporated, in the embryonic tissue of pea, in the newly formed polypeptide chain, synthesized with the CMBP population. They claimed that the CBP population may play an important role in the synthesis of the so-called stress proteins [20,21]. According to Kosowska et al. [36], there must be some mechanism of regulating translation which is activated in seeds germinating under osmotic stress that enables synthesis of specific proteins despite the total protein synthesis being reduced.

References

1. Howarth CJ, Ougham HJ. Gene expression under temperature stress. New Phytol. 1993;125(1):1–26.
2. Bhatnagar-Mathur P, Rao JS, Vadez V, Sharma KK. Transgenic strategies for improved drought tolerance in legumes of semi-arid tropics. JCRIP. 2009;24(1):92–111.
3. Szilagyi L. Influence of drought on seed yield components in common bean. Bulg J Plant Physiol. 2003;(special issue):320–330.
4. Gepts P. Legumes as a model plant family. Genomics for food and feed report of the cross-legume advances through genomics conference. Plant Physiol. 2005;137(4):1228–1235.
5. Urbano G, López-Jurado M, Frejnagel S, Gómez-Villalva E, Porres JM, Frías J, et al. Nutritional assessment of raw and germinated pea (*Pisum sativum* L.) protein and carbohydrate by in vitro and in vivo techniques. Nutrition. 2005;21(2):230–239.
6. Vidal-Valverde C, Frias J, Hernandez A, Martin-Alvarez

PJ, Sierra I, Rodriguez C, et al. Assessment of nutritional compounds and antinutritional factors in pea (*Pisum sativum*) seeds. J Sci Food Agric. 2003;83(4):298–306.

7. Hsiao TC. Rapid changes in levels of polyribosomes in zea mays in response to water stress. Plant Physiol. 1970;46(2):281–285.
8. Mason HS, Mullet JE, Boyer JS. Polysomes, messenger RNA, and growth in soybean stems during development and water deficit. Plant Physiol. 1988;86(3):725–733.
9. Uesono Y. Transient inhibition of translation initiation by osmotic stress. J Biol Chem. 2002;277(16):13848–13855.
10. Rhodes PR, Matsuda K. Water stress, rapid polyribosome reductions and growth. Plant Physiol. 1976;58(5):631–635.
11. Slaymaker DH, Hoppey CM. Reduced polysome levels and preferential recruitment of a defense gene transcript into polysomes in soybean cells treated with the syringolide elicitor. Plant Sci. 2006;170(1):54–60.
12. Kawiak J, Mirecka J, Olszewska M, Warchoła J. Podstawy cytofizjologii. Warszawa: Polish Scientific Publishers PWN; 1997.
13. Davies E, Fillingham BD, Abe S, Ito Y. Polyribosomes from peas. VII. Initial characterization of a population of polysomes associated with the cytoskeleton. Cell Biol Int Rep. 1991;17:331–340.
14. Davies E, Fillingham B, Ito Y, Abe S. Evidence for the existence of cytoskeleton-bound polysomes in plants. Cell Biol Int Rep. 1991;15(10):973–981.
15. Davies E. Intercellular and intracellular signals and their transduction via the plasma membrane-cytoskeleton interface. Semin Cell Biol. 1993;4(2):139–147.
16. Hesketh JE, Pryme IF. Interaction between mRNA, ribosomes and the cytoskeleton. Biochem J. 1991;277:1–10.
17. Nicchitta CV, Lerner RS, Stephens SB, Dodd RD, Pyhtila B. Pathways for compartmentalizing protein synthesis in eukaryotic cells: the template-partitioning model. Biochem Cell Biol. 2005;83(6):687–695.
18. Adams A, Fey EG, Pike SF, Taylorson CJ, White HA, Rabin BR. Preparation and properties of a complex from rat liver of polyribosomes with components of the cytoskeleton. Biochem J. 1983;216(1):215–226.
19. Davies E, Abe S, Larkins BA, Clore AM, Quatrano RS, Weidner S. The role of the cytoskeleton in protein synthesis. In: Bailey-Serres J, Gallie DR, editors. A look beyond transcription: mechanisms determining mRNA stability and translation in plants. Rockville MD: Amer Society of Plant Physiologists; 1998. p. 115–124.
20. Weidner S, Łukaszewicz D, Amarowicz R. Significant role for polysomes associated with the cytoskeleton in the control of protein synthesis during germination of triticale caryopses in the presence of abscisic acid. Acta Physiol Plant. 2000;22(2):185–193.
21. Weidner S, Frączek E, Romanowska M, Amarowicz R, Abe S, Davies E. The influence of abscisic acid on different polysomal populations in embryonal tissue during pea seeds germination. Acta Physiol Plant. 2003;25(1):5–12.
22. Abe S, Davies E. Isolation of F-actin from pea stems. Protoplasma. 1991;163(1):51–61.
23. Bailey-Serres J. Selective translation of cytoplasmic mRNAs in plants. Trends Plant Sci. 1999;4(4):142–148.
24. Hummel M, Rahmani F, Smeekens S, Hanson J. Sucrose-mediated translational control. Ann Bot. 2009;104(1):1–7.
25. Gong P, Wilke BM, Strozzi E, Fleischmann S. Evaluation and refinement of a continuous seed germination and early seedling growth test for the use in the ecotoxicological assessment of soils. Chemosphere. 2001;44(3):491–500.
26. Davies E, Abe S. Methods for isolation and analysis of polyribosomes. Method Cell Biol. 1995;50:209–222.
27. Abe S, Ito Y, Davies E. Co-sedimentation of actin, tubulin andmembranes in the cytoskeleton fractions from peas and mouse 3T3 cells. J Exp Bot. 1992;43(7):941–949.
28. Bradford MM. A rapid and sensitive method for the quantitation of microgram quantities of protein utilizing the principle of protein-dye binding. Anal Biochem. 1976;72(1-2):248–254.
29. Laemmli UK. Cleavage of structural proteins during the assembly of the head of bacteriophage T4. Nature. 1970;227(5259):680–685.
30. Hamayuni M, Khan SA, Shinowari ZK, Khani AL, Ahmad N, Lee IJ. Effect of polyethylene glycol induced drought stress on physio-hormonal attributes of soybean. Pak J Bot. 2010;42:977–986.
31. Türkan I, Bor M, Özdemir F, Koca H. Differential responses of lipid peroxidation and antioxidants in the leaves of drought-tolerant *P. acutifolius* Gray and drought-sensitive *P. vulgaris* L. subjected to polyethylene glycol mediated water stress. Plant Sci. 2005;168(1):223–231.
32. Brosowska-Arendt W, Weidner S. Effect of osmotic stress on the formation of a population of polysomes and their stability in pea (*Pisum sativum* L.) seeds. Acta Physiol Plant. 2011;33(4):1475–1482.
33. Gill PK, Sharma AD, Singh P, Bhullar SS. Osmotic stress-induced changes in germination, growth and soluble sugar content of *Sorghum bicolor* (L.) moench seeds. Bulg J Plant Physiol. 2002;28:12–25.
34. Gomes FE, Eneas FJ, Tarquínio PJ. Effects of osmotic stress on growth and ribonuclease activity in *Vigna unguiculata* (L.) Walp. seedlings differing in stress tolerance. Rev Bras Fisiol Veg. 1996;8:51–57.
35. Khayatnezhad M, Gholamin R, Jamaati-e-Somarin S, Zabihi-e-Mahmoodabad R. Effects of peg stress on corn cultivars (*Zea mays* L.) at germination stage. World Appl Sci J. 2010;11:504–506.
36. Kosowska M, Frączek E, Amarowicz R, Karama M, Abe S, Weidner S. Water-deficit-induced changes in cytoskeleton-bound and other polysomal populations in embryonic tissue during triticale caryopsis germination. Acta

Physiol Plant. 2004;26(1):67–74.

37. Murillo-Amador B, Lopez-Aguilar R, Kaya C, Larrinaga-Mayoral J, Flores-Hernandez A. Comparative effects of nacl and polyethylene glycol on germination, emergence and seedling growth of cowpea. J Agron Crop Sci. 2002;188(4):235–247.

38. Kawaguchi R, Williams AJ, Bray EA, Bailey-Serres J. Water-deficit-induced translational control in *Nicotiana tabacum*. Plant Cell Environ. 2003;26(2):221–229.

39. Klyachko N. Plant polysome binding to the actin cytoskeleton as a target for physiological regulation. Cell Biol Int. 2003;27(3):217–218.

40. Abe S, Azama K, Sugimoto H, Davies E. Protein accumulation in the maize endosperm: role of polyribosomes and the cytoskeleton. Plant Physiol Biochem. 2003;41(2):125–131.

41. Klyachko N. Interaction of plant polysomes with the actin cytoskeleton. Cell Biol Int. 2000;24(6):351–358.

42. Weidner S, Każarnowicz M, Frączek E, Amarowicz R, Karamać M. Exogenous abscisic acid increases stability of polysomes in embryos of triticale caryopses during germination. Acta Physiol Plant. 2006;28(6):627–634.

43. Dai H, Lo YS, Lin YH, Ruddat M, Chiang KS. In vitro polysome translation analysis of heat shock proteins in higher plants. Bot Bull Acad Sin. 1996;37:261–264.

44. Branco-Price C. Genome-wide analysis of transcript abundance and translation in arabidopsis seedlings subjected to oxygen deprivation. Ann Bot. 2005;96(4):647–660.

45. Kawaguchi R, Girke T, Bray EA, Bailey-Serres J. Differential mRNA translation contributes to gene regulation under non-stress and dehydration stress conditions in *Arabidopsis thaliana*. Plant J. 2004;38(5):823–839.

46. Nicolai M. Large-scale analysis of mrna translation states during sucrose starvation in *Arabidopsis* cells identifies cell proliferation and chromatin structure as targets of translational control. Plant Physiol. 2006;141(2):663–673.

47. Shenton D. Global translational responses to oxidative stress impact upon multiple levels of protein synthesis. J Biol Chem. 2006;281(39):29011–29021.

48. Perras M, Sarhan F. Polysome metabolism during cold acclimation in wheat. Plant Cell Physiol. 1990;31(8):1083–1089.

49. Morelli JK, Shewmaker CK, Vayda ME. Biphasic stimulation of translational activity correlates with induction of translation elongation factor 1 subunit [alpha] upon wounding in potato tubers. Plant Physiol. 1994;106(3):897–903.

50. Spriggs KA, Stoneley M, Bushell M, Willis AE. Reprogramming of translation following cell stress allows IRES-mediated translation to predominate. Biol Cell. 2008;100(1):27–38.

51. Morelli JK. Actin depolymerization affects stress-induced translational activity of potato tuber tissue. Plant Physiol. 1998;116(4):1227–1237.

12

Reduced pollen viability and achene development in *Solidago ×niederederi* Khek from Poland

Grzegorz Migdałek[1]*, Jolanta Kolczyk[2], Artur Pliszko[3], Maria Kościńska-Pająk[2], Aneta Słomka[2]

[1] Institute of Biology, Pedagogical University of Cracow, Podchorążych 2, 30-084 Cracow, Poland

[2] Department of Plant Cytology and Embryology, Institute of Botany, Jagiellonian University, Gronostajowa 9, 30-387 Cracow, Poland

[3] Department of Plant Taxonomy, Phytogeography and Herbarium, Institute of Botany, Jagiellonian University, Kopernika 27, 31-501 Cracow, Poland

Abstract

Pollen and achene characters of natural interspecific hybrid *Solidago ×niederederi* Khek were analyzed and compared with putative parental species *S. virgaurea* L. and *S. canadensis* L. to estimate the level of disturbances in generative reproduction resulting from its hybrid nature. Pollen viability (stainability) of *Solidago ×niederederi* from one newly discovered locality in NE Poland was evidently reduced to ~65% in both viability tests (acetocarmine and Alexander). The diameter of viable pollen (median 21.11 μm) fell between *S. canadensis* (median 19.52 μm) and *S. virgaurea* (median 23.48 μm). Both parental species produced normally developed achenes with high frequency (~90%) whereas in the hybrid, the seed set was dramatically low (6%). The results clearly indicated that sexual reproduction of hybridogenous taxon *S. ×niederederi* is disturbed, and its potential impact as an invasive species depends mainly on vegetative propagation.

Keywords: pollen grains; pollen viability; acetocarmine test; Alexander test; achene size; *Solidago*; hybrid

Introduction

The phenomenon of plant hybridization between alien species and their native congeners that support biotic homogenization significantly reflects global changes in native floras. This process is a major threat to genetic integrity of indigenous plant species [1] and leads to an increase in the number of new taxa which should be characterized as alien plants [2]. The spontaneous occurrence of alien-native hybrids can be very helpful in circumscribing the invasiveness of alien plant species involved in such hybridization. Unfortunately, hybridogenous specimens are often overlooked during the field studies because of the morphological resemblance to their parental species.

In Europe, *Solidago ×niederederi* Khek, a natural hybrid between the alien *S. canadensis* L. and the native *S. virgaurea* L. [3–5] has rarely been recorded in Austria, Denmark, Sweden, Norway, the United Kingdom, Germany, and in Poland [3–13]. Its occurrence is undoubtedly an indirect result of naturalization and invasion of North American *S. canadensis* [14,15]. It is usually intermediate in the shape of inflorescence, the size of the capitulum, and the shape and venation of its leaf [4,5]. It has either restricted fertility, able to produce only a few well-developed achenes [4], or is completely sterile [6]. However, this plant seems to be a long-lived perennial, which forms characteristic clumps of stems like those of *S. canadensis*. There is a need for further research on its vegetative propagation as a way to become established [5]. *Solidago ×niederederi* grows in open disturbed areas (e.g., abandoned arable fields, quarries, clay pits, railway and river embankments), usually together with the parental species [4–12].

The aim of this work is to confirm the hybrid nature of specimens of *S. ×niederederi* collected from a newly discovered locality in Poland [5] using pollen grain and achene characters. We hypothesized that male meiosis in a hybrid between two *Solidago* species that are not closely related, becomes disturbed due to the presence of non-homologous genomes. This leads to significantly decreased pollen viability and seed production as compared to its parental species *S. canadensis* and *S. virgaurea*.

Materials and methods

Origin of plant material

Plants in blooming and fruit development of *S. virgaurea* and *S. ×niederederi* were collected in September 2011 and identified by Dr. Artur Pliszko in a newly found locality in Mieruniszki village (54°10′43″ N, 22°32′04″ E) in the Zachodniosuwalskie Lakeland in NE Poland [5]. The distance between the hybrid plants and parental species did not exceed

* Corresponding author. Email: gmigd@up.krakow.pl

Handling Editor: Beata Zagórska-Marek

two meters. Individuals of *S. canadensis* were collected in the same season in Balice (50°05′10.3″ N, 19°47′01.9″E) near Kraków (southern Poland) and identified by Dr. Maria Kościńska-Pająk. All analyzed *Solidago* taxa formed compact belowground rhizome systems developing many genets. For each taxon, pollen grains and achenes were isolated from dry inflorescences of three plants (probably genets).

Estimation of pollen stainability (viability)

For acetocarmine and Alexander tests, dry pollen grains from 32 flowers of each parental species and from 40 flowers of hybrid *S. ×niederederi* were shed on slides with drops of 1% acetocarmine or Alexander's dye [16]. In the acetocarmine test cytoplasm stains red in viable pollen and remains transparent in nonviable pollen, while in the Alexander test viable pollen grains appear red and nonviable pollen grains stain green. Stamens of each flower isolated from inflorescence were divided into two equal parts. Pollen from each part was stained with acetocarmine or with Alexander.

Pollen grain diameter

Diameter of 500 pollen grains from 32 flowers of each parental species and from 40 flowers of hybrid *S. ×niederederi* was measured under Eclipse E400 optical microscope (Nikon) using NIS Elements (ver. 4.0.) program. Measurements of each grain were performed including exine along polar and equatorial axis.

Achene size

One hundred achenes of each taxon were used to estimate disturbances in achene development. Length of achenes without pappus was determined on 50 normally developed achenes of both parental species and on 6 well developed achenes of hybrid *S. ×niederederi*. Measurements were carried out under Opta-Tech SK series stereoscopic microscope using OptaView (ver. 7.3.1.7.) program.

Scanning electron microscopy (SEM)

Dried pollen grains shed directly from inflorescences were dusted onto stubs with SPI carbon-conductive double-sided adhesive discs, sputtered with gold (SPI SUPPLIES ion-sputtering system), and observed with a scanning electron microscope (HITACHI S-4700).

Statistics/data analysis

Comparison of achene size, pollen size, pollen viability between taxa, and both viability tests were conducted using Kruskal–Wallis ANOVA with Wilcoxon signed rank tests with Bonferroni correction post hoc. Statistical analyses were performed in R program (ver. 3.1.1.).

Results and discussion

Interspecific hybrids, especially between species that are not closely related, usually have disturbed generative reproduction due to abnormalities in male (microsporogenesis) and female (macrosporogenesis) meiosis. Such abnormalities result from the presence of non-homologous chromosomes in the hybrid genome as was observed in several interspecific *Viola* hybrids [17–20].

Disturbances might also accompany female and male gametophyte embryo and endosperm development, leading to reduced seed set or even total sterility [17–22]. Two putative parental species, *S. canadensis* and *S. virgaurea*, which belong to different sections – *Unilaterales* and *Solidago*, respectively (cf. Pliszko[5] and references therein), are diploid, karyologically uniform and have a somatic chromosome number 2n = 18 [23]. *Solidago ×niederederi* described from many localities in Europe [3–13] and found in one locality in Poland [5] is most likely diploid with the same chromosome number 2n = 18 as putative parental species and is partly or completely sterile [4,6]. The hybrid nature of specimens from Poland was estimated from morphological features [5]. To confirm the hybrid origin of specimens newly found in NE Poland site, pollen viability (stainability), diameter, exine sculpture and achene characters were analyzed. The indirect and quick method to estimate meiotic disturbances is pollen viability (stainability) established by histochemical tests [24,25]. Non-stained, degenerated, dwarf, differing in size (very small and giant) pollen clearly indicates abnormal male meiosis [24]. Reduced pollen viability and pollen grains with a wide range of diameter were observed in hybrids between different *Viola* species [17–20]. Frequency of stainable pollen grains is not enough to estimate pollen viability and should be combined with pollen sizes (diameter) because stained pollen grains might be of different size, thus indicating that they are cytologically imbalanced [26].

Based on two different tests, acetocarmine and Alexander, we established that both putative parental species, *S. canadensis* and *S. virgaurea* had highly stainable pollen with median values 98.45% and 98.09%, respectively in acetocarmine test and 99.15% and 99.33% in Alexander staining. Both tests indicated evidently reduced pollen stainability in natural

Tab. 1 Pollen viability (stainability) of *Solidago* taxa by two staining tests.

Taxon	No. of flowers/ No. of pollen grains analyzed	Pollen viability (%)						
		Acetocarmine			Alexander dye			
		min	max	median	min	max	median	*P*
S. canadensis	32/6209	93.80	99.59	98.45[a]	97.66	100.00	99.15[a]	0.084
S. virgaurea	32/8242	94.87	98.89	98.09[a]	96.84	100.00	99.33[a]	0.375
S. ×niederederi	40/4947	52.34	72.26	64.55[b]	36.33	74.71	64.91[b]	0.019

Values bearing different letters along one column differ significantly and indicate differences among taxa at $P = 0.05$. P values are given for tests' comparisons.

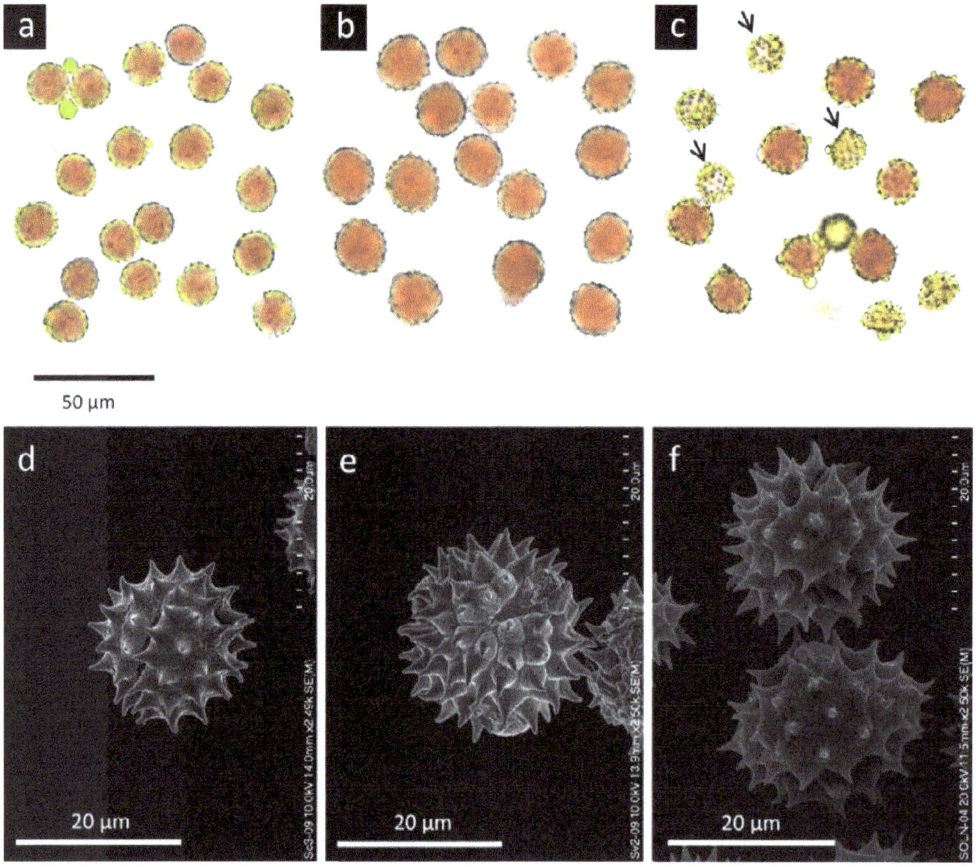

Fig. 1 Pollen grains of *Solidago* taxa stained with acetocarmine (**a–c**) and in the SEM images (**d–f**). *Solidago canadensis* (**a,d**), *S. virgaurea* (**b,e**), *S. ×niederederi* (**c,f**). Non-stainable pollen grains marked with arrows (**c**). Bar in **a** corresponds to **b** and **c**.

hybrid *S. ×niederederi* (median value 64.55% – acetocarmine test, 64.91% – Alexander staining; Tab. 1, Fig. 1a–c). There was a significant difference in pollen size (diameter) between two putative parental species. Pollen grains of *S. canadensis* were smaller (19.52 µm) than *S. virgaurea* (23.48 µm) but similar in exine sculpture as observed in SEM (Tab. 2, Fig. 1a,b,d,e). Hybrid pollen size (diameter) ranged from 15.30 µm to 25.78 µm (median 21.11 µm; Tab. 2, Fig. 1c,f). Pollen grains of all examined taxa observed in SEM were isopolar, 3-aperturate with circular equatorial outline and echinate ornamentation. *Solidago canadensis* grains had sunken apertures, while those of *S. virgaurea* had colporate apertures. Hybrid pollen grains seemed to have rather sunken apertures [27].

In the inflorescences of *S. canadensis* and *S. virgaurea* the frequency of normally developed achenes was very high reaching 90% and 89%, respectively but they differed in size. Achenes of *S. canadensis* reached 1.52 mm, in *S. virgaurea* 3.14 mm (Tab. 3, Fig. 2). Achene development in hybrids was dramatically reduced, in that only 6 achenes of 100 analyzed were normal, while the rest were empty and deformed (Fig. 2d). The length of 6 measured achenes ranged from 1.11 mm to 2.47 mm (Tab. 3).

Tab. 2 Pollen size of *Solidago* taxa.

Taxon	No. of flowers/ No. of pollen grains analyzed	Pollen diameter (µm)		
		min	max	median
S. canadensis	32/500	17.11	23.24	19.52^{a}
S. virgaurea	32/500	19.59	26.61	23.48^{b}
S. ×niederederi	40/500	15.30	25.78	21.11^{a}

Values bearing different letters differ significantly and indicate differences among taxa at $P = 0.05$.

Tab. 3 Achene size of *Solidago* taxa.

Taxon	No. of achenes analyzed/ % of normal achenes	No. of achenes measured	Achene length (mm)		
			min	max	median
S. canadensis	100/90	50	0.99	1.52	1.27^{a}
S. virgaurea	100/89	50	2.43	3.14	2.80^{b}
S. ×niederederi	100/6	6	1.11	2.47	2.15^{c}

Values bearing different letters differ significantly and indicate differences among taxa at $P = 0.05$.

Fig. 2 Achenes without pappus of *Solidago canadensis* (**a**), *S. virgaurea* (**b**) and *S. ×niederederi* (**c,d**); well developed achene (**c**) and empty achene (**d**).

Conclusion

Pollen grain characters (pollen stainability, pollen diameter) and achene development confirmed the hybrid origin of specimens from the newly found locality in Poland and based on morphological characters they could be recognized as *S. ×niederederi*, a natural hybrid between *S. canadensis* and *S. virgaurea*.

Further studies on hybrids from other localities in Poland should be analyzed using similar characters, but also its hybrid origin should be confirmed using molecular markers.

Acknowledgments

Authors would like to thank Professor Elżbieta Kuta from Jagiellonian University for invaluable help in writing this article, for fruitful discussion and valuable suggestions. We are also grateful to Professor Douglas Zook from Boston University for critical reading of the manuscript and English corrections. This work was funded by Jagiellonian University financial support of young scientists (project No. K/DSC/00810).

Authors' contributions

The following declarations about authors' contributions to the research have been made: acquisition, interpretation and documentation of light and scanning electron microscopy data, statistical analysis of data, drafting of the manuscript: GM; acquisition of light microscopy data: JK; collection of plant material, drafting of the manuscript: AP; statistical analyses, critical revision of the manuscript: AS; studies conception, interpretation of data, critical revision of the manuscript: MKP. All authors made substantial contribution to studies conception and design, data acquisition, statistical analysis, drafting the manuscript and its critical revision.

References

1. Daehler CC, Carino DA. Hybridization between native and alien plants and its consequences. In: Lockwood JL, McKinney ML, editors. Biotic homogenization. New York, NY: Kluwer Academic/Plenum Publishers; 2001. p. 81–102.
2. Pyšek P, Richardson DM, Rejmánek M, Webster GL, Williamson M, Kirschner J. Alien plants in checklists and floras: towards better communication between taxonomists and ecologists. Taxon. 2004;53(1):131–143.
3. Khek E. Floristisches aus Ober-Oesterreich. Allg Bot Z Syst. 1905;11(2):21–23.
4. Nilsson A. *Spontana gullrishybrider* (*Solidago canadensis* × *virgaurea* I sverige och Danmark. Sven Bot Tidskr. 1976;70:7–16.
5. Pliszko A. A new locality of *Solidago* ×*niederederi* Khek (Asteraceae) in Poland. Biodiv Res Conserv. 2013;29(1):57–62.
6. Burton R. *Solidago* ×*niederederi* Khek in Britain. Watsonia. 1980;13(2):123–124.
7. Melzer H. Neues zur Flora der Steiermark, XXVI. Mitt Naturwiss Ver Steiermar. 1984;114:245–260.
8. Melzer H. Neues zur Flora der Steiermark, XXVII. Mitt Naturwiss Ver Steiermar. 1985;115:79–93.
9. Leute GH. Neue und bemerkenswerte Pflanzenfunde im Bereich der Landeshauptstadt Klagenfurt in Kärten II. Carinth II. 1988;96:355–396.
10. Melzer H. Neues zur Flora der Steiermark, XXIX. Mitt Naturwiss Ver Steiermar. 1987;117:89–104.
11. Sunding P. Naturalized *Solidago* (golden rod) species in Norway. Blyttia. 1989;47(1):23–27.
12. Leute GH. Neue und bemerkenswerte Pflanzenfunde im Bereich der Landeshauptstadt Klagenfurt in Kärten VI. Carinthia: Rudolphinum Jb Landesmus; 2002.
13. Bleeker W, Schmitz U, Ristow M. Interspecific hybridisation between alien and native plant species in Germany and its consequences for native biodiversity. Biol Conserv. 2007;137(2):248–253.
14. Werner PA, Gross RS, Bradbury IK. The biology of Canadian weeds: 45. *Solidago canadensis* L. Can J Plant Sci. 1980;60(4):1393–1409.
15. Semple J, Cook R. *Solidago* Linnaeus. In: Flora of North America

Committee, editor. Flora of North America. Oxford: Oxford University Press; 2006. p. 107–166.

16. Singh R. Plant cytogenetics. 2nd ed. Boca Raton, FL: CRC Press; 2003.
17. Kuta E. Cyto-embryological studies on the species of the *Viola* L. genus; Nominium Ging. section from the territory of Poland. Fragm Flor Geobot. 1978;24:23–91.
18. Kuta E. Further cyto-embryological studies on *Viola* L., section *Viola* L. Acta Biol Cracov Ser Bot. 1981;23:69–82.
19. Kuta E. Embryological observations on two Canadian species of the *Viola* L. genus (section *Plagiostigma* Godr). Acta Biol Cracov Ser Bot. 1988;30:39–50.
20. Kuta E. Biosystematic studies on the genus *Viola* L.; section *Plagiostigma* Godr. II. Embryological analysis of *V. epipsila* Ledeb.; *V. palustris* L. and their hybrids from Poland. Acta Biol Cracov Ser Bot. 1990;31:46–62.
21. Stace CA. Taksonomia roślin i biosystematyka. Warsaw: Polish Scientific Publishers PWN; 1993.
22. Stuessy TF. Plant taxonomy. New York, NY: Columbia University Press; 2008.
23. Góralski G, Lubczyńska P, Joachimiak AJ. Chromosome number database [Internet]. 2009 [cited 2014 Feb 10].
24. Dafni A, Firmage D. Pollen viability and longevity: practical, ecological and evolutionary implications. In: Dafni A, Hesse M, Pacini E, editors. Pollen and pollination. Vienna: Springer; 2000. p. 113–132.
25. Słomka A, Kawalec P, Kellner K, Jędrzejczyk-Korycińska M, Rostański A, Kuta E. Was reduced pollen viability in *Viola tricolor* L. the result of heavy metal pollution or rather the tests applied? Acta Biol Cracov Ser Bot. 2010;52(1):123–127.
26. Słomka A, Bohdanowicz J, Poznańska E, Kwiatkowska M, Pilarska M, Struebig M, et al. Usefulness and limitations of pollen characters in environmental studies based on *Viola* L. species (sect. *Melanium* Ging.). Mod Phytomorphology. 2014;5:31–34.
27. Halbritter H, Weber M, Zetter R, Frosch-Radivo A, Buchner R, Hesse M. Pollen terminology: an illustrated handbook. Vienna: Society for the Promotion of Palynological Research in Austria; 2008.

13

The uses of *Betula pendula* Roth among Hungarian Csángós and Székelys in Transylvania, Romania

Nóra Papp[1]*, Dóra Czégényi[2], Anita Hegedűs[3], Tamás Morschhauser[4], Cassandra L. Quave[5], Kevin Cianfaglione[6], Andrea Pieroni[7]

[1] Department of Pharmacognosy, University of Pécs, Rókus 2, 7624 Pécs, Hungary

[2] Department of Hungarian Ethnography and Anthropology, University of Babeş-Bolyai of Cluj-Napoca, Horea 31, 400202 Cluj-Napoca, Romania

[3] Department of Languages for Specific Purposes, University of Pécs, Rókus 2, 7624 Pécs, Hungary

[4] Department of Plant Taxonomy and Geobotany, University of Pécs, Ifjúság 6, 7624 Pécs, Hungary

[5] Center for the Study of Human Health, Emory University, 550 Asbury Circle, Candler Library 107, Atlanta, GA 30322, USA

[6] School of Biosciences and Veterinary Medicine, University of Camerino, Via Pontoni 5, 62032 Camerino (MC), Italy

[7] University of Gastronomic Sciences, Piazza Vittorio Emanuele 9, I-12060 Fr. Pollenzo, Bra (CN), Italy

Abstract

The aim of this work is to evaluate the ethnobotanical and ethnopharmacological use of *Betula pendula* Roth in various regions of Transylvania, in East and Central Romania. Silver birch is an important pioneer species in the country, which has traditionally been used in ethnomedicine, households and various customs. Among the ethnic groups, 55 informants from the Csángós and 68 from the Székelys were interviewed in six villages during field studies occurring between 2007 and 2012. Informants were asked questions about the medical and non-medical applications of *B. pendula*, focusing on the use of the leaf and woody parts in tools and customs, and on the collection and use of birch sap with its temporal change in local ethnomedicinal knowledge. Differences related to the use of birch among the studied areas were observed. The folk remedies prepared from the sap ("virics") and leaves were documented in local customs, with similar symbolic roles assigned to birch used in other countries. Sap collection was performed during interviews and by manual means. Four special section types on the trunk were presented, and data concerning the nutritional and ethnomedicinal use of the sap was recorded. Both the significance and use of birch sap have drastically decreased in Transylvania due to the appearance of new phytotherapeutical sources and to social changes caused by migration of young people. We found that this phenomenon is gradually leading to the disappearance traditional harvesting techniques, frequency of use and ethnomedicinal knowledge concerning *B. pendula* in the studied regions.

Keywords: silver birch; ethnobotany; medicinal plant; traditional knowledge; birch sap; virics

Introduction

The silver or white birch (*Betula pendula* Roth), which belongs to the Betulaceae family, is well known as a pioneer species tolerant to cold, and which is prevalent mainly in the boreal zone in Europe [1], Asia and North Africa [2], with a wide ecological amplitude, preferring moist acidic and airy soil [3,4]. Since the Paleolithic Age, the species has been well known for its use as timber, in household tools and in traditional medicine [5,6], and as a magic element in various cultures [7–9].

The word nyír (birch) is one of the oldest (5000–6000 years old) Hungarian words in historical linguistics, originating in the Ural Mountains. From this region, the ancient elements of the use of *B. pendula* have been brought to the Carpathian basin by migration of Hungarians [10]. Silver birch is linked to special names for different regions in Hungary (such as "Nyírség") based on its presence in the local vegetation. Several settlements were also named after the surrounding birch forest in Transylvania, such as "Nyíres", "Nyírfás" [11], "Nyírszeg", "Nyírkert", "Gyakornyír" and "Marosnyír" [12]. In these regions, the traditional Hungarian use of silver birch has remained in the archaic (not current) knowledge of the people, even more so than in Hungary.

The "European Pharmacopoeia VII" [13] includes only the leaf (*Betulae folium*) as the official drug part of *B. pendula* (in Romanian: "mesteacăn"), but the leaf, the bark (*B. cortex*), the bud and its essential oil (*B. gemma*, *B. gemmae aetheroleum*) are also commonly used in the official phytotherapy nowadays [14]. However, the official "Romanian Pharmacopoeia

* Corresponding author. Email: nora.papp@aok.pte.hu

Handling Editor: Łukasz Łuczaj

X" [15] has no current data on the species, but the leaf and the bark are applied in the traditional ethnomedicinal system in the country [16]. According to the earlier reported data on the traditional use of birch, this species plays a significant role in the everyday life of local people in Transylvania. The trunk and woody parts have been used in handicraft tools, as firewood and construction elements [11,12,17]. The leaf and bark have been known as colorant substances in decoctions, for dying linen and wool to yellow, green, brown or grey [11,18,19]. Specifically, the bark has been applied for tanning leathers, for making saltboxes and baskets, and for use as postcard substitutes during World War I [12,20]. In addition, leafy branches have been collected as a protective symbol against lightning [21], and to ornament the portals of churches and houses during weddings, and in front of the houses of unwed young women in May [11].

The birch sap ("succus") circulating in the woody parts before the foliage state from March to May has been described in the European ethnomedicine [22], complete with analyses of some compounds [23–25], but without official data in the present phytotherapy. The sweet sap is called "nyírvíz" (literally birch water), "virics" or "mezge", and has been documented as an ethnomedicinal remedy for various human diseases [6,10,26–30] and in ethnoveterinary medicine [31–33], as well as in hair cosmetics [31–34]. Birch water was fermented or boiled in houses for cooking virics [35], and consumed as wine, beer, syrup or vinegar, and as a soft drink in the region in springtime [14,36–38]. The traditional extraction of sap, namely "viricselés", has been described in only a few reports in Transylvania [11,31,36,39,40]. In the 18th–19th century, the birch sap was sold in barrels as an important product in the Transylvanian markets and served as a significant source of income for the local people [31].

In the present study, the main goal was to collect and document the contemporary and evanescent local practices concerning silver birch in some isolated Transylvanian villages, and to compare these data with the traditional uses of the species in the past. The fieldwork concentrated on the harvesting, use and role of the leaf, trunk and bark, and mostly on the traditional collection method and usage of birch sap.

Material and methods

Research area and population

Selection of the study areas in Transylvania were based on the prevalence of birch-woods, on the availability of pharmacy, medical and veterinary service, and on the living knowledge of the multiple uses of birch among local people. Silver birch is common in low-level humid areas in the mountains all over the country or in the vicinity of turf-moors, especially in secondary succession or near the woodland line. It can be observed mainly after the deforestation of *Picea abies* (L.) Karst. and *Fagus sylvatica* L. [41,42].

Transylvania is a multicultural part of Romania. For centuries, it has been inhabited by various ethnic groups (mainly Hungarian, German and Romanian speaking). The selected Hungarian speaking Székely and Csángó areas represent ethnic, cultural and linguistic islands according to Hammer et al. [43], living in close relationship with the forests and natural surroundings. The selected six villages in the four major areas, namely Úz-valley, Gyimes-valley, Homoród-valley and Erdővidék (Fig. 1) are isolated from major settlements (Tab. 1).

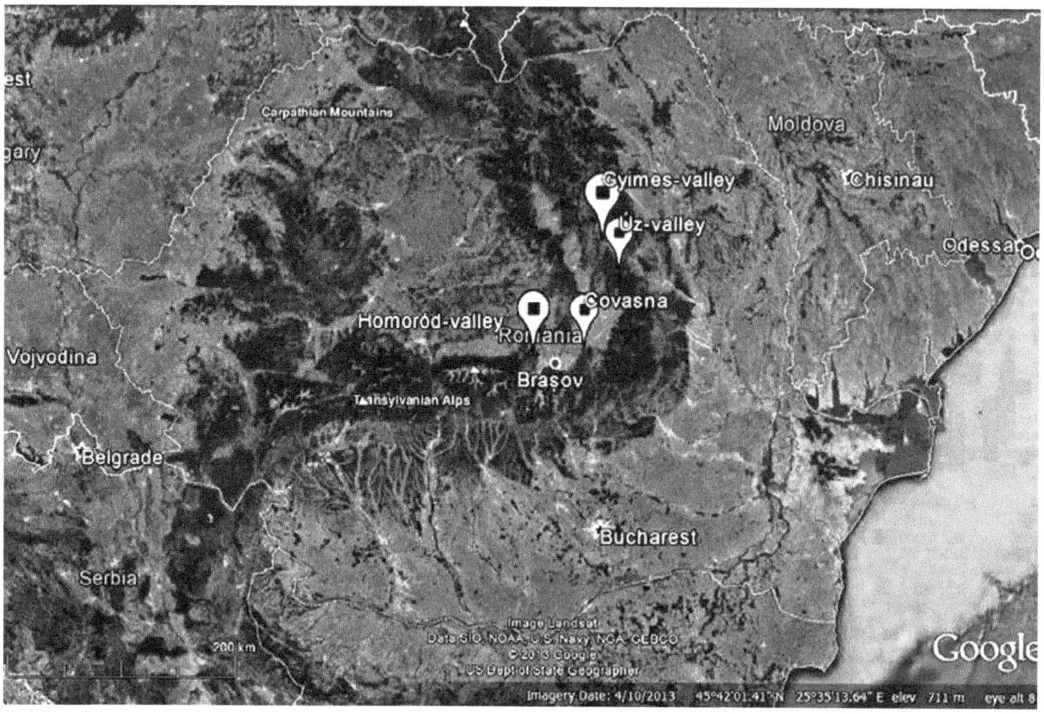

Fig. 1 Map of study areas (Google Earth 5.0, http://earth.google.com)

Tab. 1 Study area in Transylvania.

Study area	Latitude	Longitude	Informants/ Inhabitants	Confession	Medical service	Veterinary practice	Pharmacy
1. Gyimesfelsőlok	46°31′44″	25°57′33″	17/650	Catholic	+	+	+
2. Csinód	46°18′38″	26°03′53″	18/200	Catholic	-	-	-
3. Egerszék	46°20′38″	26°04′24″	20/100	Catholic	-	-	-
4. Lövéte	46°16′27″	25°29′15″	35/2900	Catholic	from a neighbouring settlement	+	+
5. Homoród-karácsonyfalva	46°11′0″	25°25′51″	15/450	Unitarian	from a neighbouring settlement	-	-
6. Nagybacon	46°5′22″	25°41′29″	18/1932	Catholic	+	+	+

Ethnic groups of the study areas. Csángós: Gyimes-valley (1), Úz-valley (2, 3); Székelys: Homoród-valley (4, 5), Erdővidék in Covasna county (6).

Local people use Hungarian as their native language in each village. In the Csángós' region in Úz-valley and Gyimes-valley, local people are subsistence farmers supplementing their income by making local dairy products and homespun clothes, and in special cases, folk costumes. In the case of acute health problems, people in Csinód (in Romanian: Cinod) and Egerszék (Egershec) of Úz-valley take a long trip to the major villages, which are provided with an own pharmacy and primary healthcare. In Gyimesfelsőlok (Lunca de Sus) in Gyimes-valley in eastern Romania, the ancestral knowledge about the native flora is represented by unique magico-mythological methods and several local folk customs in the Csángós' ethnomedicine [38]. In Székely region in southern Transylvania, Homoród-valley consists of several villages along the river Kis-Homoród. Among them, in Lövéte (Lueta), people work in agriculture and in the industry of the neighboring settlements. The Viscum Pharmacy was established in the village in 2008 with many drugs and remedies providing a daily service for the people.

Also, along the river Kis-Homoród, inhabitants of Homoródkarácsonyfalva (Crăciunel) are employed as handworkers, and work in agriculture, livestock farming and apiculture. Specialists acquired wide experience about plants, based on their observations and on inherited data, too.

Inhabitants of Nagybacon (Bătani Mari) in Erdővidék work mainly in industry. The ethnomedicinal elements are completed by methods of official medicine in this village nowadays.

Ethnobotanical collection

Informants from randomly selected households in the study communities were interviewed in the summers of 2007–2011 and in the spring of 2012, using semi-structured and unstructured interviews [44], handwritten notes and photos throughout collecting fieldtrips in the mountains. The interviews were conducted with one person at a time. In questionnaires, the visited 123 informants (Tab. 1) aged between 14 and 99 years (the ratio of female/male informants was 2:1), were asked to list the harvesting methods and time, as well as the uses of the leaf, woody parts and bark of silver birch. The medicinal uses of the plant parts contained some records arising from official sources (e.g. books, journals), which were separated from the traditional data by questions aimed at the source of the knowledge. Only the medicinal data and the use in tools and customs mentioned by more than 75% of the informants were documented in this work.

With respect to the traditional collection of the sap, appropriate details were represented orally in summers in all studied regions, and manually on the trunk of the trees presented by 8 villagers (4 men from both Lövéte and Homoródkarácsonyfalva) throughout fieldtrips in April prior to the foliage period. The documentation of the process comprised the slit form, the height and form of the cut, and the collected quantity, storage and uses of the sap. In addition, people were interviewed about the role of the species in household tools and local ceremonies, and their uses as timber and firewood for non-medicinal purposes. The original quotations of the people were written in italics between inverted commas according to the folk terminology of Csángós and Székelys. Voucher specimens collected in each village were deposited at the Department of Pharmacognosy of the University of Pécs, Hungary.

Results

Birch leaf and bark use in ethnomedicine

In all regions, the ethnomedicinal use of silver birch was mentioned mostly by women, related to their activities focusing principally on the households. Among the parts of birch, in preparations of the leaf and the bark regional variations were observed (Tab. 2). The fresh or dry leaf was used by itself or blended with other species in various home remedies applied internally and externally, too. In the Csángós' region in Gyimes-valley and Úz-valley, the fresh collected leaves were applied as a diaphoretic foment spread on the bedding. People put their painful legs into a sac for the night or into trousers filled with birch leaves during the day, or they lie among them in bed to treat mostly rheumatisms, generally four or five times daily for some hours, or all

Tab. 2 Traditional use of *Betula pendula* Roth in the studied villages and previous reports in Transylvania, and in other European countries.

	Parts used and application forms in the studied Transylvanian villages	Previous reports in Transylvania	Data from other European countries
Medicinal use	Leaf: for cold as a foment; for rheumatism and arthritis as a bath and foment (Gy) and in bed or in a sack as a foment; as a diuretic (Cs, L, H); for varicose veins in trousers and for waist pain as a foment; for kidney stones with *Alnus glutinosa*, *Origanum vulgare* and *Equisetum arvense* as a tea; for swollen legs with *Achillea millefolium* and *Lysimachia nummularia* for wounds (wound healing) chilblain, heart and liver disease; for flatulence as a tea (L), for renal pain as tea or bath (Cs); for sweating of the legs (L, H, N). Bark: for wounds (N). Sap: as a refreshing beverage earlier (Gy, H, N); for kidney disease (Cs, H); for gall stones (E); as an appetite stimulant, for stomach and liver disease; for colds (E, L); for chilblain as a foment (L).	Leaf: decoctions containing only birch leaves [18,49,50]; for rheumatism as a foment in bed and sack [11,45–48]; for stroke with garlic and for cold as a tea [47]; as a diaphoretic [61]. Bark: in veterinary practice for diarrhoea and dysentery [61]. Sap: drained from the trunk in spring [26,31,36,81]; sold in the markets [31]; vermifuge; as a milk-rennet drug in cheese [26,27]; for colds [11,12]; for anaemia [82]; for constipation and pneumonia; for a fungal skin disease called semereg or ebsemereg [12,28,29]; for abscess and pimples [30]; for herpes called ebsömöreg [8]; as a vermifuge [36,59,82]. Tar: for psoriasis and eczema as liniment, tincture and soap [59,60].	Leaf: mostly in single remedies; among ancient data: for wounds [7]; for painful body parts [51,52]; as diuretic tea (Pieroni, unpublished) [55,56,62]; as a cholagogue, for gout, rheumatism, cold [55,56,62], fever and detoxication [55,56]; to treat urinary tract infections [54–56] and kidney stone [54]; in veterinary for diarrhoea in cows, for arthritis in suint [62]. Branches: internally: rain, which has gone through a birch tree, is collected and drunk to treat kidney stones and prostatitis; externally: beaten on the affected part to treat bruises [83]. Bark: for hemorrhoids (Pieroni, unpublished); for cold, fever, rheumatisms and purifying blood [55,56]; as a diuretic; for urinary tract infections [55–58]; externally: burned, the vapours are exposed to the skin to treat skin inflammations (Pieroni, unpublished). Sap: drained from the trunk in spring [63,64]; as a milk-rennet drug in cheese [77,78]; for children with milk in the teething period, for sores on the lower legs of humans [65].
Use in household tools	Branches (Gy, Cs, E, L, H, N) and trunk (H) for broom; trunk in shaft, carriage and as firewood (L), and in tools (H).	Branches: for broom [11,46].	Branches: for broom [7,79], in torches [7].
Use in customs	Leafy branches: from Csíksomlyó blessed at Whitsunday and burnt during a storm for protection (Gy, Cs, E, H), blessed at Easter and Whitsunday, as an ornament at weddings and on May 1st for maidens (L, H), thrown in the fire during prayer (Cs).	Leafy branches: erected on May 1st for maidens [11].	Leafy branches: used in wreath ornaments; apotropaic role against witches and bad spirits [7]; sacred and magic tree in rituals [6,7]; to make magistrates' beams "that everyone fears" [7]; erected on May 1st as a gift for unwed young women [7,72] (Pieroni, unpublished data); used in herbal wreaths [84].
Other uses	Leaf: for hair loss (Gy, L), and for dyeing as a decoction (Gy, N).	Leaf: for dyeing as a decoction [11,53]. Sap: for sunspots [31], as a hair colorant with sugar [31–33], as a hair conditioner [34], as a beverage, wine, beer, syrup or vinegar [16,18,36–38].	Leaf: as a surrogate of tobacco [6], for hair loss and dandruff [54]. Sap: as a hair conditioner [85], for hair growth as rubber recently [65]; eventually sugar and raisins are added to the fresh sap, and fermented in a cool and dark place for two months - the resulting liquid is drunk and considered very "healthy" (Pieroni, unpublished); as a beverage, wine, beer, syrup [65–74] and vinegar [65–76], earlier as a fresh or fermented drink with malt and wax, wine, syrup, vinegar and confectionery; in cosmetics in lotions and shampoos; the fresh or pasteurized beverage with lemon juice, sugar and raisins; sold in shops nowadays [65].

Cs – Csinód; E – Egerszék; Gy – Gyimesfelsőlok; H – Homoródkarácsonyfalva; L – Lövéte; N – Nagybacon.

night. In addition, informants consider that this treatment is inadequate for sensitive patients, because "it can drain the strength from the muscles", causing asthenia and fatigue. In Homoród-valley in Lövéte and Homoródkarácsonyfalva, besides the independent application of the leaves, there were also reports of the use of birch leaves in combination with other taxa, such as with the herb of "egérfarkú fű" (*Achillea millefolium* L.), "piculavirág" (*Lysimachia nummularia* L.), "ezerjófű" (*Origanum vulgare* L.) and "surlófű" (*Equisetum arvense* L.), and with the leaf of "egörfa" [*Alnus glutinosa* (L.) Gaertn.] for different ailments. In Nagybacon, the fresh leaves were applied to promote heavy sweating of the legs, in the same way as in Homoród. Application of the bark was mentioned only in Nagybacon, applied to wounds as an anti-inflammatory drug, used mostly in captivity during World War II (Tab. 2).

Collection and use of birch sap

In contrast to the role of women as the main healers, men have knowledge about the growth habits, distribution and phenology of silver birch in the region. On a regional scale in Transylvania, special data were recorded on birch water (virics) and on the collecting method of the sap (viricselés). Besides the interviewed women who had no skills and knowledge about the collection, only 8 men could present the methods. The frequency of the extraction and the use of the sap is declining continuously in these days, because its consumption is replaced mostly by various refreshing beverages available in shops, and similar healing effects are achieved through the use of official remedies described in books.

In this process, young trees are bored or slit 3–4 times on average in the same spring, at about a height of one meter and mostly on the shady side of the mildly curved trunks, or on the lateral branches close to the trunk.

In Gyimes-valley, the trunk was bored in the lower part of the tree, and then an iron or copper tube was inserted in the hole, allowing the sap to flow into a jar or a pottery called "csipor". In Úz-valley, the virics was extracted by the section or boring of the trunk after breaking the thick branches.

In Székely regions, young trees were slit at the base or at the height of about one and a half meters by an axe or knife called a bicsok, resulting in the outflow of the sap from various sections. In Homoród-valley, altogether four section types were described during the collection, according to the recorded oral and manual presentations in forests (Fig. 2). Firstly, the bark can be slit in the form of "T" or "↓", 1 centimeter deep, and 15 centimeters long vertically and horizontally (Fig. 2a–b). Then the extreme parts of the slit were bent up or down beside the central section, allowing the sap to leak into a jar with the help of a funnel. In the next type, the trunk was incised in a V-form "windowy" in two directions transversally (Fig. 1c), where the sap was let out through the bottom peak of the slit, in the same way as documented in Úz-valley (Fig. 3).

In the last section type, the sap can accumulate in a deep hole, namely "lik", "göbü" or "hajk" (their preparations are called göbüzés and hajkolás), similarly to the oral data in Gyimes-valley (Fig. 2d, Fig. 4). From these holes, the sap can be drained through a hollow plant stem as in the case of *Phragmites* taxa, or collected in a jar (Fig. 5) with or without a funnel made from birch bark.

Based on the age of the tree and weather conditions, virics was observed in older trees visible as drops at the beginning of the spring, and as a continuous flow in the younger ones in the main period of the season, yielding 3–4 deciliters, sometimes 1–2 liters of fluid daily in three described section types (Fig. 2a–c), and about one deciliter in göbü (Fig. 2d). After collection, the scar of the holes and sections were smeared with mud or clay to "heal" the trunk of the tree.

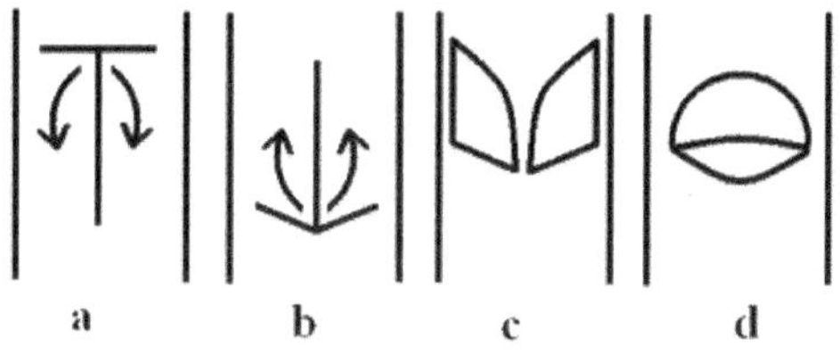

Fig. 2 Section types of the collection of birch sap on the trunk. **a** "T" type. **b** "↓" type. Arrows show the direction of the bending of the bark. **c** V-form type. **d** Göbü.

Fig. 3 V-form section in the collection of sap.

Fig. 4 Göbü in the collection of sap.

Fig. 5 Collection of sap.

The collected sap can be stored in a cool cellar or in a fridge in covered jars for two weeks, and used before it becomes cloudy. According to the informants in Homoród-valley, its consumption is healthy during the whole summer, although the excessive use leads to headache. Most common uses of virics include a drink with curative power internally, and a foment externally in all studied villages (Tab. 2).

Silver birch in household tools

Similarly to the collection of birch sap, the use of other plant parts in tools (or tool construction) was presented mostly by men (Tab. 2). In Homoród-valley, the trunk and the thick branches were used as popular firewood because of the easily ignitable bark with high calorific value, and in household tools, such as in wooden forks, shafts, shovels or axes. It was especially common in two types of broom; about 15 thin and long wattles were bound by a ring called "káva" made from the bark pelt of "magyarófa" (*Corylus avellana* L.) in the first type (Fig. 6), and short twigs were tied on a thick lateral branch called "csutakseprű" in the second one (Fig. 7). About 30–40 years ago, brooms were sold for various industries, presenting a significant livelihood activity for the local people.

Silver birch in customs

The use of *B. pendula* in local ceremonies and customs was listed both by women and men in each village. Csángós of Gyimes walk round the house with a birch broom three times at the night of St. George day (April 24th), to avoid reptiles getting into the house. During daytime, Csángó people of Úz-valley have stored the birch broom in the corner behind the door, but exclusively in the court at night as a rigorous custom in the region.

The blessing of the leafy branches is considered as a widespread traditional ceremony in Transylvania. People go on pilgrimage to Csíksomlyó at Whitsunday, praying to the Blessed Virgin at the famous catholic church of the settlement. During this trip, birch branches are blessed and taken home to say the Lord's Prayer and Ave Maria according to the number leaves on the branches. Moreover, in Csángós' regions, the blessed branches are thrown and burnt in the fire as protection from a storm, accompanied by a prayer repeated three times and by a special magic formula used as a protective symbol.

In the Transylvanian ceremonies, the leafy branches of silver birch, called "ződág" (green branch), play an important role in the ornamentation of crucifixes at roadsides, bridges and houses in weddings, and churches on the Lord's Day, at Christmas time and during confirmation. In the Homoród-valley, two branches are placed on both sides of the doorway of wedding-houses, ornamented by a bouquet for the bride, and by a jar of wine for the bridegroom. In winter, wedding-houses are decorated with the branches of spruces instead of birch. On May 1st, 2–5 meter long birch branches are erected by young men in front of the houses of unwed young women as a special gift for a week.

Discussion

The parts of silver birch were documented in various application forms in home treatments, tools and customs in the studied Csángó and Székely villages, and the data were compared with earlier records in Transylvania and some records from other European countries.

Leaf decoctions were documented not only as a single remedy, but also blended with other taxa, in contrast to the earlier Transylvanian data or records in other countries (Tab. 2). Among the ethnomedicinal use of the leaves for rheumatism, they were used in trousers, and in bed or sacks similarly to earlier data obtained in Transylvania [9,45–50], and to its use as an analgesic in Poland [51,52]. However, while the use of leaf decoctions for dyeing and for hair loss in Transylvania was reported recently and earlier [9,53], as well as in Montenegro [54], the current data for flatulence and for sweating of the legs were not found in literature sources (Tab. 2). The external use of bark for skin disorders was also recorded in Albania and Macedonia (Pieroni, unpublished), while its internal use was mentioned in Bosnia and Herzegovina and in Kosovo [55–58]. In our work, no data were recorded on the use of the tar (*Betulae pix*), in contrast to earlier reports in the country [59,60]. For diarrhoea in veterinary practice, the decoction of the bark was mentioned earlier in Transylvania [61], while the use of the leaf for the same purpose was also reported in Italy [62].

The harvesting of birch water was described in several previous reports in Transylvania, and in Germany and Russia [63,64]. In the 18th–19th century, birch sap (virics) was harvested regularly and sold in the markets, playing an important role in the livelihood of the rural classes [31]. Nowadays, traditional collection of birch sap called viricselés appears as an underrepresented and evanescent phenomenon, including mostly oral and little manual demonstration of the detailed method. During the collection, four special section types of the trunk were documented in the studied Transylvanian areas. The birch water was used in itself or with milk [65], both externally and internally (Tab. 2). It was also used as a beverage, as wine and beer, similarly to some earlier reports in the country [14,36]. As a novel data, informants have mentioned headache due to the excessive use of the sap in Homoród. The previous record for its use as vinegar (Tab. 2) was also found in Sweden, Estonia, Poland, Slovakia, Ukraine, Russia, Britain and Ireland [65–78], without any record of this documented in our field work. The significance of silver birch as a milk-rennet plant was recorded only in previous reports in the country [26,27], and in the Netherlands and Finland [77,78].

Among the uses of the leafy branches in tools, the preparation of brooms was commonly documented in each settlement (Tab. 2) similarly to some earlier records [5,9,61,79]. The common use of the woody parts as firewood was also recorded because of the easily ignitable bark of the species, while its use as a torch was reported in other countries [5]. In addition, branches were used in special protective rituals in the studied Csángós' areas on St. George's day (April 24th), and in ceremonies such as ződág in the Székelys region,

Fig. 6 First type of brooms from birch branches.

Fig. 7 Second type of brooms (csutak) from birch branches.

corresponding to the previous work in Kalotaszeg [11], and to earlier data from Netherlands, Germany, Belgium, north-eastern Italy, Central Europe and South Western Balkans [5,80] (Pieroni, unpublished data). These customs can be described as a disappearing phenomenon in Transylvania nowadays.

The blessed leafy branches are used at Whitsunday or in weddings in the studied settlements, which are ornamented with a bouquet or a jar in Homoród, and with wreaths in the Russian folklore. The protective role of the branches against bad weather (storms) was recorded mostly in the Csángó region, while against bad spirits in Russia and Sweden [7].

Conclusions

Although the traditional knowledge documented in this data is based on the memories of the ancestors and on the personal experiences of the people, a relatively low proportion of the traditional knowledge about *B. pendula* was observed among the studied ethnic groups of Csángós and Székelys from an agricultural, ethnobotanical and ethnopharmacological point of view. In addition to the existing data concerning the use of the leaves and woody parts, the use of the tar was observed as a disappeared element. Our findings add new values about birch water (virics) to the observations described earlier, but this also indicates that its consumption and multiple use has been progressively lost, similarly to the old harvesting methods in Transylvania and all over in Romania.

The ethnobotanical and ethnomedicinal data display a declining tendency in the region from day to day, largely influenced by migration trends and the employment of youth far from their ancestral lands. Moreover, the increasingly frequent use of phytotherapeutic books and other outside sources also influences the traditional knowledge of people, which is incorporated in the daily medical practices of people. Consequently, collection fieldtrips in the ethnobotanically unexplored and isolated regions of Transylvania are of primary importance nowadays, and should focus on the maintenance of ancestral harvesting techniques and ethnomedicinal elements. Such studies can also be useful for informing future phytotherapeutical research on key local ethnomedicinal resources, such as birch sap.

Acknowledgments

This work was supported by the OTKA (Hungarian Scientific Research Fund) grant (PD 108534). We are grateful for the help of the informants of the villages, and for Tamás Grynaeus†, Kata Birkás-Frendl, Mária and Mihály Péter for the instructions, and for Imre Gábos, János Gábos, Géza Szentpáli, Enikő and Mihály Benedek, as well as Judit Laki and Veronika Martos for their assistance.

Authors' contributions

The following declarations about authors' contributions to the research have been made: initiation of the work, field work and the first version of the paper: NP; literature survey: NP, DC, TM, KC, AP; participation in the writing process: DC, AH, TM, CLQ; helping to evaluate the results and discussion: AP.

References

1. Tutin TG, Burges NA, Charter AO, Edmondson JR, Heywood VH, Moore DM, et al., editors. Flora Europaea. Cambridge: Cambridge University Press; 2010. (vol 1).
2. Szabó LG. Gyógynövény-ismereti tájékoztató. Baksa: Schmidt und Co.; 2005.
3. Alejano R, Martinez Montes E, Bengoa J. *Betula pendula* en la Sierra de las Villas. Ecologia. 1993;7:203–213.
4. Gencsi L, Vancsura R. Dendrológia. Budapest: Mezőgazda; 1992.
5. Bock H. Kräuterbuch. Straßburg: 1539.
6. de Cles B, Bellomaria B, Berdini C, Pedrotti F, da Trieste F, de Cles M. Le ricette fitoterapeutiche del barone Bernardo de Cles. Camerino: Università di Camerino; 2000.
7. Cleene M, Lejeune MC. Compendium of symbolic and ritual plants in Europe: herbs. Ghent: Man & Culture; 2002. (vol 1).
8. Rácz J. Növénynevek enciklopédiája: az elnevezések eredete, a növények kultúrtörténete és élettani hatása. Budapest: Tinta; 2010.
9. Plinius GS. Naturalis historia [reprint]. Budapest: Natura; 1987.
10. Sandor K. Nyelvrokonsag es hunhagyomany. Budapest: Typotex; 2011.
11. Péntek J, Szabó A. Ember és növényvilág: kalotaszeg növényzete és népi növényismerete. Bucharest: Kriterion; 1985.
12. Tarisznyás M. A gyűjtögető gazdálkodás hagyományai Gyergyóban. In: Kós K, Faragó J, editors. Népismereti dolgozatok. Bucharest: Kriterion; 1978. p. 25–33.
13. European Pharmacopoeia. 7th ed. Strasbourg: Council of Europe; 2010.
14. Demirci B, Paper DH, Demirci F, Can Başer KH, Franz G. Essential oil of *Betula pendula* Roth. buds. Evid Based Complement Alternat Med. 2004;1(3):301–303.
15. Romanian Pharmacopoeia X. Bucharest: Medicală; 1993.
16. Borza A, Beldie A. Dicţionar etnobotanic. Bucharest: Editura Academiei Republicii Socialiste România; 1968.
17. Gazda K. Gyermekvilág Esztelneken: néprajzi monográfia. Bucharest: Kriterion; 1980.
18. Butură V. Enciclopedie de etnobotanică românească. Bucharest: Editura Ştiinţifică şi Enciclopedică; 1979.
19. Boris G. Gyógynövények népi használata a székelyföldi Lövétén [BSc thesis]. Pécs: Univeristy of Pécs; 2010.
20. Péntek J, Szabó TA. Egy háromszéki falu népi növényismerete. Ethnographia. 1976;87(1–2):203–225.
21. Szendrey Z. A növény-, állat- és ásványvilág a varázslatokban. Ethnographia. 1937;48(1):154–166.
22. Mayer JG, Saum K, Uehleke B. Kolostori gyógyászat. Budapest: Magyar Könyvklub; 2004.
23. Kallio H, Karppinen T, Holmbom B. Concentration of birch sap by reverse osmosis. J Food Sci. 1985;50(5):1330–1332.
24. Kallio H, Rine S, Pangborn RM, Jennings W. Effect of heating on the headspace volatiles of Finnish birch syrup. Food Chem. 1987;24(4):287–299.
25. Kallio H, Ahtonen S. Seasonal variations of the acids in birch sap. Food Chem. 1987;25(4):285–292.
26. Melius P. Herbárium. Bucharest: Kriterion; 1979.
27. Gunda B. Tejoltó növények a Kárpátokban. Ethnographia. 1967;78:161–175.
28. Halászné ZK. Adatok a moldvai magyarok gyógynövény-használatához. Gyógyszerészet. 1981;25:361–367.
29. Zillmann J. Népi gyógyászat Havadon. Néprajzi Látóhatár. 1997;1–2:124.

30. Keszeg V. A mezőségi Detrehemtelep népi gyógyászata. In: Népismereti dolgozatok. Bucharest: Kriterion; 1981. p. 97–117.

31. Györffy I. Viricselés a Székelyföldön. Ethnographia. 1937;48(2-3):205-220.

32. Kalašaitytė L. Liaudies medicinos priemonės. In: Vėlius N, editor. Gervėčiai. Vilnius: Mintis; 1989. p. 204–219.

33. Brøndegaard VJ. Folk og flora: Dansk etnobotanik. Copenhagen: Rosenkilde og Bagger; 1978. (vol 1-2).

34. Rácz G, Füzi J, editors. Kovászna megye gyógynövényei. Sfântu Gheorghe: Directory of Agriculture, Food and Hydrography; 1973.

35. Gunda B. Nyírvíz (szócikk). In: Magyar Néprajzi Lexikon (N-Szé). Budapest: Akadémiai Kiadó; 1987. p. 62. (vol 4).

36. Kiss L. A nyírvíz. Földgömb. 1929;1:1-6.

37. Gunda B. Irodalmi szemle. Ethnographia. 1941;52(1):154–155.

38. Rab J, Tankó P, Tankó M. Népi növényismeret Gyimesbükkön. In: János VA, Károly K, József F, editors. Népismereti dolgozatok. Bucharest: Kriterion; 1981. p. 23-38.

39. Szebeni G. A csíki juhászat. Ethnographia. 1972;73(1):54–89.

40. Bartos T. Magyar szótár: egymást magyarázó szavak és fordulatok tára. Budapest: Corvina; 2002. (vol 1).

41. Rab J. Az etnogeobotanika mint történeti ökológiai segédtudomány. In: Várkonyi RA, Kósa L, editors. Európa híres kertje. Budapest: Orpheusz; 1993. p. 223-257.

42. Babai D, Molnár Z. Népi növényzetismeret Gyimesben II.: Termőhely- és élőhelyismeret. Bot Közl. 2009;96(1-2):145-173.

43. Hammer K, Laghetti G, Pignone D, editors. Linguistic islands and plant genetic resources - the case of the Arbëreshë. Rome: Aracne; 2011.

44. Newing H. Conducting research in conservation: social science methods and practice. New York, NY: Routledge; 2011.

45. Kóczián G, Pintér I, Gál M, Szabó I, Szabó L. Etnobotanikai adatok Gyimesvölgyéből. Bot Közl. 1976;63(1):29-35.

46. Gub J. Adatok a Nagy-Homoród és a Nagy-Küküllő közötti terület népi növényismeretéhez. Néprajzi Látóhatár. 1993;1-2:95-110.

47. Gub J. Népi növényismeret a Nagy-Homoród mentén. In: Zsigmond G, editor. Növények a folklórban. Bucharest: A Magyar Köztársaság Kulturális Intézete; 2005. p. 148-162.

48. Szabó LG. Népi gyógynövény-ismeret Kalotaszegen és Gyimesvölgyében. Turán. 2002;5(4):39-52.

49. Vajkai A. Népi orvoslás a Borsavölgyében. Cluj Napoca: Erdélyi Tudományos Intézet; 1943.

50. Rab J. Újabb népgyógyászati adatok Gyimesből. Gyógyszerészet. 1982;26:325-333.

51. Kukier R. Kaszubi bytowscy: zarys monografii etnograficznej. Gdynia: Wydawnictwo Morskie; 1968.

52. Paluch A. Świat roślin w tradycyjnych praktykach leczniczych wsi polskiej. Wrocław: Wrocław University Press; 1984.

53. Miklóssy VV. Festőnövények a csíki háziiparban. In: Népismereti dolgozatok. Bucharest: Kriterion; 1978. p. 91-100.

54. Menković N, Šavikin K, Tasić S, Zdunić G, Stešević D, Milosavljević S, et al. Ethnobotanical study on traditional uses of wild medicinal plants in Prokletije Mountains (Montenegro). J Ethnopharmacol. 2011;133(1):97-107.

55. Šarić-Kundalić B, Dobeš C, Klatte-Asselmeyer V, Saukel J. Ethnobotanical study on medicinal use of wild and cultivated plants in middle, south and west Bosnia and Herzegovina. J Ethnopharmacol. 2010;131(1):33-55.

56. Šarić-Kundalić B, Dobeš C, Klatte-Asselmeyer V, Saukel J. Ethnobotanical survey of traditionally used plants in human therapy of east, north and north-east Bosnia and Herzegovina. J Ethnopharmacol. 2011;133(3):1051-1076.

57. Mustafa B, Hajdari A, Krasniqi F, Hoxha E, Ademi H, Quave CL, et al. Medical ethnobotany of the Albanian Alps in Kosovo. J Ethnobiol Ethnomed. 2012;8(1):6.

58. Mustafa B, Hajdari A, Pajazita Q, Syla B, Quave CL, Pieroni A. An ethnobotanical survey of the Gollak region, Kosovo. Genet Resour Crop Evol. 2012;59(5):739-754.

59. Csedő K, editor. Plantele medicinale şi condimentare din judeţul Harghita. Hargita megye gyógy- és fűszernövényei. Târgu Mureş: Tipografia Tirgu Mures; 1980.

60. Rácz G, Rácz-Kotilla E, Laza A. Gyógynövényismeret. Bucharest: Ceres; 1984.

61. Kóczián G, Pintér I, Szabó LG. Adatok a gyimesi csángók népi gyógyászatához. Gyógyszerészet. 1975;19:226-230.

62. Guarrera P. Usi e tradizioni della flora italiana: medicina popolare ed etnobotanica. Rome: Aracne; 2006.

63. Dieck AC. Die Birke (*Betula* L.) in der Volksmedizin Europas. Curare. 1985;3:85-94.

64. Zyryanova OA, Terazawa M, Koike T, Zyryanov VI. White birch trees as resource species of Russia: their distribution, ecophysiological features, multiple utilizations. Eurasian J Res. 2010;13:25-40.

65. Svanberg I, Sõukand R, Łuczaj Ł, Kalle R, Zyryanova O, Dénes A, et al. Uses of tree saps in northern and eastern parts of Europe. Acta Soc Bot Pol. 2012;81(4):343-357.

66. Cronquist AW. Illustrerad ordbok öfver näringsämnen och handelsartiklar. Stockholm: Hæggström; 1879.

67. Berg G. Nordskandinaviskt – nordeuropeiskt. Rig. 1933;16:118-139.

68. Berg G. Att tappa björklake. Gastron Kal. 1969:35-57.

69. Vuorela T. Suomensukuiset kansat. Helsinki: Suomalaisen Kirjallisuuden Seura; 1960.

70. Markus M. Ľudova stravá. In: Mjartan J, editor. Banícka dedina Žakarovce. Bratislava: Vydavateľstvo Slovenskej Akadémie; 1956. p. 283-324.

71. Rebane H. Voolab kasemahl. Eesti Lood. 1977;4:223-228.

72. Toren MD. Russkaya narodnaya meditsina i psikhoterapiya. St. Petersburg: Litera; 1996.

73. Allen DE, Hatfield G. Medicinal plants in folk tradition: an ethnobotany of Britain & Ireland. Portland, OR: Timber Press; 2004.

74. Łuczaj Ł. Archival data on wild food plants used in Poland in 1948. J Ethnobiol Ethnomed. 2008;4(1):4.

75. Fischerström J. Nya Swenska economiska dictionnairen. Eller Forsook til ett Allmänt och fullständigt lexicon i Swenska hushållningen och naturläran. Stockholm: Carl Stolpe; 1780. (vol 2).

76. Hupel AW. Topographische Nachrichten von Lief- und Ehstland: Nebst vollständigen Register über alle drey Bände. Riga: JF Hartknoch; 1777.

77. Hildebrand S. En holländsk beskicknings resa. Stockholm: Sune Hildegard; 1917.

78. Lange J. Løvetræer, in Kulturhistoriskt lexikon för nordisk medeltid. Malmö: Allhem; 1966. (vol 11).

79. Nedelcheva AM, Dogan Y, Guarrera P. Plants traditionally used to make brooms in several European countries. J Ethnobiol Ethnomed. 2007;3(1):20.

80. Hoffmann-Krayer E, Bächtold-Stäubli H. Handwörterbuch des deutschen Aberglauben. Berlin: Walter de Gruyter; 1927.

81. Feszt G, Rácz-Kotilla E, Fórika M, Pálffy B. A nyírfa tavaszi nedvének hatása vesekövekre "in vitro" és kísérletes nephrolithiasisban. Orvosi Szle. 1966;12(3):285-287.

82. Tita I, Mogosanu GD, Tita MG. Ethnobotanical inventory of medicinal plants from the south-west of Romania. Farmacia. 2009;57(82):141-156.

83. Pieroni A, Giusti ME, Quave CL. Cross-cultural ethnobiology in the Western Balkans: medical ethnobotany and ethnozoology among Albanians and Serbs in the Pešter plateau, Sandžak, south-western Serbia. Hum Ecol. 2011;39(3):333–349.

84. Łuczaj Ł. A relic of medieval folklore: Corpus Christi Octave herbal wreaths in Poland and their relationship with the local pharmacopoeia. J Ethnopharmacol. 2012;142(1):228–240.

85. Krylov GV, Stepanov EV. Zelenaya apteka Kuzbassa. 2nd ed. Kemerovo: Kemerovskoe knizhnoe izdatel'stvo; 1979.

14

Extracellular matrix surface network is associated with non-morphogenic calli of *Helianthus tuberosus* cv. Albik produced from various explants

Maria Pilarska[1], Marzena Popielarska-Konieczna[1]*, Halina Ślesak[1], Małgorzata Kozieradzka-Kiszkurno[2], Grzegorz Góralski[1], Robert Konieczny[1], Jerzy Bohdanowicz[2], Elżbieta Kuta[1]

[1] Department of Plant Cytology and Embryology, Jagiellonian University, Gronostajowa 9, 30-387 Cracow, Poland

[2] Department of Plant Cytology and Embryology, University of Gdańsk, Wita Stwosza 59, 80-308 Gdańsk, Poland

Abstract

Helianthus tuberosus is economically important species. To improve characters of this energetic plant via genetic modification, production of callus tissue and plant regeneration are the first steps. A new, potentially energetic cultivar Albik was used in this study to test callus induction and regeneration. Callus was produced on leaves, petioles, apical meristems and stems from field-harvested plants but was totally non-morphogenic. Its induction started in the cortex and vascular bundles as confirmed by histological analysis. The surface of heterogeneous callus was partially covered with a membranous extracellular matrix surface network visible in scanning and transmission electron microscopies. The results clearly indicate that: (***i***) the morphogenic capacity of callus in topinambur is genotype dependent, (***ii***) cv. Albik of *H. tuberosus* proved recalcitrant in in vitro regeneration, and (***iii***) extracellular matrix surface network is not a morphogenic marker in this cultivar.

Keywords: callogenesis; extracellular matrix; in vitro culture; scanning electron microscopy; topinambur; transmission electron microscopy

Introduction

Helianthus tuberosus (Jerusalem artichoke, topinambur, earth apple) is a perennial warm-season species of sunflower native to temperate regions of North America that has been known in Europe since the 17th century. It is an agricultural plant with nutritional, medicinal and energy potential. Its tubers are rich in polysaccharide inulin. The species is also cultivated for biofuels [1]. A new, cold resistance cultivar Albik was selected for further improvement (e.g., more intensive growth and cellulose production) through genetic transformation with the use of in vitro technique.

Topinambur was investigated in different aspects. Recently, phytochemical studies revealed the bioactive metabolites with medicinal value [1,2]. Microtubers induction in tissue culture was used for simulating physiological processes leading to underground storage organ formation [3]. Some efforts for the successful cryopreservation protocol were also undertaken [4]. So far the low efficiency of plant regeneration via indirect organogenesis or somatic embryogenesis was described in the leaf culture of *Helianthus tuberosus* and in anther or root culture of tetraploid interspecific hybrid *Helianthus annuus* × *Helianthus tuberosus* [5–7].

An in vitro cell and tissue culture induces different cellular responses that mediate adaptations under new environmental conditions. Cells with a morphogenetic or embryogenic potential differ from a non-regenerative callus in several features, including cell wall modifications [8]. The formation of extracellular strands, fibrils and/or a continuous layer over callus cells accompanies the induction of morphogenesis in many plants cultured in vitro. This structure, which is referred to as extracellular matrix (ECM) or extracellular matrix surface network (ECMSN), has been found during the early stages of somatic embryogenesis and organogenesis e.g., in coconut [8] and maize [9]. The exact role of this structure is still not known, but its regulatory and coordinating functions during the early stages of morphogenesis have been suggested [8,10–13].

The aim of this study was to establish protocols of callus induction and then the plant regeneration and to see whether ECMSN can be treated as a morphogenic marker of the investigated cultivar. This is the first attempt to test the morphogenetic ability of topinambur cv. Albik.

* Corresponding author. Email: m.popielarska-konieczna@uj.edu.pl

Handling Editor: Beata Zagórska-Marek

Material and methods

Culture conditions

About 100 mature plants of *H. tuberosus* cv. Albik, vegetatively propagated from tubers obtained from IHAR (Radzików, Poland), harvested in Polanka Hallera and Modlnica (experimental fields of Jagiellonian University, near Cracow, Poland), were the sources of the following explants: apical meristems, petioles (10 mm) different leaf segments (10 × 15 mm) and 2–3 mm thick stem fragments. Tuber and root segments were excluded from the study because of high contamination rate: 80–90% in preliminary tests. Plant material was surface sterilized for 70 s in 70% ethanol and for 10 min in a 50% Ace solution (commercial bleach). Media tested in the experiments were based on a modified Murashige and Skoog basal medium (MS) [14] supplied with: (*i*) plant growth regulators (PGRs): benzylaminopurine (BAP), 2,4-dichlorophenoxyacetic acid (2,4-D), indole-3-acetic acid (IAA), kinetin (KIN), 1-naphthaleneacetic acid (NAA), 2,3,5-triiodobenzoic acid (TIBA), thidiazuron (TDZ); (*ii*) other organic and inorganic substances: 2-(N-morpholin) ethanesulfonic acid (MES), potassium nitrate (KNO_3), putrescine; (*iii*) different sucrose concentrations (1, 2, 3 or 12%). Media were solidified with 8% agar (Difco, USA). In total, 36 different media were tested (Tab. 1). For statistical analysis, data from selected media (from six callus induction media and three regeneration media) were used (Tab. 2). After two or four weeks of culture, calli obtained on media 2a, b, c, 4, 8, 14 were transferred onto media 17c, 25a, b for indirect regeneration (Tab. 2). Leaf segments were cultured on 90 mm Petri dishes (5 per one Petri dish) and other explants were grown on 55 mm Petri dishes (5 petioles or stem fragments or 4 apical meristems per one Petri dish) at 26 ±3°C in darkness for the induction of calli or in a 16/8 h photoperiod under cool-white fluorescent tubes (60–90 mol photons m^{-2} s^{-1}) for regeneration. The contamination rate (bacterial or fungal) was ca. 10%.

Tab. 1 Media tested in experiments, based on MS [8] supplemented with different plant growth regulators (PGRs), additional substances and different sucrose concentration (3% in all media except for 17 and 25 b).

Medium No.	Plant growth regulators and other substances (concentration in mg l^{-1})
Callus induction	
1	without PGRs
2a, b, c	TDZ (0.25, 0.5, 0.75)
3a, b, c, d	2,4-D (0.5, 1.0, 2.5, 5.0)
4	2,4-D (0.2) + KIN (0.04)
5	2,4-D (0.2) + KIN (0.04) + KNO_3 (5000)
6	2,4-D (0.2) + KIN (0.04) + KNO_3 (5000) + MES (700)
7	2,4-D (0.2) + KIN (0.04) + putrescine (800)
8	2,4-D (2.0) + BAP (1.0)
9a, b, c, d	2,4-D (0.5, 1.0, 2.5 , 5.0) + BAP (0.5)
10	NAA (0.1) + BAP (1.0)
11	NAA (0.1) + BAP (1.0) + KNO_3 (5000)
12	NAA (0.1) + BAP (1.0) + putrescine (800)
13	NAA (1.0) + BAP (1.0)
14	NAA (0.25) + IAA (0.025) + BAP (0.25)
15	BAP (0.1) + TIBA (500)
16	BAP (2.0)
Regeneration	
17a, b, c	without PGRs, 1, 2 or 12% sucrose
18	TDZ (1.0)
19	BAP (0.5)
20	2,4-D (0.5) + KIN (1.0)
21	2,4-D (0.5) + KIN (2.0)
22	NAA (0.1) + KIN (0.5)
23	NAA (0.1) + BAP (0.2)
24	NAA (0.5) + BAP (0.1)
25a, b	BAP (1.5), 3% or 12% sucrose

Histological analysis

For histological analysis stem fragments, after 0, 6 and 10 days of culture on MS with 1.0 mg l^{-1} 2,4-D (medium 3b, Tab. 1), were used. The material was fixed in 5% glutaraldehyde in a 0.1 M phosphate buffer, pH 7.2 at room temperature for 2 h, washed four times in the same buffer and dehydrated in a graded ethanol series and kept overnight in absolute ethanol. Then, the material was embedded in Technovit 7100 (2-hydroxyethyl-methacrylate) according to manufacturer's instructions (Heraeus Kulzer). The material was sectioned into 5 μm with a rotary microtome (Microm, Adamas Instrumenten), stained with 0.1% toluidine blue O (TBO) and mounted in Entellan synthetic resin (Merck). Some sections were stained using periodic acid Schiff (PAS)/TBO double staining. The semi-thin sections were treated with 0.5% periodic acid for 10 min and rinsed in distilled H_2O for 3 min. Sections were then stained with Schiff's reagent for 30 min and rinsed in a rinsing solution (1 mol/l HCl 5 ml + 10% $Na_2S_2O_5$ 5 ml + distilled H_2O 100 ml). Then, the sections were stained with 0.1% TBO. In the PAS reaction, the total carbohydrates of insoluble polysaccharides (e.g., starch and cellulose) stain magenta to purplish red. TBO is routinely used as a counterstain in order to improve contrast. Microscopy sections were photographed with a Zeiss Axio Cam MRe digital camera. The images were processed with Zeiss Axio Vision 3.0 software.

SEM analysis

Cultured stem fragments were prefixed in 5% glutaraldehyde (in 0.1 M phosphate buffer, pH 7.2) for 2 h at room temperature. After dehydration through a graded ethanol series, the samples were dried using a CPD system (CO_2 critical point drying), sputter-coated with gold (Jeol JFC-1100 E ion-sputtering system) and observed with a Jeol JSM 5410 scanning electron microscope.

TEM analysis

Cultured stem fragments were fixed in 2.5% formaldehyde (prepared from paraformaldehyde) and 2.5% glutaraldehyde in a 0.1 M cacodylate buffer (pH 7.0) for 2 h at room

Tab. 2 Selected media and explants used for callus induction and organogenesis. Media based on MS [8] with plant growth regulators (PGRs) and different sucrose concentration (3% in all media except for 17c and 25b). Media are coded as in Tab. 1.

Medium No.	PGR (concentration in mg l^{-1})	Explant	
Callus induction			No. of explant used/No. of explants with callus induction
2a	TDZ (0.25)	apical meristem	16/16
		petioles	75/0
		leaves ventral	75/0
		leaves dorsal	25/0
		stem	25/0
2b	TDZ (0.50)	apical meristem	14/14
		petioles	75/0
		leaves ventral	75/0
		leaves dorsal	25/0
		stem	75/21
2c	TDZ (0.75)	apical meristem	16/16
		petioles	75/0
		leaves ventral	75/0
		leaves dorsal	25/0
		stem	75/19
4	2,4-D (0.2) + KIN (0.04)	apical meristem	15/15
		petioles	75/12
		leaves ventral	75/48
		leaves dorsal	75/45
		stem	75/27
8	2,4-D (2.0) + BAP (1.0)	apical meristem	16/16
		petioles	75/4
		leaves ventral	76/54
		leaves dorsal	75/57
		stem	75/69
14	NAA (0.25) + IAA (0.025) + BAP (0.25)	apical meristem	16/15
		petioles	75/6
		leaves ventral	75/45
		leaves dorsal	75/60
		stem	75/30
Regeneration			No. of explant used/No. of regenerants
17c	without PGRs + 12% sucrose	callus	110/0
25a	BAP (1.5)	callus	110/0
25b	BAP (1.5) + 12% sucrose	callus	105/0

temperature. The samples were rinsed four times in the same buffer (15 min each) and post-fixed in buffered 1% OsO_4 at 4°C overnight. After rinsing in distilled water, explants were treated with 1% uranyl acetate for 1 h, dehydrated in a graded acetone series and embedded in Spurr's resin. Ultrathin sections were cut on a Sorvall MT-2B ultramicrotome, stained with uranyl acetate and lead citrate, and examined with a Philips CM 100 transmission electron microscope. Control semi-thin sections were post-stained with 0.1% TBO.

Statistical analysis

All of the analyses and graphs were done in the R environment for statistical computing [15]. The data from six different media and five different explants (Tab. 2) were compared using a pairwise comparison between pairs of proportions with a correction for multiple testing (pairwise. prop.test function from stats library). Confidence interval values on bar charts were calculated by tests of proportion (prop.test from stats library).

Results

Callus induction and culture

Single callus cells were observed on stem discs (Fig. 1a) one week after inoculation. In subsequent days of culture, calli proliferated on callus induction and regeneration media on all explants used (Tab. 2 and Fig. 1b–d). The color of the callus depended on light conditions. In the dark, calli were white or cream, while in the light they turned green. The efficiency of callus production was explant and medium dependent. The frequency of callogenesis from apical meristems was 100% on almost all of the media. Fragments of stem and leaf blades responded with a lower frequency (Fig. 2a–c) and were not influenced by explant position on the medium (dorsal or ventral surface). The lowest response was noted when petioles (Fig. 2a–c) were used as explants. Callus tissue was produced without morphogenic response on all of the media used (Tab. 2).

Histological and ultrastructural studies

Histological analysis of callus induced from stems indicated that cell proliferation started primarily in the cortex (Fig. 3a). In subsequent days of the culture, centers of intense callus proliferation were visible in the cambial zone, phloem and xylem. The PAS reaction did not reveal starch grains in the cells of explants (data not shown). SEM observations confirmed that callus cells were induced in the cortex (Fig. 3b) and vascular bundles of stem discs. In well-developed callus, cells differed conspicuously in size and shape (Fig. 3c). Some parts of stem-derived callus (Fig. 3d) were covered with a membranous structure.

Ultrastructural studies of callus cells analyzed using TEM showed the presence of a wavy plasmalemma, profiles of a rough endoplasmic reticulum, dictyosomes, ribosomes and polysomes, which are typical for plant cells with a high level of activity (Fig. 3e–g). Fibrillar and globular structures were present in the intercellular spaces. Numerous mitochondria were located near lipid bodies. Parenchymatous, senescent cells showed a sequestration of the cytoplasm (Fig. 3e) and

Fig. 1 Explants of stem discs (**a–c**) and leaf blade (**d,e**) cultured on MS medium with 0.5 mg l^{-1} KIN and 0.1 mg l^{-1} NAA. Explant at inoculation (**a**). Callus on explant surface after 4 (**b,c**) and 3 weeks of culture (**d,e**). Scale bars: 3 mm (**a–c**); 1 mm (**d,e**).

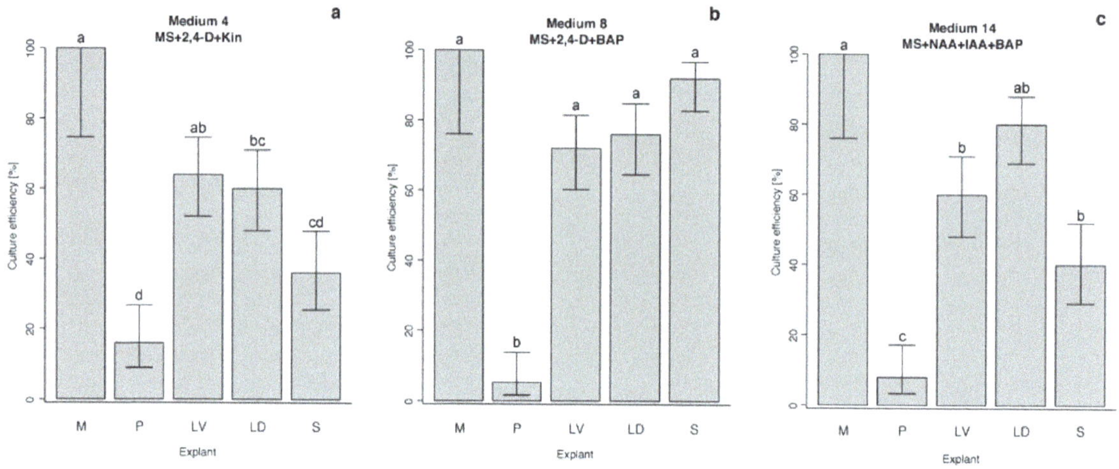

Fig. 2 **a–c** Effect of explant and medium on callus induction. Groups labelled with the same letter do not differ significantly ($P > 0.05$) by test of proportion. Error bars indicate 95% confidence intervals. Type of explants: M – meristems, P – petioles, LV – leaves ventral, LD – leaves dorsal, S – stem. Media are coded as in Tab. 1.

vacuoles containing electron-dense deposits, probably tannins (data not shown). On the surface of callus, fibers that extended from the outer cell wall were visible (Fig. 3h,i). The structure of the material covering the outer cell wall was different than the filling in the intercellular spaces. Compact material, composed of fibers and clumps, was located on the cell wall (Fig. 3g–i), whereas fibrillar and globular structures were present in the intercellular spaces (Fig. 3e,f).

Discussion

In our experiment we induced callogenesis in in vitro cultures of *H. tuberosus* cv. Albik. However, during the time of the culture, callus remained non-regenerative and no shoots, roots or somatic embryos were observed. Most of the regeneration media used in this study were previously tested with *H. tuberosus* or interspecific *Helianthus* hybrids where

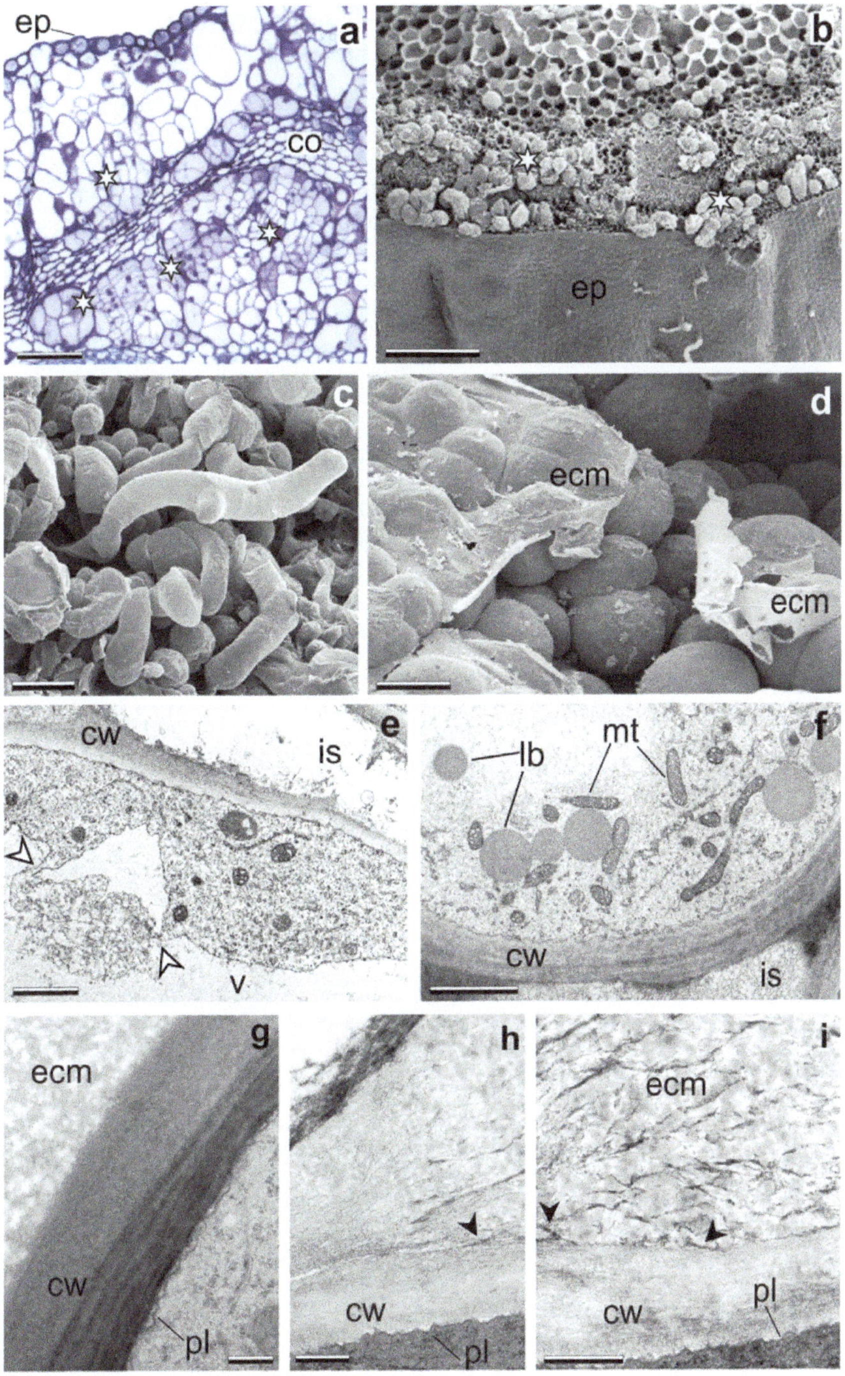

Fig. 3 Transversal section of stem explant after 10 days of culture (**a**) stained with TBO; high magnification of the cortex zone; visible epidermis (ep), cells of calli (asterisks) and remnants of collenchyma (co). SEM images of explants after 10 days (**b**) and 4 weeks (**c,d**) of culture. Cross sections of stem (**b**); visible epidermis (ep) and cells of callus (asterisks) in the cortex. Well-proliferated calli with cells differ in size and shape (**c**) and covered with ECMSN (ecm; **d**). TEM images of explants (**e–i**). Ultrastructure of meristematic cells (**f**), parenchymatic, senescent cells (**e**) and outer cell wall (**g–i**). Note wavy plasmalemma (pl), cell wall (cw), fibrillar and globular structures in the intercellular spaces (is). **e** White arrowheads indicate the sequestration of the cytoplasm and vacuole (v). **f** Visible numerous mitochondria (mt) near lipid bodies (lb). **g–i** Fragments of the surface of callus with a heterogenous ECMSN (ecm) with fibrills and clumps; black arrowheads indicate fibers extending from the outer cell wall. Visible lammellae of the cell wall (**g**). Scale bars: 50 μm (**a,c,d**); 200 μm (**b**); 1 μm (**e,f**); 0.2 μm (**g–i**).

plant regeneration by direct or indirect pathway was obtained [5–7,16–18]. In topinambur and sunflower cultures, the results highly dependent on the genotype, explant type and components of the nutrient medium [5–7,16–18]. Leaf segments of the interspecific hybrid *H. annuus* × *H. tuberosus* were successfully used for morphogenesis and somatic embryogenesis [16]. However, in previous experiments on topinambur and sunflower hybrids [5], the frequency of formation of somatic embryos was very low. Bianchi et al. [6] reported that some genotypes of the genus *Helianthus* were able to regenerate plants with a high frequency, while other showed a complete absence of morphogenetic competence. A strong genotype-dependent tissue response under in vitro conditions has been reported for some taxa of monocots and dicots [6,19,20].

Our results showed that callus was induced in the cortex and vascular bundles. It is commonly known that cambium cells are pluripotent vascular stem cells and retain the ability to divide [21,22]. A study by Rose et al. [23] on the origin of roots in leaf cultures of *Medicago truncatula*, showed that vein procambial cells stimulated by auxin gave rise to callus cells that ultimately generated root meristems. Further studies on regeneration in *M. truncatula* also revealed that somatic embryos in leaf cultures originated from procambial cells or dedifferentiated mesophyll cells that become totipotent [24]. Cell proliferation within the bundle sheath was also observed in leaf explants of *H. occidentalis* [25], leading to rhizogenesis.

The surface of non-regenerative callus derived from the cultivar Albik of *H. tuberosus* was covered with a membranous structure, similar to the ECMSN that has been reported in plant tissue cultures of different species [26–28]. Detailed TEM studies of topinambur calli have confirmed the extracellular deposition of fibrillar material and have revealed the presence of globular structures and fibers in the intercellular spaces. A similar heterogeneous ECMSN was reported in oilseed rape where osmiophilic granules were associated with fibers extending from the outer cell wall of embryogenic callus cells [29]. ECMSN was primarily linked to the acquisition of embryogenic competence [8,30] and in several species can serve as an early structural marker of somatic embryogenesis [29,31,32]. The presence of extracellular material also accompanied the induction of organogenesis [12,33]. However, in the present studies, in spite of the occurrence of ECMSN the calli were not morphogenetic. The extracellular layer covering callus cells, which is not involved in the regeneration process, has previously been reported during androgenesis in wheat [34], somatic embryogenesis in banana [35] and clover cultures [36]. The role of ECMSN cannot be generalized but it is evident that it is formed on the surface of cultured in vitro tissues regardless of its morphogenetic competence. Therefore, we agree with the suggestion that has been made by several authors [26,30] that ECMSN formation could be a stress response of plant tissues that is triggered by specific culture conditions and that covering callus with extracellular material could provide protection against external factors.

Conclusions

Based on the previous reports and the results of the present experiment, it is evident that: (*i*) the morphogenic capacity of callus in *Helianthus tuberosus* is strictly genotype dependent, (*ii*) cultivar Albik is recalcitrant in in vitro regeneration, (*iii*) ECMSN is not a morphogenic marker in this cultivar.

Acknowledgments

We would like to thank Plant Breeding and Acclimatization Institute in Radzików for providing tubers of topinambur cv. Albik. These studies were funded by the Rector of the Jagiellonian University for project AKCENT (2005. 2006).

Authors' contributions

The following declarations about authors' contributions to the research have been made: participation in in vitro culture experiments, histological technique, acquisition of in vitro data, documentation of in vitro and histological results and data interpretation, participation in drafting the article: MP; participation in in vitro culture experiments, acquisition, interpretation and documentation of scanning electron microscopy data, drafting the article: MPK; participation in in vitro culture experiments and in drafting the article: HŚ; acquisition, interpretation and documentation of transmission electron microscopy data: MKK; statistical analysis of data obtained in in vitro experiments: GG; participation in in vitro culture experiments and in drafting the manuscript: RK; acquisition, interpretation and documentation of scanning electron microscopy data, analysis and interpretation of transmission electron microscopy data, critical revision of electron microscopy part of the manuscript: JB; design of the study, contribution to interpretation of data, critical revision of the manuscript: EK.

References

1. Ma XY, Zhang LH, Shao HB, Xu G, Zhang F, Ni FT, et al. Jerusalem artichoke (*Helianthus tuberosus*), a medicinal salt-resistant plant has high adaptability and multiple-use values. J Med Plant Res. 2011;5(8):1272–1279.
2. Li XD, Miao FP, Ji NY. Two new epoxysteroids from *Helianthus tuberosus*. Molecules. 2011;16(12):8646–8653.
3. Gutierrez Pesce P, Bizzarri M, Rugini E, de Pace C. In vitro microtuberization for simulating the developmental physiology of underground storage organ in *Helianthus tuberosus*. In: Proceedings of the joint meeting AGI-SIBV-SIGA. Assisi: SIGA; 2011. p. 6A.38.
4. Volk GM, Richards K. Preservation methods for Jerusalem artichoke cultivars. HortScience. 2006;41(1):80–83.
5. Pugliesi C, Megale P, Cecconi F, Baroncelli S. Organogenesis and embryogenesis in *Helianthus tuberosus* and in the interspecific hybrid *Helianthus annuus* × *Helianthus tuberosus*. Plant Cell Tissue Organ Cult. 1993;33(2):187–193.
6. Bianchi R, Fambrini M, Pugliesi C. Morphogenesis in *Helianthus tuberosus*: genotypic influence and increased totipotency in previously regenerated plants. Biol Plant. 1999;42(4):515–523.
7. Fambrini M, Cionini G, Conti A, Michelotti V, Pugliesi C. Origin and development in vitro of shoot buds and somatic embryos from intact roots of *Helianthus annuus* × *H. tuberosus*. Ann Bot. 2003;92(1):145–151.
8. Verdeil JL, Hocher V, Huet C, Grosdemange F, Escoute J, Ferrière N, et al. Ultrastructural changes in coconut calli associated with the acquisition of embryogenic competence. Ann Bot. 2001;88(1):9–18.
9. Šamaj J, Baluška F, Bobák M, Volkmann D. Extracellular matrix surface network of embryogenic units of friable maize callus contains

arabinogalactan-proteins recognized by monoclonal antibody JIM4. Plant Cell Rep. 1999;18(5):369–374.

10. Blehová A, Bobák M, Šamaj J, Hlinková E. Changes in the formation of an extracellular matrix surface network during early stages of indirect somatic embryogenesis in *Drosera spathulata*. Acta Bot Hung. 2010;52(1):23–33.
11. Konieczny R, Bohdanowicz J, Czaplicki AZ, Przywara L. Extracellular matrix surface network during plant regeneration in wheat anther culture. Plant Cell Tissue Organ Cult. 2005;83(2):201–208.
12. Popielarska-Konieczna M, Kozieradzka-Kiszkurno M, Świerczyńska J, Góralski G, Ślesak H, Bohdanowicz J. Ultrastructure and histochemical analysis of extracellular matrix surface network in kiwifruit endosperm-derived callus culture. Plant Cell Rep. 2008;27(7):1137–1145.
13. Bobák M, Šamaj J, Hlinková E, Hlavačka A, Ovečka M. Extracellular matrix in early stages of direct somatic embryogenesis in leaves of *Drosera spathulata*. Biol Plant. 2003;47(2):161–166.
14. Murashige T, Skoog F. A revised medium for rapid growth and bioassays with tobacco tissue culture. Physiol Plant. 1962;15(3):473–497.
15. R Core Team. R: a language and environment for statistical computing [Internet]. 2012 [cited 2013 Oct 20]; Available from:
16. Fambrini M, Cionini G, Pugliesi C. Development of somatic embryos from morphogenetic cells of the interspecific hybrid *Helianthus annuus* × *Helianthus tuberosus*. Plant Sci. 1996;114(2):205–214.
17. Fambrini M, Cionini G, Pugliesi C. Acquisition of high embryogenic potential in regenerated plants of *Helianthus annuus* × *H. tuberosus*. Plant Cell Tissue Organ Cult. 1997;51(2):103–110.
18. Gamburg KZ, Vysotskaya EF, Gamanets LV. Microtuber formation in micropropagated Jerusalem artichoke (*Helianthus tuberosus*). Plant Cell Tissue Organ Cult. 1998;55(2):115–118.
19. Popielarska-Konieczna M, Kozieradzka-Kiszkurno M, Tuleja M, Ślesak H, Kapusta P, Marcińska I, et al. Genotype-dependent efficiency of endosperm development in culture of selected cereals: histological and ultrastructural studies. Protoplasma. 2013;250(1):361–369.
20. Ślesak H, Góralski G, Pawłowska H, Skucińska B, Popielarska-Konieczna M, Joachimiak AJ. The effect of genotype on a barley scutella culture. Histological aspects. Cent Eur J Biol. 2013;8(1):30–37.
21. Fukuda H. Plant cell biology: signals that control plant vascular cell differentiation. Nat Rev Mol Cell Biol. 2004;5(5):379–391.
22. Jura J, Włoch W, Kojs P, Wilczek A, Szendera W. Current trends in the structural investigations of the vascular cambium. Bull Bot Gard. 2005;14:43–47.
23. Rose RJ, Wang XD, Nolan KE, Rolfe BG. Root meristems in *Medicago truncatula* tissue culture arise from vascular-derived procambial-like cells in a process regulated by ethylene. J Exp Bot. 2006;57(10):2227–2235.
24. Wang XD, Nolan KE, Irwanto RR, Sheahan MB, Rose RJ. Ontogeny of embryogenic callus in *Medicago truncatula*: the fate of the pluripotent and totipotent stem cells. Ann Bot. 2011;107(4):599–609.
25. Šamaj J, Bobák M, Kubošníková D, Krištín J, Kolarik E, Ovečcka M, et al. Bundle sheath cells are responsible for direct root regeneration from leaf explants of *Helianthus occidentalis* L. J Plant Physiol. 1999;154(1):89–94.
26. Dubois T, Dubois J, Guedira M, Diop A, Vasseur J. SEM characterization of an extracellular matrix around somatic proembryos in roots of *Cichorium*. Ann Bot. 1992;70(2):119–124.
27. Popielarska-Konieczna M, Bohdanowicz J, Starnawska E. Extracellular matrix of plant callus tissue visualized by ESEM and SEM. Protoplasma. 2010;247(1-2):121–125.
28. Mazarei M, Al-Ahmad H, Rudis MR, Joyce BL, Stewart CN. Switchgrass (*Panicum virgatum* L.) cell suspension cultures: establishment, characterization, and application. Plant Sci. 2011;181(6):712–715.
29. Namasivayam P, Skepper J, Hanke D. Identification of a potential structural marker for embryogenic competency in the *Brassica napus* spp. *oleifera* embryogenic tissue. Plant Cell Rep. 2006;25(9):887–895.
30. Šamaj J, Bobák M, Blehová A, Krištin J, Auxtová-Šamajová O. Developmental SEM observations on an extracellular matrix in embryogenic calli of *Drosera rotundifolia* and *Zea mays*. Protoplasma. 1995;186(1-2):45–49.
31. Šamaj J, Bobák M, Blehová A, Pretová A. Importance of cytoskeleton and cell wall in somatic embryogenesis. In: Mujib A, Šamaj J, editors. Somatic embryogenesis. Berlin: Springer; 2006. p. 35–50. (vol 2).
32. Lai KS, Yusoff K, Maziah M. Extracellular matrix as the early structural marker for *Centella asiatica* embryogenic tissues. Biol Plant. 2011;55(3):549–553.
33. Konieczny R, Świerczyńska J, Czaplicki AZ, Bohdanowicz J. Distribution of pectin and arabinogalactan protein epitopes during organogenesis from androgenic callus of wheat. Plant Cell Rep. 2007;26(3):355–363.
34. Pilarska M, Czaplicki AZ, Konieczny R. Patterns of pectin epitope expression during shoot and root regeneration in androgenic cultures of two wheat cultivars. Acta Biol Crac Ser Bot. 2007;49:69–72.
35. Xu C, Zhao L, Pan X, Šamaj J. Developmental localization and methylesterification of pectin epitopes during somatic embryogenesis of banana (*Musa* ssp. AAA). PLoS ONE. 2011;6(8):e22992.
36. Pilarska M, Knox JP, Konieczny R. Arabinogalactan-protein and pectin epitopes in relation to an extracellular matrix surface network and somatic embryogenesis and callogenesis in *Trifolium nigrescens* Viv. Plant Cell Tissue Organ Cult. 2013;115(1):35–44.

15

Investigations of the capacity and strength of seed germination in *Allium victorialis* L.

Krystyna Winiarczyk[1]*, Katarzyna Skrzypczak[2], Jolanta Jaroszuk-Ściseł[3], Jan Bocianowski[4]

[1] Department of Plant Anatomy and Cytology, Maria Curie-Skłodowska University, Akademicka 19, 20-033 Lublin, Poland

[2] Faculty of Food Science and Biotechnology, University of Life Sciences in Lublin, Skromna 8, 20-704 Lublin, Poland

[3] Department of Environmental Microbiology, Maria Curie-Skłodowska University, Akademicka 19, 20-033 Lublin, Poland

[4] Department of Mathematical and Statistical Methods, Poznań University of Life Sciences, Wojska Polskiego 28, 60-637 Poznań, Poland

Abstract

The aim of the study was to examine the strength and energy of seed germination in *Allium victorialis*. Despite the normal structure of seeds containing a viable embryo and compliance with all International Seed Testing Association recommendations, no germination in *A. victorialis* was observed. Additionally, the scarification and stratification treatments applied did not improve the dynamics of germination of *A. victorialis* seeds. Microbiological analyses of soil sampled from natural localities of the plant revealed a typical composition of bacteria and fungi. The high number of fungi [4.5 $\log_{10}$CFU (colony forming units) of fungi × g^{-1} dry mass of soil] and various groups of bacteria (about 7.0 $\log_{10}$CFU of bacteria × g^{-1} dry mass of soil) were detected in the root-free-soil around garlic roots. In the interior of *A. victorialis* roots, the number of microorganisms decreased 1000 to 10 000 times but all the tested microbial groups, especially copiotrophic bacteria and fungi (1.6 and 2.2, respectively, $\log_{10}$CFU × g^{-1} dry mass of roots) were detected.

Changes in such parameters as dehydrogenase activity, pH values, and the total organic C (TOC) content in the particular parts of the rhizosphere and in comparison to the rhizosphere with root-free-soil were observed. The dehydrogenase activity and TOC content were highly positively correlated with the total number of CFU of the microorganisms.

Keywords: *Allium*; victory onion; phylogenetic analysis; ISTA; seed; embryo; tetrazolium test

Introduction

Although ontogenetic development of plants begins with zygote formation, it is assumed that seed germination is the actual first stage of development, during which a new individual starts its independent existence, often at a long distance from the parental plant. Therefore, seeds are both the end of the plant developmental cycle and the beginning.

In Angiospermae, after double fertilization, the embryo matures and becomes a seed enclosed within the fruit. The seed contains an embryo, endosperm, and a seed coat. The embryo represents a miniature plant with vegetative organ buds: the root, stem, and leaves [1]. In Mononocots, seed coats are not distinct structures being fused with the fruit wall to form a pericarp. The seed coat forms from the two integuments or outer layers of cells of the ovule, which derive from tissue from the mother plant, the inner integument forms the tegmen and the outer forms the testa. The seed is cover with the testa, which has a protective role against infections, pests, and adverse external conditions. It is usually hard and formed the mechanical layer. The cells of the seed testa become hard due to deposition of cutin, lignin, and phenolic compounds in the cell walls. The developing embryo acquires nutrients from the endosperm. This nutritive tissue is usually triploid and begins to develop immediately after fertilization through merging of the central cell in the embryo sac with the sperm nucleus. The primary endosperm nucleus usually undergoes mitotic divisions before zygote divisions. The endosperm is rich in carbohydrates, proteins, and fats; hence, the division into starch (cereals), oil (rapeseed, sunflower), and protein seeds (legume plants). Moreover, the endosperm contains hormones responsible for embryo development [2,3].

Seed germination is a set of processes leading to activation of embryos followed by initiation of seedling growth. Germination conditions are strictly defined and specific for individual species. The onset and course of the seed germination process depends on many environmental factors, e.g. minimum, maximum, and optimal temperature. Seed germination in a majority of plants is dependent on light conditions; such seeds are named photoblastic and they represent positive (light-stimulated germination) and

* Corresponding author. Email: krystyna.winiarczyk@poczta.umcs.lublin.pl

Handling Editor: Elżbieta Bednarska-Kozakiewicz

negative photoblasty (light-inhibited germination). However, light availability does not determine the process. A majority of positively photoblastic seeds germinate in darkness as well, yet light conditions have a favorable effect on germination efficiency [4]. Other environmental conditions regulating the course of seed germination include substances present in the atmosphere and the substrate and water availability. Presence of oxygen is essential for normal germination in all seeds, as lack thereof results in a slower germination rate and the developing seedlings exhibit developmental anomalies. In general, seeds are well equipped with mineral compounds by the parent plant; therefore, the content of these compounds in the substrate has no significant effect on the germination process. Nitrates, however, available in the substrate exert a significant impact on germination. As a rule, their presence enhances (sometimes several fold) the germination efficiency, which is probably a result of the oxidative activity of NO^{3-} ions. Given the nature of processes predominant in the germination process, three basic stages can be distinguished, i.e. imbibition, catabolic, and anabolic phases. The kinetics of water and oxygen uptake by germinating seeds is a factor that differentiates the individual germination phases [5]. For seed germination, some plant species require specific treatment such as scarification and stratification. Overcoming seed dormancy may involve mechanical damage to or removal of the seed coat or application of low temperature (vernalization). Additionally, each plant species needs a different duration of dormancy. Some seeds germinated immediately after harvest, while others require some dormancy period. Laboratory qualification of the seed material must be based on the recommendations included in the "International rules for seed testing" (by International Seed Testing Association – ISTA), which contain rules of the methodology of sampling and testing the sowing value of various characteristics of seeds from over 900 plant species.

Allium victorialis L. occurs on and mountain pasture and meadows, stony slopes and rocky sites. The wide geographical range of the species covers Europe, Asia, and North America [6]. In Asia, the species grows in the area between the Ural Mountains and Kamchatka [7]. It is a typical representative of alpine flora. In Asia and the Caucasus, it occurs at an altitude of 1700–2600 m a.s.l. In Central Europe, *A. victorialis* is a rare plant threatened with extinction. The species has been included in the red list of plants and fungi in Poland and regarded as endangered [8]. In the European red list of vascular plants (2002), the species has been classified as LC, i.e. least concern. In the neighboring countries, e.g. in Germany, the species occurs only in Baden-Württemberg, where it has been classified as threatened (category 1 CR – critically endangered according to the IUCN categories). In the Czech Republic, it was assigned to category C2, which corresponds to EN in the IUCN category and in Slovakia the species is included in the red list [9]. *A. victorialis* is a typical entomophilous species pollinated by *Diptera* and *Hymenoptera* insects. The number of pollen grains produced by the male flower reaches 4.8–6.0 × 104, whereas the amount of pollen produced by one plant ranges from 1.22 × 106 to 2.12 × 106. The species reproduces sexually through seeds and employs vegetative reproduction by separation of a part of the roots from the parent plant [10]. *A. victorialis* leaves are edible and the plant is a source of valuable bioactive substances used in pharmacy and gastronomy. In the past, the species was known as the so-called false mandrake, and its root successfully replaced the real *Mandragora officinarum* [11].

The aim of the study was to examine the strength and energy of *A. victorialis* germination proceeding in accordance with the ISTA guidelines and to perform a preliminary analysis of the causes of inhibited seed germination. The study also provides determination of relatedness and phylogenetic relationships between *Allium* species. Additionally, soil microflora as well as dehydrogenase activity, which catalyzes oxyreductive reactions occurring in every living cell, has been examined [12].

Material and methods

Plant material

The examined plants originated from the collections of the Maria Curie-Skłodowska University Botanical Garden in Lublin. Three *Allium* species were investigated:

(i) *Allium victorialis* L. – a critically endangered species included in the red list of plants and fungi in Poland (2006) and regarded as endangered in Poland; it was introduced from a natural locality in the Bieszczady Mountains in 1968.

(ii) *Allium angulosum* L. – a species included in the red list of plants and fungi in Poland (2006) and regarded as vulnerable in Poland (V category). The species was introduced in the UMCS Botanical Garden from Germany in 1974.

(iii) *Allium tuberosum* Roxb. – a species naturally occurring in Asia; in Europe, it is cultivated as both an ornamental and a usable plant; it was introduced from the Gatersleben Botanical Garden, Seeland, Germany in 1996. Its natural locality is Nepal.

Growth conditions

Breeding experiment was conducted in Poland in the Lublin (51°16′N, 22°30′E), 200 m a.s.l. The climate of a region is characterized by prevalence of continental features with high annual temperature amplitudes, long summers and long and cold winters. In the years 1951–2012 long-term mean annual temperature reached 8.3°C, and long-term mean annual precipitation sum achieved 550.6 mm.

Terrain relief can be described as a western slope dissected by three loess ravines with a loess podzolic soils.

Phylogenetic analyses

Genetic material was extracted from young leaves. A fragment of the internal transcribed spacer (ITS) region was amplified using primers ITS4 and ITS5. Freshly harvested young leaves of *Allium victorialis*, *A. angulosum* and *A. tuberosum* were collected from six randomly selected plants placed in Eppendorf tubes and stored at −80°C. Total DNA was extracted from leaf tissue using an Isolate Plant Mini Kit for DNA isolation according to the procedure recommended by the producer company BIOLINE. For molecular characterization and phylogenetic analyses, the regions of the nuclear ribosomal DNA gene cluster were amplified using ITS4 and ITS5 primers (Tab. 1) [13]. PCR was carried out in a 20 μl reaction mixture containing: 1 μl of each primer (20 pmol/μl; Genomed), 7 μl free RNAse water (Fermentas), 10 μl MyTaq™RedMix ready-to-use 2× (0.05 u/μl Taq polymerase; reaction buffer, 4 mM $MgCl_2$; 0.4 mM each of dNTP: dATP, dCTP, dGTP, dTTP; loading color buffer) and 1 μl DNA (100 ng/μl).

Tab. 1 Specifications of the primers.

Name	Sequence	*T* (°C)	Length	%GC*
ITS5	5'-GGAAGTAAAAGTCGTAAC AAGG-3'	51.1	22 nt	41
ITS4	5'-TCCTCCGCTTATTGAT ATGC-3'	49.7	20 nt	45

* The percentage of guanine-cytosine pairs.

Amplifications were performed in a PTC-200 thermocycler (MJ-Research, USA) under the following conditions: initial denaturation for 2 min at 94°C, 30 cycles of 1 min at 94°C, 1 min at 61°C, 2 min at 72°C, followed by 7 min at 72°C. Amplification products were separated on 1.5% agarose gel (Invitrogen) in TBE (Tris/borate/EDTA) buffer (0.178 M Tris-borate, 0.178 M boric acid, 0.004 M EDTA) and stained with ethidium bromide. Ten μl of PCR products were supplemented with 2 μl of loading buffer (0.25% bromophenol blue, 30% glycerol), prior to agarose-gel electrophoresis at 3 V cm^{-1} for about 2 h. A 100-bp DNA Ladder Plus (Fermentas) was used as a size standard. When developed, the gels were viewed under UV light and photographs were taken (Syngen UV visualizer).

After electrophoresis, extraction of amplification products from agarose gel was performed using an Isolate Gel and PCR Kit according to the procedure recommended by the producer company BIOLINE. After extraction and purification from the gel, the amplification products were prepared for sequencing according to the recommendations of Genomed. Uploaded by Genomed sequencing results were analysed in the Finch TV. All nucleotide sequences were transformed in the appropriate format-FASTA, and then read using ClustalW2.

Assessment of the strength and energy of seed germination

Determination of the germinating capacity of the seeds from the plant material analyzed was performed in accordance with the recommendations included in "International rules for seed testing" [14]. The method provides reproducible results within the limits set by the random variation in the sample. The conditions of the experiments were standardized following the ISTA guidelines to allow control and regulation.

Estimation of the seed germination capacity in each species was performed in 15 replicates using 12 seeds in a Petri dish. Calculations were based on counted numbers of germinating seeds on experimental days 7, 10, 14, and 21. The germination energy was estimated for the cumulative number of seeds germinating on day 10. The germination strength was estimated for the cumulative number of seeds germinating on experimental day 21.

Tetrazolium test for seed viability and vigor

The tetrazolium test is widely recognized as an accurate means of estimating seed viability. Tetrazolium test results can be extremely valuable for providing labeling information for immediate shipment of seed lots without waiting for completion of germination tests. It is also a valuable research technique for estimating seed viability and determining reasons for poor germination. The staining patterns reveal the live and dead areas of the embryos thus enabling determination whether seeds have the capacity to produce normal seedlings. This biochemical test can distinguish live from dead tissues of seed embryos on the basis of dehydrogenase enzyme activity (respiration enzymes). Upon seed hydration, the activity of dehydrogenase enzymes increases resulting in release of hydrogen ions, which reduce the colorless tetrazolium salt solution (2,3,5-triphenyltetrazolium chloride) into a chemical red compound called formazan. Formazan stains living cells with a red color, while dead cells remain colorless [15].

The analyses were based on the recommendations and methodology contained in the "International rules for seed testing" [14]. The method allows receiving reproducible results, within the limits set by the random variation in the sample. The conditions of the experiment were standardized to allow control and regulation.

Sixty of seeds subjected to the experiment were soaked in sterile distilled water at 20°C for 17 hours. They were opened into half and immersed in a 1% tetrazolium solution (pH 6.5). The samples were incubated for 30 minutes at 37°C. Dried seeds were observed and diagnosed under a microscope.

Scanning electron microscopy (SEM)

The biological material constituted by 20 seeds of each tested species was fixed in a 5:5:90 (v:v:v) mixture of glacial acetic acid:formalin (40%):ethanol (70%) and dehydrated in a graded acetone series (40%, 70%, 80%, 100%). Subsequently, all material was frozen by using liquid nitrogen. Samples were analyzed in the scanning electron microscope (LEO1430VP) with an accelerating potential of 15 kV.

Microbial analysis of root-free-soil, rhizosphere soil and roots

Cores of soil (brown podzolic soil developed from loess) with *A. victorialis* plants at the full maturity stage of growth were aseptically taken from randomly selected sites in the plots. At the same time, cores of root-free soil (RFS) from the same field (plot) were collected.

Intact cores were placed in plastic bags, returned to the laboratory and stored overnight at 4°C.

The root contaminated soil (RCS) was obtained through gentle shaking of cores of soil with *A. victorialis* plants. Next, the plants were carefully removed from the soil and then separated into tops (shoots) and roots.

The roots were gently shaken (3 times – 15 s) to remove adhering clumps of rhizosphere soil loosely attached to roots (RhSL). Next, the garlic roots were placed in sterile glass flasks and shaken for 15 min at 50 rpm min^{-1} in an Innova chamber. After this procedure, the roots were separated from the removed rhizosphere soil strongly attached to roots (RhSS) and aseptically transferred to sterile Erlenmayer flasks.

Colonization of the rhizoplane (RhP) was evaluated by the following procedure: samples (10 g) of roots without RhSS transferred to sterile Erlenmayer flasks with 90 ml of sterile distilled water (SDW) were shaken for 1h at 50 rpm min^{-1} in the Innova chamber. The suspension obtained was used as a source of rhizoplane microorganisms.

Colonization of the RhP and interior of roots was evaluated by the following procedure: samples (10 g) of roots without RhSS in 90 ml of SDW were homogenized in an Ultra Turrax IKA® T18 homogeniser at 20 000 rpm min^{-1} for 30 s (the homogenate was cooled in an ice bath). This treatment was repeated three times at 15-s intervals. The homogenate obtained was used as a source of microorganisms colonizing the rhizoplane and interior of the root.

Colonization of the root interior was evaluated by the following procedure: 10 g samples of roots were surface-disinfected by brief washing 3 times in SDW for 5 min at 150 rpm min^{-1} in the Innova chamber, next by soaking for 10 min in a solution of $HgCl_2$ (0.1%), then in H_2O_2 (30%), and rinsing 10 times in sterile distilled water. Subsequently, the homogenate of the surface-disinfected roots was prepared by the method described above. The homogenate obtained was used as a source of microorganisms colonizing the interior of the root.

Soil colonization – colonization of RFS and RCS as well as rhizosphere soils (RhSL and RhSS) was evaluated by placing 10 g of soil in a sterile Erlenmayer flasks with 90 ml of SDW followed by shaking for 30 min at 50 rpm min^{-1} in the Innova chamber.

The soil dilution technique was used to determine the number of bacterial and fungal colony forming units (CFU) in the soil and roots samples. Seven 10-fold dilutions in 9 ml of SDW were made from the soil and root suspensions obtained and 100 μl of each dilution was spread onto the surface of appropriate solid media in Petri dishes.

The number of copiotrophs was determined on plates with PYS [12,16] medium and the number of CFU of oligotrophs was counted on 100 times diluted PYS agar. The numbers of *Pseudomonas fluorescens* on King's medium B [17] To determine the number of *Actinobacteria*, RBSCN agar [11] was used. Martin's medium (1950) [18] was used to calculate the total number of fungal CFU.

Colonies of bacterial strains were counted after 48 h of incubation, and colonies of fungal strains were counted after 7 days of growth at 28°C. The means of three replicates were expressed as log CFU × g^{-1} of dry mass (for RFS, RCS, RhSL, RhSS) of soil or dry mass of roots (for RhP and the interior of roots).

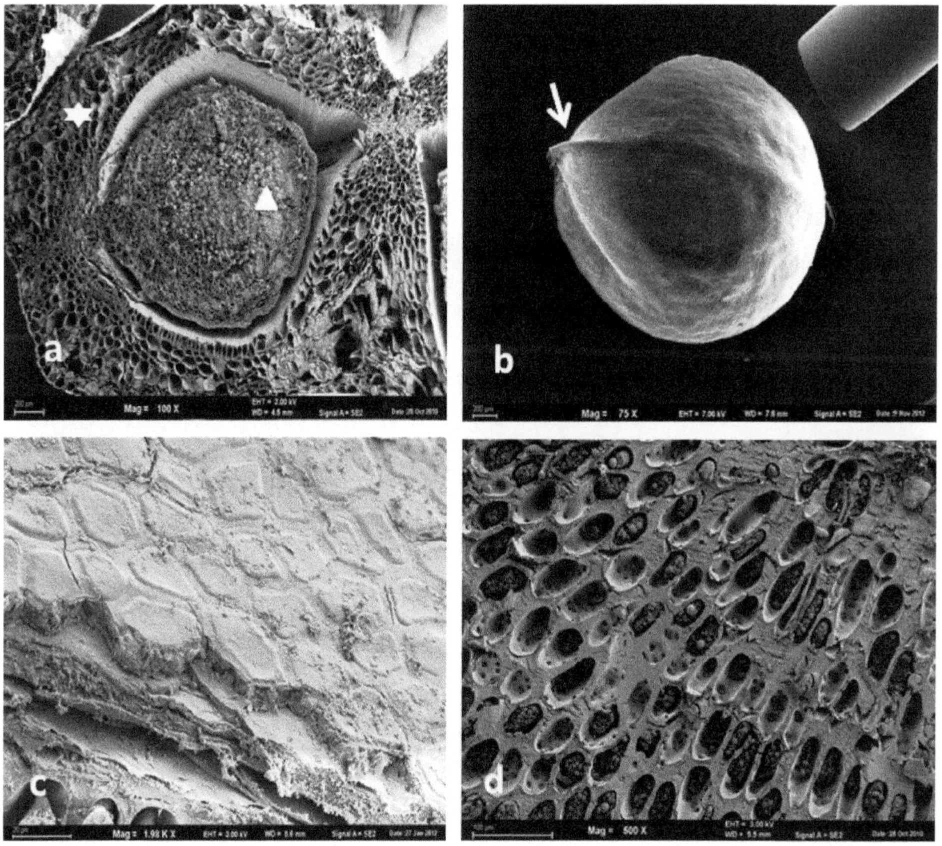

Fig. 1 *Allium victorialis*. **a** Ovary cross section with a multilayered pericarp (asterisk); central part taken up by a young seed (arrow). **b** General view of a victory onion seed; the raphe is marked by an arrow. **c** Reticulate seed coat surface pattern. **d** Seed cross section showing flatted and adhered to each other endosperm cells with starch granules. Scale bars: **a,b** 200 μm; **c** 20 μm; **d** 100 μm.

Dehydrogenase activity

Dehydrogenase activity was determined using the Tabatabai (1982) [19], Thalmann (1968) [20] method described by Alef [12]; it is based on estimation of the rate of triphenyltetrazolium chloride reduction to triphenyl formazan (TPF) in soils after incubation at 28°C for 24 h. TPF was extracted with methanol. The TPF methanol extract was filtered and the optical density of the clear supernatant was measured against the blank at 485 nm (red color). The total organic carbon content (TOC) was determined using the Tiurin method after vaporizing the sample [16]. The values of pH were measured with a potentiometer in a 1-mol dm^{-3} KCl solution and in water.

Data analysis

The numbers of microorganisms in root-free-soil, rhizosphere soil, and roots (in three replicates from each treatment) were analyzed for significant differences ($P < 0.05$) by one-way (species) analysis of variance (ANOVA) using Microsoft® Excel 2000 [21]. The CFU of the microorganisms × g^{-1} dry mass of soil/root were subjected to analysis of variance (ANOVA), and means were separated using the Neuman–Keuls test at a 5% level of significance. The coefficients of genetic similarity of the investigated species were calculated. The coefficients were used to group the species hierarchically using the unweighted pair group method of arithmetic means. The phylogenetic similarity among species was presented in the form of a dendrogram. The relationships between observed traits were analyzed by correlation coefficients.

Results

The structure of the seed in SEM

A single *A. victorialis* plant produced 23–35 flowers, which contained between 69 and 105 developing embryos. After fertilization, the trilocular ovary contained a pericarp and a young seed (Fig. 1a). A single plant from the population growing in the natural habitat produced from 14 to 70 seeds, which were black, ovoid, and their size reached approximately 3 mm (Fig. 1b). The weight of a single seed ranged between 6.0 and 6.7 mg. The *A. victorialis* seed was covered by a thick testa with a raphe visible on its surface. The seed coat was made of strongly flattened cells. The seed coat exhibited irregularly reticulate surface pattern with polygonal epidermal cells (Fig. 1c). The seed interior was filled with the endosperm – a storage tissue containing store substances for the embryo (Fig. 1d). After 21 days, an embryo located peripherally and surrounded by the nutritive tissue developed inside the seed (Fig. 2a). In the early phase of embryonic development, the endosperm contained nutrients

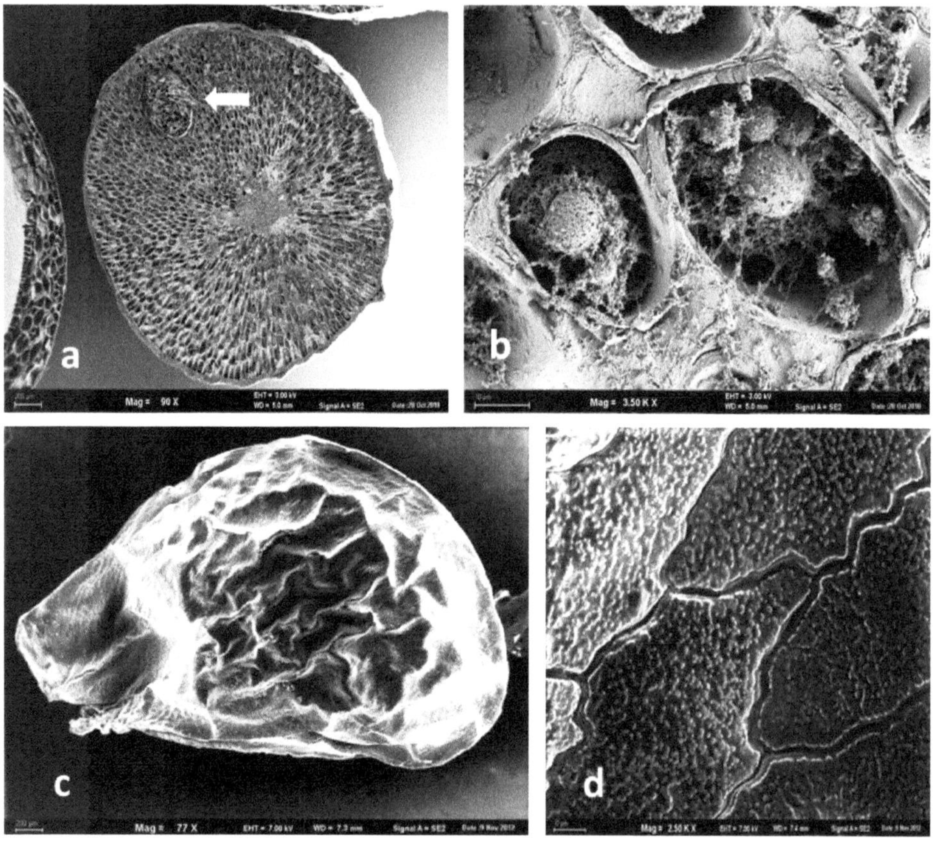

Fig. 2 *Allium victorialis* after 21 days of the in vitro experiment (**a**,**b**). **a** The seed cross section with peripheral location of a part of a young seedling (arrow). **b** Endosperm cells with protoplast remains; the cells have very thick walls; lack of intercellular spaces between them. **c** General view of the *A. tuberosum* seed. **d** Reticulate seed coat surface pattern showing irregularly polygonal testa cells of the *A. tuberosum* seed. Scale bars: **a**,**c** 200 μm; **b**,**d** 10 μm.

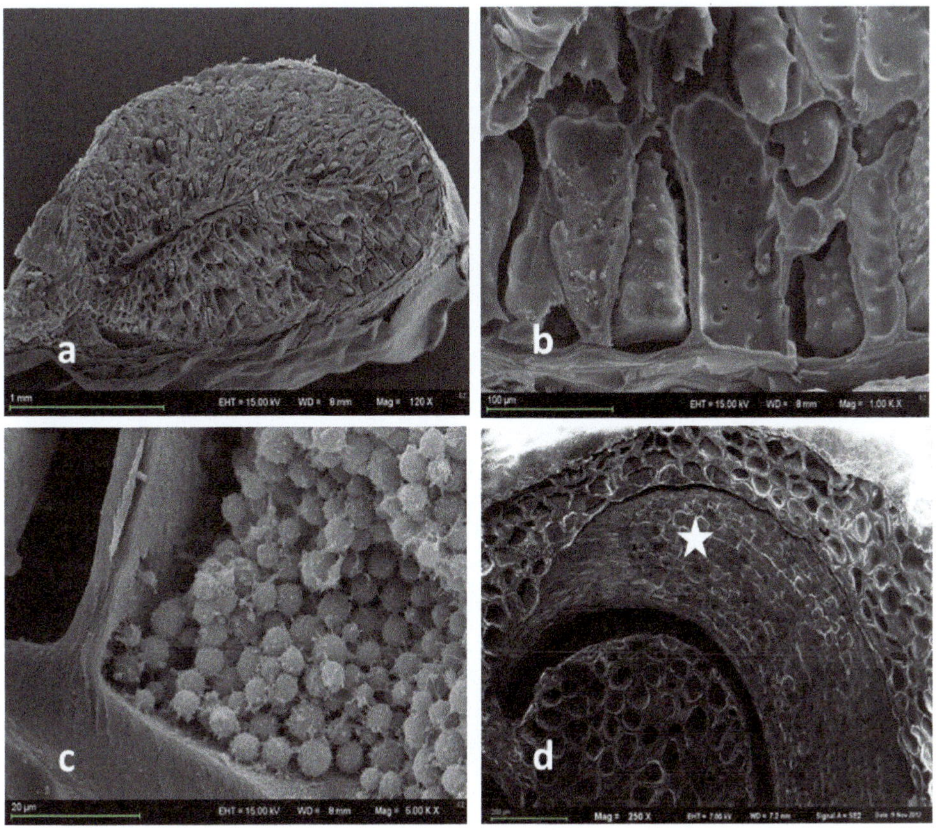

Fig. 3 **a** Cross section of the *Allium angulosum* mature seed. **b** The aleuronic palisade layer cells in the *A. angulosum* seed. There are alternately arranged pits in the cell wall. **c** Starch grains inside the endosperm cell. **d** Cross section of a young seedling in the *A. angulosum* seed (asterisk radicle).

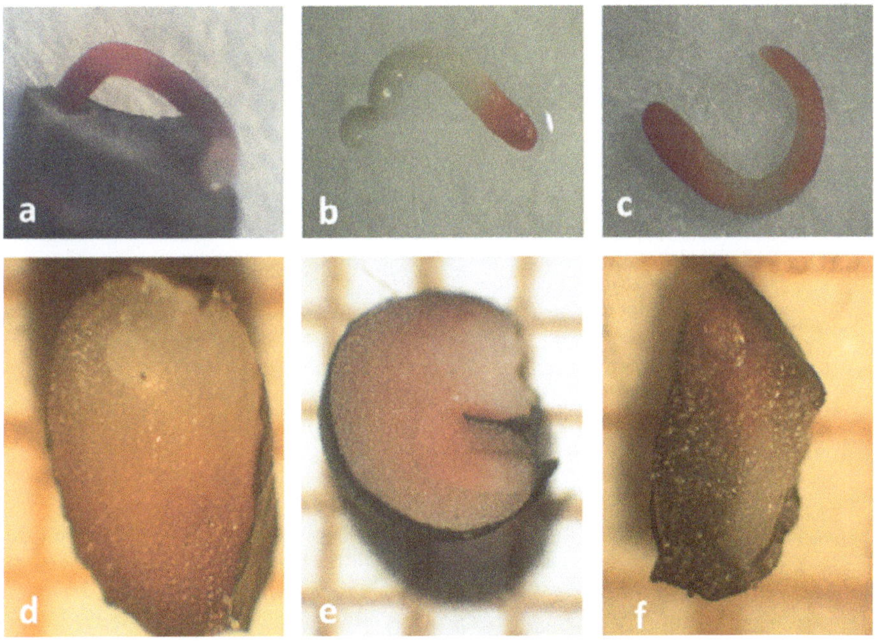

Fig. 4 Tetrazolinum test for seed viability and vigour carried out on day 21 of the experiment. **a,d** *Allium victorialis*. **b,e** *A. tuberosum*. **c,f** *A. angulosum*.

Tab. 2 Strength and energy germination of tested *Allium* species.

Allium species	Number of germinated seeds/180 seeds				The strength of germination after 10 days (%)	The energy of germination after 21 days (%)
	After 7 days	After 10 days	After 14 days	After 21 days		
A. tuberosum	60	98	113	172	54.44	95.55
A. angulosum	-	-	30	45	0	25
A. victorialis	-	-	-	-	0	0

Statistically significant differences ($P < 0.05$) were observed for all terms.

and energetic substances for the developing embryo. Protoplasts with starch grains were gradually consumed by the new developing plant – a young sporophyte (Fig. 2b). The seeds in all the *Allium* species examined differed in their size, shape, and surface sculpture (Fig. 1c,d, Fig. 2c,d, Fig. 3a). Immediately under the seed coat, there was an aleuronic layer of elongated cells arranged perpendicular to the seed coat (Fig. 3b). Isodiametric cells filled with numerous starch grains were located deeper (Fig. 3c).

After 21 days of the experiment, young developing embryos were observed in the seeds of all the species examined (Fig. 3d). The tetrazolium test revealed viability of the embryos developing inside the seeds (Fig. 4a–c).

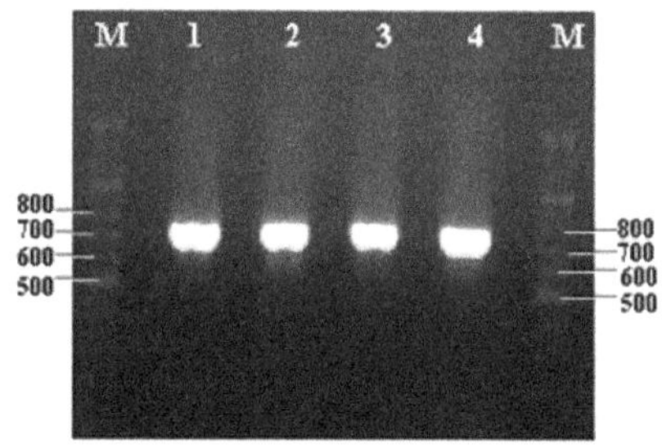

Fig. 5 Electrophoresis of PCR products amplified from leaves with ITS4 and ITS5 in agarose gel stained with ethidium bromide. Lanes: M – M-weight marker; 1 – *Allium angulosum*; 2 – *A. tuberosum*; 3 – *A. victorialis*.

Estimation of the strength and energy of germination

The experimental results presented in Tab. 2 reveal that the seed germination energy estimated after 10 days was the same (0%) in *A. angulosum* and *A. victorialis*, while in *A. tuberosum* it reached 54.16%. Moreover, *A. tuberosum* was characterized by high germination strength reaching 95.83%. After 21 days, *A. victorialis* seeds remained dormant. Statistically significant differences ($P < 0.05$) were observed for all terms.

In order to facilitate germination of the *A. victorialis* embryo, the seeds were incised with a scalpel to ensure better access of water and gases. Additionally, the seeds were stratified at 4°C for 7 days. Neither of the treatment applied activated development of the embryo, although it was proved viable by the tetrazolium test (Fig. 4d–f).

Phylogenetic analyses

After examination of germination strength and energy, an analysis was performed to assess the genetic similarity of *A. victorialis* to the other *Allium* species.

Amplification of the plant genetic material yielded amplicons of 700-bp length (Fig. 5). The phylogenetic analysis of nucleotide sequences in the DNA amplification products from *A. victorialis*, *A.uberosum* and *A. angulosum* revealed the highest similarity of the IST region between *A. victorialis* and *A. tuberosum* (Fig. 6). Despite the high homology of the primary structure of the genome fragment, the plants differ substantially in the germination strength. *Allium tuberosum* seeds were characterized by considerable germination strength and energy, whereas *A. victorialis* seeds did not exhibit any.

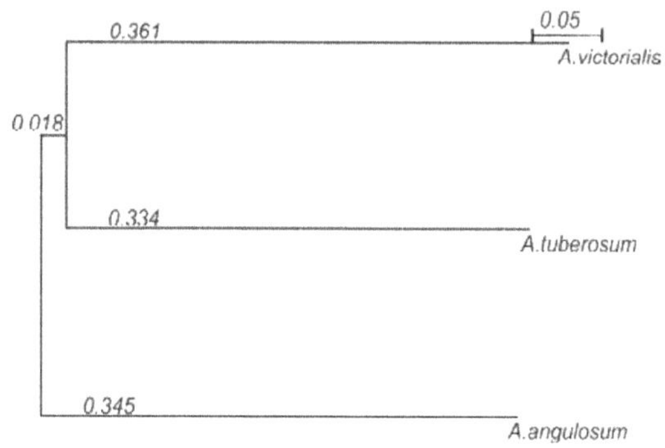

Fig. 6 Comparative phylogenetic analysis of the obtained ITS nucleotide sequences.

Changes in the number of microorganisms and some soil parameters in root-free-soil, rhizosphere soil and *Allium victorialis* roots

In order to find possible causes of lack of germination in the case of *A. victorialis* seeds, analysis of soil sampled from natural plant habitats was performed. The high number of fungi (4.5 $\log_{10}$CFU of fungi × g^{-1} dry mass of soil) and

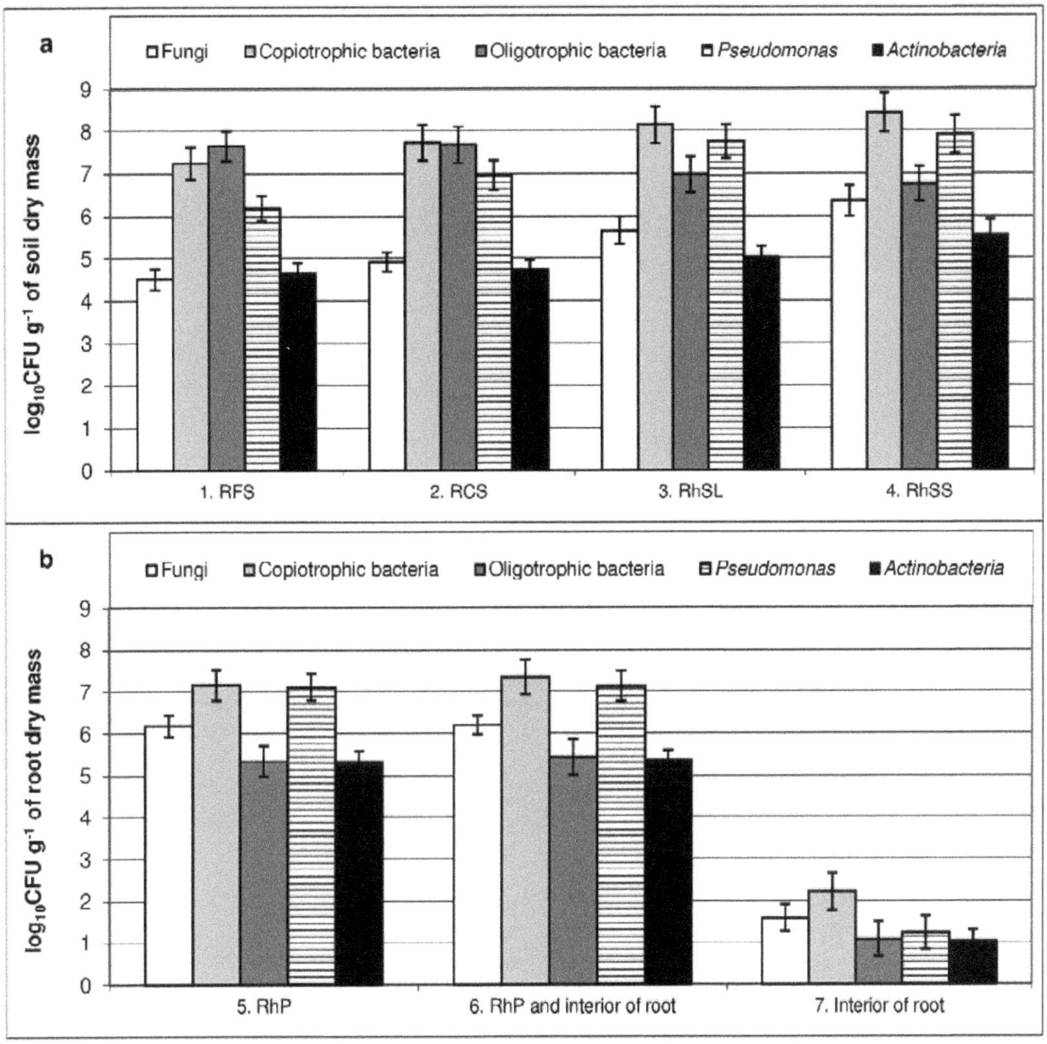

Fig. 7 Number of particular microorganisms in root- free-soil and garlic rhizosphere soil (**a**), and in garlic rhizoplane and root (**b**). RFS – root-free soil; RCS – root contaminant soil; RhSL – rhizosphere soil loosely attached to roots; RhSS – rhizosphere soil strongly attached to roots; RhP – rhizoplane.

various groups of bacteria (about 7.0 $\log_{10}$CFU of bacteria × g^{-1} dry mass of soil) were detected in the RFS around garlic roots (Fig. 7a). The number of CFU of the all tested groups of microorganisms increased in the RCS; however, the increase in the number of oligotrophic bacteria was not statistically significant. Only the number of CFU of oligotrophic bacteria decreased in the rhizosphere soil (Fig. 7b). The number of CFU of the other tested groups of microorganisms increased distinctly (about 10 fold) in the rhizosphere soil. The garlic rhizoplane was also inhabited by all the tested groups of microorganisms: especially by copiotrophic bacteria and bacteria from the genus *Pseudomonas*. The rhizoplane of garlic was colonized by above 6.0 $\log_{10}$CFU of fungi × g^{-1} dry mass of roots. In the interior of *A. victorialis* roots, the number of microorganisms decreased 1000 to 10 000 times but all the tested microbial groups, especially copiotrophic bacteria and fungi (1.6 and 2.2, respectively, $\log_{10}$CFU × g^{-1} dry mass of roots) were detected.

Changes in such parameters as dehydrogenase activity, pH values, and the TOC content in the particular parts of the rhizosphere and in comparison to the rhizosphere with root-free-soil were observed (Fig. 8). A gradual increase in dehydrogenase activity (Fig. 2a) and the TOC content (Fig. 8) from RFS to RhSS was detected as well as a gradual decrease in the pH values (Fig. 8b). The dehydrogenase activity and TOC content were highly positively correlated with the total number of CFU of the microorganisms (r = 0.918 and 0.93, respectively).

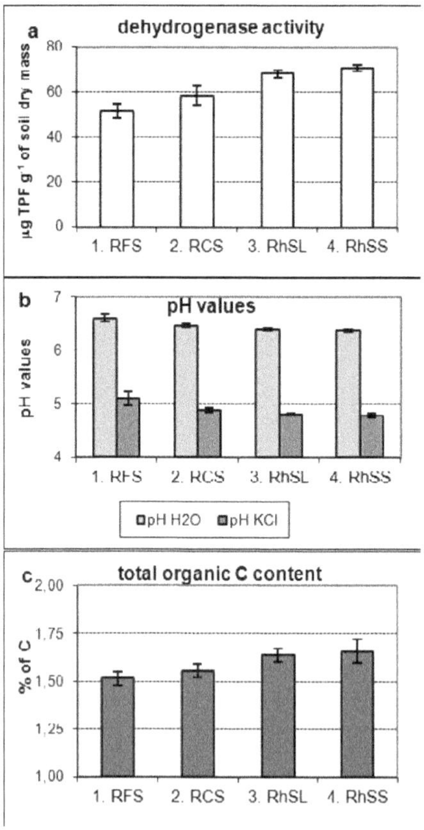

Fig. 8 Dehydrogenase activity (**a**), pH values (**b**) and total organic C content (**c**) in root-free-soil and garlic rhizosphere soil. RFS – root-free soil; RCS – root contaminant soil; RhSL – rhizosphere soil loosely attached to roots; RhSS – rhizosphere soil strongly attached to roots; RhP – rhizoplane.

Discussion

In European conditions, *A. victorialis* (victory onion, alpine leek) can be found in few, isolated localities or is introduced in botanical gardens. This implies a considerable impact of climatic conditions, geographical location, and the environment on the strength and energy of seed germination in the victory onion. The results of our scanning microscopy examinations have excluded anatomical and morphological abnormalities in the seed structure that could inhibit the germination process. All seed structures such as the embryo and endosperm developed normally, in a manner characteristic for all species from the *Allium* genus. Moreover, the tetrazolium test revealed viability of the embryos in the examined seeds. The present study consisted in tests of germination strength and energy in the laboratory under recommended optimal conditions for *Allium* germination outlined by the International Seed Testing Association [14]. Although all the recommendations were implemented, no germination was observed in *A. victorialis*. Additional scarification and stratification treatments did not improve the dynamics of *A. victorialis* seed germination. The germination strength and energy did not change and remained at a zero level, which may imply that the treatments cannot improve the generative reproduction of this species. Scarification of onion (*A. cepa*) seeds not only reduced germination parameters but also contributed to fungal infections [22].

Some authors emphasize the impact of bacteria, actinomycetes, and fungi on the rhizosphere and plant root tissue [23,24]. The above-mentioned microorganisms detoxify secondary metabolites, which in turn may inhibit seed germination in in vitro conditions [25,26]. Since young individuals developing from seeds can be found in nature, it was hypothesized that the germination process is correlated with soil microflora.

The studies on intensity of colonization of *A. victorialis* rhizosphere soil and garlic roots by particular microbial groups and dehydrogenase activity indicated that the rhizosphere effect is present in the garlic rhizosphere just as in many other plants [27,28]. The rhizosphere effect is often evaluated in terms of the *R*/*S* ratio, where *R* = the number of microbes in the rhizosphere and *S* = the number of similar microbes in bulk soil. Equally typical for plant rhizosphere, the number of fungal CFU was generally lower than the number of bacterial CFU. Literature data about microorganisms inhabiting the garlic rhizosphere and root is sparse. Garlic diseases such as basal rot disease in garlic infected by *Fusarium oxysporum* fungi [29] and white-rot disease in garlic infected by *Sclerotium cepivorum* fungi [30] are frequently described. There are also publications describing beneficial fungi (e.g. *Trichoderma harzianum*) and bacteria colonizing the root region of garlic plants and effective in biocontrol of fungal phytopathogens. Research papers provide information about arbuscular mycorrhizal fungi [31,32] rather than about other endophytic fungi colonizing garlic roots [33].

The present study and literature data do not allow identification of the causes of inhibited *A. victorialis* germination. Despite the high genetic homology between *A. victorialis* and *A. tuberosum*, the species differ markedly in their ability of seed germination. This may imply the effect of yet unknown gene mutations or varied expression of the same regions with the same genetic structure. Investigations into the physiology of seed germination have shown that some seeds will not germinate even in favorable habitat conditions, as they display absolute dormancy caused by, e.g. immaturity of the embryo or accumulation of inhibitors of growth and development [5]. An important stage in the germination process is initiation of the imbibition phase, which is only possible in conditions of high humidity and suitable temperature. These conditions were fulfilled in the present study but the seeds did not germinate; therefore, it can be concluded that other factors should be employed in overcoming seed dormancy. It seems advisable that the levels of abscisic acid and other germination inhibitors, e.g. phenols and their derivatives, should be determined, as their high concentrations may substantially inhibit *A. victorialis* germination. The similarity of 13 species and five ornamental cultivars of *Allium* was studied by Kamenetsky and Rabinowitch [34].

In order to determine unequivocally the causes of seed germination inhibition and low energy and strength, further investigations are indispensable focused on photoperiodic and temperature changes and biochemical analyses (especially of such growth regulators as auxins, gibberellins and abscisic acid). Solving the problem of inhibited *A. victorialis* germination will be of great importance for conservation of this valuable, endangered species.

Acknowledgments

This study was supported by the basic funds for maintaining research potential at the Faculty of Biology and Biotechnology of UMCS in Lublin. The authors would like to acknowledge two anonymous reviewers for constructive and helpful comments.

Authors' contributions

The following declarations about authors' contributions to the research have been made: hypothesis of the study, writing of the manuscript: KW; microscopical analysis: KW, KS; photographs: KW, KS; microbial analysis of root-free soil, rhizosphere soil and roots, dehydrogenase activity: JJŚ; data analysis: JB.

References

1. Moïse JA, Han S, Gudynaitę-Savitch L, Johnson DA, Miki BLA. Seed coats: structure, development, composition, and biotechnology. Vitro Cell Dev Biol Plant. 2005;41(5):620–644.
2. Mallek SB, Prather TS, Stapleton JJ. Interaction effects of *Allium* spp. residues, concentrations and soil temperature on seed germination of four weedy plant species. Appl Soil Ecol. 2007;37(3):233–239.
3. Beligni MV, Lamattina L. Nitric oxide stimulates seed germination and de-etiolation, and inhibits hypocotyl elongation, three light-inducible responses in plants. Planta. 2000;210(2):215–221.
4. Mathew SJ, Rembold M, Leptin M. Role for Traf4 in polarizing adherens junctions as a prerequisite for efficient cell shape changes. Mol Cell Biol. 2011;31(24):4978–4993.
5. Bewley JD. Seed germination and dormancy. Plant Cell. 1997;9(7):1055–1066.
6. Fritsch RM, Friesen N. Evolution, domestication, and taxonomy. In: Rabinowitch HD, Currah L, editors. *Allium* crop science: recent advances. Wallingford: CABI; 2002. p. 5–26.
7. Hanelt P. Alliaceae. In: Hanelt P, Büttner R, Mansfeld R, editors. Mansfeld's encyclopedia of agricultural and horticultural crops (except ornamentals). Vienna: Springer; 2001. p. 2250–2269.
8. Mirek Z, Piękoś-Mirkowa H, Zając A, Zając M, editors. Flowering plants and pteridophytes of Poland – a checklist. Cracow: W. Szafer Institute of Botany, Polish Academy of Sciences; 2002. (Biodiversity of Poland; vol 1).
9. Bilz M, Kell SP, Maxted N, Lansdown RV. European red list of vascular plants. Luxembourg: Publications Office of the European Union; 2011.
10. Kawano S, Nagai Y. Life-history monographs of Japanese plants. 4: *Allium victorialis* L. ssp. *platyphyllum* (Makino) Hulten (Alliaceae) Syn. *Allium victorialis* L. var. *platyphyllum* Makino; *A. latissimum* Prokh. Plant Species Biol. 2005;20(3):219–225.
11. Rätsch C. The encyclopedia of psychoactive plants: ethnopharmacology and its applications. Rochester, NY: Inner Traditions/Bear; 2012.
12. Alef K. Dehydrogenase activity. In: Alef K, Nannipieri P, editors. Methods in applied soil microbiology and biochemistry. London: Academic Press; 1995. p. 228–231.
13. Hirschegger P, Jakse J, Trontelj P, Bohanec B. Origins of *Allium ampeloprasum* horticultural groups and a molecular phylogeny of the section *Allium* (*Allium*: Alliaceae). Mol Phylogenet Evol. 2010;54(2):488–497.
14. International Seed Testing Association. International rules for seed testing. Bassersdorf: International Seed Testing Association; 2010.
15. Graber ER, Harel YM, Kolton M, Cytryn E, Silber A, David DR, et al. Biochar impact on development and productivity of pepper and tomato grown in fertigated soilless media. Plant Soil. 2010;337(1–2):481–496.
16. Lityński T, Jurkowska H, Gorlach E. Analiza chemiczno-rolnicza: przewodnik metodyczny do analizy gleby i nawozów. Warsaw: Polish Scientific Publishers PWN; 1976.
17. King EO, Ward MK, Raney DE. Two simple media for the demonstration of pyocyanin and fluorescin. J Lab Clin Med. 1954;44(2):301–307.
18. Martin JP. Use of acid, rose bengal, and streptomycin in the plate method for estimating soil fungi. Soil Sci. 1950;69(3):215–232.
19. Tabatabai MA. Soil enzymes. In: Page AL, Keeney DR, editors. Methods of soil analysis. Part 2. Chemical and microbiological properties. Madison, WI: American Society of Agronomy, Soil Science Society of America; 1982. p. 903–948.
20. Thalmann A. Zur Methodik der Bestimmung der Dehydrogenasеaktivität im Boden mittels Triphenyltetrazoliumchlorid (TTC). Landwirtsch Forsch. 1968;21:249–258.
21. Armitage P, Berry G, Matthews JNS. Statistical methods in medical research. Oxford: Blackwell; 1987.
22. Sardi P, Saracchi M, Quaroni S, Petrolini B, Borgonovi GE, Merli S. Isolation of endophytic *Streptomyces* strains from surface-sterilized roots. Appl Env Microbiol. 1992;58(8):2691–2693.
23. Azaizeh HA, Marschner H, Römheld V, Wittenmayer L. Effects of a vesicular-arbuscular mycorrhizal fungus and other soil microorganisms on growth, mineral nutrient acquisition and root exudation of soil-grown maize plants. Mycorrhiza. 1995;5(5):321–327.
24. Schroth MN, Hildebrand DC, Starr MP. Phytopathogenic members of the genus *Pseudomonas*. In: Starr MP, Stolp H, Trüper HG, Balows A, Schlegel HG, editors. The Prokaryotes. Berlin: Springer Berlin Heidelberg; 1981. p. 701–718.
25. Leifert C, Murphy K, Lumsden P. Mineral and carbohydrate nutrition of plant cell and tissue cultures. Crit Rev Plant Sci. 1995;14(2):83–109.
26. Kuzyakov Y. Review: factors affecting rhizosphere priming effects. J Plant Nutr Soil Sci. 2002;165(4):382–396.
27. Raaijmakers JM, Paulitz TC, Steinberg C, Alabouvette C, Moënne-Loccoz Y. The rhizosphere: a playground and battlefield for soilborne pathogens and beneficial microorganisms. Plant Soil. 2009;321(1–2):341–361.
28. Fatawi ZD, Wiyono H, Widono S. Analysis of rhizosphere bacterial community in suppressive and conducive soils to basal rot of garlic based on PCR-RISA. Biomirror. 2012;3(2):9–12.
29. Avila Miranda ME, Herrera Estrella A, Peña Cabriales JJ. Colonization of the rhizosphere, rhizoplane and endorhiza of garlic (*Allium sativum* L.) by strains of *Trichoderma harzianum* and their capacity to control allium white-rot under field conditions. Soil Biol Biochem. 2006;38(7):1823–1830.
30. Rajeshkumar S, Nisha MC. Diversity of arbuscular mycorrhizal fungi associated with some garlic varieties in Nilgiri District. Herb Tech Ind. 2012;3:1–3.
31. Thangavelu M, Tamilselvi V. Occurrence and morphology of endorhizal fungi in crop species. Trop Subtrop Agroecosyst. 2010;12(3):593–604.
32. Rodriguez RJ, White JF, Arnold AE, Redman RS. Fungal endophytes: diversity and functional roles. New Phytol. 2009;182(2):314–330.
33. Krzymińska A, Gawłowska M, Wolko B, Bocianowski J. Genetic diversity of ornamental *Allium* species and cultivars assessed with isozymes. J Appl Genet. 2008;49(3):213–220.
34. Kamenetsky R, Rabinowitch HD. Floral development in bolting garlic. Sex Plant Reprod. 2001;13(4):235–241.

Investigation on the pollen morphology of traditional cultivars of *Prunus* species in Sicily

Anna Geraci*, Vincenza Polizzano, Pasquale Marino, Rosario Schicchi
Department of Environmental Biology and Biodiversity, University of Palermo, Via Archirafi, 38-90123, Italy

Abstract

In this study pollen grains of 13 cultivars and 3 rootstocks belonging to 5 species (*P. armeniaca*, *P. domestica*, *P. dulcis*, *P. persica*, *P. avium*) of the genus *Prunus* collected from North-East Sicily were examined for the micromorphological characterization through the scanning electron microscopy (SEM). The length of polar axis (*P*) and the equatorial diameter (*E*) of grain, *P/E* ratio, the length of colpi (*C*), diameter of perforations (*DP*) and the number of perforations in 25 μm² (*PN*), the width of muri (*WM*), the distance between muri (*DM*) and their number in 25 μm² (*MN*), the width of grooves (*WG*) were measured and their variation was compared among studied taxa. Moreover multivariate statistical analysis was carried out to distinguish morphometric information from measured parameters. All pollen grains are trizonocolpate, isopolar, medium-large sized and their shape varies from prolate to perprolate. Regarding outline pollen grains are subtriangular in polar view and elliptic in equatorial view. Exine sculpturing is striate with perforations on grain surface. The arrangement of ridges appears roughly parallel but too sloped (sometimes curved) compared to polar axis, or branched and oriented in different directions, or perfectly parallel or more irregular with bifurcated ridges often sinuous. The analyses showed a great variability (particularly in *P. domestica* cultivars) related in some cases to the diversity in the morphological features of the leaves and the fruits of the investigated entities.

Keywords: pollen morphology, SEM, Sicilian *Prunus*, species, cultivars

Introduction

The cultivation of fruit trees and shrubs in the traditional agrosystems of Sicily played a significant role until the years after World War II. Its products were a primary source of nutrition for rural people but also a consistent integration of their income. It was a non-specialized fruit-growing based on mixed systems in which the richness of species and cultivars was primarily aimed at providing fresh and durable fruits for their family consumption.

Since the eighth century BC, Sicily has been subjected to various dominations (Phoenician, Greek, Roman, Arab, Norman) who have contributed to the introduction of different tree species and fruit trees. The wide range of varieties selected over time by the rural communities, closely linked to the economy of the past and the local agricultural practices, has been preserved until half of the twentieth century in the context of traditional agrosystems [1].

* Corresponding author. Email: anna.geraci@unipa.it

Prunus belongs to the Rosaceae family, subfamily Prunoideae. This genus includes several both wild and cultivated species, mainly deciduous shrubs and trees, distributed in tropical, temperate and sometimes in semiarid regions of the northern hemisphere. It represents an important economical resource because many species produce edible fruits (e.g. plums, peaches, apricots, cherries and almonds), oil, wood and some species are employed as ornamental plants [2].

Prunus is one of the most problematic genera in the family of Rosaceae and its taxonomy is very complicated because of the polymorphism and the wide ecological tolerance of the species, as well as the presence of an enormous number of cultivars [3].

For the identification of plant material, morphological characteristics such as plant vigour and leaf, flower, and fruit parameters could be utilized [4]. In addition to these features, differences in the size and the surface ultrastructure of pollen grain are a useful tool for distinguishing species and cultivars of fruit trees. Using pollen morphology in identification has become more important since the advent of the scanning electronic microscope (SEM) [5–13].

Taxonomists and botanists have recognized the importance of pollen morphology in clarifying the classification of many plants: in particular Hebda et al. [10] showed that exine sculpturing, aperture and aperture zone structure, grain shape and grain size are all useful features to distinguish genera and even species of rosaceous pollen; Fogle [5,6] demonstrated species-specific variation in pollen size and in exine sculpturing among the fruit-crop of the genera *Malus*

Mill., *Prunus* L. (peaches and cherries), and *Pyrus* L. [14]; Moore et al. [15] have emphasized a very large differentiation of pollen morphology of the taxa of Rosaceae family, even among the populations within the same species. They thought the pollen size to be the least reliable feature and they related it to the comparatively frequent occurrence of hybrid and polyploid species in this family. In the subgenera *Sorbus*, *Chamaespilus* and *Torminaria* belonging to the genus *Sorbus*, pollen morphology corresponds with their systematic classification [16], while in the genus *Rosa*, a recent study showed that the exine sculpture can be helpful in distinguishing groups of species although only slightly corroborates division of the genus in subgenera and sections [17].

The purpose of this study is to increase knowledge of the aspects related to the variability of the genus *Prunus* fruit trees in order to facilitate the emergence of initiative aimed at the characterization and preservation of the germoplasm. For this reason SEM analysis was conducted to characterize the pollen morphology of some Sicilian *Prunus* cultivars and to evaluate the utility of pollen features in order to develop an additional alternative plant material identification tool.

Tab. 1 The species and cultivars investigated in this work.

Species	Cultivar – rootstock	ID
P. armeniaca	"Arancinu"	A1
	"Pirmintìu"	A2
	rootstock	A3
P. domestica	"Sittimmirinu"	B1
	"Rapparinu"	B2
	"Occhi di voi niuri"	B3
	"Occhi di voi biancu"	B4
	"Minacore"	B5
P. avium	"Cappuccia niura"	C1
	"Maiulina"	C2
P. dulcis var. *dulcis*	"Cavalera"	D1
	"Rappa"	D2
	rootstock	D3
P. dulcis var. *amara*	rootstock	D4
P. persica subsp. *persica*	"Pirmintìu"	E1
P. persica subsp. *nucipersica*	"Spaccarelli"	E2

Material and methods

In this study the pollen grains coming from 13 cultivars and 3 rootstocks (plants obtained from seeds) belonging to four species of the genus *Prunus*; in particular *P. armeniaca* L. (apricot), *P. persica* (L.) Batsch (peach), *P. domestica* L. (plum), *P. dulcis* (Miller) D. A. Webb (almond) were examined (Tab. 1).

The plant samples were collected from fruit-trees grown in orchards or in abandoned fields, both situated in different localities of Madonie Mts. in North-Eastern side of Sicily. The flower buds of cultivars belonging of *Prunus* species were taken at the balloon stage (just before the opening of the anthers) in order to collect pollen; sepals and petals of the buds were removed, and anthers were isolated, put into petri dishes and kept at room temperature for 24 h. Afterwards, the pollen was sprinkled on the surface. The pollen was further dried at room temperature for 6–8 h and put into vials with silica gel until require for SEM evaluation. Dry pollen grains were sifted onto transparent double-sided tape on the disc surface of polished aluminum stabs. The sample on each stub was sputter-coated with a gold layer ca. 200 A thickness in a vacuum evaporator under certain pressure.

After that, for SEM evaluation, the pollen grains were observed and photographed at 446–2000× for the whole grain and at 6000–9000× for sectioning the exine pattern. For each cultivars and rootstocks, 35 pollen grains were employed to determine the pollen size and the following morphological traits were measured: (***i***) length of pollen along the polar axis (*P*); (***ii***) length of pollen along the equatorial diameter (*E*); (***iii***) ratio of polar axis to equatorial diameter (*P/E*); (***iv***) diameter of perforations (*DP*); (***v***) number of perforations in 5 µm × 5 µm (*PN*); (***vi***) width of muri (*WM*); (***vii***) number of muri in 5 µm × 5 µm (*MN*); (***viii***) distance between muri (*DM*); (***ix***) width of grooves (*WG*); (***x***) length of colpi (*C*). The terminology follows that of Erdtman [18,19] and Hesse et al. [20].

In order to find out the number of perforations and the number of muri on the exine surface, perforations and muri in 25 μm^2 were counted in 5–10 different locations on the surface. In order to calculate the diameter of the perforations, the width and the distance between muri and the width of grooves, for each of these parameters 25–35 different measures were conducted.

Statistical analyses of principal component analysis (PCA) component and matrix of similarity were realized using PAST software [21].

Results

General pollen characteristics

Pollen grains of all investigated species are symmetric, isopolar, monad, trizonocolpate (with three germinal furrows or colpi), medium-large sized. Regarding outline, in polar view, pollen grains are subtriangular and in equatorial view are elliptic (Fig. 1, Fig. 2, Tab. 2). The mean length of the polar axis varies from 38.54 µm in B3 (*P. domestica* "Occhi di voi niuru") to 63.46 µm in E1 (*P. persica* "Pirmintìu"). The mean of equatorial diameter varies from 20.39 µm in B3 (*P. domestica* "Occhi di voi niuru") to 32.41 µm in D1 (*P. dulcis* "Cavalera"). The shape of dried pollen grains is mainly prolate (P/E = 1.33–2) except in A1 (*P. armeniaca* "Arancinu") and E1 (*P. persica* "Pirmintìu"), both with perprolate pollen grains (P/E > 2). The mean P/E ratio ranges from 1.58 in D1 (*P. dulcis* "Cavalera") to 2.02 in E1 (*P. persica* "Pirmintì") and A1 (*P. armeniaca* "Arancinu"). The smallest mean of pollen grains length was found in B3 and B4 (*P. domestica* "Occhi di voi niuru" and "Occhi di voi biancu") and the largest mean pollen length occurs in E1 (*P. persica* "Pirmintìu").

Regarding exine sculpture, the ornamentations (muri) are described as striate type and they are separated by grooves, which appear longer in B1 (*P. domestica* "Sittimmirinu") and shorter in A2 (*P. armeniaca* "Pirmintiù"). The surface of the pollen shows the largest perforations in B3 (*P. domestica* "Occhi di voi niuru") and the smallest in D2 and D3 (*P. dulcis* "Rappa" and rootstock of *P. dulcis* var. *dulcis*) and the largest number of perforations is found in E1 (*P. persica* "Pirmintìu") and the smallest in rootstock of *P. dulcis* var. *amara* (Fig. 3, Fig. 4, Tab. 2).

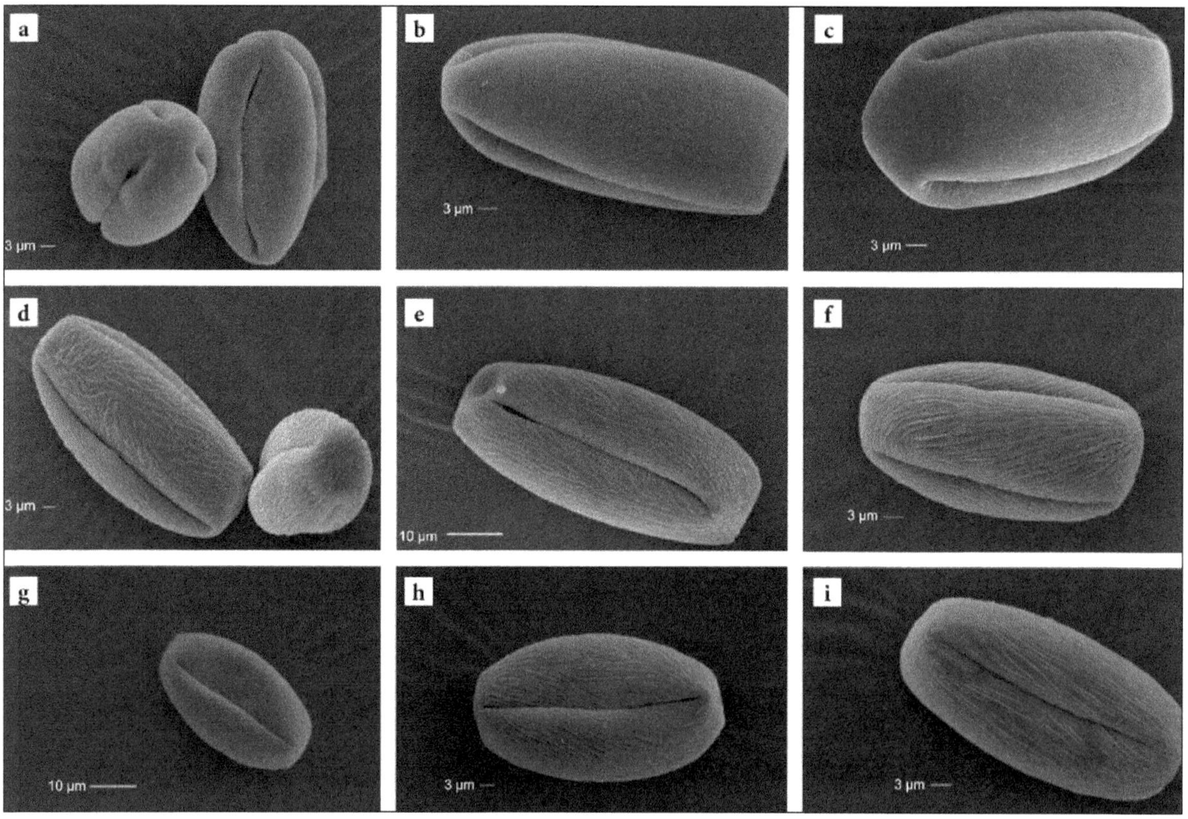

Fig. 1 Pollen shape of *Prunus* species and cultivars. **a–c** *P. armeniaca*: **a** "Arancinu"; **b** "Pirmintìu"; **c** Rootstock. **d–h** *P. domestica*: **d** "Sittimmirinu"; **e** "Rapparinu"; **f** "Occhi di voi niuri"; **g** "Occhi di voi biancu"; **h** "Minacore". **i** *P. avium* "Cappuccia niura".

The species investigated: pollen morphological characterization

P. ARMENIACA TYPE

Two cultivars and one rootstock of *P. armeniaca* (Tab. 1) were examined. All the pollen grains are large-sized (Fig. 1a–c), while the shape varies from prolate to perprolate (Fig. 1a–c). The mean of the polar diameter ranges between 52.92 and 54.30 µm and the width varies from 26.64 to 28.27 µm. The *P/E* ratio ranges between 1.92 and 2.02. The mean length of colpi changes from 46.36 to 48.38 µm (Tab. 2).

Concerning the exine sculpturing, the muri appear roughly parallel but too sloped compared to the polar axis and sometimes curved at the end near the colpi, ± bifurcated and often intersecting (Fig. 3a–c). On the pollen surface small-sized perforations are distinguishable (0.29–0.30 µm), which are dense in A1 (20.70 ±1.93 in 25 µm^2), and more reduced and sparsely distributed in A2 (15.00 ±2.70) and in the rootstock (17.60 ±3.43). The width of muri ranges between 0.29 and 0.35 µm (relatively narrow), while the arbitrary distance between them varies from 0.38 to 0.45 µm. The number of muri calculated in 25 µm^2 ranges between 16.20 and 21.00, while the width of grooves changes from 0.18 to 0.23 µm (very narrow).

P. DOMESTICA TYPE

Five cultivars of this species were investigated (Tab. 1). The pollen grains show variable sizes (Fig. 1d–h, Tab. 2) from medium (B3 and B4) to large (B1, B2, B5), while the shape is prolate. The mean length of polar axis ranges between 38.54 µm and 56.69 µm and the length of equatorial diameter varies from 20.39 µm to 29.52 µm. The length/width ratio changes from 1.88 to 1.97. The length of germinal furrows varies from 31.13 to 48.58 µm (Fig. 5).The arrangement of muri shown by the five cultivars is classifiable in the three following arbitrary ways (Fig. 3d–h):

(*i*) In B1 ("Sittimmirinu"), the muri are very well marked, branched, often intersecting and oriented in different directions (irregular pattern; Fig. 3d). The large muri (0.52 ±0.11 µm) are separated by long intervals (*DM* = 0.64 ±0.18 µm) and by large grooves (0.50 ±0.12 µm) with small-sized perforations (0.30 ±0.08 µm). The number of perforations in 25 µm^2 is 17.80 ±3.12, while the number of muri in 25 µm^2 is 11.60 ±1.71.

(***ii***) In B2 ("Rapparinu"), the relatively large (0.47 ±0.07 µm) muri appear roughly parallel but too sloped respect to polar axis, sometimes curved (the exine arrangement is similar to *P. armeniaca*, but more irregular), often intersecting and ± bifurcated with very abundant (*PN* = 23.20 ±1.80) small-sized perforations situated in relatively short grooves (*WG* = 0.36 ±0.06 µm). The number of muri in 25 µm^2 is 14.00 ±0.88 and the distance between them is 0.56 ±0.08 µm (Fig. 3e).

(***iii***) In B3 ("Occhi di voi niuru"), the muri are comparable to B2 but arrangement is more regular and separated by very narrow grooves (0.28 ±0.06 µm) with relatively large perforations (0.36 ±0.09 µm; Fig. 3f). A similar arrangement of muri was observed in B4 ("Uocchi ri voi biancu") and B5 ("Minacore"; Fig. 3g,h) too, but the former presents relatively large perforations (0.35 ±0.10 µm), and relatively narrow muri (0.38 ±0.07µm) separated by shorter grooves (0.35 ±0.11 µm), the latter shows small perforations (0.29 ±0.07 µm), large muri (like as B3 "Occhi di voi niuru") separated by large grooves (*WG* = 0.42 ±0.09 µm). The number of perforations in 25 µm^2

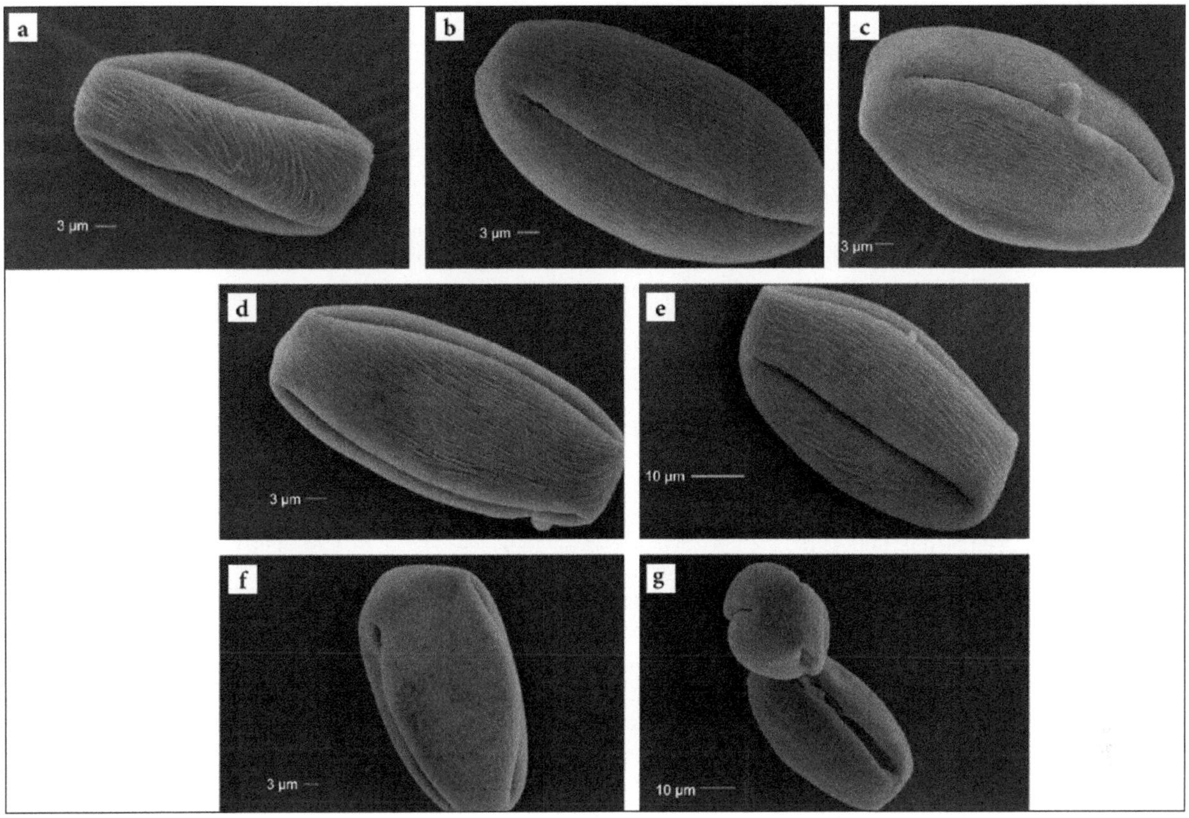

Fig. 2 Pollen shape of *Prunus* species and cultivars. **a** *P. avium* "Maiulina". **b–d** *P. dulcis* var. *dulcis*: **b** "Cavalera"; **c** "Rappa"; **d** rootstock. **e** *P. dulcis* var. *amara* rootstock. **f** *P. persica* subsp. *persica* "Pirmintiu". **g** *P. persica* subsp. *nucipersica* "Spaccarelli".

is not variable in B4 and B5 (15.20, 15.40), while the number of muri in 25 µm^2 is 13.60 ±1.00 and 11.20 ±1.69, respectively. The distance between the muri is longer in B5 ("Minacore") (0.64 ±0.11 µm; Fig. 5).

P. AVIUM TYPE

Regarding this species, only two cultivars were examined (Tab. 1), both showing medium sized pollen grains and prolate shape (Fig. 1i, Fig. 2a, Tab. 2). The length of polar axis ranges between 45.74 and 48.20 µm, while the width of grain varies from 23.55 to 24.78 µm. The length/width ratio is similar (1.95 and 1.96). The length of colpi changes from 37.60 to 40.07 µm. The exine pattern (Fig. 3i, Fig. 4a) is not variable in arrangement compared to those observed in *P. armeniaca* and in *P. domestica*. The muri appear relatively narrow (0.34, 0.35 µm) like in apricot, separated by relatively short grooves (0.36 µm). The perforations are very small (0.26, 0.28 µm) and sparsely distributed (13.00, 14.20 in 25 µm^2). The arbitrary distance between the muri is not very different (0.55, 0.56 µm), such as the muri number calculated in 25 µm^2 (12.80, 13.20).

P. DULCIS TYPE

For this species two cultivars and two rootstocks were compared (Tab. 1). All the examined pollen grains are large sized and prolate shaped. The length of polar axis varies from 50.48 to 58.42 µm (rootstock of *P. dulcis*), while the width of the grain ranges between 29.18 and 32.41 µm. The length/width ratio varies from 1.58 to 1.98 (Fig. 2b–e, Tab. 2).

The exine pattern is very regular (Fig. 4b–e); in fact the muri are parallel to the colpi, very bifurcated, sometime fused. The diameter of perforations ranges between 0.24 and 0.32 µm (from very small to medium-sized), while their mean number counted in 25 µm^2 varies from 11.00 to 17.40 (very few in D4 *P. dulcis* var. *amara*). The width of muri changes from 0.31 to 0.37 µm (relatively narrow) and the arbitrary distance between muri ranges from 0.57 to 0.64 µm (rather wide). The number of muri counted in 25 µm^2 results 10.40–15.20, while the width of grooves varies from 0.34 to 0.40 µm. (medium in D2 and in the two rootstocks, large in D1).

P. PERSICA TYPE

Regarding this species the pollen from two cultivars belonging two different subspecies (Tab. 1) was observed. The examined samples show large-sized pollen grains but their shape varies from prolate to perprolate (Tab. 2). The length of polar axis (equatorial view) is 52.90 and 63.46 µm and the width of equatorial diameter is 30.21 and 31.49 µm. The length/width ratio is 1.76 and 2.02. The length of colpi is 46.64 and 55.07. The diameter of perforations is not variable (0.30 and 0.31 µm, small-sized), while the number of perforations in 25 µm^2 is respectively 19.20 and 25.60 (the pits are very abundant in E1). The width of muri varies from 0.40 to 0.46 µm and their arbitrary distance is 0.58 and 0.62 µm. The number of muri in 25 µm^2 is respectively 10.40 and 12.60 (few), while the width of grooves is 0.39 and 0.40 µm. Concerning exine arrangement we found some differences: in E1 ("Pirmintiu"), the exine pattern appears more irregular with bifurcated muri, often sinuous and sometime merged (Fig. 4f); in E2 ("Spaccarelli") the exine pattern is regular (Fig. 4g) with muri oriented like in *P. dulcis*.

Tab. 2 Details of examined pollen characters of *Prunus* species and cultivars.

ID	Species/cultivar/ rootstock	*P*	*E*	*P/E*	*DP*	*PN*	*WM*	*MN*	*DM*	*WG*	*C*
	P. armeniaca										
A1	"Arancinu"	48.06–58.40	24.31–28.65	1.68–2.25	0.09–0.45	19.00–24.50	0.18–0.55	15.00–19.00	0.27–0.64	0.18–0.36	44.52–50.32
		53.73 ±2.34	26.64 ±0.86	2.02 ±0.12	0.30 ±0.07	20.70 ±1.93	0.35 ±0.06	17.00 ±1.39	0.45 ±0.07	0.23 ±0.04	48.19 ±2.28
A2	"Pirmintìu"	46.30–57.28	26.15–31.87	1.70–2.05	0.12–0.59	10.00–18.00	0.24–0.47	19.00–23.00	0.29–0.47	0.12–0.24	42.62–49.57
		54.30 ±1.91	28.27 ±1.03	1.92 ±0.07	0.29 ±0.09	15.00 ±2.70	0.30 ±0.08	21.00 ±1.64	0.38 ±0.04	0.18 ±0.03	46.36 ±2.21
A3	rootstock	49.33–56.66	24.45–29.21	1.83–2.18	0.13–0.50	14.00–24.00	0.25–0.38	14.00–19.00	0.27–0.55	0.13–0.38	46.25–50.00
		52.92 ±1.38	26.77 ±1.08	1.98 ±0.09	0.29 ±0.07	17.60 ±3.43	0.29 ±0.03	16.20 ±1.90	0.38 ±0.55	0.23 ±0.05	48.38 ±1.43
	P. domestica										
B1	"Sittimmirinu"	50.46–55.77	24.49–29.03	1.74–2.09	0.12–0.59	14.00–23.00	0.35–0.88	9.00–14.00	0.35–1.18	0.24–0.82	41.33–50.67
		52.96 ±0.96	26.91 ±0.89	1.97 ±0.07	0.30 ±0.08	17.80 ±3.12	0.52 ±0.11	11.60 ±1.71	0.64 ±0.18	0.50 ±0.12	47.20 ±2.82
B2	"Rapparinu"	47.81–55.86	24.63–33.42	1.63–2.09	0.18–0.55	20.00–25.00	0.36–0.73	13.00–15.00	0.36–0.73	0.18–0.55	43.20–49.60
		53.82 ±1.49	28.74 ±1.67	1.88 ±0.09	0.30 ±0.08	23.20 ±1.80	0.47 ±0.07	14.00 ±0.88	0.56 ±0.08	0.36 ±0.06	47.04 ±2.35
B3	"Uocchi ri voi	31.28–41.16	16.62–23.47	1.74–2.04	0.18–0.55	20.00–29.00	0.36–0.73	11.00–16.00	0.36–0.82	0.18–0.36	27.50–35.00
	nivuri"	38.54 ±1.80	20.39 ±1.25	1.90 ±0.07	0.36 ±0.09	24.60 ±3.20	0.46 ±0.07	13.20 ±1.90	0.59 ±0.09	0.28 ±0.06	31.13 ±2.75
B4	"Uocchi ri voi	33.05–42.89	17.96–24.52	1.35–2.19	0.18–0.64	13.00–20.00	0.18–0.55	12.00–15.00	0.36–0.82	0.18–0.73	32.31–40.00
	biancu"	39.62 ±1.95	21.22 ±1.38	1.96 ±0.15	0.35 ±0.10	15.20 ±2.43	0.38 ±0.07	13.60 ±1.00	0.51 ±0.09	0.35 ±0.11	36.00 ±3.10
B5	"Minacore"	53.28–62.79	26.74–32.69	1.68–2.20	0.18–0.55	13.00–18.00	0.36–0.55	9.00–14.00	0.36–1.00	0.18–0.64	44.52–51.85
		56.69 ±1.94	29.52 ±1.29	1.96 ±0.12	0.29 ±0.07	15.40 ±1.59	0.46 ±0.06	11.20 ±1.69	0.64 ±0.11	0.42 ±0.09	48.58 ±2.66
	P. avium										
C1	"Cappuccia niura"	45.25–50.47	20.51–27.04	1.83–2.26	0.12–0.47	10.00–17.00	0.24–0.47	11.00–16.00	0.35–0.71	0.24–0.71	36.00–40.00
		48.20 ±1.02	24.78 ±1.15	1.95 ±0.07	0.28 ±0.07	13.00 ±2.70	0.35 ±0.04	13.20 ±1.69	0.55 ±0.06	0.36 ±0.10	37.60 ±1.92
C2	"Maiolina"	42.48–48.04	18.71–26.03	1.81–2.33	0.12–0.47	10.00–20.00	0.24–0.41	11.00–14.00	0.35–0.76	0.24–0.59	35.00–44.00
		45.74 ±1.21	23.55 ±1.44	1.96 ±0.10	0.26 ±0.08	14.20 ±3.30	0.34 ±0.04	12.80 ±1.14	0.56 ±0.07	0.36 ±0.09	40.07 ±2.87
	P. dulcis* var. *dulcis										
D1	"Cavalera"	46.19–56.61	27.19–38.51	1.32–2.02	0.09–0.45	7.00–16.00	0.18–0.55	12.00–14.00	0.36–0.91	0.18–0.64	40.53–50.00
		50.48 ±2.14	32.41 ±1.91	1.58 ±0.14	0.28 ±0.07	13.00 ±3.10	0.37 ±0.07	12.40 ±0.78	0.60 ±0.10	0.40 ±0.08	45.68 ±3.32
D2	"Rappa"	49.37–54.39	27.20–34.37	1.52–1.97	0.06–0.53	14.00–21.00	0.24–0.41	14.00–17.00	0.41–1.00	0.24–0.47	41.18–45.88
		51.50 ±1.02	29.84 ±1.61	1.74 ±0.10	0.24 ±0.09	17.40 ±2.53	0.31 ±0.04	14.80 ±1.14	0.57 ±0.11	0.34 ±0.05	44.24 ±1.56
D3	rootstock	55.97–61.13	26.65–31.65	1.79–2.08	0.13–0.38	14.00–20.00	0.25–0.38	13.00–16.00	0.38–0.88	0.13–0.75	48.46–56.54
		58.42 ±1.04	29.55 ±1.14	1.98 ±0.08	0.24 ±0.06	16.00 ±2.06	0.32 ±0.04	15.20 ±1.14	0.64 ±0.11	0.35 ±0.10	53.31 ±2.71
	P. dulcis* var. *amara										
D4	rootstock	49.95–58.76	25.76–32.33	1.68–2.22	0.13–0.63	6.00–16.00	0.25–0.50	9.00–12.00	0.44–1.13	0.13–0.69	45.00–52.00
		55.86 ±1.79	29.18 ±1.07	1.92 ±0.10	0.32 ±0.11	11.00 ±3.16	0.36 ±0.06	10.40 ±1.00	0.64 ±0.12	0.34 ±0.10	49.90 ±2.49
	P. persica* subsp. *persica										
E1	"Pirmintìu"	58.50–67.98	29.34–33.22	1.88–2.14	0.09–0.73	20.00–28.00	0.36–0.55	11.00–14.00	0.36–0.73	0.18–0.73	51.98–58.53
		63.46 ±1.55	31.49 ±0.81	2.02 ±0.05	0.30 ±0.12	25.60 ±2.81	0.40 ±0.04	12.60 ±1.33	0.58 ±0.08	0.40 ±0.10	55.07 ±2.07
	P. persica subsp. nucipersica										
E2	"Spaccarelli"	46.51–59.80	26.81–33.09	1.55–2.02	0.09–0.55	13.00–23.00	0.36–0.73	9.00–12.00	0.36–0.91	0.27–0.45	44.17–50.10
		52.90 ±2.44	30.21 ±1.15	1.76 ±0.11	0.31 ±0.08	19.20 ±3.47	0.46 ±0.08	10.40 ±1.00	0.62 ±0.12	0.39 ±0.04	46.64 ±1.89

C – length of colpi; *DM* – distance between muri; *DP* – diameter of perforations; *E* – length of equatorial diameter; *MN* – number of muri in 25 μm^2; *P* – length of polar axis; *P/E* – ratio of polar axis to equatorial diameter; *PN* – number of perforations in 25 μm^2; *WG* – width of grooves; *WM* – width of muri.

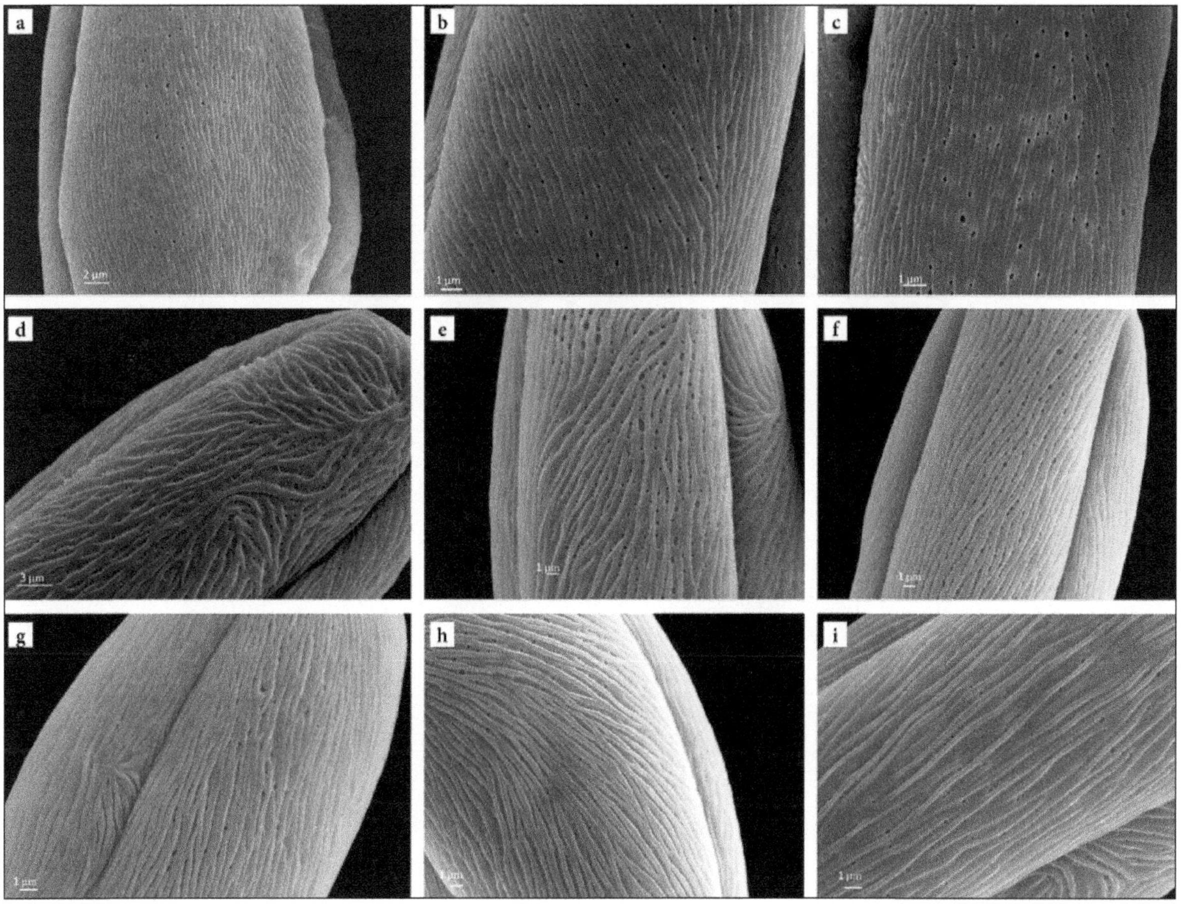

Fig. 3 Exine patterns of *Prunus* species and cultivars. **a–c** *P. armeniaca*: **a** "Arancinu"; **b** "Pirmintìu"; **c** rootstock. **d–h** *P. domestica*: **d** "Sittimmirinu"; **e** "Rapparinu"; **f** "Occhi di voi niuri"; **g** "Occhi di voi biancu"; **h** "Minacore". **i** *P. avium* "Cappuccia niura".

Statistical and cluster analysis

The resultant data of the morphological characteristics of pollen grains were statistically analysed.

Statistical analysis was conducted using PCA. Analysing the correlation matrix, three principal components were needed to account for the 80.46% of the total variation, whose 64.32% was represented by the first two components (Tab. 3) consisting in length of polar axis and length of equatorial diameter.

The principal component (*PC*) score, associated to each variable (Tab. 4) on the three principal components, identifies the variables that mostly define them. The *PC*1 defined the 33.58% of cumulative variance and this is positively correlated with following variables: length of polar axis (*P*), width of equatorial diameter (*E*), width of muri (*WM*), width of grooves (*WG*), distance between muri and length of colpi. The *PC*2 signified 30.74% of cumulative variance and this is positively correlated with: *P*, *E*, *P/E* ratio, *MN* (number of muri in 25 μm^2) and *C*. The *PC*3 defined 16.14% of cumulative variance and it is positively correlated with all variables except *E*, *DM* and *WM*. The projections of the *PC* loadings defined by the first two principal components allow us to visualize the position of *Prunus* cultivars and rootstocks as shown in score plot (Fig. 6).

In addition, the differences among the samples analysed on the basis of parameters measured on their pollen grains, were verified by the cluster analysis using UPGMA algorithm on the basis Euclidean distance matrix. The analysis of the UPGMA tree (Fig. 7; constructed on the strenght of quantitative traits of pollen grains) revealed three principal groups of morphological similarity: (***i***) B5 (*P. domestica* "Minacore"), rootstocks of *P. dulcis* (D3 and D4); D1 (*P. dulcis* var. *dulcis* "Cavalera"), rootstock of *P. armeniaca* (A3); A1 (*P. armeniaca* "Arancinu"); B2 (*P. domestica* "Rapparinu"); B1 (*P. domestica* "Sittimmirinu"); E2 (*P. persica* subsp. *nucipersica* "Spaccarelli"); D2 (*P. dulcis* var. *dulcis* "Rappa") and A2 (*P. armeniaca* "Pirmintìu"); (***ii***) C1 and C2 (*P. avium* "Cappuccia nìura" and "Maiulina"); (***iii***) B3 and B4 (*P. domestica* "Occhi di voi niuri" and "Occhi di voi biancu"). E1 (*P. persica* subsp. *persica* "Pirmintìu") resulted in a distinct clade.

Discussion

The genus of *Prunus* s.l. comprises more domesticated and cultivated species of temperate fruits than the other genera in the family of Rosaceae. One of the obvious reasons for the abundant domestication might be the coincidence between the location of the centre of variability of *Prunus* and/or of the places of the first high civilization of human history [22]. Many species such as plum, peach, apricot, almond, have attracted the human interest for their edible fruits, while others are planted also for their timber and their ornamental value like *P. avium* (cherry).

In this paper, the detailed analysis of morphological characters of the pollen grains showed that in all the examined

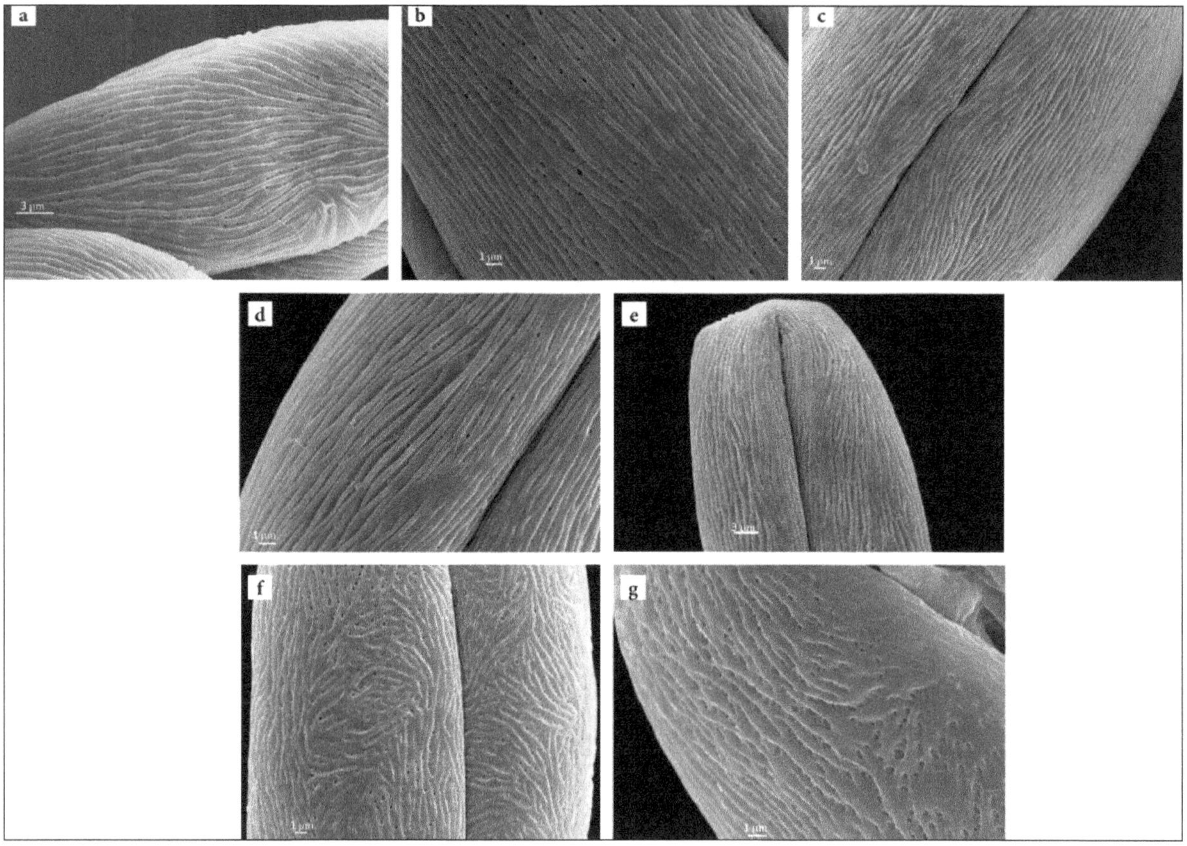

Fig. 4 Exine patterns of *Prunus* species and cultivars. **a** *P. avium* "Maiulina". **b–d** *P. dulcis* var. *dulcis*: **b** "Cavalera"; **c** "Rappa"; **d** rootstock. **e** *P. dulcis* var. *amara* rootstock. **f** *P. persica* subsp. *persica* "Pirmintìu". **g** *P. persica* subsp. *nucipersica* "Spaccarelli".

Tab. 3 Percentage variance contributions by principal components for the dimension and characteristics of *Prunus* pollen grains.

	Total variance explained		
PCs	Eigenvalue	Variance (%)	Cumulative variance (%)
1	3.36	33.58	33.58
2	3.07	30.74	64.32
3	1.61	16.14	80.46
4	0.92	9.20	89.66
5	0.52	5.22	94.88
6	0.25	2.52	97.40
7	0.15	1.53	98.93
8	0.07	0.67	99.60
9	0.04	0.39	99.98
10	0.00	0.02	100

PCs – principal components.

Tab. 4 Principal component analysis loadings and contributions of the principal components.

	PCA loadings		
Variables	*PC1*	*PC2*	*PC3*
P	0.31	0.42	0.28
E	0.37	0.37	–0.03
P/E	–0.15	0.03	0.50
DP	–0.14	–0.39	0.31
PN	0.00	–0.09	0.62
WM	0.28	–0.38	0.31
MN	–0.39	0.33	0.09
DM	0.45	–0.23	–0.11
WG	0.46	–0.22	–0.07
C	0.30	0.42	0.25

*PC*1–3 – principal components.

samples they are symmetric, isopolar, monad, trizonocolpate (with three germinal furrows or colpi), medium-large sized. Regarding outline, in polar view, pollen grains are subtriangular and in equatorial view are elliptic. In the general evaluation of pollen surface, all the pollen grains show striate type of exine sculpture with perforations in the grooves confirming the general characteristics for the pollen of Rosaceae [23–27].

In fact an useful feature for distinguishing the species in the Rosaceae family consists in examining exine sculptures, the most variable morphological character of the pollen surface. Different exine sculptures were observed in some genera of

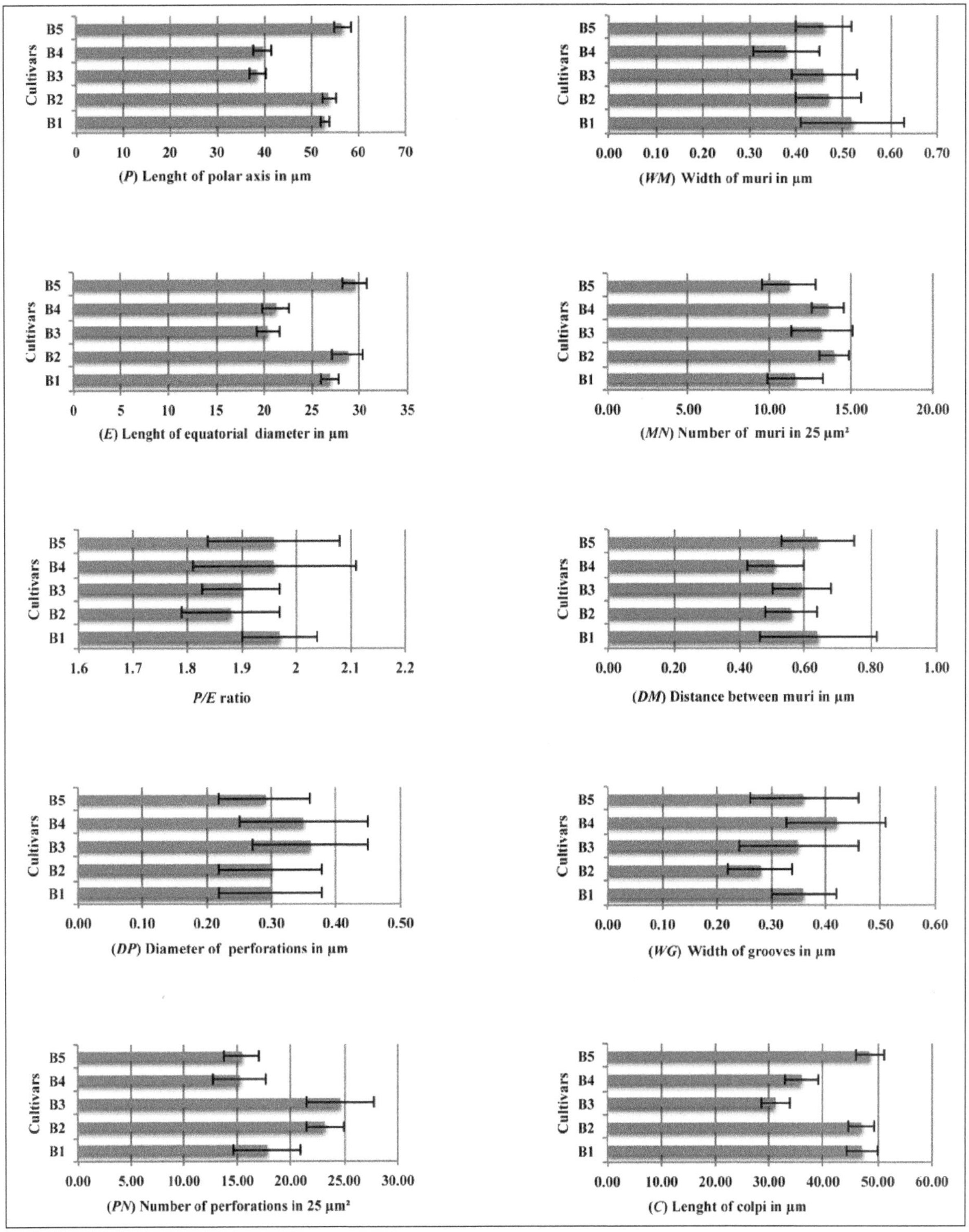

Fig. 5 Graphics of pollen morphology features in *P. domestica*.

Rosaceae family including Turkish Crataegus [28] and some members of rosaceous and Prunoideae in Canada [24–26]. Our results showed similar arrangement of the muri organisation in *P. armeniaca*, *P. avium* and some cultivars of *P. domestica* in which they were roughly parallel but too sloped compared to the polar axis and sometimes curve; in *P. dulcis* and in *P. persica* subsp. *nucipersica* the muri are regular and perfectly parallel. The muri pattern observed in surface of B1 (*P. domestica* "Sittimmirinu") pollen grain, was markedly branched, while in *P. persica* subsp. *persica* "Pirmintìu" (E1) it was very irregular and sinuous. Irregular pattern of exine ornamentation was observed in one Italian cultivar ("Maria Emilia") of *P. persica* cultivated in Argentina [29], that presented a globular structure. Regarding to the dimensional parameters, *P. persica* was

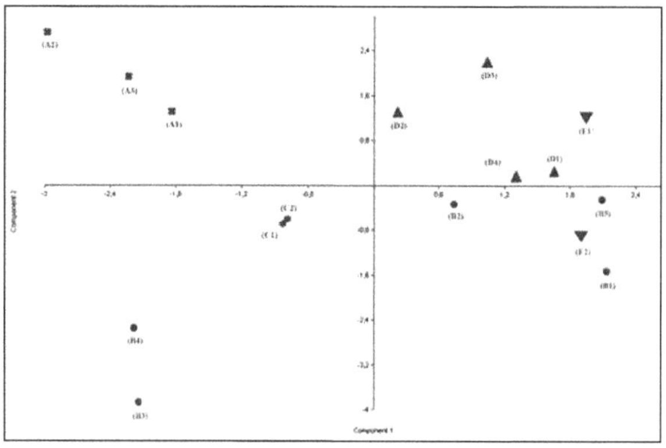

Fig. 6 Scatter diagram of *Prunus* species and cultivars in the first and second principal components.

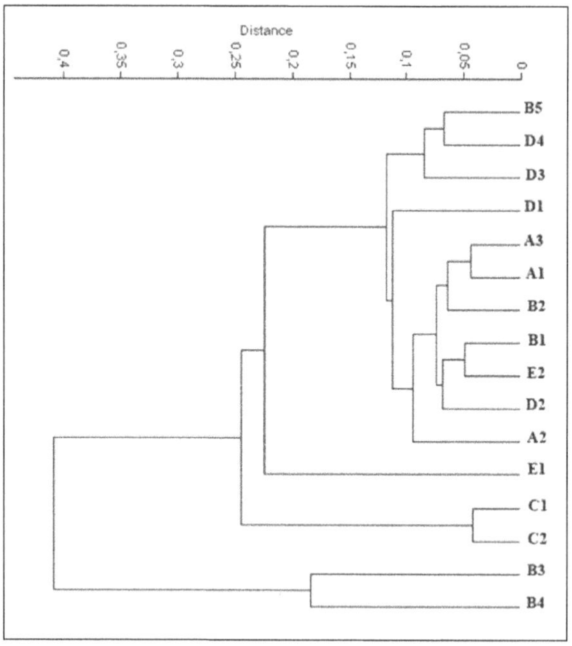

Fig. 7 UPGMA tree of *Prunus* species and cultivars constructed on the basis of the Euclidean distance matrix. Follow Tab. 1 for species and cultivar shortcuts explanations.

ranked first in terms of pollen length and width (mean value 58.18 × 38.85 μm among two cultivars examined), followed by *P. dulcis* (mean value 54.07 × 30.25 μm). Our data agreed with those reported by Evrenosoğlu & Misirli [4] for pollen size and for *P/E* ratio mean value in peach and almond and for the diameter of perforations in almond; concerning the width of muri in peach, moreover, they obtained similar values to those shown by Gilani et al. [30]. The average pollen length and width, as well as *P/E* and the width of muri mean value in *P. armeniaca*, obtained in our study, are similar to the data observed by Arzani et al. [31], while Gilani et al. [30] reported similar range value as to *DM* (distance between muri) and *WM*. Otherwise, from the observations on almond from Iran reported by Vadafar et al. [32], pollen grains of *P. dulcis* s.l. (D1–D4) presented small and medium sized perforations; their number decreases in *P. dulcis* var. *amara* (D4).

The measurements made on *P. domestica* pollen grains showed a clear variability in the most morphological investigated characteristics. However, our dimensional range data (*P* × *E*, *DP*, *DM*, *WM*) can be only partially compared to those described by Evrenosoğlu and Misirli [4] and Gilani et al. [30]. In the current study, pollen size in *P. avium* is larger than the samples analysed by Gilani et al. [30], while the range of mean values of *DM* and *WM* was comparable. In order to analyse the variance, the correlation matrix shows the variables that mostly identify and define the three principal components: *PC*1 is strongly correlated with *WG* (0.46) variable, *PC*2 with *C* (0.42) variable and *PC*3 with *PN* (0.62).

The cluster analysis shows a great morphological affinity among nine cultivars and three rootstocks, while the two cultivars of *P. avium* (C1 and C2) were different due to the medium-sized pollen grains (45.74–48.20 μm × 23.55–24.78 μm) and the very small diameter of perforations (0.26–0.28 μm, respectively), which were sparsely distributed. Similar to cherry, plum cultivars B3 and B4 ("Occhi di voi niuri" and "Occhi di voi biancu") were characterized by medium-sized pollen but smaller than cherry (38.54–39.62 μm × 20.39–21.22 μm) and the large diameter of perforations. Perforations on the exine surface of the *P. persica* "Pirmintiu" pollen were very abundant (25.60 in 25 μm^2) and small-sized, while the size of pollen grains was the largest (63.46–31.49 μm).

In conclusion, the fine observations on pollen surface allowed us to characterize further the different cultivars and in particular to point out the great variability found in the fruits and also in other morphological traits [1]. In fact, particularly, examining *P. domestica* cultivars we discovered a correlation of variability between the pollen grains structures and the fruits shapes. This latter species, shows the highest diversity of the observed parameters, probably because of its hybrid origin according to the hypothesis of Zeven & De Wet [33]. Moreover, in the subfamily of Prunoideae the exine structure could be considered an auxiliary diagnostic character, as also reported for the genus *Rosa* [17], to increase the number of the taxonomical characters that can be utilized for discrimination of the species.

Other studies are in progress to analyse and to preserve this important tree-patrimony realized by the human activities in several centuries.

Acknowledgements

We are very grateful to Dr. Carmela Di Liberto of the same University Department for providing SEM photos.

References

1. Schicchi R, Marino P, Raimondo FM. Individuazione, valutazione e raccolta del germoplasma delle specie arboree da frutto di prevalente interesse negli agrosistemi tradizionali della Sicilia. Palermo: Dipartimento Azienda Regionale Foreste Demaniali; 2008.
2. Lee S, Wen J. A phylogenetic analysis of *Prunus* and the Amygdaloideae (Rosaceae) using ITS sequences of nuclear ribosomal DNA. Am J Bot. 2001;88(1):150–160.
3. Dönmez AA, Yildirimli S. Taxonomy of genus *Prunus* L. (Rosaceae) in Turkey. Turk J Bot. 2000;24:187–202.

4. Evrenosoğlu Y, Misirli A. Investigations on the pollen morphology of some fruit species. Turk J Agric For. 2009;33:181–190.
5. Fogle HW. Identification of clones within four tree fruit species by pollen exine patterns. J Amer Soc Hort Sci. 1977;102:552–560.
6. Fogle HW. Identification of tree fruit species by pollen ultrastructure. J Amer Soc Hort Sci. 1977;102:548–551.
7. Maas JL. Pollen ultrastructure of strawberry and other small fruitcrops. J Amer Soc Hort Sci. 1977;77:548–555.
8. Cargnello G, Carraro L, Lombardo G, Gerola FM. Pollen morphology of Picolit grown in different Italian regions. Vitis. 1980;19:2016.
9. Galletta GJ. Pollen and seed management. In: Moore JN, Janick J, editors. Methods in fruit breeding. West Lafayette IN: Purdue University Press; 1983. p. 23–47.
10. Hebda RJ, Chinnappa CC, Smith BM. Pollen morphology of the Rosaceae of Western Canada. Grana. 1988;27(2):95–113.
11. Merev N. Doğu Karadeniz Bölgesi'nde doğal olarak yetişen kuşburnu (*Rosa* L.) taksonlarının palinolojik yönden incelenmesi. Kuşburnu Sempozyumu Bildiriler Kitabı, Gümüşhanev. Ahi Evran Üniversitesi, Ziraat Fakültesi, Bahçe Bitkileri Bölümü: Kirşehir; 1996. p. 47–54.
12. Jones GD. Pollen as indicators of source areas and foraging resources [Internet].
13. Perveen A, Qaiser M. Pollen flora of Pakistan - XVII. Illecebraceae. Pak J Bot. 2003;35:141–144.
14. Chung KS, Elisens W, Skvarla J. Pollen morphology and its phylogenetic significance in tribe Sanguisorbeae (Rosaceae). Plant Syst Evol. 2010;285(3):139–148.
15. Moore P, Webb JA, Collinson ME. Pollen analysis. 2nd ed. Oxford: Blackwell Scientific Publications; 1991.
16. Bednorz L, Maciejewska-Rutkowska I, Wronska-Pilarek D, Fujiki T. Pollen morphology of the polish species of the genus *Sorbus* L. Acta Soc Bot Pol. 2005;74(4):315–322.
17. Wrońska-Pilarek D, Jagodziński A. Systematic importance of pollen morphological features of selected species from the genus *Rosa* (Rosaceae). Plant Syst Evol. 2011;295(1):55–72.
18. Erdtman G. An introduction to pollen analysis. Stockholm: Waltham Mass; 1943.
19. Erdtman G. On terminology in pollen and spore morphology. Grana. 1958;1(3):6–9.
20. Hesse M, Halbritter H, Zetter R, Weber M, Buchner R, Frosch-Radivo A, et al. Pollen terminology: an illustrated handbook. Vienna: Springer; 2009.
21. Hammer Ø, Harper DAT, Ryan PD. PAST: paleontological statistics software package for education and data analysis. Palaeontol Electron. 2001;4(1):1–9.
22. OECD. Stone fruits (*Prunus* spp.). OECD Agriculture & Food. 2008;2006(21):628–660.
23. Ueda Y, Tomita H. Morphometric analysis of pollen exine patterns in roses. J Japan Soc Hort Sci. 1989;58(1):211–220.
24. Hebda RJ, Chinnappa CC. Studies on pollen morphology of Rosaceae in Canada. Rev Palaeobot Palynol. 1990;64(1-4):103–108.
25. Hebda RJ, Chinnappa CC. Pollen morphology of the Rosaceae of western Canada. III. Geum. Can J Bot. 1990;68(6):1369–1378.
26. Hebda RJ, Chinnappa CC, Smith BM. Pollen morphology of the Rosaceae of western Canada. IV. Luetkea, Oemleria, Physocarpus, Prunus. Can J Bot. 1991;69(12):2583–2596.
27. Zhou LH, Wei ZX, Wu ZY. Pollen morphology of Prunoideae of China (Rosaceae). Acta Bot Yunnan. 1999;21:207–211.
28. Dönmez E. Pollen morphology in Turkish Crataegus (Rosaceae). Bot Helv. 2008;118(1):59–70.
29. Radice S, Ontivero M, Giordani E, Bellini E. Morphology and physiology of pollen grains of Italian *Prunus persica* (L.) Batsch cultivars grown in Argentina. Adv Hort Sci. 2003;17(2):93–96.
30. Gilani SA, Qureshi RA, Khan AM, Potter D. Morphological characterization of the pollens of the selected species of genus *Prunus* Linn. from Northern Pakistan. Afr J Biotechnol. 2010;9(20):2872–2879.
31. Arzani K, Nejatian MA, Karimzadeh G. Apricot (*Prunus armeniaca*) pollen morphological characterisation through scanning electron microscopy, using multivariate analysis. N Z J Crop Hort Sci. 2005;33(4):381–388.
32. Vafadar M, Attar F, Maroofi H, Mirtadzadini M. Pollen morphology of *Amygdalus* L. (Rosaceae) in Iran. Acta Soc Bot Pol. 2010;79(1):63–71.
33. Zeven AC, de Wet J. Dictionary of cultivated plants and their regions of diversity: excluding most ornamentals, forest trees and lower plants. Dehra Dun: International Book Distributors; 1993.

17

Primary results of 2-dimensional electrophoresis for protein studies of *Gentiana kurroo* Royle somatic embryos derived from long-term embryogenic cell suspensions

Agnieszka Niedziela[1], Jan Jarosław Rybczyński[2*]

[1] Department of Plant Physiology and Biochemistry, Plant Breeding and Acclimatization Institute – National Research Institute, Radzików, 05-870 Błonie, Poland

[2] Department of Experimental Plant Biology, Botanical Garden – Center for Biological Diversity Conservation of the Polish Academy of Sciences in Powsin, Prawdziwka 2, 02-973 Warsaw, Poland

Abstract

Two-dimensional gel electrophoresis was used to compare protein expression profiles between various stages of *Gentiana kurroo* Royle somatic embryos. Seven distinct stages (I–VII) were pointed out and measured from long-term embryogenic cell suspension. Isoelectric focusing was done in the pH intervals 3–10, and the second dimension was carried out with 13% SDS-polyacrylamide gel electrophoresis. Dependent on the stage from about 400 (stage IV) to more than 700 (stage II) protein spots were totally detected. The molecular weight of abundant proteins range from 12 to 70 kDa, however, majority of proteins were located between 20–49 kDa spots on the gels. The highest difference in the number of spots appeared in the case of globular embryo (stage I) and elongated cotyledonary stage (stage VII) with differences being about 130 spots. The relevance of embryogenic cell suspension choice for proteomic analysis as well as expediency of the increasing number of particular embryo stages is discussed.

Keywords: *Gentiana kurroo*; two-dimensional gel electrophoresis; protein studies; embryogenic cell suspension

Introduction

Gentianaceae family consists about 1700 species, but only 28 are the object of plant tissue culture and biotechnology. The first papers summarizing achievements of gentian plant tissue cultures were published in 1988 and 1991 [1,2]. Since then, considerable progress in the biochemistry and biotechnology of the genus has been achieved. The very first paper showed possibilities to establish plant multiplication of *Gentiana kurroo* by culturing the shoot tips and nodal segments on Murashige and Skoog (MS) medium supplemented with BAP (benzylaminopurine) and NAA (naphtaleneacetic acid) [3]. For most effective rooting, individual shoots with 3–4 nodes were implanted on hormone-free agar MS medium supplemented only with 6% sucrose [3]. Randomly amplified polymorphic DNA and karyotypic analysis showed the lack of the variation and genetic stability of regenerants and confirmed efficacy of the protocol for micropropagation of *G. kurroo* over a ten-year period [4]. The significant progress in vegetative plant cell manipulation was recognized when the somatic embryogenesis of primary explants for a few gentian species was described [5]. Among the studied gentians, *G. kurroo* appeared very embrygenic with tremendous morphogenic potential since seedling and leaf explants [6] were used for culture initiation. Furthermore, once established cell suspensions [7] appeared to be an excellent source of embryogenic cells and their protoplasts [8]. Cell suspensions also created embryogenic aggregates and somatic embryos [9]. Somatic embryogenesis is a complex developmental process that offers multidisciplinary studies of embryogenesis and great potential in plant propagation. Since two-dimensional electrophoresis (2-DE) was developed the application of the method has been increasingly used as a tool for studies of somatic embryogenesis [10–12]. It should be stressed that proteomic study in gentians has not been published until now, with the exception of one describing isoenzymes and protein patterns in leaves of in vitro micropropagated plantlets of *G. lutea* in the presence of various combination of plant growth regulators [13].

The aim of this paper is to show the primary results on the newly discovered proteomic changes between various stages of *G. kurroo* somatic embryos.

Material and methods

Somatic embryos were selected from hypocotyl derived suspension culture (Fig. 1a) described previously [7] and used for protein isolation and detection (Fig. 2). Different

* Corresponding author. Email: jjryb@obpan.pl

Handling Editor: Elżbieta Bednarska-Kozakiewicz

stages of development were regenerated spontaneously in one conical flask (Fig. 1b,c). The scale of typical four stages of zygotic embryogenesis was replaced with seven in vitro stages scale. The following stages were identified: I – globular embryo; II – heart with thick epidermis and differentiated rootlet; III – an elongated embryo with well-differentiated primary root; IV – cotyledonary embryo with closed cotyledons; V – cotyledonary embryo with detached cotyledons 2–3 mm in size; VI – cotyledonary embryo 3–4 mm; VII – cotyledonary embryos larger than 5 mm in size.

Protein extraction from 10 mg of tissue was performed using an extraction buffer (0.7 M sucrose, 0.5 M Tris, 30 mM HCl, 50 mM EDTA, 10 mM KCl, 2.0 mM PMSF, 13 mM DTT) and a resolving buffer (9.0 M urea, 4% Nonidet NP-40, 2% Servalyte, pH 2–4. Two-dimensional gel electrophoresis performed according to Hochstrasser et al. [14]. In the first dimension, isoelectrofocusing (IEF), 1 × 150 mm capillary was used for polymerization a polyacrylamide column gel with Servalyte pH 3–10. The IEF was conducted at room temperature for 18 h (50–1000 V). Protein were separated in the second dimension (SDS-PAGE) in 13% polyacrylamide slabs (1 × 150 × 150 mm). After electrophoresis, the peptides were stained with silver nitrate according to Heukeshoven and Dernick [15]. Protein spots on gels were measured with Image Master 2-D Elite LKB software.

Results and discussion

The paper presents second field of our experimental activity with the application both embryogenic cell suspension and its proteomic studies. Two-dimensional gel electrophoresis has been successfully engaging in proteome studies of encapsulated embryogenic cell suspension during its adaptation to osmotic stress in cryopreservation protocol of *Gentiana cruciata* [16]. In these studies, more attention was paid for description of dynamics of the electrophoretic proteomic profile changes considering proteins that showing upstream growth of activity and other proteins being in the decreasing activity.

Fig. 1 Established embryogenic cell suspension of *Gentiana kurroo*: 80-ml cell suspension carried on in 200-ml conical flask – the source of studied material (**a**); embryos in globular (**b**), and cotyledon stage (indicated by arrows; **c**).

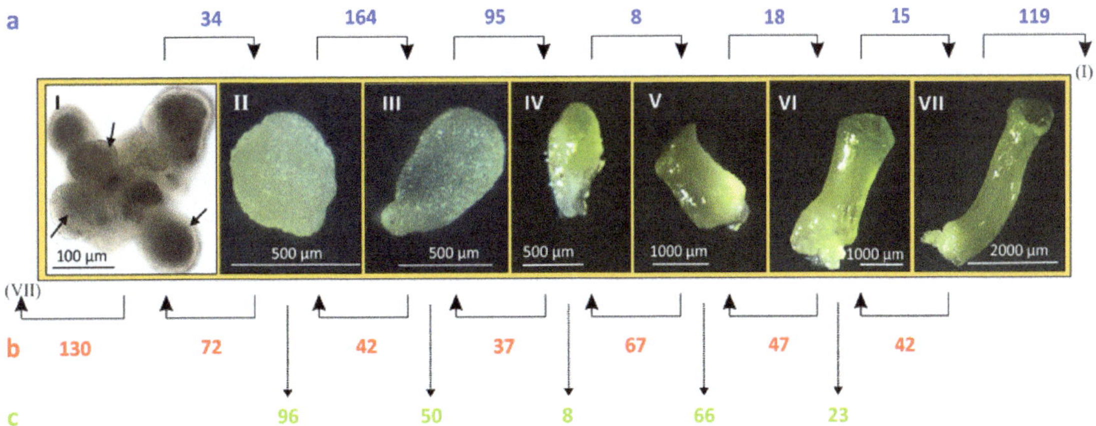

Fig. 2 Seven selected stages (I–VII) of studied embryos in embroygenic cell suspension of *Gentiana kurroo* and their characteristic number of protein spots. Number of protein spots not observed in the following stage (blue color; **a**), in previous stage (red color; **b**), and characteristic for particular stage (green color; **c**).

Number of protein spots produced by two-dimensional electrophoresis

Total number of protein spots produced by 2-DE was different and characteristic for the seven stages of *G. kurroo* somatic embryogenesis analyzed (Fig. 2). The differences ranged from the lowest number of spots presented by stage IV to the highest number of spots presenting in the stage II with over 300 different protein spots. We did not observe the unique differences in the number of produced protein spots between stage III, VI and VII with the number about 545. In the case of *Cyclamen persicum*, studies revealed that the differing number of spots between particular stages is not large, and globular somatic embryo stage was characterized by 451 protein spots while the torpedo stage was characterized by 460 protein spots [11].

More attention was paid to recognize the number of protein spots characterizing particular stages. Looking on data from the next stage to the previous stage, the highest difference appeared, of course, in the case of globular embryo (stage I) and elongated cotyledonary stage (stage VII) with differences being about 130 spots (Fig. 2). Other stages were characterized by the difference ranging from 37 to 72 protein spots between following and previous stage. However, the focusing on the number of protein spots which differed between previous stage to following one gave the other perspective on the possibilities to recognize the protein markers of a particular somatic embryo stage of *G. kurroo*. The highest number of not observed protein spots was recognized between stages VII and I stage, over one hundred, but the lowest number of protein spot differences was observed between IV and V stage, just about 10. The Fig. 3 presents the results obtained in case of the stage I, III and VII with the criterion of not find protein spots between the compared stages.

Proteomics of somatic embryogenesis

Proteomic patterns have been used for various purposes, however, the comparison between particular stages of embryogenesis, embryogenic and non-embryogenic callus during somatic embryogenesis [17] or plant ontogenesis [18] appeared most frequently analyzed in a limited number of previously published papers. In the case of the H99 inbred maize line, the proteome analysis of embryogenic and non-embryogenic callus revealed the existence of 42 proteins spots that were differentially expressed [17]. The protein expression in rice was used for the description of embryogenic and non-embryogenic cultures, as well as isolated embryos from which cultures were initiated. In both types of cultures, the strongest signals were for 56, 54 and 36 kDa proteins, when in case of initial embryos the signals were not present. Only one protein of 54 kDa appeared to be a marker of embryogenesity, and it was not expressed in non-embryogenic tissue [19]. For the pea, two proteins of 45 kDa and 70 kDa were described being the markers of embryogenesity [20]. In the case of cassava, proteome analysis using extended stages of plant development showed that the only a few of the identified proteins are unique to somatic embryos, shoots, adventitious roots and tubers [18].

With the help of 2-DE, the proteins of seven consecutive stages of somatic embryos of *G. kurroo* were analyzed (Fig. 2). These stages of somatic embryos differ from those usually described in references because of untypical environmental condition of liquid culture in the presence of both plant growth regulators, auxin (2,4-D) and cytokinin (Kinetin). In the case of the last three stages (V, VI, VII) the size (length of embryos) was considered. The Fig. 1 presents the morphological evidences of a particular stage from proembryo to cotyledonary somatic embryo. With the help of the first dimension, the isoelectrofocusing points of isolated proteins from embryos were established. Compare two most distinct stages: globular (I) and cotyledonary (VII) the major difference in the distribution of IEF proteins has been recognized. The first stage of embryo development located studied proteome in majority between pH values ranging from 6 to 9, when the total scale was pH value ranging from 3 to 10. For the oldest studied stage, the majority of proteins were located between pH value ranging 6.5 and 8.5. For both

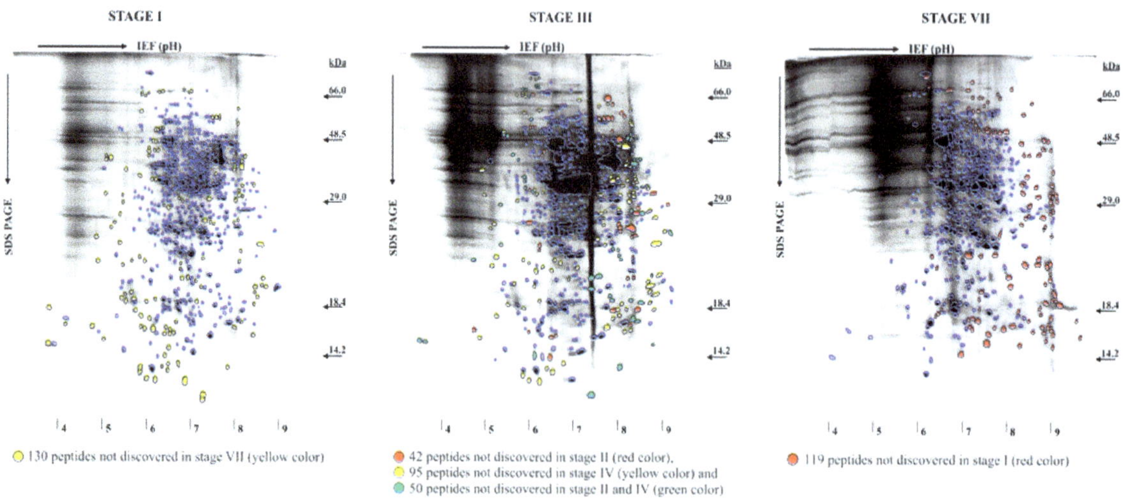

Fig. 3 Number of protein spots and their distribution in 2-DE gel electrophoresis of different stages of somatic embryos. **a** 130 peptides not discovered in stage VII (yellow color). **b** 42 peptides not discovered in stage II (red color), 95 peptides not discovered in stage IV (yellow color) and 50 peptides not discovered in stage II and IV (green color). **c** 119 peptides not discovered in stage I (red color).

extreme stages, some proteins had acid character just about pH 4.0 or a base character just between pH values ranging from 9 to 10. Globular and torpedo somatic embryos of *C. persicum* presented IEF of proteins ranging pH 5–8, only [11].

SDS-PAGE gel indicated that the molecular weight of abundant proteins spots for *G. kurroo* is located between 12 and 70 kDa, however, majority of proteins were characterized by kDa from 20 to 49. In the embryogenic callus of *Crocus sativus*, many high abundant protein spots indicated the range of 14–20 and 30–43 kDa [21]. Describing only three stages: globular, torpedo and cotyledonary of somatic embryos of *Coffea arabica* two-dimensional electrophoresis analysis revealed a wide range of protein spots ranging between 10 and 160 kDa, when pH value ranging from 3 to 10 [10]. All these studies confirm earlier published results indicating quantitative and qualitative protein variation between proembryo and well-developed cotyledonary stage because of mechanisms involved in differentiation against development and maturation, respectively [12].

Our results concerning protein analysis of *G. kurroo* somatic embryos have shown that proposed seven stages of embryogenesis (I–VII stages) can be utilized for estimation of the correlation between protein expression and gene involved in the development of somatic embryos. Moreover, they could be used for supporting the progress in studies of genetic and physiological controls of somatic embryogenesis in such model plants as *Arabidopsis thaliana* and *Medicago truncatula*.

Acknowledgments
Presented studies were sponsored by Polish State Committee of Scientific Research 3PO4C03723.

Authors' contributions
The following declarations about authors' contributions to the research have been made: designed research: AN, JJR; performed experiments: AN, JJR; wrote the manuscript: AN, JJR.

References
1. Barešová H. *Centaurium erythraea* Rafn.: Micropropagation and the production of secoiridoid glucosides. In: Bajaj YPS, editor. Medicinal and aromatic plants I. Berlin: Springer Berlin Heidelberg; 1988. p. 350–366. (Biotechnology in agriculture and forestry; vol 4).
2. Miura H. *Swertia* spp. In: Bajaj YPS, editor. Medicinal and aromatic plants III. Berlin: Springer Berlin Heidelberg; 1991. p. 451–463. (Biotechnology in agriculture and forestry; vol 15).
3. Sharma N, Chandel KPS, Paul A. In vitro propagation of *Gentiana kurroo* – an indigenous threatened plant of medicinal importance. Plant Cell Tissue Organ Cult. 1993;34(3):307–309.
4. Kaur R, Panwar N, Saxena B, Raina R, Bharadwaj SV. Genetic stability in long-term micropropagated plants of *Gentiana kurroo* – an endangered medicinal plant. J New Seeds. 2009;10(4):236–244.
5. Mikuła A, Rybczyński JJ. Somatic embryogenesis of *Gentiana* genus I. The effect of the preculture treatment and primary explant origin on somatic embryogenesis of *Gentiana cruciata* (L.), *G. pannonica* (Scop.), and *G. tibetica* (King). Acta Physiol Plant. 2001;23(1):15–25.
6. Fiuk A, Rybczyński JJ. Morphogenic capability of *Gentiana kurroo* Royle seedling and leaf explants. Acta Physiol Plant. 2008;30(2):157–166.
7. Fiuk A, Rybczyński JJ. Factors influencing efficiency of somatic embryogenesis of *Gentiana kurroo* (Royle) cell suspension. Plant Biotechnol Rep. 2008;2(1):33–39.
8. Fiuk A, Rybczyński JJ. The effect of several factors on somatic embryogenesis and plant regeneration in protoplast cultures of *Gentiana kurroo* (Royle). Plant Cell Tissue Organ Cult. 2007;91(3):263–271.
9. Rybczyński JJ, Borkowska B, Fiuk A, Gawrońska H, Śliwińska E, Mikuła A. Effect of sucrose concentration on photosynthetic activity of in vitro cultures *Gentiana kurroo* (Royle) germlings. Acta Physiol Plant. 2007;29(5):445–453.
10. Tonietto Â, Sato JH, Teixeira JB, Souza EM, Pedrosa FO, Franco OL, et al. Proteomic analysis of developing somatic embryos of *Coffea arabica*. Plant Mol Biol Rep. 2012;30(6):1393–1399.
11. Bian F, Zheng C, Qu F, Gong X, You C. Proteomic analysis of somatic embryogenesis in *Cyclamen persicum* Mill. Plant Mol Biol Rep. 2010;28(1):22–31.
12. Sallandrouze A, Faurobert M, El Maataoui M, Espagnac H. Two-dimensional electrophoretic analysis of proteins associated with somatic embryogenesis development in *Cupressus sempervirens* L. Electrophoresis. 1999;20(4–5):1109–1119.
13. Petrova M, Stoilova T, Zagorska N. Isoenzyme and protein patterns of in vitro micropropagated plantlets of *Gentiana lutea* L. after application of various growth regulators. Biotechnol Biotechnol Equip. 2006;20(1):15–19.
14. Hochstrasser DF, Harrington MG, Hochstrasser AC, Miller MJ, Merril CR. Methods for increasing the resolution of two-dimensional protein electrophoresis. Anal Biochem. 1988;173(2):424–435.
15. Heukeshoven J, Dernick R. Simplified method for silver staining of proteins in polyacrylamide gels and the mechanism of silver staining. Electrophoresis. 1985;6(3):103–112.
16. Domżalska L, Kędracka-Krok S, Rybczyński JJ. Proteomic changes in *Gentiana cruciata* L. cell suspension during adaptation to osmotic stress in cryopreservation protocol. In: Pukacki PM, editor. Proceeding of the 17th cold hardiness seminar in Poland. Kórnik: Institute of Dendrology of the Polish Academy of Sciences; 2011. p. 44–52.
17. Sun L, Wu Y, Zou H, Su S, Li S, Shan X, et al. Comparative proteomic analysis of the H99 inbred maize (*Zea mays* L.) line in embryogenic and non-embryogenic callus during somatic embryogenesis. Plant Cell Tissue Organ Cult. 2013;113(1):103–119.
18. Li K, Zhu W, Zeng K, Zhang Z, Ye J, Ou W, et al. Proteome characterization of cassava (*Manihot esculenta* Crantz) somatic embryos, plantlets and tuberous roots. Proteome Sci. 2010;8(1):10.
19. Chen LJ, Luthe DS. Analysis of proteins from embryogenic and non-embryogenic rice (*Oryza sativa* L.) calli. Plant Sci. 1987;48(3):181–188.
20. Stirn S, Jacobsen HJ. Marker proteins for embryogenic differentiation patterns in pea callus. Plant Cell Rep. 1987;6(1):50–54.
21. Sharifi G, Ebrahimzadeh H, Ghareyazie B, Gharechahi J, Vatankhah E. Identification of differentially accumulated proteins associated with embryogenic and non-embryogenic calli in saffron (*Crocus sativus* L.). Proteome Sci. 2012;10(1):3.

18

Determination of potential glacial refugia and possible migration routes of *Campylocentrum* (Vandeae, Orchidaceae) species through the Darién Gap

Marta Kolanowska*

Department of Plant Taxonomy and Nature Conservation, University of Gdańsk, Wita Stwosza 59, 80-308 Gdańsk-Oliwa, Poland

Abstract

The possible glacial refugia of four *Campylocentrum* species reported from the Darién Gap are estimated based on the ecological niche modeling. The potential migration routes of the studied species are presented. The current ranges of all species designated based on the available floristic data are compared with their potential habitats distribution.

Keywords: *Campylocentrum*; Darién Gap; ENM; migration; glacial refugia

Introduction

The Darién Gap (Spanish Tapón del Darién, Fig. 1) is an area of swampland and tropical forest separating Panamian province of Darién from Colombia. The Isthmus is about 50 km wide, from the Carribean Sea to the North Pacific Ocean, and 84 km long from Yarviza (Panama) to Lomas Aisiadas (Colombia). Darién is one of the least known and most diverse regions of the World. The Panamian part of the Isthmus is covered with the rainforest, while the Colombian side is dominated by the flat marshland created in the delta of the Atrato River. In terms of floristic diversity Darién Gap is one of the least recognized region of the World, partially due to the lack of the roads in this area, which makes the field studies difficult to conduct. In fact, this is the only missing part of the Pan-American Highway, which crosses through the entirety of North, Central, and South America [1].

Undoubtedly, apart from the insufficient herbarium material collected in the Darién Gap, orchids are important part of the regional flora. Unfortunately, the exact number of the species is difficult to estimate. While Williams et al. [2] provided information only about 14 orchids occurring in the province of Darién and Cárdenas-López [3] listed 33 species found in the Colombian part of the Isthmus, the preliminary research conducted in the herbaria MO and AMES resulted in a list of over 250 species belonging to 11 tribes occurring in the Darién Gap.

One of the most interesting orchids are plants classified under the tribe Vandeae, which is best represented in the Paleotropics. Within this group, only the species of the subtribe Angraecinae are found also in Neotropics. Those New World plants belong to one of three genera: *Campylocentrum* Benth., *Dendrophylax* Rchb.f. or *Harrisella* Fawc. & Rendle.

Proposed in 1881 [4] the genus *Campylocentrum* is the only representative of the tribe Vandeae in Central America. Species constituting the genus are epiphytic, monopodial, either leafy or leafless plants producing lateral inflorescence with non-resupinate, usually inconspicuous flowers. The variously 3-lobed or entire lip is always elongated into a prominent spur [5]. The geographical range of the genus extends from Florida through Mexico to Brazil. About 60 species have been described so far, just four of them have been reported from the Darién Gap: *C. brenesii* Schltr., *C. micranthum* (Lindl.) Rolfe, *C. panamense* Ames and *C. tyrridion* Garay & Dunst. [6–8].

So far no research about the migration of Orchidaceae through the Isthmus was conducted. In this paper the maximum entropy method was applied to determine localization of potential refugia during last glacial maximum (LGM, 26 500–19 000 years ago) [9] and to estimate the possible postglacial migration routes of *Campylocentrum* species based on the occurrence data of their populations. Additionally, the similarities between potential ranges of the studied species are discussed.

Material and methods

List of localities

A total of over 170 herbarium specimens were studied to prepare a list of localities of the studied species populations. Moreover, the information from the electronic database of the Missouri Botanical Garden (available at http://www.

* Email: martakolanowska@wp.pl

Handling Editor: Beata Zagórska-Marek

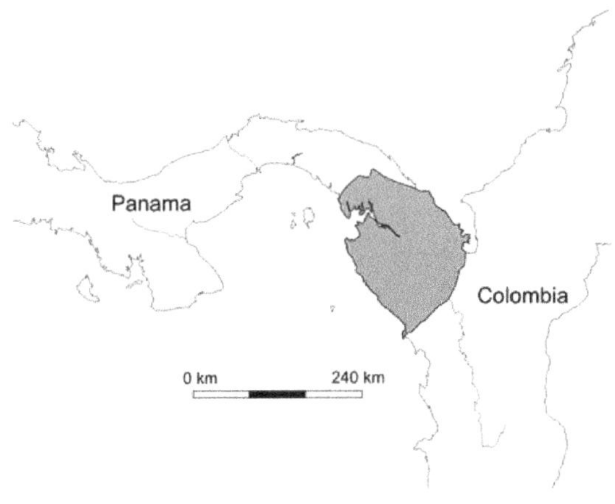

Fig. 1 Location of the Darién Gap.

tropicos.org) was used. The complete list of the examined specimens is presented in Tab. S1.

Dried herbarium specimens were examined according to the standard procedures. Every studied sheet was photographed and the data from the labels were taken. Both vegetative and generative characters of each plant were examined under a stereomicroscope. The comparison of the studied material with the designated type specimens and protologues assured the correctness of identifications. The material was examined in, or borrowed from the following institutions: AMES, AMO, COL, JAUM, K, MO, UGDA and VALLE. Moreover, the digitalized materials of AAU and RENZ were used.

Tab. 1 Variables used in the modeling.

Code	Variable
bio1	Annual mean temperature
bio2	Mean diurnal range = Mean of monthly (max temp – min temp)
bio3	Isothermality (bio2/bio7) (× 100)
bio4	Temperature seasonality (standard deviation × 100)
bio5	Max temperature of warmest month
bio6	Min temperature of coldest month
bio7	Temperature annual range (bio5 – bio6)
bio8	Mean temperature of wettest quarter
bio9	Mean temperature of driest quarter
bio10	Mean temperature of warmest quarter
bio11	Mean temperature of coldest quarter
bio12	Annual precipitation
bio13	Precipitation of wettest month
bio14	Precipitation of driest month
bio15	Precipitation seasonality (coefficient of variation)
bio16	Precipitation of wettest quarter
bio17	Precipitation of driest quarter
bio18	Precipitation of warmest quarter
bio19	Precipitation of coldest quarter
alt	Altitude

Maximum entropy analysis

To estimate the potential niche for present and past, the maximum entropy analysis was made. Only the localities, which could be precisely placed on the map, were used for the analysis – a total of 183 different occurrence data, which is more than the minimum number of records required by MaxEnt to obtain trustable predictions [10].

Potential niche modeling was carried out using Maxent version 3.3.2 [11–13]. As input data 19 climatic variables in 2.5 arc minutes developed by Hijmans et al. [14], as well as the altitudinal data (Tab. 1) were used. To assess maximum specificity of the modeling, the maximum iterations was set to 10 000 and convergence threshold to 0.00001. For each run 20% of the data were used to be set aside as test points [15]. Moreover, the "random seed" option, which provided random test partition and background subset for each run, was used. The run was performed as a bootstrap with 1000 replicates, and the output was set to logistic. The analogical settings were used to estimate distribution of the suitable niches during last glacial maximum. In this modeling the data developed by the Paleoclimate Modelling Intercomparison Project Phase II [16] were applied.

Results

Range – floristic data

The maps presenting geographical ranges of the studied *Campylocentrum* species were prepared based on the floristic data only. While the ranges of *C. brenesii* and *C. panamense* seem to be restricted to the Central America, *C. micranthum* is widely distributed through the Neotropics. Most of *C. tyrridion* populations were found in Central America, but there are few records of this species from South America (Ecuador and Venezuela; Fig. 2).

Ecological niche modeling

All repeated ecological niche models for the present time received high AUC scores of 0986–0.999 (Tab. 2) that indicate great reliability of the analysis.

The results of ecological niche modeling for the present time as well as for LGM are consistent in the contributions of the analyzed climatic variables to the distribution of all studied species (Tab. 3). The crucial climatic variables deciding on the distribution of *C. brenesii* is the annual precipitation and the altitude. The altitude is also important in the modeling of the potential niches of *C. tyrridion*, but for this species the principal variable was the min. temperature of the coldest month. The temperature seasonality gave a significant contribution in both models made for *C. micranthum* and *C. panamense*. For the first one it was the most important variable together with the annual range of the temperature. For the second species the precipitation of the coldest quarter was essential.

The analyses indicated Cordillera de Talamanca and north part of Cordillera de San Blas as areas of suitable habitats for *C. brenesii* (Fig. 3a). Some additional potentially available niches are also distributed in Colombian Sierra Nevada de Santa Marta and northern Andes. In the model created

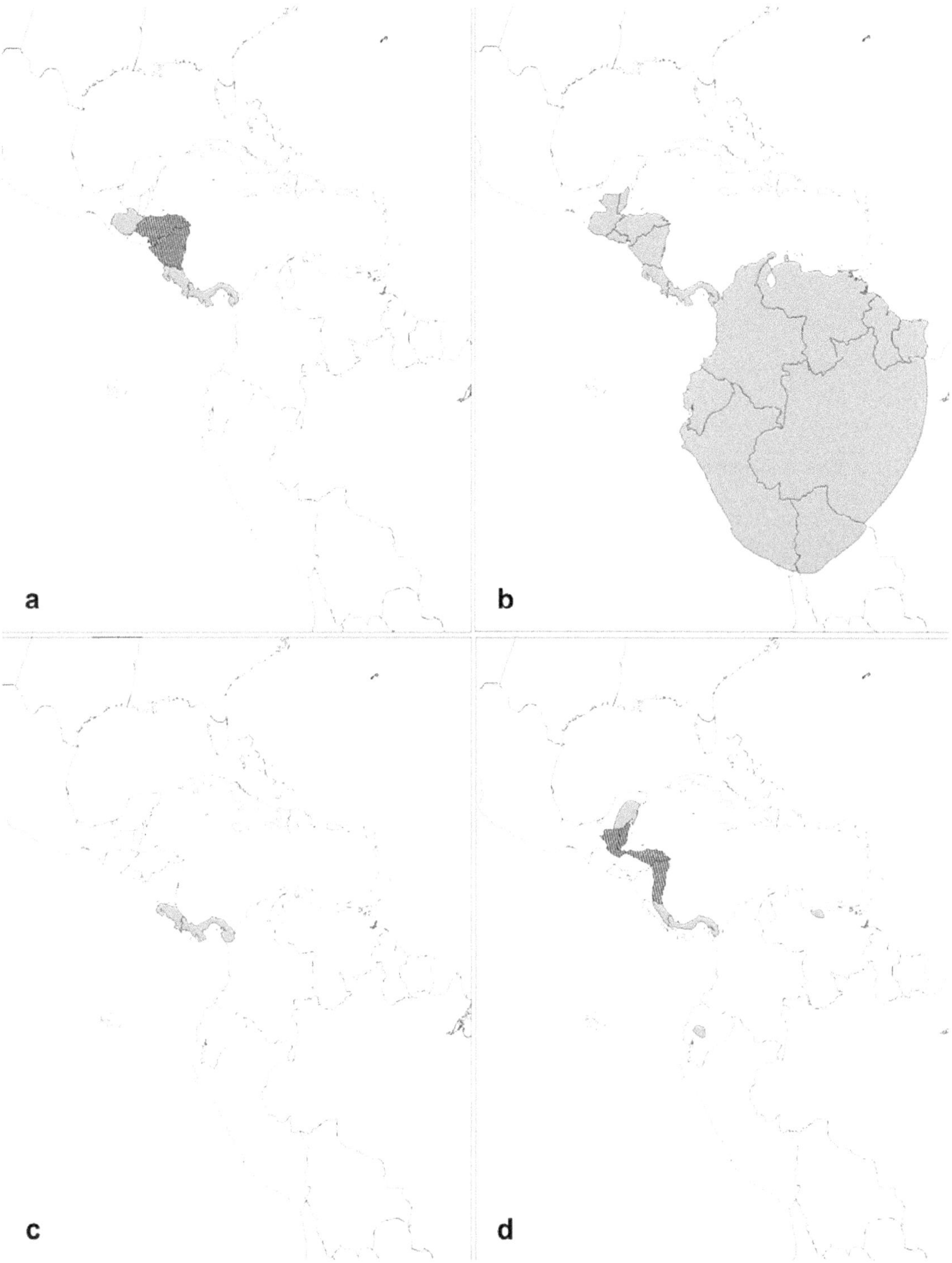

Fig. 2 Current geographical ranges of *C. brenesii* (**a**), *C. micranthum* (**b**), *C. panamense* (**c**) and *C. tyrridion* (**d**). The pattern fill was used for the unconfirmed parts of the ranges.

for *C. micranthum* (Fig. 3b) the most suitable habitats of this species are located along the coast of Mesoamerica as well as along the Pacific coast of Colombia, lower Orinoco and Brazilian Atlantic forests. The niches of *C. panamense* (Fig. 3c) are distributed along the cost of Panamian Isthmus and Brazilian Atlantic coast. Relatively extensive potential range was also indicated in the analysis of *C. tyrridion* (Fig. 3d). The niches suitable for this species are associated with the lowland areas of Neotropics, such as Atlantic coast, Orinoco lowland and Amazon region.

Tab. 2 The AUC scores for the repeated models for different analysis per species.

	C. brenesii	*C. micranthum*	*C. panamense*	*C. tyrridion*
Model for the present time	0.999	0.993	0.996	0.987
Model for LGM	0.999	0.993	0.989	0.986

Tab. 3 The climatic variables with the highest contribution to the potential niches distribution models.

Species	*C. brenesii*	*C. micranthum*	*C. panamense*	*C. tyrridion*
Present time				
Var_1 (% contribution)	Bio12 (35.8)	Bio4 (18.1)	Bio19 (23.6)	Bio6 (49.4)
Var_2 (% contribution)	Alt (19.2)	Bio7 (15.8)	Bio4 (15.9)	Alt (21.8)
LGM				
Var_1 (% contribution)	Bio12 (35.3)	Bio4 (16.9)	Bio19 (30.3)	Bio6 (55.6)
Var_2 (% contribution)	Alt (19.8)	Bio7 (15.0)	Bio4 (23.6)	Alt (21.0)

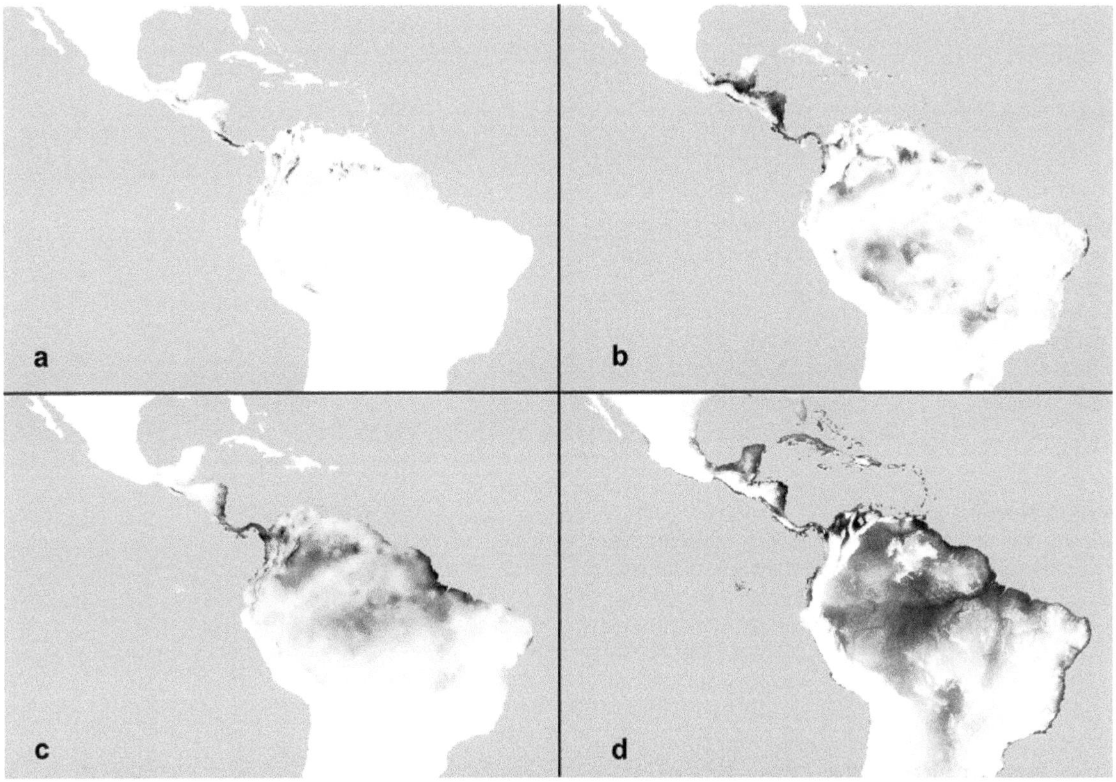

Fig. 3 Potential niche modeling for present time for *C. brenesii* (**a**), *C. micranthum* (**b**), *C. panamense* (**c**) and *C. tyrridion* (**d**).

The niche modeling for last glacial maximum shows a different distribution of suitable habitats for each studied species. For *C. brenesii* (Fig. 4a), the most appropriate climatic conditions included Mesoamerican Sierra Cuchumatanes and Cordillera de Talamanca as well as the northern part of the Colombian Andes. Moreover, some proper niches were located in the lower Orinoco. The most suitable niches for *C. micranthum* (Fig. 4b) existed in the coastal Central America as well as in the Andes from Colombia to Peru. Some appropriate habitats were also located in the eastern

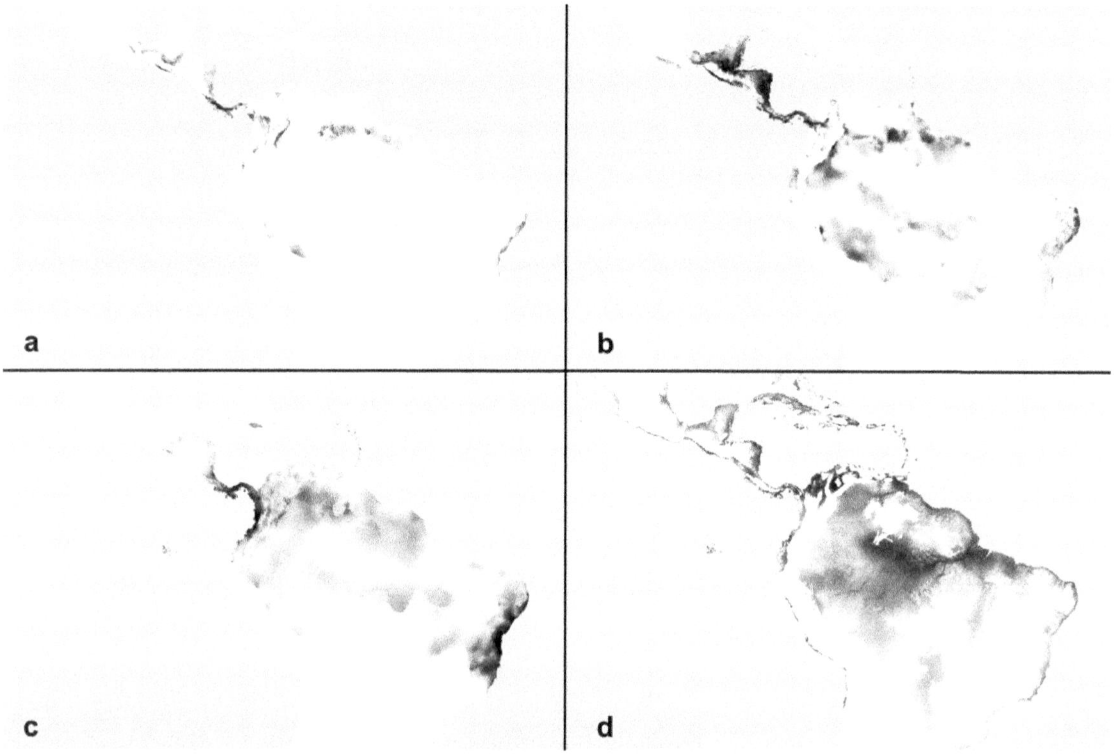

Fig. 4 Habitat fitness simulation for LGM for *C. brenesii* (**a**), *C. micranthum* (**b**), *C. panamense* (**c**) and *C. tyrridion* (**d**).

Andean slope, lower Orinoco and todays' Brazilian Atlantic forests. Along the Pacific coast of Colombia extended habitats suitable for *C. panamense* (Fig. 4c) and *C. tyrridion* (Fig. 4d) were distributed. The proper niches for the first one were also located in the Brazilian Atlantic coast and for the latter some habitats were also available in the Caribbean region.

Discussion

Current species ranges

The potential ranges estimated based on the maximum entropy analysis are significantly wider than those drawn based on the floristic data only. The latter may be underestimated due to insufficient herbarium material. While plants of *Campylocentrum* are rather common element of epiphytic flora of the Neotropics, especially leafless species are difficult to notice during the field works. There are also records of the sympatric occurrence of some species in Central America [8], which, due to the superficial similarity of many species, may be the reason of unrepresentative herbarium collections of *Campylocentrum*.

Possible refugia

Based on the results of the predictive habitat distribution modeling, the most possible distribution of the refugia was estimated. The most probable refugial areas of *C. brenesii* included Sierra Madre de Chiapas, Cordillera de Talamanca and Cordillera de San Blas.

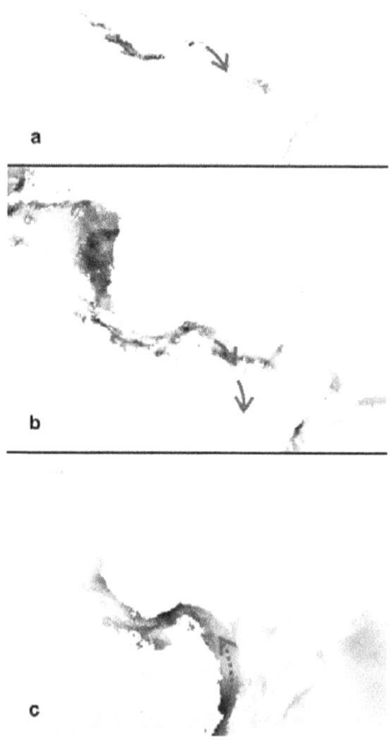

Fig. 5 Possible migration routes of *C. brenesii* (**a**), *C. micranthum* (**b**) and *C. panamense* (**c**) through the Darién Gap.

Refugia of *C. micranthum* were located in Chiapas Highlands, Cordillera Chontaleña, Cordillera de Talamanca, Serranía de Tabasará, Cordillera de San Blas and Serranía del Darién in Mesoamerica, as well as in Colombian Serranía de Abibe, Serranía de San Jéronimo, Serranía de Ayapar. Region of Bocas del Toro, Panama Canal and Colombian Pacific Coast excluding Serranía del Baudó probably were refugial areas of *C. panamense*.

The potential refugia of *C. tyrridion* were located in the coast of Gulf of Panama, Serranía Del Sapo, coasts of Gulf of Morrosquilo, Cartagena Bay, Maracaibo Lake as well as Ecuadorian coast of Bay of Sardinas.

According to the Olsons' maps, all the areas selected here as a potential refugia corresponded during the LGM to tropical rainforest and tropical woodland [17].

Estimated migration routes

Based on the distribution of the possible refugia during the last glacial maximum and the present ranges of the studied *Campylocentrum* species, their migration routes through the Darién Gap were estimated and graphically presented (Fig. 5). Due to the lack of any herbarium material from Colombia and the Carribean, the migration routes of *C. tyrridion* could not be estimated.

The most probable origin of the southernmost population of *C. brenesii* found in the valley between Pirre and next most southerly peak sloping hillside was Cordillera de San Blas (Fig. 5a). There were probably several independent sources of *C. micranthum*, currently the widest spread through the Neotropics (Fig. 5b). In case of *C. panamense* the regression of its range is observed (Fig. 5c).

Conclusions

The maximum entropy analysis algorithms applied in the presented research seems to be a useful tool for studying the possible refugia of the plants which habitat requirements are poorly known. Moreover, the modeling of the suitable niches for LGM shows that the results of the analysis accompanied with the floristic data may indicate the possible migration routes of the species.

In both approaches, the models should be interpreted with an extreme care, taking into account the representativeness of the floristic material and factors limiting species distribution not included in the analysis, such as concurrence and the pollinator range.

Acknowledgments

This article was prepared thanks to the grants from European Commission's Research Infrastructure Action via the SYNTHESYS Project (AT-TAF-2483, GB-TAF-2445) and funding obtained from Faculty of Biology, University of Gdańsk (538-L150-B065-13). The curators and staff of the cited herbaria are thanked for their kind hospitality and assistance during visits and for making specimens available on loan.

Competing interests

No competing interests have been declared.

Supplementary material

The following supplementary material for this article is available online at http://pbsociety.org.pl/journals/index.php/asbp/rt/suppFiles/asbp.2014.030/0:

1. Tab. S1: list of *Campylocentrum* specimens included in the ENM analysis.

References

1. Kanning M. Darién Gap. In: McColl RW, editor. Encyclopedia of world geography. New York, NY: Infobase Publishing; 2005. p. 234–235.
2. Williams LO, Allen PH, Dressler RL. Orchids of Panama: a facsimile reprint of "The Orchidaceae", "Flora of Panama". Saint Louis, MO: Missouri Botanical Garden; 1980. (Monographs in systematic botany; vol 4).
3. Cárdenas-López D. Inventario florístico en el Cerro del Cuchillo, Tapón del Darién Colombiano. Caldasia. 2003;25(1):101–117.
4. Bentham G. Notes on Orchidaceae. J Linn Soc Bot. 1881;18:281–361.
5. Dressler RL. *Campylocentrum*. In: Hammel BE, Grayum MH, Herrera C, Zamora Villalobos N, editors. Manual de plantas de Costa Rica. Saint Louis, MO: Missouri Botanical Garden; 2003. p. 38–41. (Monographs in systematic botany; vol 3).
6. Dressler RL. Field guide to the orchids of Costa Rica and Panama. New York, NY: Cornell University Press; 1993.
7. Bogarín D, Pupulin F. The genus *Campylocentrum* (Orchidaceae: Angraecinae) in Costa Rica: some critical questions and a few answers. In: Pridgeon AM, Suarez JP, editors. Proceedings of the second scientific conference on Andean orchids. Loja: Universidad Técnica Particular de Loja; 2009. p. 32–45.
8. Bogarín D, Pupulin F. The genus *Campylocentrum* (Orchidaceae: Angraecinae) in Costa Rica: a Revision. Harv Pap Bot. 2010;15(2):353–414.
9. Clark PU, Dyke AS, Shakun JD, Carlson AE, Clark J, Wohlfarth B, et al. The last glacial maximum. Science. 2009;325(5941):710–714.
10. Pearson RG, Raxworthy CJ, Nakamura M, Townsend Peterson A. Predicting species distributions from small numbers of occurrence records: a test case using cryptic geckos in Madagascar. J Biogeogr. 2006;34(1):102–117.
11. Elith J, Phillips SJ, Hastie T, Dudík M, Chee YE, Yates CJ. A statistical explanation of MaxEnt for ecologists: Statistical explanation of MaxEnt. Divers Distrib. 2011;17(1):43–57.
12. Phillips SJ, Dudík M, Schapire RE. A maximum entropy approach to species distribution modeling. In: ICML '04: Proceedings of the twenty-first international conference on machine learning. New York, NY: ACM Press; 2004. p. 655–662.
13. Phillips SJ, Anderson RP, Schapire RE. Maximum entropy modeling of species geographic distributions. Ecol Model. 2006;190(3–4):231–259.
14. Hijmans RJ, Cameron SE, Parra JL, Jones PG, Jarvis A. Very high resolution interpolated climate surfaces for global land areas. Int J Clim. 2005;25(15):1965–1978.
15. Urbina-Cardona JN, Loyola RD. Applying niche-based models to predict endangered-hylid potential distributions: are neotropical protected areas effective enough. Trop Conserv Sci. 2008;1(4):417–445.
16. Braconnot P, Otto-Bliesner B, Harrison S, Joussaume S, Peterchmitt J-Y, Abe-Ouchi A, et al. Results of PMIP2 coupled simulations of the mid-holocene and last glacial maximum. Part 1: experiments and large-scale features. Clim Past. 2007;3(2):261–277.
17. Olson JS, Allison LJ, Watts JA. Carbon in live vegetation of major world ecosystems. Report DOE/NBB-0037. Washington, DC: United States Department of Energy; 1983.

19

The progressive and ancestral traits of the secondary xylem within *Magnolia* clad – the early diverging lineage of flowering plants

Magdalena Marta Wróblewska*

Department of Developmental Plant Biology, Institute of Experimental Biology, University of Wrocław, Kanonia 6/8, 50-328 Wrocław, Poland

Abstract

The qualitative and quantitative studies, presented in this article, on wood anatomy of various species belonging to ancient *Magnolia* genus reveal new aspects of phylogenetic relationships between the species and show evolutionary trends, known to increase fitness of conductive tissues in angiosperms. They also provide new examples of phenotypic plasticity in plants.

The type of perforation plate in vessel members is one of the most relevant features for taxonomic studies. In *Magnolia*, until now, two types of perforation plates have been reported: the conservative, scalariform and the specialized, simple one.

In this paper, are presented some findings, new to magnolia wood science, like exclusively simple perforation plates in some species or mixed perforation plates – simple and scalariform in one vessel member. Intravascular pitting is another taxonomically important trait of vascular tissue. Interesting transient states between different patterns of pitting in one cell only have been found. This proves great flexibility of mechanisms, which elaborate cell wall structure in maturing tracheary element. The comparison of this data with phylogenetic trees, based on the fossil records and plastid gene expression, clearly shows that there is a link between the type of perforation plate and the degree of evolutionary specialization within *Magnolia* genus.

Keywords: *Magnolia*; basal angiosperms; secondary xylem; perforation plate

Introduction

Magnolias are classified as the magnoliids group that represent 3% of all angiosperms species. They diverged from the angiosperm ancestral line before the separation of the two major evolutionary lines of monocotyledones and dicotyledones [1]. They have both the ancestral features and those depicting specialization, therefore they lately became intensively studied. Morphological and anatomical studies gave the evidence for ancestral character and early rise of this group taking into account the number of original features such as undifferentiated perianth, numerous spirally arranged generative parts, construction of the pollen and presence of the oil cells [2–5].

To the Magnoliales order belongs family Magnoliaceae. First detailed classification of this family, created by James E. Dandy, was based on the flower and fruit morphology. This classification was widely accepted over the years, just with minor modifications, until Noteboom's study. Noteboom has reorganized the taxonomy taking into account "tricks of nature", such as homoplasies (the same features of different origin) in fruit structure [6,7]. Subgenera of magnolias are classified based on the morphology of flowers on flowering time and structure of leaves. With the development of DNA sequencing a new method became available and the *ndhF* chloroplast gene was first to be examined in Magnoliaceae. The results proved polyphyletic origin of *Magnolia* and established new connections between minor sections [8]. Molecular data confirmed also early origin of magnoliids [9].

In addition to flower structure the taxonomists use wood anatomy in various classifications systems and anatomical keys. The wood (secondary xylem) is a secondary conducting tissue in perennial woody plants, which develops due to activity of cambium – secondary meristem. Vessels, present only in angiosperm wood, are most significant elements with traits of taxonomic value. The units of the vessel, the cells called vessel members, function as dead elements. They are produced by cambial fusiform initials through their periclinal divisions. Vessel elements differentiate increasing in diameter, thus their length is more or less the same as the length of producing them cambial initials. Their length's mean value increases in early ontogeny of a tree (in accordance with Sanio rule for non-storied cambium). They are composed of the cell wall only and are interconnected vertically by perforation plates to form a pipe conducting water. The cell wall pattern is characteristic and can be diverse, its formation is driven by the expression of specific genes involved in biosynthesis of polysaccharides, lignins,

* Email: magdalena.turczyn@gmail.com

Handling Editor: Beata Zagórska-Marek

proteins etc. [10]. The development of a perforation plate is especially well-studied process. The perforation is created by the primary wall enzymatic breakdown on the border between the two adjacent elements, in the areas not covered by the secondary cell wall. Important role in this process is played by cellulase, which degrades some cellulose microfibrils. The remaining fibrils are either carried away by a transpiration stream [11,12] or some of them stay forming thread-like structures in a plate, this feature is found in less specialized vessel members [13]. The perforation plate can be set obliquely or transversely, with regard to the axis of the vessel. The plate may have only one large opening and then it is called a simple perforation plate (Fig. 1a) or several small perforations. Their arrangement allows distinguishing three basic types of plates: scalariform (Fig. 1b), which is composed of flattened openings arranged parallel to each other, reticulate (Fig. 1c) where the openings form a network and foraminate (Fig. 1d) with the series of circular openings. In the course of evolution the perforation plates specialized from scalariform to the simple plate [14], evident is also gradual shortening of vessel members [13].

Vessels contact with other, adjacent xylem elements through the pits situated on the lateral walls. Pits may form defined patterns. When they are horizontally elongated and form a ladder-like vertical series with the lateral length of each pit comparable to the vessel width, their arrangement is called scalariform pitting (Fig. 1a). In transitional pitting (Fig. 1b) some pits are shorter, they occur in laterally adjacent groups of 2–3 pits and these groups alternate longitudinally with long pits. Pseudoscalariform pattern (Fig. 1b) occurs when the shorter pits appear in between two longer ones, disrupting the ladder-like structure. The lateral series of round or elliptical in outline pits are present in a pattern of opposite pitting (Fig. 1c). The pits are arranged on the vessel wall in horizontal and vertical rows. In another variant of this pattern the pits in consecutive horizontal rows are laterally shifted, which leads to formation of diagonal rows instead of vertical (Fig. 1d). This, most common in dicotyledonous plants type of pitting, is called an alternate pattern [15]. The pit packing in alternate pattern is much more efficient than in opposite one, accommodating more pits on the wall surface of the same size. This naturally increases the lateral transport of water and thus should be treated as an advantageous, evolutionary novelty.

Magnolia's wood is generally described as diffuse – porous, where vessels are of the same diameter and evenly distributed within a growth ring, either solitary or in short radial multiples containing 3–4 adjacent vessels. Vessels in such ancient plants should be considered as a progressive trait, because there are some magnoliids like Trochodendron and Tetracentron, which are completely vesselless. Their wood resembles more wood of gymnosperms than of flowering plants. In *Magnolia* wood the annual rings are distinct and limited by the marginal band of tracheids and/or parenchyma cells. Rays are hetero- and homocellular. Vessel members in *Magnolia* have spiral thickenings on the walls [16,17]. Wood features, which are used to distinguish the individual species within Magnoliaceae are mainly the vessel member traits, like the type of perforation plate and pattern of pitting. Therefore these elements are of special interest. In *Magnolia* the presence of both simple and scalariform type of plate has been described. The species with exclusively simple perforation plates are for example: *Magnolia acuminata* (L.) L., *M. stellata* (Sieb. & Zucc.) Maxim., *M. obovata* Thumb. (previous name: *M. hypoleuca*), *M. offiicinalis* Rehd. & Wilson, *M. kobus* DC, *M. liliiflora* Desr. [prev.: *M. quinquepeta*

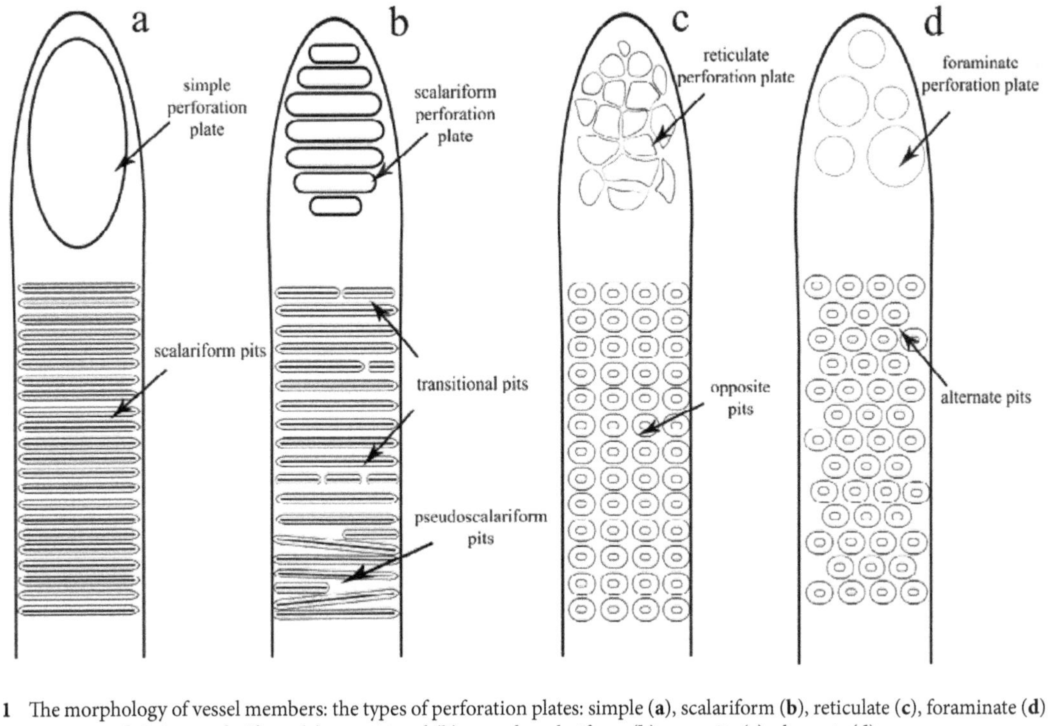

Fig. 1 The morphology of vessel members: the types of perforation plates: simple (**a**), scalariform (**b**), reticulate (**c**), foraminate (**d**) and different patterns of pitting: scalariform (**a**), transitional (**b**), pseudoscalariform (**b**), opposite (**c**), alternate (**d**).

(Buchoz) Dandy], the species with scalariform perforations are *M. sieboldii* K. Koch, *M. cylindrica* Wilson, *M. tripetala* (L.) L. Interestingly, there are some species with both types of perforation plates, like M· denudata Desr. [prev.: *M. heptapeta* (Buchoz) Dandy] [18].

In this paper it is shown that variation in wood architecture described in the literature occurs in magnolias growing in Poland in humid continental climate (Köppen–Geiger classification) [19] and that quantifying these variations, provides new information on their taxonomic significance. Interesting modifications in the perforation plate morphology have been noted as an example of flexibility in this group of flowering plants. Since the simple perforation plate is considered to be a progressive and characteristic feature of phylogenetically younger species, its presence was examined more closely and showed the connection between the timing of species emergence in the tree of life and its wood micromorphology. The results were compared with widely available information on magnolias geographic distribution, fossil records and their plastid and nuclear genes sequences analysis.

Material and methods

Wood samples were collected from mature branches of trees and shrubs. Into the study only solidly confirmed species were included, excluding hybrids, varieties and trees, which were decoratively trimmed. All examined individuals, representing the *Magnolia* species, grow in the Botanical Garden of the University of Wrocław (Wrocław, Poland) and in WULS-SGGW Arboretum in Rogów (Rogów, Poland; Tab. 1).

Tab. 1 Species of the *Magnolia* from WULS-SGGW Arboretum in Rogów (Rogów, Poland) and Botanical Garden of the University of Wrocław (Wrocław, Poland) used in the present study.

Section	Species	Number of examined trees	Age of branches for each tree
Yulania	*Magnolia acuminata*	4 (2)	10/9/9/8
	Magnolia cylindrica	1	6
	Magnolia denudata	2 (1)	8/9
	Magnolia kobus	3 (1)	9/8/7
	Magnolia liliiflora	4 (1)	9/8/8/7
	Magnolia salicifolia	3 (1)	9/9/9
	Magnolia stellata	2 (1)	10/8
Rytidospermum	*Magnolia obovata*	3 (1)	10/9/7
	Magnolia officinalis	3 (1)	9/8/7
	Magnolia sieboldii	4 (2)	9/9/8/8
	Magnolia tripetala	3 (1)	9/8/7
	Magnolia wilsonii	2	10/9
Magnolia	*Magnolia virginiana*	2	8/8

In parentheses is given the number of individuals coming from the Botanical Garden of the University of Wrocław.

Wood anatomy was analyzed with use of two techniques: wood sections and macerates. Initially wood discs from 6–10 year old branches were boiled in the distilled water with glycerol for 30 hours to soften the material. Then, out of pretreated wood blocks, 20 μm thick sections (transverse, radial and tangential) were cut using a sliding microtome Leica S14 2000R (Leica Instruments, GmbH, Germany). Permanent microscope slides were prepared as follows: the sections mounted on the slides had been bleached with use of sodium hypochlorite (NaClO), stained with Malachite Green and dehydrated with ethanol and isopropanol series. Afterwards they were embedded in Euparal [20,21].

The wood macerates were prepared from the outermost growth ring. Each wood sample was treated overnight with acetic acid and then boiled in 3% hydrogen peroxide for 6 hours or until visible separation of cells had occurred. Some samples were also stained with alcian blue for better contrast [20,21]. The pictures were taken with Olympus BX50 Microscope, Olympus Camera DP71 and Cell B Software (Olympus Optical Co., Warsaw, Poland).

To compare the frequencies of perforation plate types in *Magnolia*, 120 vessel members for each species were analyzed. 30 measurements of the length and width from vessel elements for 13 species, separately for each individual, were taken using Cell B Software Arbitrary Measure Tool. For these species in which more than one type of perforation plate occurred, 30 measurements were taken separately for the cells with exclusively scalariform, simple and mixed perforation plates. To correlate different features we performed statistical tests. Normal distribution was analyzed with the Shapiro–Wilk test and the comparisons of mean vessel length and diameter between *Magnolia* species with the Welch Two Sample *t*-test. Statistical data were obtained with the R program [22].

Results

General wood morphology

Blocks of wood, collected from mature branches, were analyzed in three planes: transverse, radial and tangential to show all the anatomical features of a secondary xylem. The anatomical features common for all *Magnolia* species, studied here, are listed as follows. The wood was diffuse-porous. Vessels of the same growth ring had at the average the same diameter and were arranged either solitary or in 2–4 radial multiples (Fig. 2a), rarely in clusters. The terminal layer (Fig. 2b) was distinct, formed by flattened tracheids, 3–4 in each row. Rays were homogeneous and usually 1–4 seriate (Fig. 2c,d). In the vessels both simple and scalariform perforation plates and various patterns of pits were observed (Fig. 2c). Tertiary thickenings, very thin, were present in some vessel elements. Better description of the morphology of a single and entire vessel member was possible while observing macerates rather than microtome sections.

Perforation plates

Perforation plates were examined qualitatively and quantitatively in the wood of 13 *Magnolia* species with use of macerates. Three types of vessel elements with different

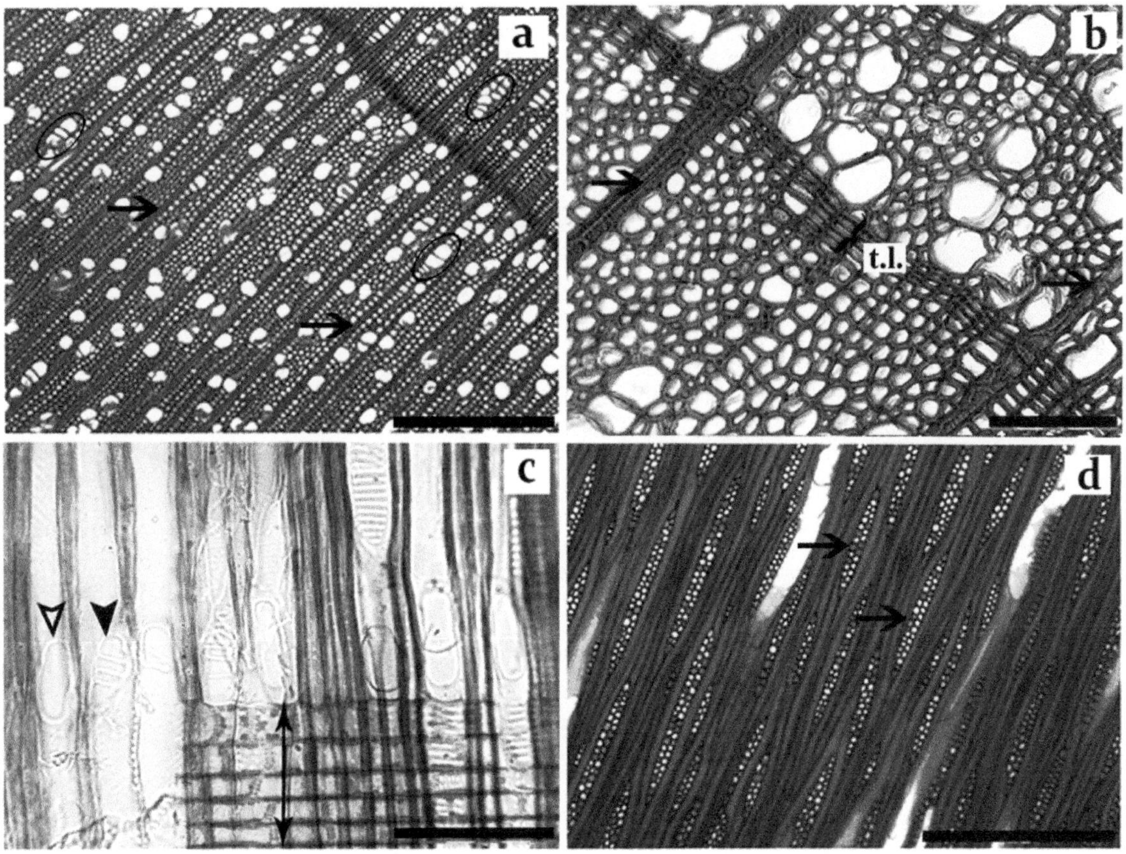

Fig. 2 Details of *Magnolia* wood anatomy. Diffuse-porous wood of *M. sieboldii* on transverse sections (**a,b**), the wood rays indicated by arrows, vessels in radial multiples marked with the circles (**a**); flattened tracheids in terminal layer (t.l.; **b**); simple (white arrowhead) and scalariform (black arrowhead) perforation plates in vessels and part of homogeneous ray marked with double arrow on the radial section of *M. stellata* wood (**c**); 2–3- seriate rays on tangential section of *M. stellata* wood – pointed by arrows (**d**). Scale bars: 500 μm (**a**); 100 μm (**b,c**); 200 μm (**d**).

perforation plates were documented: exclusively scalariform, exclusively simple or mixed, with two different perforation plates at the opposite ends of one vessel member. Exclusively scalariform perforation plates were characteristic of four examined species: *M. virginiana* (Fig. 3a), *M. wilsonii* (Fig. 3b), *M. sieboldii* (Fig. 3c) and *M. tripetala* (Fig. 3d). Vessel members with exclusively simple perforation plates were observed in *M. acuminata*, *M. salicifolia* and *M. kobus* (Fig. 3e–g, respectively). In the species: *M. obovata* (Fig. 4a), *M. liliiflora* (Fig. 4b), *M. cylindrica* (Fig. 4e), *M. officinalis* (Fig. 4f), *M. stellata* (Fig. 4h), *M. denudata* (Fig. 4k) vessel members with both types of perforation plate in one cell were encountered among others. Except for the two groups of species with exclusively simple or scalariform plates, the frequency of different types of plates, for the specimens analyzed in this paper, was species specific. In *M. obovata* (Fig. 4a), *M. stellata* (Fig. 4h,i), *M. officinalis* (Fig. 4f,g), *M. denudata* (Fig. 4j,k) only simple and mixed perforation plates occurred, whereas *M. liliiflora* (Fig. 4b,c) and *M. cylindrica* (Fig. 4d,e) had all three types of vessel elements with simple, scalariform and mixed perforation plates. The frequency of specific vessel member types is shown in Tab. 2. Results (the ratio of plate types) for individuals from one species analyzed from two places were the same. Within the species that have simple and mixed perforation plates the greater percentage of cells had simple type of plate, more than 70%. The same was true for *M. liliiflora*, which has all three types of plates. In this species the vessel members with mixed and exclusively scalariform plates together were still in minority, of 30% only. However, in *M. cylindrica* the ratio of simple, mixed and scalariform types of plates was as 1:1:2. Spiral thickenings of the vessel member wall were visible in *M. cylindrica* (Fig. 4d,e), *M. stellata*, *M. denudata* (Fig. 4i,k; respectively)

Vessel members with exclusively scalariform plates – on both ends of one cell, had at the average 12 bars in the plate. The highest number of bars (20) was found in *M. sieboldii* and in *M. wilsonii*. Species with mixed perforation plates usually had plates with only 1–4 bars. In *M. sieboldii*, *M. tripetala* and *M. wilsonii* the modification of plate structure was found, such as bifurcations of the bars (Fig. 5a). Usually there was only one bifurcation per plate but there were cases where even 3 bifurcating bars were observed. Other interesting aberrations resulted from the evidently

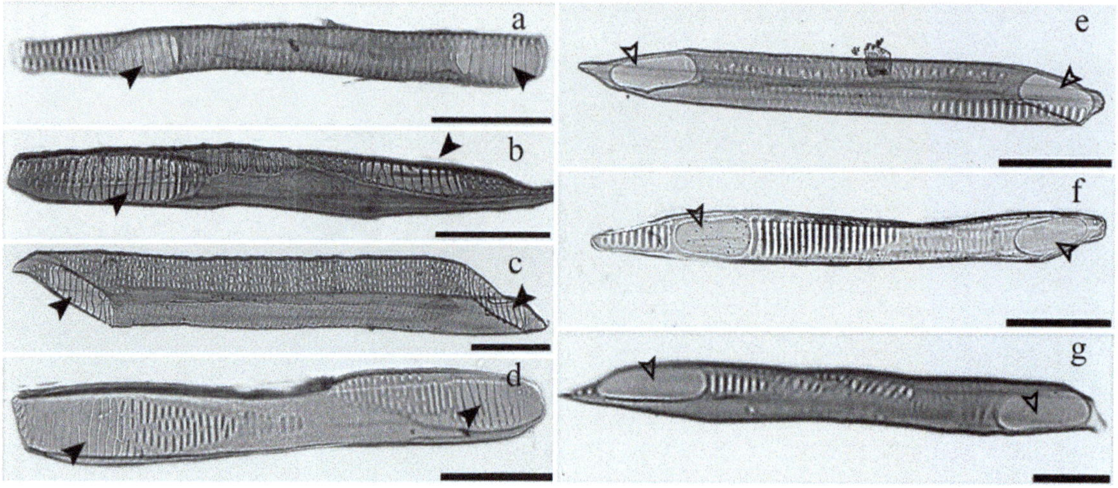

Fig. 3 The *Magnolia* species with exclusively scalariform perforation plates (**a–d**; pointed by black arrowheads) and with exclusively simple perforation plates (**e–g**; pointed by white arrowheads). *M. virginiana* (**a**); *M. wilsonii* (**b**); *M. sieboldii* (**c**); *M. tripetala* (**d**); *M. acuminata* (**e**); *M. salicifolia* (**f**); *M. kobus* (**g**). Scale bar: 100 μm.

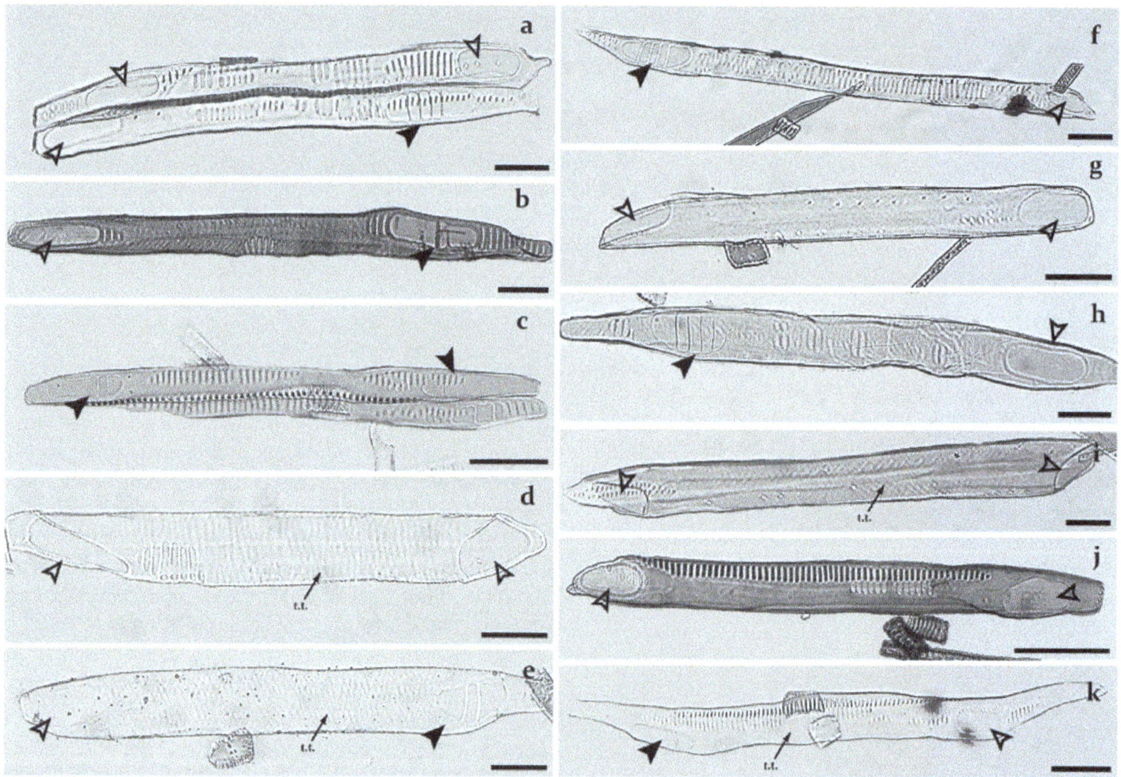

Fig. 4 The *Magnolia* species with more than one type of perforation plate. **a** *M. obovata* – two vessel members: one with exclusively simple perforation plates (above), the second with mixed perforation plates (below). **b** *M. liliiflora* – vessel member with two types of perforation plates. **c** *M. liliiflora* – vessel member with scalariform perforation plates having two bars only. **d,e** *M. cylindrica* – vessel member with only simple perforation plates (**d**) and mixed type of plate (**e**). **f,g** *M. officinalis* – vessel member with mixed perforation plates (**f**) and with exclusively simple plates (**g**). **h,i** *M. stellata* – vessel member with mixed perforation plates (**h**) and with two simple perforation plates (**i**). **j,k** *M. denudata* – vessel member with exclusively simple perforation plate (**j**) and mixed perforation plates (**k**). In vessel members delicate tertiary thickenings (t.t.) of the cell wall are visible – pointed by arrows (**d,e,g,i**); the scalariform perforation plates pointed by black arrowheads and simple by white arrowheads. Scale bar: 100 μm.

Tab. 2 Frequency of various perforation plate types present in analyzed *Magnolia* species.

	Perforation plate		
Species*	**Simple**	**Scalariform**	**Mixed**
Magnolia sieboldii	----	100%	----
Magnolia tripetala	----	100%	----
Magnolia virginiana	----	100%	----
Magnolia wilsonii	----	100%	----
Magnolia acuminata	100%	----	----
Magnolia kobus	100%	----	----
Magnolia salicifolia	100%	----	----
Magnolia denudata	86%	----	13%
Magnolia obovata	90%	----	10%
Magnolia officinalis	74%	----	26%
Magnolia stellata	85%	----	15%
Magnolia cylindrica	24%	51%	25%
Magnolia liliiflora	72 %	10%	18%

* Species are listed in alphabetical order within each group having specific pattern of perforation plate types.

imperfect plate formation in the maturing vessel member. In *M. obovata* the structures resembling very short bars in otherwise simple perforation plates were noted (Fig. 5b). In *M. denudata* (Fig. 5c) partially developed bars occurred: the plate structure where a single bar ended in the middle of a plate was observed. This was in a sense a negative image of the bifurcated bar – a bifurcated perforation. In *M. salicifolia* two simple perforation plates on one side of a cell have been found (Fig. 5d).

Intravascular pitting

The scalariform type of intravascular pitting (Fig. 6a) present on the lateral cell walls of vessel members was next characteristic feature for all examined species. In some cases, like in *M. liliiflora*, a continuous transitions between three types of pitting was observed: reduction of one of the pits lateral dimension, resulted in a change from scalariform to transitional pattern and then further to opposite one (Fig. 6b). In *M. tripetala* the oval pits of similar dimensions in the opposite arrangement were present. In this case instead of a single, laterally elongated pit two, three or even four shorter ones occurred in one row (Fig. 6c). Additionally, in *M. liliiflora* and *M. denudata* pseudoscalariform pits had been found (Fig. 6d). In this pattern some pits were shorter than others and not arranged in the lateral series. They looked like wedges pushing apart neighboring lateral pits.

Vessel cells dimensions

Measurements of the lengths and widths of vessel members in all examined species did not reveal significant intraspecific differences between individuals. There were, however, interesting differences among the species (Fig. 7). The longest and narrowest vessel elements were documented for *Magnolia denudata* in the category of mixed-plate cells, and the shortest (but not the narrowest) in *M. virginiana* where the plates were exclusively scalariform. The highest diameter was scored for *M. acuminata* having exclusively simple plates. The statistical tests were performed to evaluate the significance of scored dimensions. There was no correlation within the species between vessel member dimensions and the perforation plate types. The only statistically significant differences in vessel cell lengths (upper part of a plot in the Fig. 7), occurred between *M. virginiana* and the group with exclusively scalariform perforation plates, and also between *M. salicifolia* and the species with exclusively simple perforation plates.

Considering vessel widths, three groups of statistically similar species can be indicated. The first one consists of the species with scalariform perforation plate: *M. liliiflora*, *M. sieboldii*, *M. virginiana* and *M. wilsonii*, to the second, with mixed perforation plate, *M. cylindrica* and *M. liliiflora* belong and to the third, with simple perforation plate, *M. cylindrica*, *M. denudata*, *M. kobus*, *M. liliiflora* and *M. obovata*.

Discussion

The Magnoliaceae family is distributed natively in temperate, subtropical and tropical regions of northern part of South America, Central America and North America to the

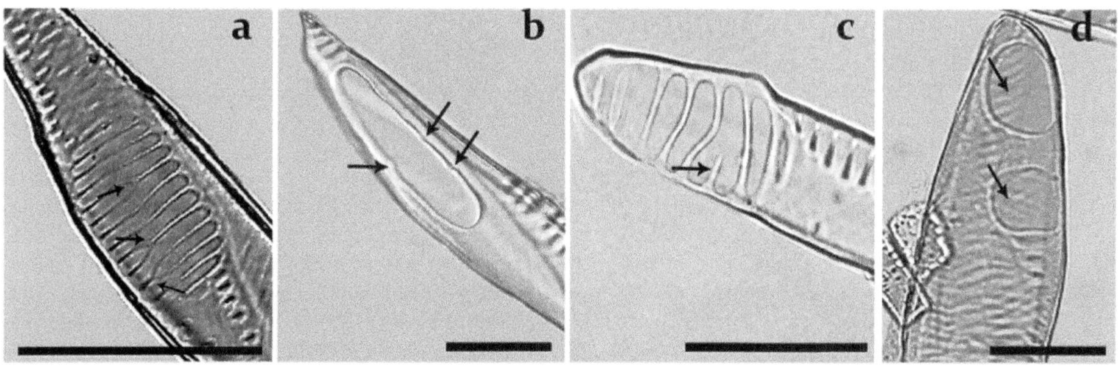

Fig. 5 Aberrant perforation plates. **a** Three bar bifurcations in one plate (pointed by arrows) in *M. sieboldii*. **b** The structures resembling developing but "unfinished" bars (indicated by arrows) in otherwise simple perforation plate of *M. obovata*. **c** A scalariform perforation plate with partially developed bar (see arrow) in *M. denudata*. **d** Tip of the vessel element with two perforation plates formed close by in *M. salicifolia*. Scale bar: 100 μm.

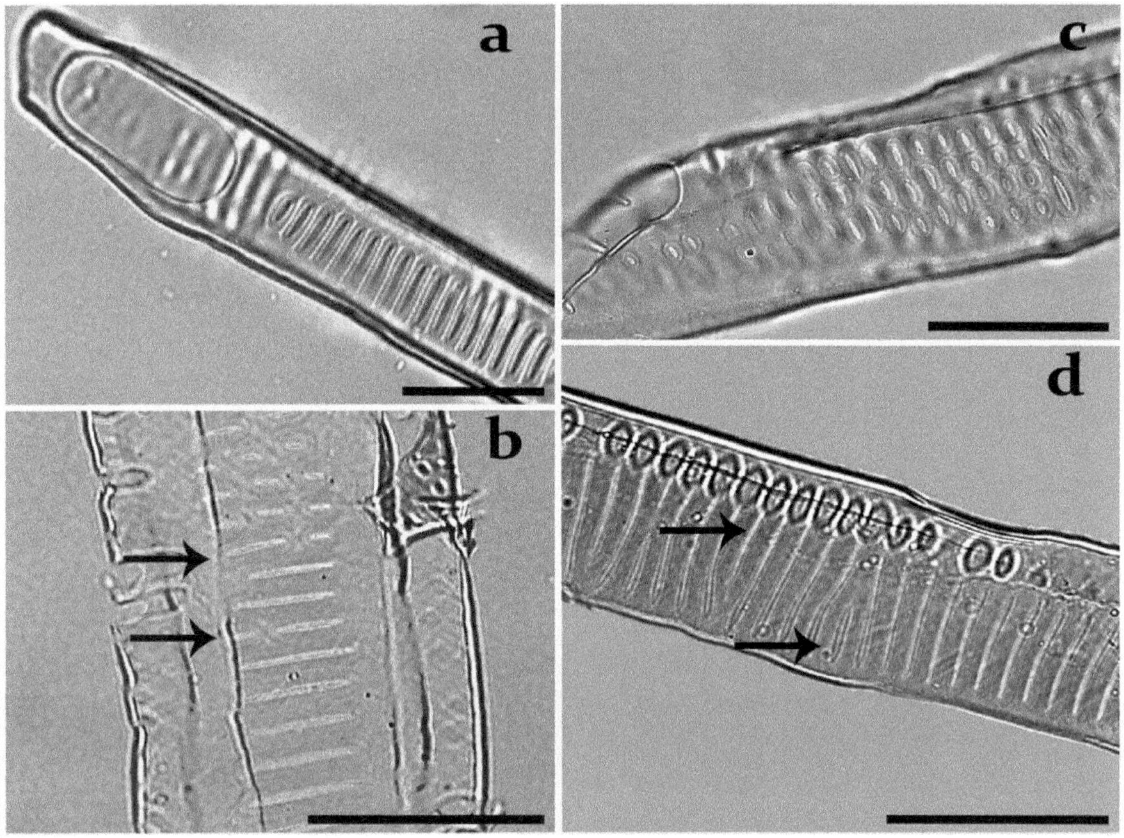

Fig. 6 Patterns of lateral cell wall pitting in vessel members. a Progressively simple perforation plate and ancestral scalariform pitting of the cell wall in *M. acuminata*. **b** Transitional pitting in the area denoted by arrows in *M. liliiflora*. **c** Opposite pits in *M. tripetala*. d Pseudoscalariform pitting in *M. denudata*, arrows point out the sites of shorter pits insertion. Scale bar: 50 μm.

Great Lakes and southern and eastern Asia [23]. Taxonomists and evolutionists based their research on magnolia wood considering only the genets (individuals) from these regions. In Poland, like elsewhere in Europe, magnolias extinct due to the Pleistocene glaciations, were reintroduced, thus their properties as of plants growing in different than native climatic conditions seemed to be of interest, especially in light of arguments for environmental impact on wood anatomy provided by "ecological wood anatomy" [24]. The *Magnolia* wood analyzed in this study was sampled from the trees growing in Poland and apparently was not much different from that described in literature with the exceptions discussed below.

The material used in this study was the wood from *Magnolia* branches. These parts of a tree or a shrub are more accessible than the main trunk and no specific drills are necessary. Therefore, in this research the main attention has been focused on that type of plant material, which is readily available for taxonomical analyzes. It is possible that besides the potential impact of environmental cues this was the reason why some characteristics of this wood were different from the literature data based on the wood sampled from the main stem.

Significant differences were found in the perforation plates. The data presented in the literature depicts the species like *M. officinalis*, *M. liliiflora* as those with exclusively simple perforation plates [18]. Data presented in these paper show that, beside simple plates also have mixed perforation plates. Furthermore, in *M. liliiflora* the cells with exclusively scalariform perforation plates were found. *Magnolia obovata* is described by Chen group as a species with exclusively simple perforation plates and by the InsideWood Working Group (IWG) as a species with simple and scalariform perforation plates [25,26]. In this study mostly simple but also mixed perforation plates have been shown. *Magnolia cylindrica* described in the literature as the species with exclusively scalariform type [18] in this study showed the vessel elements with all possible types of perforation plates: exclusively simple, scalariform and mixed. For *M. denudata* the presence of exclusively scalariform perforation plates reported by Chen group or simple and scalariform perforation plates described by IWG was not confirmed. The majority of vessel members in this species had simple or mixed perforation plates. Except for above differences the results of analyses performed on stem wood, available in the literature, are similar to those obtained in this study for the wood sampled from branches.

In three species with exclusive scalariform perforation plates: *M. sieboldii*, *M. tripetala* and *M. wilsonii* a feature known in literature as forking bars [27] was noticed. This

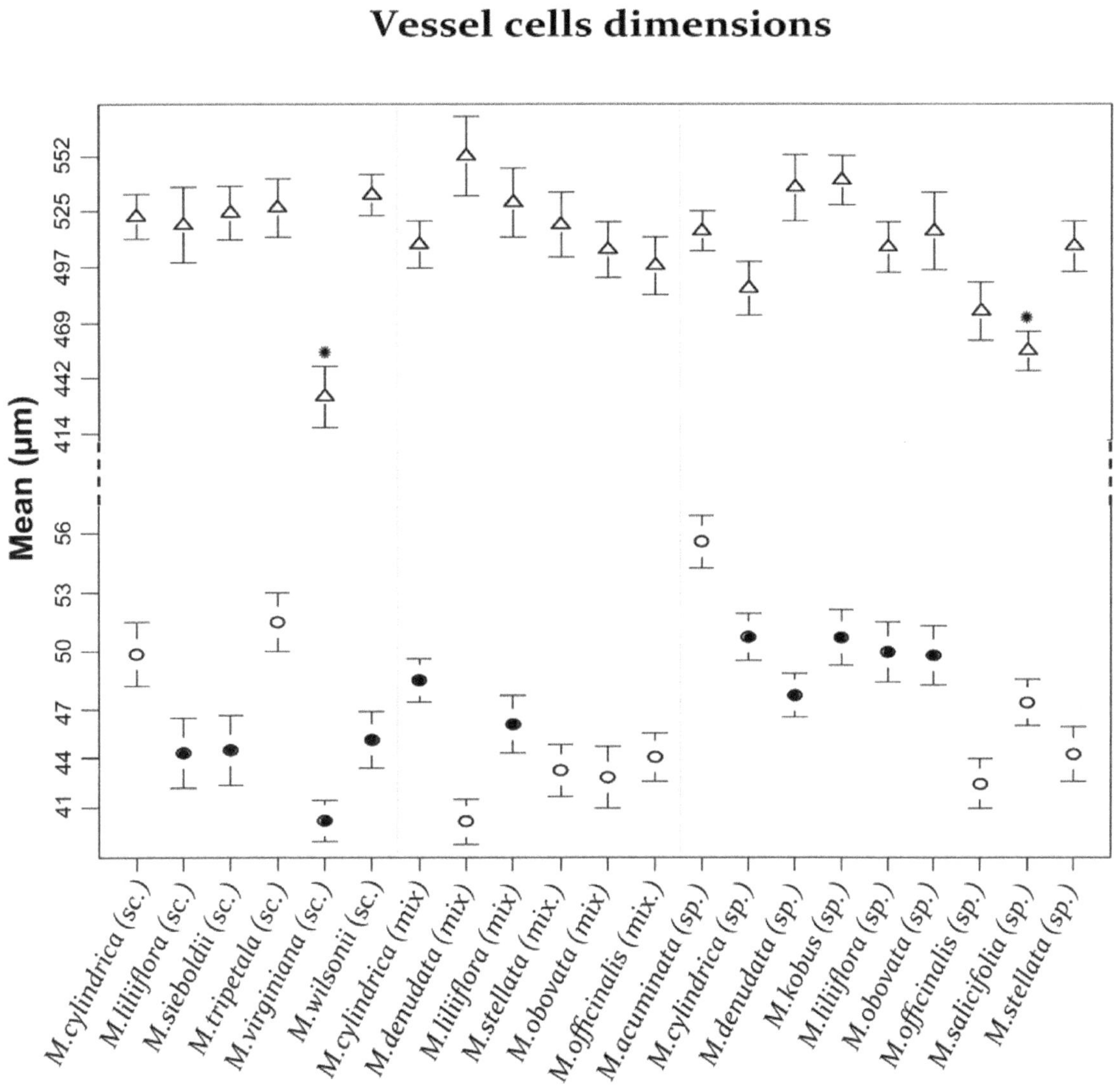

Fig. 7 Vessel cells dimensions. Mean (µm) ±std. err. of the vessel member length (upper part of the plot) and width (lower part of the plot) shown in groups of magnolia species distinguished by perforation plates, listed in alphabetical order within each group. (sc.) – vessel cells with exclusively scalariform plate; (mix.) – cells with mixed perforation plates; (sp.) – cells with exclusively simple perforation plates. Asterisks in the upper part of the plot show the species with significant differences in cells length and black dots in lower part indicate the species with no significant differences between mean values of the width.

phenomenon occurs in many angiosperms and sometimes it leads to the identification of a plate as a scalariform type, especially when most of the bars in one plate is forked, like for example in *Miodocarpus fraxinifolius* [28]. In this study maximum three forkes per one plate were documented (Fig. 5a). Interestingly, bifurcations occurred often on both perforation plates of the vessel.

In the literature there is a variety of molecular and fossil data important and useful in reconstructing magnolias evolution [8,29,30]. The researchers speculate that the evolutionary trend in wood evolution leads from scalariform perforation plate to the simple one [31]. Therefore, the occurrence of plate types within analyzed *Magnolia* species was correlated in this study with the time of species divergence. New morphological data were compared here with the information from the cladograms created by independent groups [8,29,30]. The combined cladogram (Fig. 8) allowed concluding that the time of species disjunction is indeed reflected in the vessel morphology. Younger species like *M. acuminata* have exclusively simple perforation plate while the older ones, like *M. virginiana*, exclusively scalariform. The results of comparison mentioned above are supported also by literature data. Scalariform perforation plates were reported in *Magnolia* species (prev. *Talauma* genus) from Borneo in tropical Asia. This clad is known as ancestral lineage within Magnoliaceae family [32]. Azuma group has noticed that there is a need to compare once again the morphological characters between *M. tripetala* and section

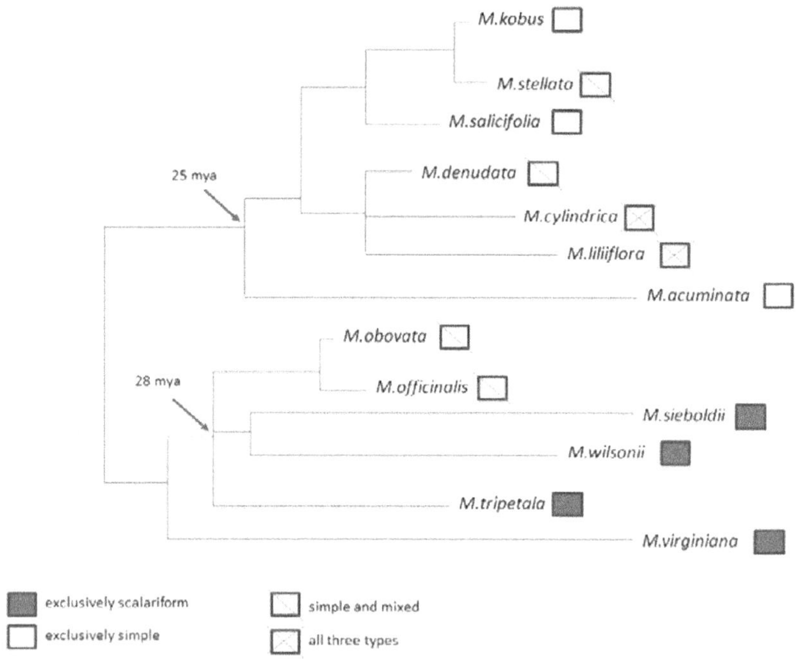

Fig. 8 Combined cladogram of our data on the types of perforation plates with the data known from literature on magnolia species divergence (Azuma et al. [29], modified; Kim et al. [8], modified; Nie et al. [30], modified). Arrows indicate estimated time of separation given in million years (mya). Squares show the type of perforation plate occurring in the species.

Oyama, which includes *M. sieboldii* and *M. wilsonii* to confirm or disprove the phylogenetic relationship based on molecular data. The presence of exclusively scalariform perforation plate in both section *Oyama* and *Magnolia tripetala* found in this study could be the missing morphological feature sustaining the molecular relationships. This close association is in opposition to *M. virginiana* (section *Magnolia*) in which also exclusively scalariform perforation plates occur but the statistical analyses suggest that this species has shorter and narrower vessel members. Evolutionary trends are shortening of tracheary elements of cambial origin and changing the ancestral scalariform perforation plates to the simple ones. *Magnolia virginiana* with its short vessel members and exclusively scalariform perforation plates is an example of heterobatmy – coexistence of progressive and ancestral traits in the same organism in the result of different rate of their evolutionary specialization. Opposite condition seems to occur in *M. denudata* where the vessel elements are quite long (ancestral trait) whereas perforation plates are specialized – mixed or simple.

Analysis of general vessel morphology: perforation plates, pits, vessel length and width are invaluable for better understanding phylogenetic relationships and evolution of the plant hydraulic system [14]. There are studies that show high correlation between vessel members' length and diameter [33] while others report that there is no relationship between those traits [34]. In this paper no correlations have been found between the dimensions of *Magnolia* vessel elements. Vessel diameter is correlated with environmental conditions and also with the stem age and size [14,35] and vessel length is related to cambial initial cell length [36]. In roots the vessel elements are wider in diameter but not always longer than in the stems [37]. Because the magnolias in this research were collected from the areas with similar climate [38], and the material was taken from the same organs – 6–12 years old branches – their general wood architecture as well as vessel cell dimensions were similar.

There is available a very limited set of morphological and anatomical data for mature branches. To make proper taxonomic investigations the knowledge about wood structure and anatomy not only from the main stem but also from other parts of the plant is important. Results from this paper show that using branch material can be quite advantageous and may provide satisfactory, sufficient data for phylogenetic and taxonomic analysis.

In conclusion, this paper confirmed that wood anatomy is a great source of traits that may be used in the studies of evolutionary relationships within flowering plants. Thorough phenotypic analysis of these plants, which represent the ancestral lines, is very important for better understanding the evolutionary trends and different opportunities for development. Those studies show that within *Magnolia* genus phylogenetic progressions are already visible. Wood modifications, similar to those observed in evolution of younger lineages of flowering plants, like shortening of cambial cells and vessel members or progression from scalariform to simple perforation plate, make this group quite advanced, but also provide a good example of convergent evolution.

Acknowledgments
This work was supported by the University of Wrocław (grant No. 1068/S/IBE/13).

Competing interests
No competing interests have been declared.

References
1. Soltis PS, Soltis DE. The origin and diversification of angiosperms. Am J Bot. 2004;91(10):1614–1626.
2. Canright JE. The comparative morphology and relationships of the Magnoliaceae. I. trends of specialization in the stamens. Am J Bot. 1952;39(7):484.
3. Canright JE. The comparative morphology and relationships of the Magnoliaceae. II. The significance of the pollen. Phytomorphology. 1953;3:355–365.
4. Zagórska-Marek B. O przyczynach zmienności planu budowy kwiatu. Postępy Biol. Komórki. 27(1):97–109.
5. Zagórska-Marek B. *Magnolia* flower – the living crystal. Magnolia. 2011;89:11–21.
6. Nooteboom HP. Notes on Magnoliaceae, with a revision of *Pachylarnax* and *Elmerrillia* and the Malesian species of *Manglietia* and *Michelia*. Blumea. 1985;31:65–121.
7. Figlar RB, Nooteboom HP. Notes on Magnoliaceae IV. Blumea. 2004;49(1):87–100.
8. Kim S, Park CW, Kim YD, Suh Y. Phylogenetic relationships in family Magnoliaceae inferred from *ndhF* sequences. Am J Bot. 2001;88(4):717–728.
9. Mathews S, Donoghue MJ. Basal angiosperm phylogeny inferred from duplicate phytochromes *a* and *c*. Int J Plant Sci. 2000;161(6 suppl):S41–S55.
10. Demura T, Fukuda H. Transcriptional regulation in wood formation. Trends Plant Sci. 2007;12(2):64–70.
11. Yata S, Itoh T, Kishima T. Formation of perforation plates and bordered pits in differentiating vessel elements. Wood Res. 1970;50:1–11.
12. Plomion C, Leprovost G, Stokes A. Wood formation in trees. Plant Physiol. 2001;127(4):1513–1523.
13. Carlquist S. Xylem heterochrony: an unappreciated key to angiosperm origin and diversifications. Bot J Linn Soc. 2009;161(1):26–65.
14. Herendeen PS, Wheeler EA, Baas P. Angiosperm wood evolution and the potential contribution of paleontological data. Bot Rev. 1999;65(3):278–300.
15. Carlquist S. Lateral wall pitting of vessels. In: Timell TE, editor. Comparative wood anatomy systematic, ecological, and evolutionary aspects of dicotyledon wood. Berlin: Springer Berlin Heidelberg; 2001. p. 74–82.
16. Canright J. The comparative morphology and relationships of the Magnoliaceae. IV. Wood and nodal anatomy. J Arnold Arbor. 1955;36:119–143.
17. Kubitzki K, Rohwer J, Bittrich V. Flowering plants – Dicotyledons. Magnoliid, Hamamelid and Caryophyllid families. In: Kadereit JW, editor. Families and genera of vascular plants. Berlin: Springer; 1993. p. 391–402. (vol 2).
18. Baas P, Liang CB, Wheeler EA, Shuming W. Wood anatomy of trees and shrubs from China. VI. Magnoliaceae. IAWA J. 1993;14(4):391–412.
19. Peel MC, Finlayson BL, McMahon TA. Updated world map of the Köppen–Geiger climate classification. Hydrol Earth Syst Sci. 2007;11(5):1633–1644.
20. Johansen DA. Plant microtechnique. New York, NY: McGraw-Hill Book Co.; 1940.
21. Gerlach D. Zarys mikrotechniki botanicznej. Warsaw: PWRiL; 1972.
22. R Development Core Team. R: a language and environment for statistical computing. Vienna: R Foundation for Statistical Computing; 2011.
23. Frodin D, Govaerts R. World checklist and bibliography of Magnoliaceae. Kew: Royal Botanic Gardens; 1996.
24. Carlquist S. Ecological strategies of xylem evolution. Berkeley, CA: University of California Press; 1975.
25. InsideWood [Internet]. 2004 [cited 2014 Oct 2].
26. Wheeler EA. InsideWood – a web resource for hardwood anatomy. IAWA J. 2011;32(2):199–211.
27. Meylan BA, Butterfield BG. Occurrence of simple, multiple, and combination perforation plates in the vessels of New Zealand woods. N Z J Bot. 1975;13(1):1–18.
28. Carlquist SJ. Scalariform perforation plate variants. In: Timell TE, editor. Comparative wood anatomy: systematic, ecological, and evolutionary aspects of dicotyledon wood. Berlin: Springer; 1988. p. 68–72.
29. Azuma H, Garcia-Franco JG, Rico-Gray V, Thien LB. Molecular phylogeny of the Magnoliaceae: the biogeography of tropical and temperate disjunctions. Am J Bot. 2001;88(12):2275.
30. Nie ZL, Wen J, Azuma H, Qiu YL, Sun H, Meng Y, et al. Phylogenetic and biogeographic complexity of Magnoliaceae in the Northern Hemisphere inferred from three nuclear data sets. Mol Phylogenet Evol. 2008;48(3):1027–1040.
31. Frost FH. Specialization in secondary xylem of dicotyledons. II. Evolution of end wall of vessel segment. Bot Gaz. 1930;90(2):198.
32. Takahashi A. Wood anatomical report of some Magnoliaceae from Borneo. J Phytogeogr Taxon. 1989;37:93–100.
33. Lens F, Sperry JS, Christman MA, Choat B, Rabaey D, Jansen S. Testing hypotheses that link wood anatomy to cavitation resistance and hydraulic conductivity in the genus *Acer*. New Phytol. 2011;190(3):709–723.
34. Wheeler JK, Sperry JS, Hacke UG, Hoang N. Inter-vessel pitting and cavitation in woody Rosaceae and other vesselled plants: a basis for a safety versus efficiency trade-off in xylem transport. Plant Cell Environ. 2005;28(6):800–812.
35. Zhang S, Baas P. Wood anatomy of trees and shrubs from China. III. Rosaceae. IAWA J. 1992;13:21–91.
36. Kitin P. Variations in the lengths of fusiform cambial cells and vessel elements in *Kalopanax pictus*. Ann Bot. 1999;84(5):621–632.
37. Jacobsen AL, Pratt RB, Tobin MF, Hacke UG, Ewers FW. A global analysis of xylem vessel length in woody plants. Am J Bot. 2012;99(10):1583–1591.
38. Ziernicka-Wojtaszek A. Weryfikacja rolniczo-klimatycznych regionalizacji Polski w świetle współczesnych zmian klimatu. Acta Agrophys. 2009;13(3):803–812.

20

Protocol for acclimatization of in vitro cultured *Potamogeton praelongus* – aspect of plantlet size and type of substrate

Romana Prausová[1], Zuzana Kozelková[1]*, Lenka Šafářová[2]

[1] Department of Biology, Faculty of Science, University of Hradec Králové, 500 02 Hradec Králové, Czech Republic

[2] East Bohemian Museum in Pardubice, Zámek 2, 530 02 Pardubice, Czech Republic

Abstract

The aim of the experiment was to determine suitable substrate type and optimal plant size for transfer of plantlets from in vitro to ex vitro under experimental outdoor conditions. Tests focused on the effect of substrate type (muddy and sandy) and starting size of plantlets gained through in vitro seed germination (0–3, 3.1–5,5.1–6, 6.1–10 cm) on plant growth. Three parameters (fresh weight, length, and the number of leaves) were compared to evaluate growth. Basic water parameters in experimental water tanks were regularly measured (pH, temperature, electrical conductivity, shadow intensity) and controlled to reach similar conditions to those in the natural habitat of this species. Overwintering was studied in a cellar with newly defined size categories (≤6, 6.1–8, 8.1–10, 10.1–12, 12.1–15 cm).

Both substrate type and starting size of plantlets significantly impacted growth. Plantlets grew better in the muddy substrate while a 100% success rate of rooting was gained with a starting size of 6.1–10 cm in both substrates. The biggest increase in fresh weight was observed with a starting size of 3.1–5 cm and 5.1–6 cm in both substrates. The greatest increase in fresh weight was observed in plants with a starting size of 3.1–5 cm in the muddy substrate (more than 95% increase). The best overwintering results were gained in the 6.1–8 cm size category.

Keywords: *Potamogeton praelongus*; plant tissue culture; growth experiment

Introduction

Potamogeton praelongus is a rare plant all over the world, being found in northern, mildly suboceanic and circumpolar regions. It grows primarily in the northern part of Europe and in similar latitudes in Asia and North America [1] in running waters, lakes and water pools. It occurs predominantly in clear, still, up to moderately flowing water, preferably mezotrophic, with a muddy or clay-sandy bed, often also with a layer of sapropel or an admixture of gravel or stones [2].

In the Czech Republic, it is considered critically endangered and currently is only found in two rather poor micropopulations in oxbows of the Orlice River near Hradec Králové [2], where it grows in a layer of organic mud [3]. Only one of these micropopulations is native – in the temporarily protected area (TPA) "Rameno u Stříbrného rybníku". In the past, *P. praelongus* was abundant in the CR from lowlands to highlands.

* Corresponding author. Email: zuzana.kozelkova@uhk.cz

Handling Editor: Beata Zagórska-Marek

The last native Czech micropopulation of *P. praelongus* (in TPA) is endangered by deposition of nutrient-rich sediment which covers the leaf surfaces with fine particles [4,5] and causes growth of metaphytic filamentous green algae. Disturbance of plants and micropopulations due to flood events is also of potential concern.

Genetic analyses detected a low level of intra- and inter-population genetic diversity in the Czech micropopulations [6] due to clonal growth, as known with other aquatic plant populations [7]. In nature, the achenes of *P. praelongus* are probably only used for colonization of new sites [8]. Present results indicate that the genetic variability of the micropopulations can be considerably increased by sexual reproduction in rescue cultures. Experimental studies reported a rapid spread of immigrant genomes within inbred populations due to heterosis [9–11]. Small amounts of artificial genes that have flowed into natural populations quickly reduced inbreeding depression and fitness reductions from a fixed genetic load [12–14].

Potamogeton praelongus prefers vegetative to generative reproduction in its whole range [2]. Germination tests were conducted during 2007–2013 to determine how to use the generative reproductive ability of the plant, which is strongly restricted by dormancy [8,15]. A *P. praelongus* tissue culture was set up in Ostrava (Czech Republic) and 30 clones are maintained there at present. Plantlets from this tissue culture were provided to the University of Hradec Králové for this experiment.

Since 2003, the Czech Republic has been running a conservation programme aimed at preserving *P. praelongus* through establishing new micropopulations in appropriate original locations of this species. Re-introductions of *P. praelongus* have only been successful twice although many potential original locations have been tried. The aim of this experiment was to determine how to increase the success rate of further re-introductions by choosing the appropriate substrate and determining the best plantlet starting size for acclimatization of the minute and vulnerable tissue cultured plantlets. These plantelets would then have to be grown until they reach a sufficient size and vitality for their successful plantation into suitable natural sites. Incorporating genetically variable material from 30 different clones should also help to improve the genetic variability.

Material and methods

Material from plant tissue culture

Seeds harvested in the rescue culture of *P. praelongus* at the Institute of Botany in Třeboň (Czech Republic) in September 2009 were used for preparation of a sterile in-vitro culture in Ostrava [5]. After two months of post-germination growth, seedlings were transplanted into a fresh liquid medium of a higher concentration (half strength Gamborg B5 medium with 500 mg l^{-1} KNO_3, with 2.5 % sucrose, pH 5.7 or 6.5) [16]. The volume of renewed culture medium was 50–70 ml in 350-ml flasks or 300 ml in 0.5-l flasks. Sterile plantlets were cultivated at 21°C, with artificial lighting of 20–30 PAR coming from two linear fluorescent tubes for 12 hours a day. Cultivation was conducted without shaking [5].

Seedlings rapidly showed vigorous growth after transfer into the new medium. The plants outgrew the flasks after 2–3 months (about 8–15 shoot apices were produced) and exhausted the medium (its final pH was 4.65–6.25). At this point, 1–3 apical shoot segments were transplanted into new flasks [5] to remain in the rescue culture and the rest of the plants were provided to the University of Hradec Králové for a growth experiment.

Starting the growth experiment

The experiment was carried out in an experimental culture in Býšť in artificially established water tanks with site conditions comparable to those of the last habitat of this species in the CR. At the end of June 2012, 488 plantlets of 23 clones of *P. praelongus* from the plant tissue culture in Ostrava (Fig. 1) were sorted into four size categories (plant length 0–3, 3.1–5, 5.1–6, 6.1–10 cm). On 1.7.2012, these plantlets were planted into plastic containers of 0.25–0.75 l volume (0.25 l for 0–3 cm plants, 0.3 l for 3.1–5 cm plants, 0.5 l for 5.1–6 cm plants, 0.75 l for 6.1–10 cm plants) and placed into 3 parallel water tanks (A, B, C) filled to the top edge with unchlorinated well water. The dimensions of the tanks were (length, width, depth): A: 0.70 × 0.70 × 0.28 m (135 l); B: 0.70 × 0.70 × 0.28 m (135 l); C: 1.00 × 0.70 × 0.26 m (175 l). The plants were grown in two types of substrate, muddy (245 plants) and sandy (243 plants). Plastic containers with either muddy or sandy substrate were put into each tank, e.g. water conditions were similar in all water tanks. Both substrates contained sand (pH of the sand infusion in distilled water was 6.77) and were enriched with a small amount of clay (1/6 of total volume), which acted as a source of Ca^{2+} ions thus influencing the final pH of the substrate and water. The mud added into the muddy substrate consisted of undecomposed black organic material with a pH of 7.4, and nutrient contents of NH_4^+: 23 mg kg^{-1}, NO_3^-: 30 mg kg^{-1}, P_{total}: 420 mg kg^{-1}. The proportion of mud and sand in the muddy substrate was 1:1. Both the mud and the sand were taken from the standing oxbow in the TPA (mud from its filled parts, sand from the sandy alluvium below the mouth of Stříbrný stream into the oxbow).

Fig. 1 Plantlets from sterile plant tissue culture (clone No. 29), scale in cm.

The starting length and fresh weight of every plantlet were determined (Tab. 1). The fresh weight was measured immediately after blotting the plantlets with filter paper, using laboratory scales with an accuracy of 0.0001 g. Fresh weight instead of dry weight was used to assess growth in order not to destroy the plants.

Tab. 1 Survey of number of plants in each size category.

	Mud		Sand	
Plant length (cm)	Number of plants	Mean weight (g)	Number of plants	Mean weight (g)
0–3	118	0.0677	117	0.0720
3.1–5	80	0.1737	79	0.1897
5.1–6	27	0.2599	27	0.2245
6.1–10	10	0.3663	10	0.3319
Total	245	0.2169	243	0.2045

End of experiment

The growth experiment finished on 24.10.2012. The plants were carefully washed with tap water and the final fresh weight, plant length and number of leaves on the stalk were determined immediately after blotting the fresh plants with filter paper (Fig. 2). All plants from this growth experiment were placed into a cellar for overwintering from November to May. Water temperature in the overwintering water tank did not drop below 0°C. Every weekend, the water was continuously aerated with an aerator device. Newly defined size categories (≤6, 6.1–8, 8.1–10, 10.1–12, 12.1–15 cm) were used for studying overwintering success.

Fig. 2 Plants at the end of the growth experiment (before measurement), scale in cm.

Continuous measurement of parameters of the experiment

Basic water parameters (pH, temperature, electrical conductivity, and shadow intensity) were regularly measured in the water tanks. Electrical conductivity and pH were measured by a Gryf 107 L portable combined device. Air and water temperatures were measured by a Multi-Thermo portable thermometer with range from −50°C to 300°C. Light conditions in particular water tanks were measured using a Voltcraft LX-1108 digital photometer. Shadow intensity was calculated as the ratio of actual lighting of the water tank and lighting of an unshaded space, and expressed in percentage of shading. Long-term monitoring of the temperature course was made with a Minikin data logger (Environmental Measuring Systems, Brno) placed in water tank C (representative tank) and in the water tank with overwintering plants. Chemical analyses of samples of surface water and muddy substrate were carried out to compare the conditions in the experimental culture in Býšť and in the native locality of *P. praelongus* in TPA; the analyses were conducted by the water management laboratories in the state enterprise of Povodí Labe, s.p.

Furthermore, a parallel all-day measurement of water temperature, shadow intensity, pH, and electrical conductivity was carried out with portable devices in both localities on 26.7.2012 (Fig. 3) so to compare the conditions in the representative water tank and TPA.

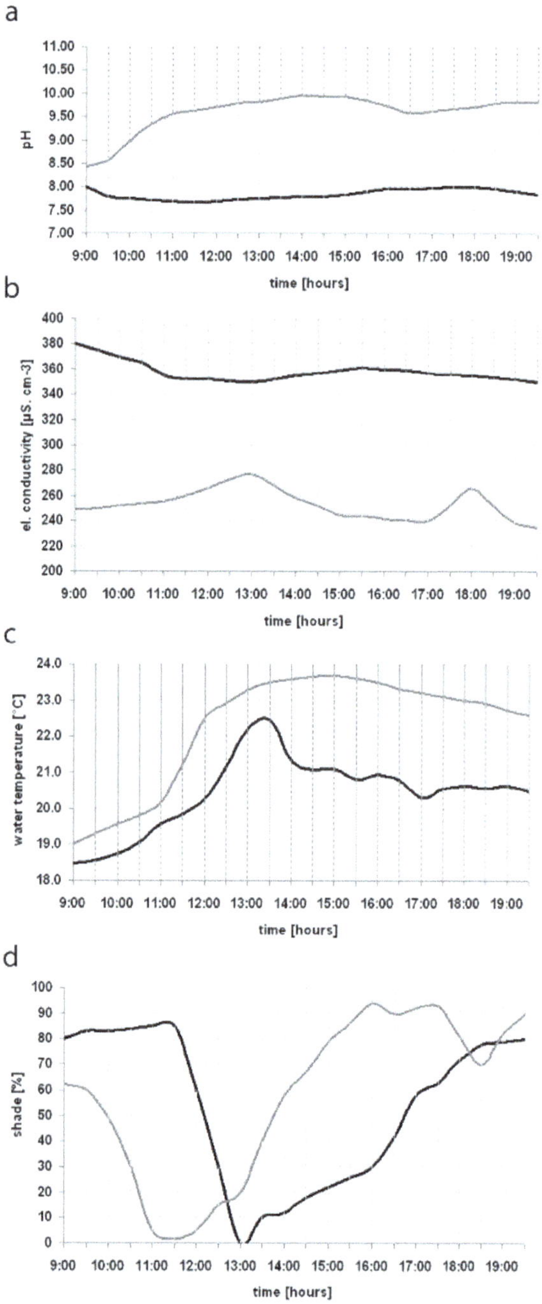

Fig. 3 All-day measurement of pH, water temperature, light conditions (shade) and electrical conductivity in TPA (black) and the experimental culture (grey).

Statistical evaluation

All statistical analyses were performed using NCSS 2001 [17]. The size categories of 5–6 and 6–10 cm had to be combined (5–10 cm), because of the small number of plants in each of these categories. The effect of the substrate type (mud and sand) on plant survival was evaluated by a Chi-Squared test. *T*-tests were run to test the effect of the substrate type (mud and sand) on the final fresh weight in particular size categories. The non-parametric Kruskal–Wallis test was used to compare the final fresh weight and number of leaves in the mud and sand substrates.

Results

pH values were higher than 8 in the water tanks, surpassing the values measured in TPA by 1–2 degrees (Fig. 3a). In both the experimental culture and TPA, precipitated $CaCO_3$ created white, rough layers on the leaf surface. The electrical conductivity in the tanks with the experimental culture was lower (23–27 mS m^{-1}) than in TPA (35–37 mS m^{-1}; Fig. 3b). Temperature in the water tanks was comparable to that in TPA, being higher by 2°C at most than in the original locality during only a short period of the day (Fig. 3c). Also the shadow intensity in the experimental culture and TPA were similar (Fig. 3d).

One-time samples of water from TPA and the experimental culture showed differences in COD_{Mn}, NO_2, NH_4, NO_3, N_{total}, PO_4, Ca, and Mg contents (Tab. 2). Although the soil substrates used in the experimental culture came from TPA, their chemical properties were different when used in the water tanks.

Substrate type in the experimental culture significantly affected the growth and number of surviving plants originating from a plant tissue culture. On the whole, 318 of 488 plants survived till the end of the experiment. The survival rate was higher in the muddy substrate (188 plants, 36.89%) than in the sandy one (130 plants, 26.64%; Tab. 3).

The survival rate was lowest in the 3.1–5 cm size category in both types of substrate. A 100% success rate of rooting was observed in the 6.1–10 cm size category in both types of substrate. The substrate type significantly impacted the number of surviving plants in the 3.1–5 cm and 5.1–10 cm size categories (Tab. 4).

The final fresh weight was higher in plants grown in the muddy substrate. The impact of substrate type on the fresh weight of the surviving plants was statistically significant in all size categories (0–3, 3–5, 5–10 cm; Tab. 4, Fig. 4).

The starting size of the tissue-cultured plantlets also significantly impacted the growth and fresh weight production of the plants (Tab. 5, Fig. 5). Plants grown in the muddy substrate had greater increases in fresh weight than those growing in the sand in all size categories. The best results were reached in the size category 0–3 cm where the fresh weight increased 14.5× in the mud and 5× in the sand. The second highest increase of fresh weight was observed with plants in the 3.1–5 cm size category in mud (11.7×) while the lowest increase (1.7×) was found in the sand. Fresh weight increased 5.5× in the mud and 2.4× in the sand in the 5.1–10 cm size category.

Tab. 2 Comparison of water parameters in the experimental culture and in TPA (sampling June 2013).

Water parameters	Unit	Experimental culture in Býšť	TPA
el. conductivity	mS m^{-1}	23.4	32.3
pH		9.8	7.6
COD_{Mn}	mg l^{-1}	17	16.80
NO_2	mg l^{-1}	0.013	0.072
NH_4	mg l^{-1}	0.06	0.04
NO_3	mg l^{-1}	<0.5	10.2
N_{total}	mg l^{-1}	1.1	2.7
PO_4	mg l^{-1}	0.33	0.16
Ca	mg l^{-1}	28.7	49.7
Mg	mg l^{-1}	4.2	5.5

Tab. 3 Number of planted and successfully grown plants of *P. praelongus*, and survival rates in mud and sand.

Plant length (cm)	Situation	Mud	Sand
0–3	Planted	118	117
	Successfully grown	86	76
	Survival rate (%)	72.9	64.9
3.1–5	Planted	80	79
	Successfully grown	56	24
	Survival rate (%)	70.0	30.4
5.1–6	Planted	37	37
	Successfully grown	36	20
	Survival rate (%)	97.3	54.05
6.1–10	Planted	10	10
	Successfully grown	10	10
	Survival rate (%)	100	100
Planted altogether		245	243
Successfully grown altogether		188	130
Total survival rate (%)		76.7	53.5

Tab. 4 Chi-Squared statistics: associations between the substrate type and plants´ survival in particular size categories. Significant results ($P \leq 0.05$) are in bold.

Size category (cm)	Test statistics	d.f.	*P*
0–3	1.72	1	0.189
3.1–5	24.96	1	**<0.001**
5.1–10	17.59	1	**<0.001**

The initial size category significantly affected the final number of leaves on the stalk (Tab. 5). The maximum number of leaves was reached by plants of starting size 3.1–5 cm in the mud (25 leaves), and with plants of starting size 0–3 cm in the sand (10 leaves; Fig. 6). In the 6.1–10 cm size category, the final number of leaves did not noticeably differ from the initial one. However, leaf surface visually increased in this category.

Tab. 5 Kruskal–Wallis test: associations between final fresh weight and number of leaves in the mud and sand substrates. Significant results ($P < 0.05$) are in bold.

Size category (cm)	Test statistics	d.f.	*P*
0–3	6.29	1	**<0.001**
3.1–5	3.45	1	**0.001**
5.1–10	2.31	1	**0.024**

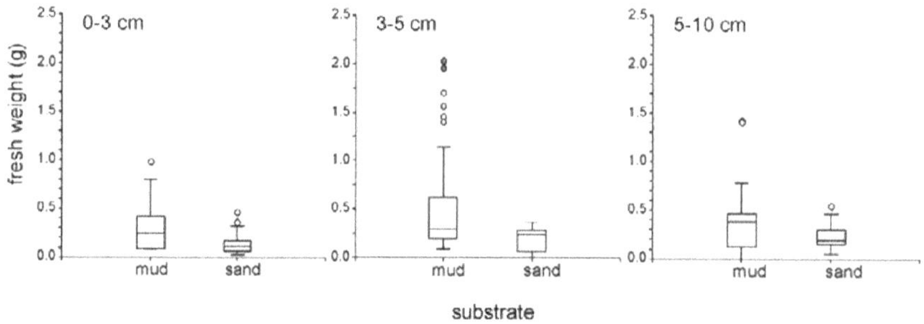

Fig. 4 Final fresh weight (g) of plants in the size categories 0–3, 3.1–5 and 5.1–10 cm.

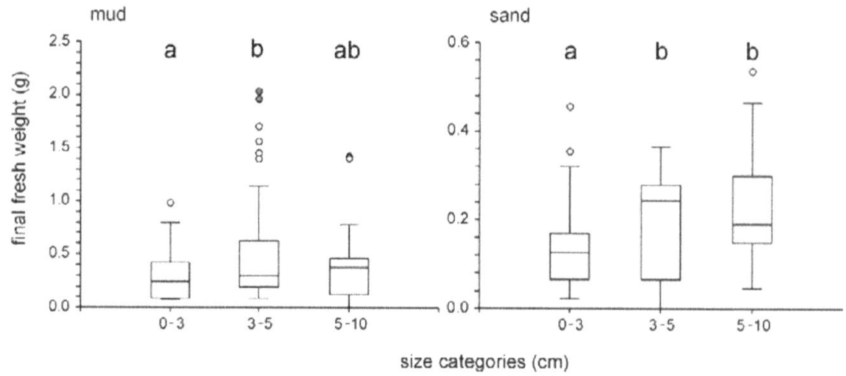

Fig. 5 Final fresh weight (g) of plants grown in the mud and sand substrates. Significant differences in medians (Dunn's test with $P = 0.05$) between the size categories are marked by different letters above the diagrams.

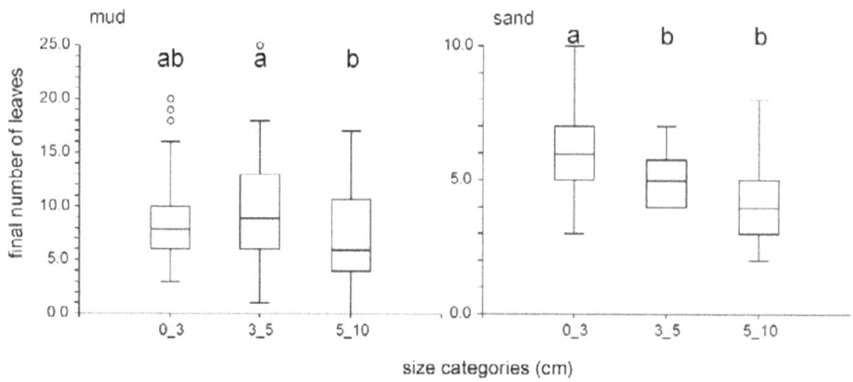

Fig. 6 Final leaf number on plants grown in the mud and sand substrates. Significant differences in medians (Dunn's test with $P = 0.05$) between the size categories are marked by different letters above the diagrams.

Tab. 6 Kruskal–Wallis test: associations between final fresh mass in the mud and in the sand. Significant results ($P \leq 0.05$) are in bold.

Substrate	Test statistics	d.f.	*P*
mud	7.09	2	**0.029**
sand	18.13	2	**<0.001**

A total of 56.38% of plants successfully overwintered (Tab. 6), with the best results occurring in the 6.1–8 cm size category (100%).

Discussion

Both substrates used in this growth experiment were taken from the last original locality of *P. praelongus* in CR, TPA "Rameno u Stříbrného rybníku". The muddy black substrate with a high share of undecomposed organic material (fragments of leaves, branches, etc.) contains a high proportion of nitrogen in the forms NO_3^- and NH_4^+, phosphorus, and Ca^{2+} and Mg^{2+} cations. These nutrients have a positive effect on stalk growth, leaf formation, and underground organ formation. Plants grown in the sandy substrate were smaller, had less vitality, and lower survival rate. This confirms earlier results showing that *P. praelongus* has high demands for nutrients, which are mainly gained from the muddy sediment [2,3]. This species is found in relatively calcium rich water, on humic sandy soils and mud.

Potamogeton praelongus is very efficient photosynthetically and is able to efficiently use HCO_3^- ions as a source of carbon for photosynthesis, which gives it an advantage when growing in hard water [18–20].

Potamogeton praelongus is mostly found in weakly acidic up to weakly alkaline water with a moderate content of calcium [21–24] and in acidic to neutral substrates [25,26]. Research of the Czech localities showed that *P. praelongus* best thrives where pH values range from 7.1 to 8.4, electrical conductivity is 11–41 mS m^{-1} and overall alkalinity is higher than 1.2 mekv l^{-1} [5]. Both the water pH and electrical conductivity values in the experimental culture were in the optimal ranges for the species. The content of nitrogenous materials in the experimental culture was lower than in TPA, because the water in the Orlice River is being constantly enriched with nitrogenous material washed off from the whole river basin. But the amount of PO_4^- was higher than in TPA, probably coming from the garden where the water tanks with *P. praelongus* were situated.

In CR, *P. praelongus* prefers water depth of 30–150 cm [2]. Outside the CR, it grows to a depth of several metres in lakes with high water transparency [22,23,26,27]. The species does not prosper in shallow and warmed water, where the stalks become brown and perish at the water surface in extremely hot temperatures. This was documented by observations of reserve micropopulations of this species in revitalized pools in the landscape park (LP) Kokořínsko [5]. Because the water tanks with the experimental culture in Býšť were shallow, with depth under 0.3 m, shadow intensity and water aeration were constantly controlled so that the plants were not overheated and damaged. Thus, the water temperature in the tanks was similar to that in TPA, although it was slightly higher during a short period of the day but by only 2°C at most. The shadow intensity in the experimental culture was managed to reach 0–80%, which was in compliance with the natural conditions in TPA. 100% lighting of the water surface was limited to 1 hour per day at maximum in the water tanks as well as in TPA. Ellenberg [28] mentioned that *P. praelongus* is a light-demanding species only rarely growing in conditions with relative lighting under 40%. However, in shallow waters in the CR, it prefers moderate shading by bank vegetation or floating plants that do not compete with it. *P. praelongus* was negatively affected by *P. natans*, *Calliergonella cuspidata* and *Chara* sp. in pools in LP Kokořínsko [5].

Unlike with previous plantings of *P. praelongus*, this method used 30 clones that were raised from seeds and thus should lead to increased genetic variability of the surviving Czech population. Kitner et al. [6] noticed that the current state of genetic variability of the Czech population of *P. praelongus* is alarming, because intrapopulation variability is higher than interpopulation variability. Within the scope of the rescue program for *P. praelongus*, the only successful attempts were repatriations into TPA, reintroduction into the Kašparovo oxbow in the flood plain of the Orlice River, and reintroduction into the Ploučnice River near Heřmaničky. These successful attempts used vital plants of about 0.5 m stalk size with well-developed rhizome systems from the rescue culture of the Institute of Botany of AS CR in Třeboň, and the reserve populations in LP Kokořínsko. An important condition of rooting for the tissue-cultured plants in a natural environment is that they grow to a sufficient size and vitality before planting. Such important information was gained by this growth experiment, which showed the necessity of growing the tissue-cultured plantlets in a rich, alkaline substrate and in particular conditions (concerning the physical and chemical properties of the water and site). Also, this experiment confirmed that plant tissue culture is a good source of *P. praelongus* for subsequent repatriations and re-introductions. In this way, *P. praelongus* plants to be planted in an experimental culture reach a sufficient size in a much shorter time than through germination in the laboratory.

For planting, it is necessary that *P. praelongus* plants have 2–3 leaves. However, roots are formed mostly with plants that have 4–5 leaves. The presence and progressive extension of the root system is a helpful, but not necessary, condition for successful growth. Plant growth is noticeably accelerated after transfer into a soil substrate rich in nutrients. They then activate the growth of rootlets and rhizome. For planting into an experimental culture, plantlets of at least 3 cm length are sufficient, but for planting into a natural site, vital stalks of 20–30 cm (preferably with a rhizome) are necessary.

According to our experience with plantlets obtained from a tissue culture, sufficiently strong plants will be obtained in a rescue or experimental culture after 3 growing seasons, when the plants may start to be fertile. We are now testing plantlets from germination tests and suppose that their growth up to a sufficient size will take longer. It is important to successfully overwinter as many plants as possible. An outdoor deep water tank or a shallow water tank in a cellar, where the water temperature does not drop below 0°C, can be used. Our overwintering experiment was carried out in a shallow water tank in a cellar with 56.38% survival.

Future growth experiments will focus on changes in underground organs during the growing and hibernal seasons. Current efforts in the frame of the rescue program for *P. praelongus* in the CR [29] aim to use the most recent findings about dormancy breaking in *P. praelongus* achenes [8] to gain genetically variable material for repatriations and re-introductions. The aim of this experiment was to find an alternative, possibly more efficient, method to rescue this critically endangered species in the CR and Central Europe.

Acknowledgments
The authors would like to thank M. Jůzová for translation into English and native speaker K. Edwards for English correction. The study was supported by funds awarded by EHP/Norway and the Ministry of Environment of the Czech Republic in 2009–2010, and by the funds of specific research financed by Ministry of Education of the Czech Republic in 2011–2012 No. 2121/2011, 2122/2012.

Authors' contributions
The following declarations about authors' contributions to the research have been made: research designing: RP; conducting experiments: RP, ZK; writing the manuscript: RP, ZK; statistical evaluation: LŠ.

Competing interests
No competing interests have been declared.

References
1. Vőge M. Die Entwicklung von *Potamogeton praelongus* im Großensee bei Hamburg. Tuexenia. 1992;12:275–284.
2. Kaplan Z. *Potamogeton* L. – rdest. In Chrtek J (jun.), Kaplan Z, Štěpánková J, editors. Květena České republiky. Prague: Academia; 2010. p. 330–378. (vol 8).
3. Šumberová K. Vegetace vodních rostlin zakořeněných ve dně (Potametea). In: Chytrý M, Šumberová K, Hájková P, Hájek M, Hroudová Z, Navrátilová J, et al., editors. Vegetace České republiky – vodní a mokřadní vegetace. Prague: Academia; 2011. p. 100–247. (vol 3).
4. Prausová R, Janová J. Současný stav výskytu rdestu dlouholistého (*Potamogeton praelongus*) v České republice. Příroda. 2010;27:155–168.
5. Prausová R, Janová J, Adamec L. Rescue of the critically endangered long-stalked pondweed (Potamogeton praelongus) in the Czech Republic. Acta Biol Slov. 2011;54:43–54.
6. Kitner M, Prausová R, Adamec L. Present status of genetic diversity of *Potamogeton praelongus* populations in the Czech Republic. Phyton. 2013;53(1):73–86.
7. Waycott M. Assessment of genetic variation and clonality in the seagrass *Posidonia australis* using RAPD and allozyme analysis. Mar Ecol Prog Ser. 1995;116:289–295.
8. Prausová R, Janová J, Šafářová, L. Testing achene germination of *Potamogeton praelongus* Wulfen. Cent Eur J Biol. 2013;8(1):78–86.
9. Ebert D, Haag C, Kirkpatrick M, Riek M, Hottinger JW, Pajunen VI. A selective advantage to immigrant genes in a *Daphnia* metapopulation. Science. 2002;295:485–488.
10. Richards CM. Inbreeding depression and genetic rescue in a plant metapopulation. Am Nat. 2000;155:383–394.
11. Saccheri IJ, Brakefield PM. Rapid spread of immigrant genomes into inbred populations. Proc R Soc Lond B. 2002;269:1073–1078.
12. Madsen T, Shine R, Olsson M, Wittzell H. Conservation biology: restoration of an inbred adder population. Nature. 1999;402:34–35.
13. Stokstad E. "Genetic rescue" helps panthers but puts researchers on the spot. Science. 2005;309:1162–1162.
14. Westemeier RL, Brawn JD, Simpson SA, Esker TL, Jansen RW, Walk JW, et al. Tracking the long-term decline and recovery of an isolated population. Science. 1998;282:1695–1698.
15. Prausová R, Sikorová P, Šafářová L. Generative reproduction of long stalked pondweed (*Potamogeton praelongus* Wulfen) in the laboratory. Aquat Bot. 2014;120:268–274.
16. Adamec L, Pásek K. Medium optimization for growing *Aldrovanda vesiculosa* in vitro. Carniv Plant Newslett. 2000; 29:122–124.
17. Hintze, J. NCSS – Number Cruncher Statistical System. NCSS: Kaysville, UT; 2001.
18. Hough RA, Wetzel RG. Photosynthetic pathways of some aquatic plants. Aquat Bot. 1977;3:297–313.
19. Husák Š, Kaplan Z. Studium a záchrana vybraných ohrožených druhů rodu *Potamogeton*. I. *Potamogeton praelongus* (Rameno u Stříbrného rybníka). Report for Agency for Nature Protection of the Czech Republic. Třeboň: Institute of Botany of the Academy of Science of the Czech Republic in Třeboň and Průhonice; 1997.
20. Maberly SC, Spence DHN. Photosynthetic inorganic carbon use by freshwater plants. J Ecol. 1983;71:705–724.
21. Berg C, Dengler J, Abdank A, Isermann M, editors. Die Pflanzengesellschaften Mecklenburg. Vorpommernsundihre Gefährdung – Textband. Jena: Weissdorn, Verlag; 2004.
22. Burnett JH, editor. The vegetation of Scotland. Edinburgh: Oliver & Boyd; 1964.
23. Doll R. Die Pflanzengesellschaften der stehender Gewässer in Mecklenburg – Vorpommern. Teil I. 3. Potamogetonetea Tx. et Prsg. 42. Laichkrautgesellschaften. Feddes Repert. 1991;102:217–317.
24. Nowak A, Nowak S. Czerniawska-Kusza I. Rare and threatened pondweed communities in anthropogenic waterbodies of Opole Silesia (SW Poland). Acta Soc Bot Pol. 2007;76(2):151–163.
25. Miljan A. Vegetations untersuchungen an Naturwiesen und Seen im Otepääschen Moränengebiet Estlands. Acta Commentat. 1933;25:1–139.
26. Vőge M. Ecological studies on water plants of 14 sites around Kangerlussuaq, southern West Greenland, with special regard to *Potamogeton* [Internet]. 2002 [cited 2014 Apr 23]. Available from:
27. Fajer M, Waga JM, Rzetala M, Szymczyk A, Nita M, Machowski R, et al. The Late Vistulian and Holocene evolution of Jezioro Lake: a record of environmental change in southern Poland found in deposits and landforms. J Paleolimnol. 2012;48:651–667.
28. Ellenberg H, Weber HE, Düll R, Wirth Y, Werner W, Paulißen D. Zeigerwerte der Gefässpflanzen Mitteleuropas. Scripta Geobot. 1991;18:1–248.
29. Prausová R, Rybka V, Adamec L, Husák Š, Rydlo J. Záchranný program pro rdest dlouholistý. Depon. In: Report for Ministry of Environment of CR. Pardubice: Agency for Nature Protection of the Czech Republic; 2003.

21

The evolution of land plants: a perspective from horizontal gene transfer

Qia Wang[1,2], Hang Sun[1], Jinling Huang[3]*

[1] Key Laboratory for Plant Diversity and Biogeography of East Asia, Kunming Institute of Botany, Chinese Academy of Sciences, Kunming 650201, China

[2] University of Chinese Academy of Sciences, Beijing 10049, China

[3] Department of Biology, East Carolina University, Greenville, NC 27858, USA

Abstract

Recent studies suggest that horizontal gene transfer (HGT) played a significant role in the evolution of eukaryotic lineages. We here review the mechanisms of HGT in plants and the importance of HGT in land plant evolution. In particular, we discuss the role of HGT in plant colonization of land, phototropic response, C_4 photosynthesis, and mitochondrial genome evolution.

Keywords: mechanisms of HGT; plant evolution; mitochondrial genome; adaptation

Introduction

Horizontal gene transfer (HGT) refers to the movement of genetic information between distinct species. Being different from vertical genetic transmission from parent to offspring, HGT circumvents normal mating barriers and may introduce novel genes into recipient genomes. As such, HGT may not only accelerate the genome evolution of recipient organisms, but also allow recipients to explore new resources and niches [1–4].

HGT was once thought to be frequent only in prokaryotes and certain unicellular eukaryotes [2,5]. During the past decade, horizontally acquired genes have been found in all major multicellular eukaryotic lineages (plants, animals, and fungi) [6–9], suggesting that HGT has occurred throughout the history of eukaryotic evolution and is critical for the adaptive evolution of some multicellular lineages. Here, we review cases of HGT in land plants, and discuss the genetic mechanisms and evolutionary importance of HGT in plant evolution. In particular, we focus on acquired genes of adaptive significance in plants, rather than genes that are selectively neutral or nearly neutral [10,11]. Because intracellular gene transfer (IGT) from organelles (mitochondria and plastids) to the nucleus has been discussed in multiple other articles [12–14], it will not be considered here.

* Corresponding author. Email: huangj@ecu.edu

Handling Editor: Andrzej Bodył

Mechanisms of HGT in plants

A complete and successful HGT requires that a foreign gene first enters cells of the recipient organism, be integrated into the recipient genome and then transmitted to the offspring [7]. Because the physical isolation of germ cells may prevent foreign genes from being transmitted to the offspring [15,16], plants and animals were traditionally regarded to be resistant to HGT [3,15]. Nevertheless, the occurrence of HGT in all major eukaryotic lineages suggests that no barrier to HGT is insurmountable in eukaryotes [7]. Specifically for plants, their meristems, which represent the germline equivalent but can be formed through de-differentiation of somatic cells, are less protected and may result in a higher susceptibility to HGT [16]. Several mechanisms have also been proposed regarding how HGT may occur in plants [6,7,16–18] (Fig. 1).

INTIMATE PHYSICAL ASSOCIATION. Sustained contact such as symbiosis, parasitism, epiphytism, pathogen infection or plant grafting may offer opportunities to gene transfer. In particular, parasitism has been reported in several cases of HGT in plants [6,18–23]. In most of these documented examples, genes were transferred from host to parasitic plants. For instance, multiple mitochondrial and nuclear genes of Rafflesiaceae, a family of parasitic plants, were reportedly acquired from their host *Tetrastigma* (Vitaceae) [19–21]. Similar cases have also been observed in other parasitic plants such as *Cistanche deserticola* (Orobanchaceae) [23], *Orobanche aegyptiaca* (Orobanchaceae) and *Cuscuta australis* (Convolvulaceae) [18], all of which acquired genes from their hosts. Considering the main direction of nutrient transmission from host to parasitic plants, it is possible that genetic material may be carried within, thus offering opportunities for HGT to parasitic plants.

Fig. 1 Mechanisms of HGT in land plants. Several mechanisms may facilitate HGT in plants, some of which are shown here. **a** The intimate physical association of the parasite *Cuscuta* with host plants offers opportunities for gene transfer. **b** Pollen germination on stigma of *Cyananthus* (as shown by image of fluorescence labeled callose. Photo courtesy of Yang Niu, Kunming Institute of Botany). Genetic materials may be delivered to egg cells via elongated pollen tubes. **c** Aphids may serve as a gene transfer agent between two plant species (photo courtesy of Jianwen Zhang, Kunming Institute of Botany). **d** Horizontal transfer of entire genomes between green plants may be explained by mitochondrial fusion. **e** Weakly protected zygote may serve as an entry point for foreign genes into the moss genome according to the weak-link model.

However, even if gene transfer in the opposite direction (from parasitic to host plants) may be rare, there are still cases that have been reported. For instance, Mower and his colleagues described two HGT cases of the mitochondrial atp1 gene from parasitic flowering plants, *Cuscuta* and *Bartsia* (Orobanchaceae), to their host plants of *Plantago* (Plantaginaceae) [22]. Therefore, the link between HGT direction and nutrient flow is not universal.

ILLEGITIMATE POLLINATION. This hypothesis presumes that pollen grains of one flowering plant species may germinate on the stigma of another species and, after pollination, some genes from the former may happen to be integrated into the latter species [16,24,25]. Such genetic materials may have been delivered to egg cells via elongated pollen tubes and, in some cases or under laboratory conditions, whole plastid and/or mitochondrial genomes are involved [24,26]. Based on available data, HGT due to illegitimate pollination may happen between closely related species such as different *Alloteropsis* grasses [25], or even members from different genera of the grass family Poaceae [27,28].

TRANSFER AGENTS. Vectors such as viruses, bacteria, fungi or insects could transfer genes between organisms. Although concrete evidence is still scarce by now, the possibility that a vector with a wide host range may serve as a gene transfer agent between two unrelated plant species should not be dismissed prematurely [6,16].

MITOCHONDRIAL FUSION. HGT in land plant mitochondria is frequent and sometimes involves large pieces of DNA or even whole genomes (discussed below). Mitochondrial fusion has been suggested as a mechanism to explain these observations [29]. The mitochondrial genomes of green plants may be incompatible with those of fungi or animals, thus limiting the donors of mitochondrial fusion events to other green plants [29].

THE WEAK-LINK MODEL. To explain how foreign genes may overcome the isolation of germ cells and reach the offspring, Huang recently proposed a weak-link model [7]. This model suggests that foreign genes may enter recipient cells at weakly protected unicellular or early developmental stages of the lifecycle (e.g. zygotes, embryos or spores). These stages represent the weak link in recipient organisms for the entry of foreign genes. Subsequent cell proliferation and differentiation may spread foreign genes to other tissues, including germ cells. As such, foreign genes could ultimately be transmitted to the offspring and then fixed in the population through drift or positive selection on newly acquired functions [7].

Some of the above mechanisms (e.g., illegitimate pollination) are only applicable to plants, whereas others (physical association and weak-link model) may explain HGT from miscellaneous donors such as viruses, fungi and prokaryotes. These mechanisms also are not mutually exclusive. For instance, the entry of weakly protected unicellular or early developmental stages of land plants may be facilitated by transfer agents or physical associations. In other words, transfer agents, intimate physical association, and illegitimate pollination concern how foreign genes may be carried from donor to recipient organisms, while the weak-link model focuses on how the barrier of germ cells may be circumvented and foreign genes be integrated into recipient genomes. Even boundaries of these mechanisms may not be clear-cut. For example, a common host plant could serve as an intermediate vector for horizontally transferred genes between different parasites [16,30].

Currently, our knowledge about how HGT happened in land plants is still limited. Future work and new technologies are needed in this research area.

The impacts of HGT on land plant evolution

HGT in plant colonization of land

Fossil data and molecular evidence suggest that land plants emerged from a pioneer green algal ancestor about 480–500 million years ago [31]. Compared with aquatic environments, terrestrial habitats brought many challenges to early land plants (e.g., desiccation, UV radiation, and microbial attack). During their colonization of land, plants evolved some complex regulatory systems, body plans, and other phenotypic novelties that facilitated their adaptation and radiation in a hostile terrestrial environment [32,33]. The origin and evolution of these novelties were aided through acquisition of genes from other organisms [32–34].

The phenylpropanoid pathway is responsible for the production of compounds such as flavonoids and lignin. The first and essential step of the phenylpropanoid pathway is catalyzed by phenylalanine ammonia lyase (PAL) [35,36]. At 2009, Emiliani et al. reported that plants acquired the *PAL* gene from soil bacteria or fungi during their early stages of land colonization [32]. Given the role of lignin in xylem formation and flavonoids in protecting plants from microbial infection and/or UV radiation, the acquisition of *PAL* might have been crucial for plant adaptation to terrestrial environments. A more comprehensive genome analysis of the moss *Physcomitrella patens* identified 57 gene families that were transferred from prokaryotes, fungi or viruses to the most recent common ancestors of land plants or green plants [33]. These genes are involved in some essential or plant-specific activities such as xylem formation, plant defense or nitrogen recycling [33,34]. Nineteen of these identified gene families also appear to be specific to mosses, suggesting that recent HGT events may not be rare. Further analyses also show that the origin of plant auxin biosynthesis was shaped by HGT [37]. Since these acquired genes were identified based on analyses of a single genome using stringent phylogenomic approaches, it is reasonable to believe that more horizontally acquired genes will be found in land plants [34].

The origin of vascular tissues is a significant event in land plant evolution. By ensuring long-distance transport of water, nutrients and organic compounds and increasing the mechanical support for plants, vascular tissues enable land plants to overcome some of the major challenges in terrestrial habitats (e.g., anchorage of plants in the ground, absorption of water and nutrients, support for upright growth) [33,38]. Several horizontally acquired genes are involved in the development of vascular tissues. For instance, the vein patterning 1 (*VEP1*) gene in land plants, presumably acquired from bacteria, is involved in vascular strand formation [33,39]. Most recently, Yang et al. investigated the evolution of TAL-type transaldolase (*TAL*) gene and found that land plant *TAL* genes are derived from actinobacteria [40]. *TAL* is ubiquitous in land plants and positively selected [40]. Transgenic experiments in rice showed that *TAL* is specifically expressed in vascular tissues. Importantly, knockdown of *TAL* expression leads to fewer, smaller and immature vascular bundles in culms, suggesting that the acquisition of *TAL* genes have played a pivotal role in plant vascular development and, therefore, adaptation to land environments.

HGT in phototropic response and C_4 photosynthesis

Millions of years after the initial plant invasion of land, vascular plants arose (during Silurian) [41]. Ferns dominated the earth's land surface antecedently, and their dominance was then replaced by angiosperms. Surprisingly, ferns did not die out. Instead, they mostly adapted to a shade-dwelling habit created by angiosperms-dominated forest canopies and then went through another round of diversification, which led to the emergence of the majority of living ferns. Neochrome, a novel chimeric photoreceptor derived from the fusion of red-sensing phytochrome and blue-sensing phototropin modules [42,43], is considered as a key innovation that allows ferns to thrive under low-light conditions [44,45]. In order to understand the origin of fern neochrome, Li and his colleagues [46] performed some comprehensive phylogenetic analyses of neochrome and related gene families. After ruling out other causes (e.g., phylogenetic artifacts and independent origins), they concluded that neochrome originated in hornworts and then was transferred to ferns horizontally. This finding suggests that HGT may have had a profound evolutionary impact on the diversification of ferns and, without it, the fern lineages we see today would likely be dramatically different.

C_4 photosynthesis evolved multiple times from C_3 ancestors independently [47]. HGT also has affected the evolution of C_4 photosynthesis in Poaceae [25]. *Alloteropsis*, a genus of the grass family that includes five C_4 species and one C_3 subspecies (*A. semialata* subspecies *ecklonian*a), is an excellent example for studying the evolution of C_4 photosynthesis [48,49]. Christin and his colleagues [25] reported that two essential genes of the C_4 pathway in *Alloteropsis* grasses were acquired from other C_4 taxa within the same family (Poaceae). A reasonable mechanism for the gene transfer events in *Alloteropsis* grasses was illegitimate pollination. This study also demonstrates that, even without sustained physical associations like parasitism or epiphytism, occasional contact such as pollination may offer opportunities for HGT in flowering plants.

HGT in plant organellar genomes

Although HGT to plastid genomes has rarely been reported (but see [50,51]), mitochondrial genes of land plants are subject to widespread, sometimes in a large scale, gene transfers [24,52,53]. For instance, up to 41% of mitochondrial genes in the parasitic family Rafflesiaceae were reportedly subject to HGT events [21]. Such massive mitochondrial gene transfers in Rafflesiaceae are not entirely unexpected, considering the frequent reports of HGT in plant mitochondrial genes and in parasitic plants.

An unprecedented case of HGT in flowering plants is related to *Amborella trichopoda* (Amborellaceae), a basal flowering plant. The mitochondrial genome of *Amborella* reportedly contains six genome equivalents of foreign mitochondrial DNA, including entire mitochondrial genomes from three green algae and one moss [29]. What makes *Amborella* so rich in foreign mitochondrial genes? The authors postulated that direct contact between *Amborella* and epiphytes, meristem regeneration through wounding, mitochondrial fusion, as well as a lower rate of DNA loss might have contributed to the large number of foreign genes in *Amborella* mitochondrial genome [17,29]. On the other hand, the distinctive phylogenetic position of *Amborella* (it diverged the earliest from other lineages of flowering plants [54]) allows researchers to confidently detect genes acquired from other plants [17]. The massive HGT in *Amborella* also highlights the tendency of plant mitochondrial genomes to take up foreign DNA, for there is no evidence of HGT in the chloroplast genome of *Amborella* [17,29].

HGT of mitochondrial genes is traditionally considered neutral or nearly neutral and, therefore, insignificant for plant adaptive evolution. However, because acquired mitochondrial genes can be transferred to the nucleus or recombine with endogenous organellar DNA fragments, they may increase genetic variation of the recipient plant taxon [14,55,56]. Indeed, it has been demonstrated that gene acquisition and subsequent recombination/gene conversion led to some mosaic mitochondrial genes in plants [53,55,56]. Therefore, it is premature to dismiss the functional implications of mitochondrial HGT.

Conclusions

The foregoing cases suggest that HGT happened in all major land plant lineages (mosses, ferns and flowering plants). The acquired genes may also be derived from different donor groups including viruses, bacteria, fungi, and other plants [6,33,57,58]. Many other cases of HGT have been reported in land plants, but cannot be detailed here [37,52,59–64]. To sum up, while HGT in plants may not be as frequent as in prokaryotes or unicellular eukaryotes, it still plays an important role in some essential or plant-specific activities and has had a significant impact on the evolution of land plants. In the end, it should be noted that most of the HGT cases discussed above were detected through analyses of whole genomes or transcriptomic data. In view of the increasing amount of genomic data that become available, we will have a better understanding of the frequency and the role of HGT in land plants.

Acknowledgments

We thank three reviewers for their comments and suggestions. This work is supported by a NSF "Assembling the tree of life" grant (DEB 0830024), an NSFC Oversea, Hong Kong, Macao collaborative grant (31328003), and the CAS/SAFEA "International partnership program for creative research teams".

Authors' contributions

The following declarations about authors' contributions to the research have been made: contributed to the writing of this review: QW, HS, JH.

Competing interests

No competing interests have been declared.

References

1. Ochman H, Lawrence JG, Groisman EA. Lateral gene transfer and the nature of bacterial innovation. Nature. 2000;405(6784):299–304.
2. Gogarten JP, Doolittle WF, Lawrence JG. Prokaryotic evolution in light of gene transfer. Mol Biol Evol. 2002;19(12):2226–2238.

3. Andersson JO. Lateral gene transfer in eukaryotes. Cell Mol Life Sci. 2005;62(11):1182–1197.
4. Keeling PJ, Palmer JD. Horizontal gene transfer in eukaryotic evolution. Nat Rev Genet. 2008;9(8):605–618.
5. Andersson JO. Gene transfer and diversification of microbial eukaryotes. Annu Rev Microbiol. 2009;63:177–193.
6. Richardson AO, Palmer JD. Horizontal gene transfer in plants. J Exp Bot. 2006;58(1):1–9.
7. Huang J. Horizontal gene transfer in eukaryotes: the weak-link model. Bioessays. 2013;35(10):868–875.
8. Dunning Hotopp JC. Horizontal gene transfer between bacteria and animals. Trends Genet. 2011;27(4):157–163.
9. Boto L. Horizontal gene transfer in the acquisition of novel traits by metazoans. Proc Biol Sci. 2014;281(1777):20132450.
10. Gogarten JP, Townsend JP. Horizontal gene transfer, genome innovation and evolution. Nat Rev Microbiol. 2005;3(9):679–687.
11. Huang J, Gogarten JP. Concerted gene recruitment in early plant evolution. Genome Biol. 2008;9(7):R109.
12. Selosse MA, Albert B, Godelle B. Reducing the genome size of organelles favours gene transfer to the nucleus. Trends Ecol Evol. 2001;16(3):135–141.
13. Timmis JN, Ayliffe MA, Huang CY, Martin W. Endosymbiotic gene transfer: organelle genomes forge eukaryotic chromosomes. Nat Rev Genet. 2004;5(2):123–135.
14. Kleine T, Maier UG, Leister D. DNA transfer from organelles to the nucleus: the idiosyncratic genetics of endosymbiosis. Annu Rev Plant Biol. 2009;60(1):115–138.
15. Kurland CG, Canback B, Berg OG. Horizontal gene transfer: a critical view. Proc Natl Acad Sci USA. 2003;100(17):9658–9662.
16. Bock R. The give-and-take of DNA: horizontal gene transfer in plants. Trends Plant Sci. 2010;15(1):11–22.
17. Bergthorsson U, Richardson AO, Young GJ, Goertzen LR, Palmer JD. Massive horizontal transfer of mitochondrial genes from diverse land plant donors to the basal angiosperm *Amborella*. Proc Natl Acad Sci USA. 2004;101(51):17747–17752.
18. Zhang D, Qi J, Yue J, Huang J, Sun T, Li S, et al. Root parasitic plant *Orobanche aegyptiaca* and shoot parasitic plant *Cuscuta australis* obtained Brassicaceae-specific strictosidine synthase-like genes by horizontal gene transfer. BMC Plant Biol. 2014;14(1):19.
19. Davis CC, Wurdack KJ. Host-to-parasite gene transfer in flowering plants: phylogenetic evidence from Malpighiales. Science. 2004;305(5684):676–678.
20. Xi Z, Bradley RK, Wurdack KJ, Wong K, Sugumaran M, Bomblies K, et al. Horizontal transfer of expressed genes in a parasitic flowering plant. BMC Genomics. 2012;13:227.
21. Xi Z, Wang Y, Bradley RK, Sugumaran M, Marx CJ, Rest JS, et al. Massive mitochondrial gene transfer in a parasitic flowering plant clade. PLoS Genet. 2013;9(2):e1003265.
22. Mower JP, Stefanović S, Young GJ, Palmer JD. Plant genetics: gene transfer from parasitic to host plants. Nature. 2004;432(7014):165–166.
23. Li X, Zhang TC, Qiao Q, Ren Z, Zhao J, Yonezawa T, et al. Complete chloroplast genome sequence of holoparasite *Cistanche deserticola* (Orobanchaceae) reveals gene loss and horizontal gene transfer from its host *Haloxylon ammodendron* (Chenopodiaceae). PLoS ONE. 2013;8(3):e58747.
24. Sanchez-Puerta MV, Cho Y, Mower JP, Alverson AJ, Palmer JD. Frequent, phylogenetically local horizontal transfer of the *cox1* group I intron in flowering plant mitochondria. Mol Biol Evol. 2008;25(8):1762–1777.
25. Christin PA, Edwards EJ, Besnard G, Boxall SF, Gregory R, Kellogg EA, et al. Adaptive evolution of C_4 photosynthesis through recurrent lateral gene transfer. Curr Biol. 2012;22(5):445–449.
26. Svab Z, Maliga P. Exceptional transmission of plastids and mitochondria from the transplastomic pollen parent and its impact on transgene containment. Proc Natl Acad Sci USA. 2007;104(17):7003–7008.
27. Diao X, Freeling M, Lisch D. Horizontal transfer of a plant transposon. PLoS Biol. 2006;4(1):e5.
28. Vallenback P, Jaarola M, Ghatnekar L, Bengtsson BO. Origin and timing of the horizontal transfer of a *PgiC* gene from *Poa* to *Festuca ovina*. Mol Phylogenet Evol. 2008;46(3):890–896.
29. Rice DW, Alverson AJ, Richardson AO, Young GJ, Sanchez-Puerta MV, Munzinger J, et al. Horizontal transfer of entire genomes via mitochondrial fusion in the angiosperm *Amborella*. Science. 2013;342(6165):1468–1473.
30. Park JM, Manen JF, Schneeweiss GM. Horizontal gene transfer of a plastid gene in the non-photosynthetic flowering plants *Orobanche* and *Phelipanche* (Orobanchaceae). Mol Phylogenet Evol. 2007;43(3):974–985.
31. Sanderson MJ, Thorne JL, Wikström N, Bremer K. Molecular evidence on plant divergence times. Am J Bot. 2004;91(10):1656–1665.
32. Emiliani G, Fondi M, Fani R, Gribaldo S. A horizontal gene transfer at the origin of phenylpropanoid metabolism: a key adaptation of plants to land. Biol Direct. 2009;4:7.
33. Yue J, Hu X, Sun H, Yang Y, Huang J. Widespread impact of horizontal gene transfer on plant colonization of land. Nat Commun. 2012;3:1152.
34. Yue J, Hu X, Huang J. Horizontal gene transfer in the innovation and adaptation of land plants. Plant Signal Behav. 2013;8(5).
35. Dixon RA, Achnine L, Kota P, Liu CJ, Reddy MSS, Wang L. The phenylpropanoid pathway and plant defence – a genomics perspective. Mol Plant Pathol. 2002;3(5):371–390.
36. Dixon RA, Paiva NL. Stress-induced phenylpropanoid metabolism. Plant Cell. 1995;7(7):1085–1097.
37. Yue J, Hu X, Huang J. Origin of plant auxin biosynthesis. Trends Plant Sci. 2014;19(12):764–770.
38. Ligrone R, Duckett JG, Renzaglia KS. Major transitions in the evolution of early land plants: a bryological perspective. Ann Bot. 2012;109(5):851–871.
39. Jun JH, Ha CM, Nam HG. Involvement of the *VEP1* gene in vascular strand development in *Arabidopsis thaliana*. Plant Cell Physiol. 2002;43(3):323–330.
40. Yang Z, Zhou Y, Huang J, Hu Y, Zhang E, Xie Z, et al. Ancient horizontal transfer of transaldolase-like protein gene and its role in plant vascular development. New Phytol. 2014 (in press).
41. Cai C, Ouyang S, Wang Y, Fang Z, Rong J, Geng L, et al. An Early Silurian vascular plant. Nature. 1996;379(6566):592–592.
42. Möglich A, Yang X, Ayers RA, Moffat K. Structure and function of plant photoreceptors. Annu Rev Plant Biol. 2010;61:21–47.
43. Kawai H, Kanegae T, Christensen S, Kiyosue T, Sato Y, Imaizumi T, et al. Responses of ferns to red light are mediated by an unconventional

photoreceptor. Nature. 2003;421(6920):287–290.

44. Schneider H, Schuettpelz E, Pryer KM, Cranfill R, Magallón S, Lupia R. Ferns diversified in the shadow of angiosperms. Nature. 2004;428(6982):553–557.
45. Schuettpelz E, Pryer KM. Evidence for a Cenozoic radiation of ferns in an angiosperm-dominated canopy. Proc Natl Acad Sci USA. 2009;106(27):11200–11205.
46. Li FW, Villarreal JC, Kelly S, Rothfels CJ, Melkonian M, Frangedakis E, et al. Horizontal transfer of an adaptive chimeric photoreceptor from bryophytes to ferns. Proc Natl Acad Sci USA. 2014;111(18):6672–6677.
47. Osborne CP, Beerling DJ. Nature's green revolution: the remarkable evolutionary rise of C_4 plants. Philos Trans R Soc Lond B Biol Sci. 2006;361(1465):173–194.
48. Ibrahim DG, Burke T, Ripley BS, Osborne CP. A molecular phylogeny of the genus *Alloteropsis* (Panicoideae, Poaceae) suggests an evolutionary reversion from C_4 to C_3 photosynthesis. Ann Bot. 2009;103(1):127–136.
49. Christin PA, Freckleton RP, Osborne CP. Can phylogenetics identify C_4 origins and reversals? Trends Ecol Evol. 2010;25(7):403–409.
50. Rice DW, Palmer JD. An exceptional horizontal gene transfer in plastids: gene replacement by a distant bacterial paralog and evidence that haptophyte and cryptophyte plastids are sisters. BMC Biol. 2006;4:31.
51. Moszczynski K, Mackiewicz P, Bodyl A. Evidence for horizontal gene transfer from bacteroidetes bacteria to dinoflagellate minicircles. Mol Biol Evol. 2012;29(3):887–892.
52. Bergthorsson U, Adams KL, Thomason B, Palmer JD. Widespread horizontal transfer of mitochondrial genes in flowering plants. Nature. 2003;424(6945):197–201.
53. Mower JP, Stefanović S, Hao W, Gummow JS, Jain K, Ahmed D, et al. Horizontal acquisition of multiple mitochondrial genes from a parasitic plant followed by gene conversion with host mitochondrial genes. BMC Biol. 2010;8(1):150.
54. Soltis PS, Soltis DE. Angiosperm phylogeny: a framework for studies of genome evolution. In: Greilhuber J, Dolezel J, Wendel JF, editors. Plant genome diversity. Vienna: Springer; 2013. p. 1–11. (vol 2).
55. Hao W, Palmer JD. Fine-scale mergers of chloroplast and mitochondrial genes create functional, transcompartmentally chimeric mitochondrial genes. Proc Natl Acad Sci USA. 2009;106(39):16728–16733.
56. Hao W, Richardson AO, Zheng Y, Palmer JD. Gorgeous mosaic of mitochondrial genes created by horizontal transfer and gene conversion. Proc Natl Acad Sci USA. 2010;107(50):21576–21581.
57. Maumus F, Epert A, Nogué F, Blanc G. Plant genomes enclose footprints of past infections by giant virus relatives. Nat Commun. 2014;5:4268.
58. Richards TA, Soanes DM, Foster PG, Leonard G, Thornton CR, Talbot NJ. Phylogenomic analysis demonstrates a pattern of rare and ancient horizontal gene transfer between plants and fungi. Plant Cell. 2009;21(7):1897–1911.
59. Won H, Renner SS. Horizontal gene transfer from flowering plants to Gnetum. Proc Natl Acad Sci USA. 2003;100(19):10824–10829.
60. Yoshida S, Maruyama S, Nozaki H, Shirasu K. Horizontal gene transfer by the parasitic plant *Striga hermonthica*. Science. 2010;328(5982):1128–1128.
61. Intrieri MC, Buiatti M. The horizontal transfer of *Agrobacterium rhizogenes* genes and the evolution of the genus *Nicotiana*. Mol Phylogenet Evol. 2001;20(1):100–110.
62. Nickrent DL, Blarer A, Qiu YL, Vidal-Russell R, Anderson FE. Phylogenetic inference in Rafflesiales: the influence of rate heterogeneity and horizontal gene transfer. BMC Evol Biol. 2004;4(1):40.
63. Schönenberger J, Anderberg AA, Sytsma KJ. Molecular phylogenetics and patterns of floral evolution in the Ericales. Int J Plant Sci. 2005;166(2):265–288.
64. Davis CC, Anderson WR, Wurdack KJ. Gene transfer from a parasitic flowering plant to a fern. Proc Biol Sci. 2005;272(1578):2237–2242.

22

Helicosporidia: a genomic snapshot of an early transition to parasitism

Yukun Sun, Jean-Francois Pombert*

Department of Biological and Chemical Sciences, Illinois Institute of Technology, Chicago, IL 60616, USA

Abstract

Helicosporidia are gut parasites of invertebrates. These achlorophyllous, non-photosynthetic green algae are the first reported to infect insects. Helicosporidia are members of the green algal class Trebouxiophyceae and are further related to the photosynthetic and non-photosynthetic genera *Auxenochlorella* and *Prototheca*, respectively, the latter of which can also turn to parasitism under opportunistic conditions. Molecular analyses suggest that Helicosporidia diverged from other photosynthetic trebouxiophytes less than 200 million years ago and that its adaptation to parasitism is therefore recent. In this minireview, we summarize the current knowledge of helicosporidian genomics. Unlike many well-known parasitic lineages, the *Helicosporidium* sp. organelle and nuclear genomes have lost surprisingly little in terms of coding content aside from photosynthesis-related genes. While the small size of its nuclear genome compared to other sequenced trebouxiophycean representatives suggests that *Helicosporidium* is going through a streamlining process, this scenario cannot be ascertained at this stage. Genome expansions and contractions have occurred independently multiple times in the green algae, and the small size of the *Helicosporidium* genome may reflect a lack of expansion from a lean ancestor state rather than a tendency towards reduction.

Keywords: *Helicosporidium*; green algae; comparative genomics; entomopathogen

Introduction

Parasitism comes in many guises, transient or obligate, extracellular or intracellular. Transient parasites are free-living or symbiotic organisms that are turning to parasitism upon a set of favorable opportunistic conditions. These organisms do not rely exclusively on parasitism for their survival nor are they always well-equipped to the task. Obligate parasites, on the other hand, cannot thrive without hosts and, by necessity, came up over the course of their evolution with the strategies and the tools to facilitate infection [1]. While the modes of action (extracellular or intracellular) and levels of specialization differ between various parasites, the goal remains the same: to benefit at the expense of the host.

The road to parasitism is littered with acquisitions and losses [2–6]. Intuitively, functions and strategies that can help avoid, disrupt or even disable the hosts' defenses are valuable targets for acquisition by parasites, even more so when their survival depends on it. What is not developed intrinsically over time can be acquired sometimes quickly by horizontal transfer of foreign genetic material. Parasites can gain new functions by capturing useful genes from their hosts (e.g. [7–10]), from co-pathogens or co-symbionts (e.g. [11,12]), from other species during transit (e.g. [13]), or from other various sources (e.g. [14–17]). Conversely, functions that are no longer necessary are often discarded. This is especially true for intracellular parasites, whose switch from external to internal environments shifted many functions from necessary to accessory. Given their internalization, intracellular parasites are constantly pressured to adapt to the responses of their host(s), which in turn may lead to accelerated rates of evolution compared to their free-living relatives [18–20].

Due to their reduced and/or highly adapted nature, the physiological features of parasites are often cryptic, which hinders taxonomical classification. Two well-known examples are the malaria-causing agent *Plasmodium* and the once enigmatic Microsporidia, which have undergone several rounds of taxonomical revisions [21,22]. Helicosporidia, the subject of this minireview, is another protist lineage whose position within the tree of life has long remained elusive. First characterized in 1921 [23], Helicosporidians were in consequence over time classified as sporozoan, myxosporidean, fungal and protozoan parasites before their position within the green algae was first hinted at nearly 80 years later based on morphological similarities with *Prototheca wickerhamii*, a non-photosynthetic trebouxiophyte [24]. This affiliation was soon corroborated by molecular phylogenetic inferences based on nuclear-encoded actin and rRNA genes [25] and further confirmed with additional plastid [26,27], mitochondrial [28] and nuclear data [29–31].

Helicosporidians are the first reported green algae that infect insects [23,24]. These entomopathogens are found

* Corresponding author. Email: jpombert@iit.edu

Handling Editor: Andrzej Bodył

in the outside environment as cysts surrounded by a thin pellicle and composed of three ovoid cells wrapped within a helicoidal filamentous one [24]. The relatively small helicosporidian cysts (sometimes referred as spores in the older literature), up to 6 µm wide, are ingested per os and dehisces in the host's gut at the start of the invasive stage to liberate the four cells within [32]. The long filamentous cell is then uncoiled and attaches itself to the host epithelium with the help of the barbed protuberances located at its extremities. Upon successful breach of the epithelial barrier, the ovoid cells invade the hemolymph wherein they will replicate during the vegetative stage [24]. Despite their mode of replication within insect hosts, helicosporidian vegetative cells can also be cultured in vitro on limited nutrients media suggesting that, unlike many parasites, they have retained the metabolic pathways required for saprobic growth. For more information on the morphology and lifecycle of Helicosporidia, we refer to the readers to the educated description by Boucias et al. [24] and the thorough review by Tartar [32]. In this minireview, we will focus on the genomic features of Helicosporidia.

Helicosporidians are of particular interest for our understanding of the genetic changes that occur during the gradual transition to parasitism. This lineage emerged recently and is part of a green algal lineage that encompasses free-living, symbiotic and parasitic organisms [31]. Helicosporidia are members of the trebouxiophycean order Chlorellales, whose best-known genera include *Nannochloris*, the type species *Chlorella* and the achlorophyllous *Prototheca* [33]. The latter has been reported to sometimes infect humans and other mammals under opportunistic conditions [34]. The Chlorellales order originated circa 350 million years ago (mya), within which the *Prototheca* and *Chlorella* genera diverged somewhere between 200 to 350 mya [35]. Helicosporidians are closely related to the genus *Prototheca*, but as can be seen in Fig. 1, their surprisingly fast rate of evolution renders molecular clock inferences based on 18S rDNA somewhat unreliable. Nevertheless, Helicosporidia are bound to have arisen after the late Paleozoic/early Mesozoic eras, likely within the last 200 million years or so, making them one of the youngest parasitic lineages known. In comparison, the well-known parasitic lineages from *Giardia*, Microsporidia and *Plasmodium* arose 2.2, 1.2 and 1 bya, respectively [36,37]. The recent transition from free-living to insect pathogen in Helicosporidia thus offers the opportunity to shed some light onto the early stages of parasitism in this lineage.

The *Helicosporidium* organelle genomes are present and conserved

Organelles are often the target of major modifications in parasites. Severe alterations to organelles can force organisms towards alternate lifestyles while changes in lifecycles can impact organelles to the point that these can become barely recognizable over time. Following internalization, the chloroplast is often victim of severe atrophy, for the capability to harness sunlight and convert it to energy is limited, if not absent within the confines of the host. Keeping a functional photosynthetic apparatus in this environment is rarely beneficial if not deleterious. In the apicomplexan *Plasmodium*, the relict organelle corresponding to the plastid is so derived that it took decades before it was finally identified as a remnant of a photosynthetic organelle [38–40]. The functions encoded within have been greatly diminished and the apicoplast genome has retained a limited set of genes allowing for its replication and maintenance [41,42]. It is unclear, however, if the *Plasmodium* plastid was already reduced prior to its adaptation to parasitism or if its reduction is a consequence of its pathogenic lifestyle. The blood vessels carrying the erythrocytes that these parasites infect are iron-rich environments favorable to heme synthesis but are almost systematically shrouded in darkness, such that a photosynthetic *Plasmodium* species would have little reason to exist.

Not surprisingly, the entomopathogenic helicosporidians are also non-photosynthetic. Losing photosynthetic capability however, by itself, does not define a parasite. Many free-living lineages never had this ability and the loss of atmospheric CO_2 fixation capability occurred independently and recurrently across many photosynthetic lineages, reverting the organisms to heterotrophy. The closest known relatives of Helicosporidia from the *Prototheca* genus (Fig. 1) are also non-photosynthetic and, while sometimes parasitic, they can do so only under opportunistic conditions [43]. The *Helicosporidium* sp. plastid genome has been streamlined to a greater extent than that of *Prototheca wickerhamii* (Junbiao Dai, personal communication) but not as much as that of the apicoplast from *Plasmodium falciparum* [27,38]. As one would expect, the *Helicosporidium* plastid gene losses mainly affect those associated with the fixation of atmospheric carbon dioxide, i.e. none of the genes coding for products involved in it has been retained, but it also lacks all genes coding for ATP synthases that are present in *Prototheca* ([27,44]; Fig. 2) and four ribosomal proteins (Rpl23, Rps2, Rps9 and Rps18) that are found in other green algae. The *Helicosporidium* plastid genome is compact and arrayed with a single replication origin (inferred by GC-skew analyses [27]), suggesting an optimization towards efficiency and a reduction in the amount of energy spent per replication, but it is unknown if its streamlining is still actively ongoing or if it has reached a certain equilibrium.

Modifications to the mitochondrion can also have a drastic, much more potent impact on parasites. As the principal energy factory of the cell, any evolutionary trend that leads to the loss of ATP production will severely affect the parasite and increase its host dependency. In many parasites, the mitochondrion has been severely overhauled, losing its genome in the process. Those highly reduced organelles, referred to as mitosomes or hydrogenosomes, are involved in only a few of the original mitochondrial pathways [45,46]. Mitosomes are involved principally in iron–sulfur cluster assembly [47] and are incapable of oxygenic respiration, such that mitosome-bearing parasites like Microsporidia or *Giardia* have to rely either on the alternate glycolysis and pentose phosphate pathways for ATP production or on the energy that they hitchhike from their hosts to power all of their biological functions [45,48]. Hydrogenosomes, like those found in *Trichomonas*, are similarly reduced but are further distinguished by their ability to generate molecular hydrogen [49].

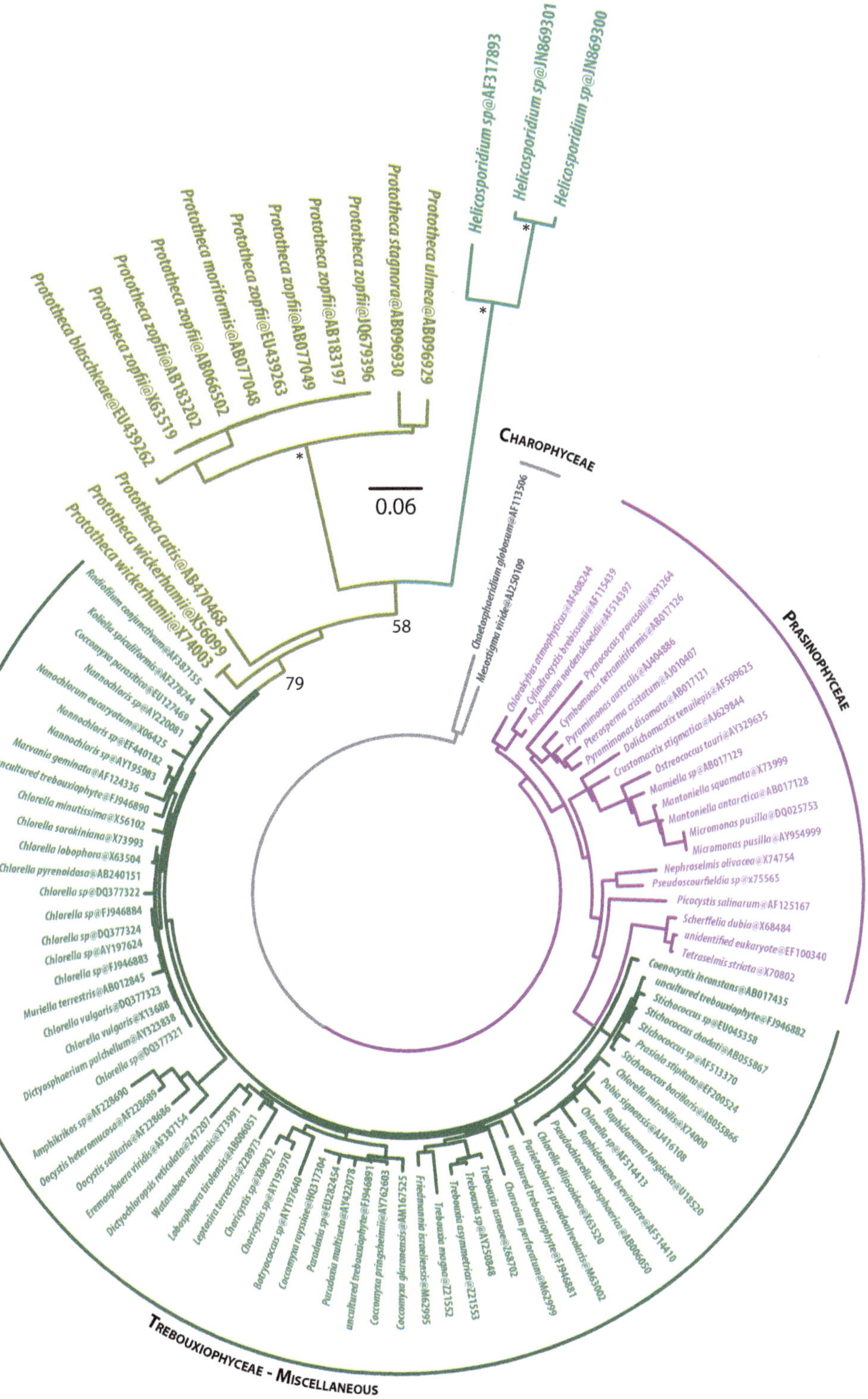

Fig. 1 Phylogenetic niche of *Helicosporidium* within the Trebouxiophyceae. The best 18S rDNA maximum likelihood tree shown here was computed with PhyML 3.1 [62] under the general time reversible model of nucleotide substitution, as implemented in SeaView 4 [63]. The tree dataset is derived from de Wever et al. [35], to which *Helicosporidium* and *Prototheca* sequences were added. Analyses were run with and without sequences from the Ulvophyceae and Chlorophyceae (not shown), with Prasinophyceae (in purple) and the charophycean algae *Mesostigma* and *Chaetosphaeridium* (in gray) as outgroups. The Trebouxiophyceae are shown is shades of green. The grouping of *Helicosporidium* with *Prototheca* was recovered in both analyses. For brevity, the tree shown here does not include the Ulvophyceae and Chlorophyceae. Bootstrap values for the *Prototheca*/*Helicosporidium* clades above 50% are indicated over the corresponding nodes; asterisks indicate nodes recovered in all bootstrap replicates. Branch lengths are drawn to scale.

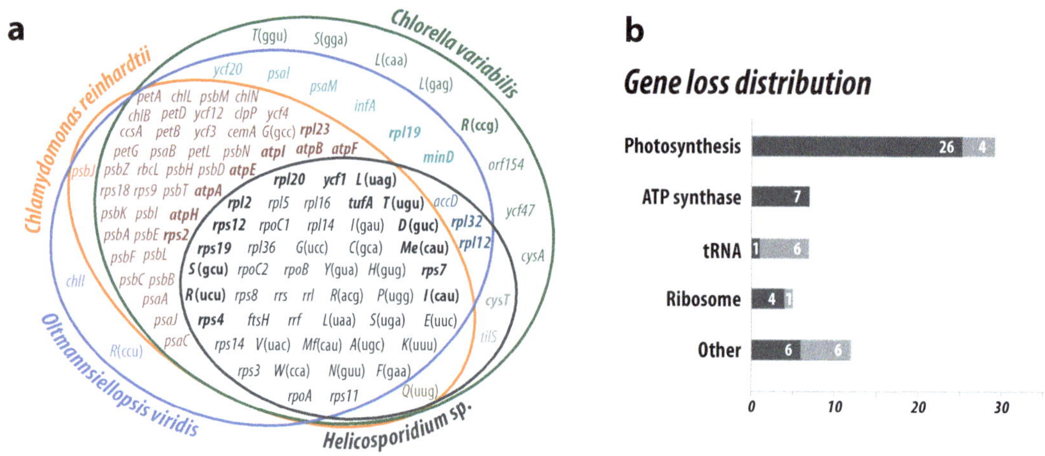

Fig. 2 Distribution of gene losses in the *Helicosporidium* plastid genome. **a** Venn representation of the genes that are shared between *Helicosporidium* and selected representatives of the Ulvophyceae (purple), Chlorophyceae (orange) and Trebouxiophyceae (green). Genes found in the *Prototheca wickerhamii* partially-sequenced plastid genome [44] are highlighted in bold. **b** Losses distributed by functional categories. Darker shades indicate gene losses in *Helicosporidium* that are shared between the Ulvophyceae, Chlorophyceae and Trebouxiophyceae (UTC). Lighter shades indicate the number of genes that are present in one or more UTC but not in *Helicosporidium*. None of the proteins that were lost in the *Helicosporidium* plastid genome were found encoded in the nucleus.

This is not the case for Helicosporidia. Helicosporidians can use oxygen as a terminal electron acceptor, as indicated by their successful growth in oxygenic conditions on nutrients-limited petri dishes [24], and have retained a mitochondrion with a full genome similar to other free-living, early-diverging green algal lineages [28]. In fact, if we make abstraction of the Chlorophyceae, a later-diverging lineage with mid- to heavily-modified mitochondrial genomes [50], one could describe the circular, 49.3 kbp-long *Helicosporidium* mitochondrial chromosome as a typical green algal mitogenome. Perhaps the most salient feature in the otherwise somewhat unremarkable *Helicosporidium* mitochondrial genome is the presence of a trans-spliced group I intron. Trans-spliced group I introns are rare (e.g. [51] and references within) but have little to do with parasitism.

The *Helicosporidium* nuclear genome is compact

We often think about parasite genomes in terms of reduction, for the evolutionary pressures applied on them to stay small is intuitive, whether to facilitate replication or minimize energetic expenses. A caveat of that way of thinking is that we sometimes make involuntary shortcuts when comparing with free-living organisms. Quite often, when the genome of the parasite is smaller, we say that it has been reduced. This is true for many instances, but is it always the case? In the green algae, judging by size alone is not enough. Most genome sizes were estimated from C-values (Fig. 3), which are not always accurate. For example, the *Ostreococcus* genome was thus once estimated at around 100 Mbp [52], or one order of magnitude larger than its real size [53]! Even if the sizes are accurate, many free-living algae harbor genomes that are bloated with repeated elements (e.g. [54,55]) that can propagate rapidly upon the right conditions and which have little effect on the overall metabolic profile of the species. Comparisons based on the total number of genes are therefore preferable but not necessarily accurate either. Automated gene prediction algorithms can struggle with their proper detection, sometimes grossly under- or overestimating their presence by the thousands [55,56], and we lack the manpower to manually curate all of the incoming onslaught of genomic data. Furthermore, a greater gene count is not necessarily synonymous with a greater metabolic potential, as duplications have occurred many times throughout evolution. Without looking at the genes themselves, inferences about parasitic reduction, free-living expansion or a mixture of both processes cannot be made without prior knowledge of the state that was present in the common ancestor.

By glancing quickly at sizes and gene numbers alone (Fig. 4), the *Helicosporidium* nuclear genome (12.4 Mbp assembled; 17 Mbp estimated [31]) appears to have shrunken down to almost a third of that of the free-living *Coccomyxa* (49 Mbp; [57]) and the symbiotic *Chlorella* (46 Mbp; [58]) and encodes a gene set that is about 40% smaller than that of its relatives. However, when looking at their respective metabolic profiles (Fig. 4), the picture that emerges is quite different. The core green algal metabolic pathways are all present in *Helicosporidium* and the only clear reduction has been the loss of its photosynthetic ability (Fig. 4). But even here the reduction is not complete and *Helicosporidium* has retained all of the genes involved in the storage of sugars as starchy polymers. The lower gene count in *Helicosporidium* can be attributed at least in part to the presence of fewer paralogous gene copies and to the absence of non-essential alternate metabolic branches synthesizing compounds that are either accessory or produced via other pathways. Other factors that contribute to the small size of the *Helicosporidium* genome are the limited number of introns, about on par

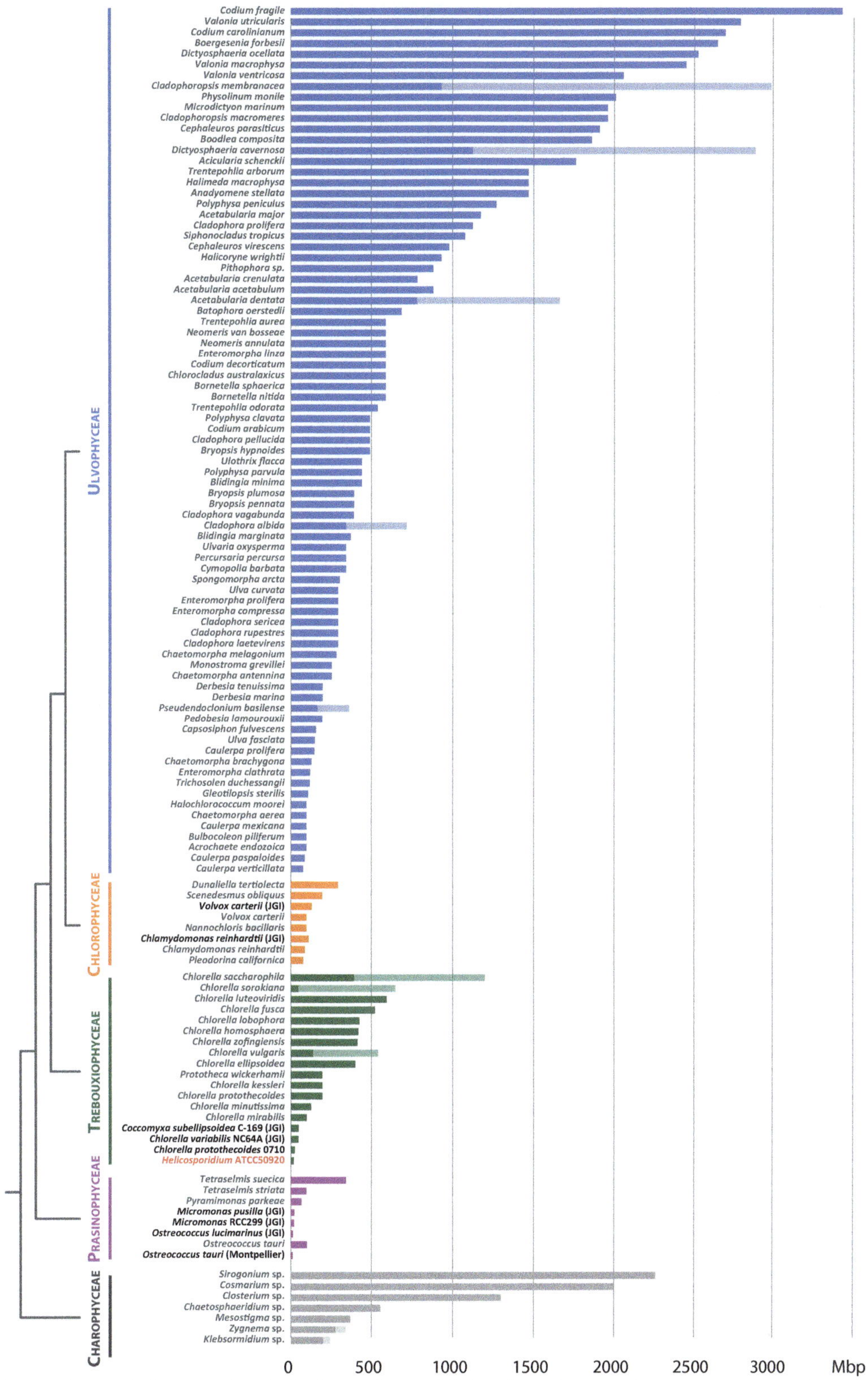

Fig. 3 Estimated green algal genome sizes in the Streptophyta and Chlorophyta. Genomes that were sequenced are lettered in black, with *Helicosporidium* lettered in red. Genomes whose sizes were estimated based on their 1C-values are lettered in gray. Estimated 1C-values for algae from the Streptophyta (Charophyceae) were inferred from Kapraun ([52]; only the smallest representatives were selected) whereas others from the Chlorophyta (Prasinophyceae, Trebouxiophyceae, Chlorophyceae, Ulvophyceae) were retrieved from the RBG Kew Algae DNA C-values database (http://data.kew.org/cvalues). Dark and light shades in the histogram bars correspond to lower and upper 1C-value size estimates. The histogram is color-coded according to the underlying cladogram depicting the branching order between these lineages.

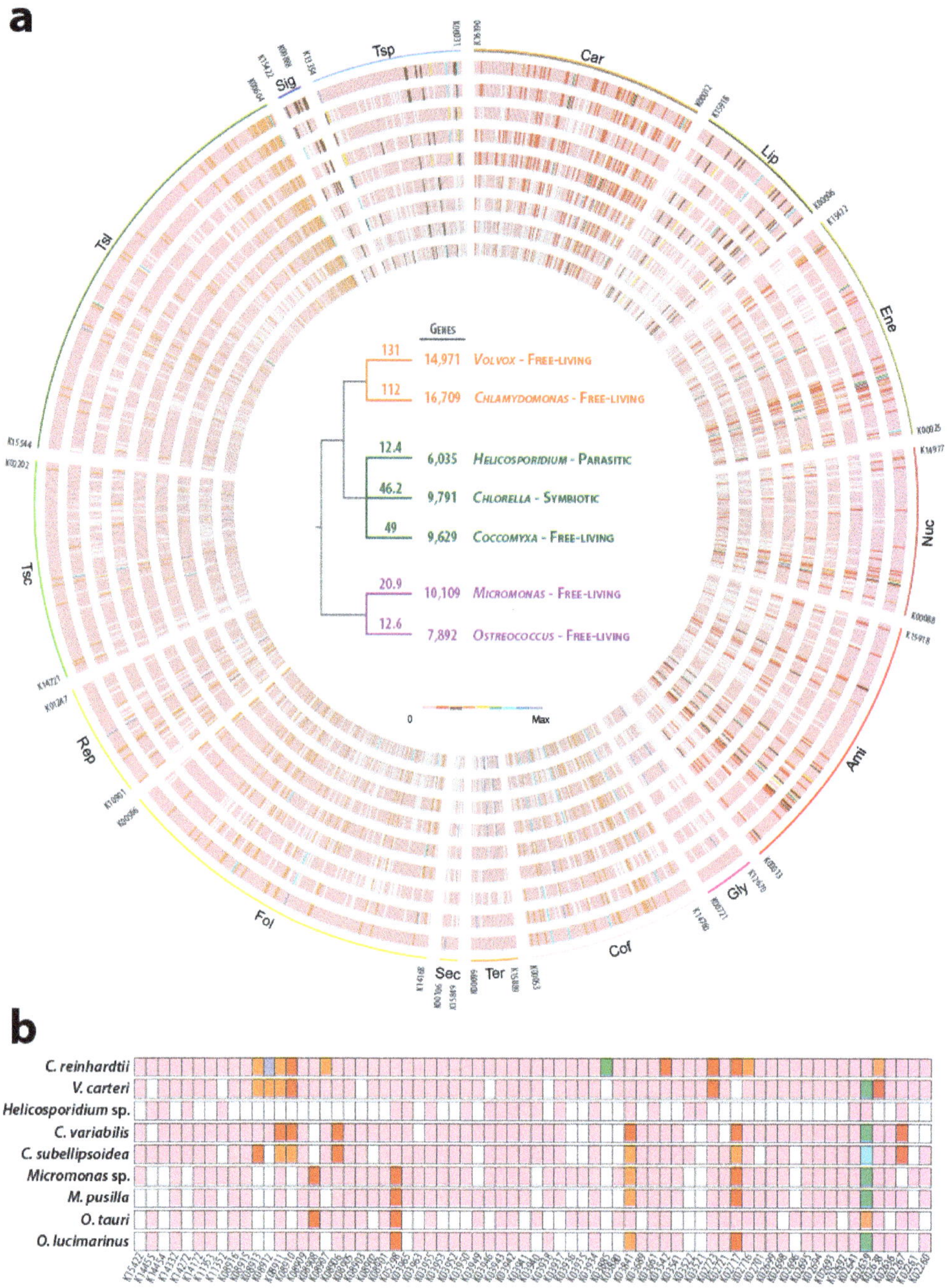

Fig. 4 Distribution of KEGG metabolic pathways among sequenced green algal nuclear genomes. **a** The concentric rings, from the outside to the inside, are: *Chlamydomonas reinhardtii*, *Volvox carteri* f. *nagariensis*, *Helicosporidium* sp., *Chlorella variabilis* NC64A, *Coccomyxa subellipsoidea* C-169, *Micromonas* sp. RCC299, *Micromonas pusilla*, *Ostreococcus tauri*, *Ostreococcus lucimarinus*. Each bar in the concentric rings corresponds to a unique protein in the overarching pathways. KEGG orthology identifiers (KO) indicated at the start and end of each arc correspond to the order in Tab. S1. Color gradients indicate the respective number of each protein-coding gene in the corresponding organisms. The gradient is a heat map standardized per pathway for visual balance. The absence of a protein-coding gene is indicated by a white color, whereas the pink to purple gradient (pink, red, brown, orange, gold, green, cyan, blue, purple) indicate the relative number of gene copies per pathway. The KEGG metabolic pathways are: Ami: amino acids; Car: carbohydrates; Cof: cofactors and vitamins; Ene: energy; Fol: folding, sorting and degradation; Gly: glycan; Lip: lipids; Nuc: nucleotides; Rep: replication and repair; Sec: secondary metabolites; Sig: signal transduction; Ter: terpenoids and polyketides; Tsc: transcription; Tsl; translation; Tsp; transport. The cladogram in the center schematizes the relationships between the prasinophytes (in purple), trebouxiophytes (in green) and chlorophycean algae (in orange). Assembled genome sizes (in Mbp) are indicated above the corresponding branches; values for picoprasinophytes are from *Ostreococcus tauri* and *Micromonas* sp. RCC299. **b** Zoom-in of the major *Helicosporidium* losses in the KEGG energy-related pathways. The corresponding KEGG orthology identifiers (KO) indicated under the corresponding blocks are further described in Tab. S1.

with picoprasinophytes and roughly three times less than in the other Trebouxiophyceae and Chlorophyceae, and the dearth of transposable elements [31]. *Helicosporidium* lacks the Dicer and Argonaute proteins hypothesized to suppress the propagation of transposable elements (TEs) [59], and it is unclear if the RNA interference machinery was lost because its function was accessory.

So, did the *Helicosporidium* genome shrink over time, was it the other trebouxiophycean genomes that have expanded or perhaps a bit of both? Almost concurrently with the publication of the *Helicosporidium* genome, Gao and co-authors reported the genome of another free-living Trebouxiophyceae, *Chlorella protothecoides*, assembled at 22.9 Mbp with a maximum estimated size of 27.6 Mbp [60]. The *C. protothecoides* genome is also quite compact and displays features that are intermediates between the *Helicosporidium* and *C. variabilis*/*Coccomyxa* genomes. In particular, like *Helicosporidium*, multi-copy genes are fewer in *C. protothecoides* than in the other trebouxiophycean genomes, suggesting that expansion by duplication may have occurred in this lineage [60]. The variation in size observed in the Trebouxiophyceae is not limited to this lineage and all green algal classes feature a mixture of species with small, medium or large genomes (Fig. 3). The problem is that we actually don't know what the genomic state of the green algal ancestor was, but here we infer that it was likely small. The genomes of prasinophytes occupying the most basal branches of the Chlorophyta phylum tend to be on the smaller scale and, while charophycean genomes in the Streptophyta have been somewhat stigmatized as humongous (many of the early ones investigated were indeed quite large [61]), species with genomes in the 100 Mbp range also exists [52] and, again, tend to branch at the base of the clade. We cannot rule out that the ancestral genome of the green algae may have been bloated and then reduced repeatedly and independently over time, but the opposite scenario is more parsimonious and thus, we find it more likely. Considering the above, we think that the *Helicosporidium* genome did not really experience much reduction, aside from a few clear photosynthetic and other gene losses, and that it has retained a genome whose features are probably closer to the ancestral ones than that of *Coccomyxa* and *C. variabilis*.

Our picture of the evolution of Helicosporidia is far from complete however. The *Helicosporidium* genome reported corresponds to the one with the shortest branch in our 18S rDNA tree (Fig. 1) and may thus represent one of the less derived helicosporidians. Looking at the other species may reveal greater levels of specialization and/or host dependency and help identify the components that are important for their infection strategies. *Helicosporidium* sp. was found to harbor numerous chitinases that can help breach its invertebrate host gut epithelium but also featured over 850 unique predicted proteins that were found to be expressed and for which no homology was found [31]. It is unclear if these transcripts are functional or the result of transcriptional noise, but their unicity in some cases may reflect novel or derived-beyond-recognition components potentially involved in pathogenicity. In that regard, sequencing the genomes of the closest known relatives of Helicosporidia from the genus *Prototheca* would most likely provide interesting insights, considering that these related species can also parasitize various hosts upon favorable conditions, potentially with the help of shared genetic components.

Acknowledgments

This work was supported by funds from the Illinois Institute of Technology.

Authors' contributions

The following declarations about authors' contributions to the research have been made: performed the phylogenetic analysis: YS; wrote the mini-review: YS, JFP.

Competing interests

No competing interests have been declared.

References

1. Walker DM, Oghumu S, Gupta G, McGwire BS, Drew ME, Satoskar AR. Mechanisms of cellular invasion by intracellular parasites. Cell Mol Life Sci. 2014;71(7):1245–1263.
2. Jackson AP. Genome evolution in trypanosomatid parasites. Parasitology. 2014 (in press).
3. Wijayawardena BK, Minchella DJ, DeWoody JA. Hosts, parasites, and horizontal gene transfer. Trends Parasitol. 2013;29(7):329–338.
4. Corradi N, Selman M. Latest progress in microsporidian genome research. J Eukaryot Microbiol. 2013;60(3):309–312.
5. Raffaele S, Kamoun S. Genome evolution in filamentous plant pathogens: why bigger can be better. Nat Rev Microbiol. 2012;10:417–430.
6. Blouin NA, Lane CE. Red algal parasites: models for a life history evolution that leaves photosynthesis behind again and again. Bioessays. 2012;34(3):226–235.
7. Selman M, Pombert JF, Solter L, Farinelli L, Weiss LM, Keeling P, et al. Acquisition of an animal gene by microsporidian intracellular parasites. Curr Biol. 2011;21(15):R576–R577.
8. Pombert JF, Selman M, Burki F, Bardell FT, Farinelli L, Solter LF, et al. Gain and loss of multiple functionally related, horizontally transferred genes in the reduced genomes of two microsporidian parasites. Proc Natl Acad Sci USA. 2012;109(31):12638–12643.
9. Anderson MT, Seifert HS. Opportunity and means: horizontal gene transfer from the human host to a bacterial pathogen. mBio. 2011;2(1):e00005-11.
10. Kishore SP, Stiller JW, Deitsch KW. Horizontal gene transfer of epigenetic machinery and evolution of parasitism in the malaria parasite *Plasmodium falciparum* and other apicomplexans. BMC Evol Biol. 2013;13(1):37.
11. Lee SC, Weiss LM, Heitman J. Generation of genetic diversity in microsporidia via sexual reproduction and horizontal gene transfer. Commun Integr Biol. 2009;2(5):414–417.
12. Weigel LM, Clewell DB, Gill SR, Clark NC, McDougal LK, Flannagan

SE, et al. Genetic analysis of a high-level vancomycin-resistant isolate of *Staphylococcus aureus*. Science. 2003;302(5650):1569–1571.

13. Nakabachi A, Nikoh N, Oshima K, Inoue H, Ohkuma M, Hongoh Y, et al. Horizontal gene acquisition of *Liberibacter* plant pathogens from a bacteriome-confined endosymbiont of their psyllid vector. PLoS ONE. 2013;8(12):e82612.
14. Grant JR, Katz LA. Phylogenomic study indicates widespread lateral gene transfer in *Entamoeba* and suggests a past intimate relationship with parabasalids. Genome Biol Evol. 2014;6(9):2350–2360.
15. Richards TA, Soanes DM, Jones MDM, Vasieva O, Leonard G, Paszkiewicz K, et al. Horizontal gene transfer facilitated the evolution of plant parasitic mechanisms in the oomycetes. Proc Natl Acad Sci USA. 2011;108(37):15258–15263.
16. Gardiner DM, Kazan K, Manners JM. Cross-kingdom gene transfer facilitates the evolution of virulence in fungal pathogens. Plant Sci. 2013;210:151–158.
17. Soanes D, Richards TA. Horizontal gene transfer in eukaryotic plant pathogens. Annu Rev Phytopathol. 2014;52(1):583–614.
18. Slamovits CH, Fast NM, Law JS, Keeling PJ. Genome compaction and stability in microsporidian intracellular parasites. Curr Biol. 2004;14(10):891–896.
19. Cuomo CA, Desjardins CA, Bakowski MA, Goldberg J, Ma AT, Becnel JJ, et al. Microsporidian genome analysis reveals evolutionary strategies for obligate intracellular growth. Genome Res. 2012;22(12):2478–2488.
20. Burki F, Corradi N, Sierra R, Pawlowski J, Meyer GR, Abbott CL, et al. Phylogenomics of the intracellular parasite *Mikrocytos mackini* reveals evidence for a mitosome in rhizaria. Curr Biol. 2013;23(16):1541–1547.
21. Rich SM, Xu G. Resolving the phylogeny of malaria parasites. Proc Natl Acad Sci USA. 2011;108(32):12973–12974.
22. Keeling P. Five questions about microsporidia. PLoS Pathog. 2009;5(9):e1000489.
23. Keilin D. On the life-history of *Helicosporidium parasiticum*, n.g., n.sp., a new type of protist parasitic in the larva of *Dasyhelea obscura* Winn. (Diptera, Ceratopogonidae) and in some other arthropods. Parasitology. 1921;13(02):97.
24. Boucias DG, Becnel JJ, White SE, Bott M. In vivo and in vitro development of the protist *Helicosporidium* sp. J Eukaryot Microbiol. 2001;48(4):460–470.
25. Tartar A, Boucias DG, Adams BJ, Becnel JJ. Phylogenetic analysis identifies the invertebrate pathogen *Helicosporidium* sp. as a green alga (Chlorophyta). Int J Syst Evol Microbiol. 2002;52(1):273–279.
26. Tartar A, Boucias DG, Becnel JJ, Adams BJ. Comparison of plastid 16S rRNA (rrn16) genes from *Helicosporidium* spp.: evidence supporting the reclassification of *Helicosporidia* as green algae (Chlorophyta). Int J Syst Evol Microbiol. 2003;53(6):1719–1723.
27. de Koning AP, Keeling PJ. The complete plastid genome sequence of the parasitic green alga *Helicosporidium* sp. is highly reduced and structured. BMC Biol. 2006;4(1):12.
28. Pombert JF, Keeling PJ. The mitochondrial genome of the entomoparasitic green alga *Helicosporidium*. PLoS ONE. 2010;5(1):e8954.
29. de Koning AP, Keeling PJ. Nucleus-encoded genes for plastid-targeted proteins in *Helicosporidium*: functional diversity of a cryptic plastid in a parasitic alga. Eukaryot Cell. 2004;3(5):1198–1205.
30. de Koning A. Expressed sequence tag (EST) survey of the highly adapted green algal parasite, *Helicosporidium*. Protist. 2005;156(2):181–190.
31. Pombert JF, Blouin NA, Lane C, Boucias D, Keeling PJ. A lack of parasitic reduction in the obligate parasitic green alga *Helicosporidium*. PLoS Genet. 2014;10(5):e1004355.
32. Tartar A. The non-photosynthetic algae *Helicosporidium* spp.: emergence of a novel group of insect pathogens. Insects. 2013;4(3):375–391.
33. Leliaert F, Smith DR, Moreau H, Herron MD, Verbruggen H, Delwiche CF, et al. Phylogeny and molecular evolution of the green algae. Crit Rev Plant Sci. 2012;31(1):1–46.
34. Lass-Florl C, Mayr A. Human protothecosis. Clin Microbiol Rev. 2007;20(2):230–242.
35. de Wever A, Leliaert F, Verleyen E, Vanormelingen P, van der Gucht K, Hodgson DA, et al. Hidden levels of phylodiversity in Antarctic green algae: further evidence for the existence of glacial refugia. Proc Biol Sci. 2009;276(1673):3591–3599.
36. Hedges SB. The origin and evolution of model organisms. Nat Rev Genet. 2002;3(11):838–849.
37. Escalante AA, Ayala FJ. Evolutionary origin of *Plasmodium* and other *Apicomplexa* based on rRNA genes. Proc Natl Acad Sci USA. 1995;92(13):5793–5797.
38. Gardner MJ, Hall N, Fung E, White O, Berriman M, Hyman RW, et al. Genome sequence of the human malaria parasite *Plasmodium falciparum*. Nature. 2002;419(6906):498–511.
39. McFadden GI, Reith ME, Munholland J, Lang-Unnasch N. Plastid in human parasites. Nature. 1996;381(6582):482–482.
40. Kohler S. A plastid of probable green algal origin in apicomplexan parasites. Science. 1997;275(5305):1485–1489.
41. Lim L, McFadden GI. The evolution, metabolism and functions of the apicoplast. Philos Trans R Soc Lond B Biol Sci. 2010;365(1541):749–763.
42. Nair SC, Striepen B. What do human parasites do with a chloroplast anyway? PLoS Biol. 2011;9(8):e1001137.
43. Mayorga J, Barba-Gómez JF, Verduzco-Martínez AP, Muñoz-Estrada VF, Welsh O. Protothecosis. Clin Dermatol. 2012;30(4):432–436.
44. Knauf U, Hachtel W. The genes encoding subunits of ATP synthase are conserved in the reduced plastid genome of the heterotrophic alga *Prototheca wickerhamii*. Mol Genet Genomics. 2002;267(4):492–497.
45. Makiuchi T, Nozaki T. Highly divergent mitochondrion-related organelles in anaerobic parasitic protozoa. Biochimie. 2014;100:3–17.
46. Martin WF, Müller M. Origin of mitochondria and hydrogenosomes. New York, NY: Springer; 2007.
47. Goldberg AV, Molik S, Tsaousis AD, Neumann K, Kuhnke G, Delbac F, et al. Localization and functionality of microsporidian iron-sulphur cluster assembly proteins. Nature. 2008;452(7187):624–628.
48. Vávra J, Lukeš J. Microsporidia and "the art of living together". In: Rollinson D, editor. . New York, NY: Elsevier; 2013. p. 253–319. (vol 82).
49. Schneider RE, Brown MT, Shiflett AM, Dyall SD, Hayes RD, Xie Y, et al. The *Trichomonas vaginalis* hydrogenosome proteome is highly reduced relative to mitochondria, yet complex compared with mitosomes. Int J Parasitol. 2011;41(13-14):1421–1434.
50. Smith DR, Hamaji T, Olson BJSC, Durand PM, Ferris P, Michod RE, et al. Organelle genome complexity scales positively with organism size in volvocine green algae. Mol Biol Evol. 2013;30(4):793–797.
51. Pombert JF, Otis C, Turmel M, Lemieux C. The mitochondrial genome of the prasinophyte *Prasinoderma coloniale* reveals two trans-spliced group I introns in the large subunit rRNA gene. PLoS ONE. 2013;8(12):e84325.

52. Kapraun DF. Nuclear DNA content estimates in green algal lineages: Chlorophyta and Streptophyta. Ann Bot. 2006;99(4):677–701.

53. Derelle E, Ferraz C, Rombauts S, Rouze P, Worden AZ, Robbens S, et al. Genome analysis of the smallest free-living eukaryote *Ostreococcus tauri* unveils many unique features. Proc Natl Acad Sci USA. 2006;103(31):11647–11652.

54. Prochnik SE, Umen J, Nedelcu AM, Hallmann A, Miller SM, Nishii I, et al. Genomic analysis of organismal complexity in the multicellular green alga *Volvox carteri*. Science. 2010;329(5988):223–226.

55. Godman J, Balk J. Genome analysis of *Chlamydomonas reinhardtii* reveals the existence of multiple, compartmentalized iron-sulfur protein assembly machineries of different evolutionary origins. Genetics. 2008;179(1):59–68.

56. Yandell M, Ence D. A beginner's guide to eukaryotic genome annotation. Nat Rev Genet. 2012;13(5):329–342.

57. Blanc G, Agarkova I, Grimwood J, Kuo A, Brueggeman A, Dunigan DD, et al. The genome of the polar eukaryotic microalga *Coccomyxa subellipsoidea* reveals traits of cold adaptation. Genome Biol. 2012;13(5):R39.

58. Blanc G, Duncan G, Agarkova I, Borodovsky M, Gurnon J, Kuo A, et al. The *Chlorella variabilis* NC64A genome reveals adaptation to photosymbiosis, coevolution with viruses, and cryptic sex. Plant Cell. 2010;22(9):2943–2955.

59. Chung WJ, Okamura K, Martin R, Lai EC. Endogenous RNA interference provides a somatic defense against *Drosophila transposons*. Curr Biol. 2008;18(11):795–802.

60. Gao C, Wang Y, Shen Y, Yan D, He X, Dai J, et al. Oil accumulation mechanisms of the oleaginous microalga *Chlorella protothecoides* revealed through its genome, transcriptomes, and proteomes. BMC Genomics. 2014;15(1):582.

61. Kapraun DF. Nuclear DNA content estimates in multicellular green, red and brown algae: phylogenetic considerations. Ann Bot. 2005;95(1):7–44.

62. Guindon S, Dufayard JF, Lefort V, Anisimova M, Hordijk W, Gascuel O. New algorithms and methods to estimate maximum-likelihood phylogenies: assessing the performance of PhyML 3.0. Syst Biol. 2010;59(3):307–321.

63. Gouy M, Guindon S, Gascuel O. SeaView version 4: a multiplatform graphical user interface for sequence alignment and phylogenetic tree building. Mol Biol Evol. 2010;27(2):221–224.

23

New records of *Chara connivens* P. Salzmann ex A. Braun 1835 – an extremely rare and protected species in Polish brackish waters

Paulina Brzeska[1]*, Adam Woźniczka[2], Mariusz Pełechaty[3], Irmgard Blindow[4]

[1] Department of Aquatic Ecology, Maritime Institute in Gdańsk, Długi Targ 41/42, 80-830 Gdańsk, Poland

[2] Research Station in Świnoujście, National Marine Fisheries Research Institute, Plac Słowiański 11, 72-600 Świnoujście, Poland

[3] Department of Hydrobiology, Faculty of Biology, Adam Mickiewicz University in Poznań, Umultowska 89, 61-614 Poznań, Poland

[4] Biological Station of Hiddensee, University of Greifswald, Biologenweg 15, 18565 Kloster, Germany

Abstract

The stonewort *Chara connivens* was rediscovered in the Vistula Lagoon in 2011, almost 35 years after its last record. In 2012, the species was recorded for the first time in the Szczecin Lagoon. *Chara connivens* occurred at shallow (0.5–1.2 m) sandy-muddy and muddy bottoms of small embayments. In the Vistula Lagoon, the stonewort was represented by single small specimens, while in the Szczecin Lagoon, it formed dense and extensive patches.

Keywords: *Chara*; rare species; Vistula Lagoon; Szczecin Lagoon; Baltic

Introduction

Charophytes are a group of macroscopic green algae represented by 314 species [1] assigned to the family Characeae within the phylum Chlorophyta; alternatively, they are treated as a separate phylum, the Charophyta [2]. The occurrence of most of the Characeae species in Europe is limited to clear, freshwater lakes of low fertility [3–8]. In response to increasing trophic status and the associated light limitation, many charophytes have therefore become rare, and are treated as sensitive bioindicators of water quality [6,9]. Although the number of sites with abundant charophyte vegetation has decreased, these macroalgae are widely distributed in different types of aquatic environments throughout the world. Due to their ability for ion regulation and osmotic adjustment, a set of charophyte species, commonly known from freshwaters, can occur in brackish coastal lagoons, as shown for the Baltic Sea [10].

In brackish waters, charophytes experience changes in osmotic pressure and mineral composition [7]. A brackish environment unique on a global scale is the Baltic Sea, the world's largest brackish water body [11]. The variety of the Baltic coast types gives rise to diverse and specific biocoenoses. Among submerged algae, a number of freshwater and brackish charophyte species were identified to represent five out of six genera within the family of Characeae [10]. Among the rarest brackish charophytes, *Chara connivens* P. Salzmann ex A. Braun 1835 exemplifies those species whose distribution has been reported to have greatly declined [12,13].

Chara connivens is widely distributed, with records known from Europe, Africa and Northern Asia [6,12,14–16]. Within Europe, *C. connivens* was recorded in the western European maritime regions, along the coasts of the Mediterranean and the Baltic Sea, and in inland saline waters of central and southern Europe [4,6,12]. In Poland, historical records of the species are available for the Gulf of Gdańsk [4] and for the Vistula Lagoon [17]. Currently, *C. connivens* has a status of a strictly protected species. In the "Red list of plants and fungi in Poland", published in 2006 [18], the species was assigned to the category Ex, i.e., extinct or probably extinct species.

The present paper reports new records of *C. connivens* from the Vistula and the Szczecin Lagoons. The Vistula Lagoon observation is a rediscovery of *C. connivens* after almost 35 years of the previous record, while in the Szczecin Lagoon this species has never been noted so far.

Material and methods

The occurrence of *C. connivens* was evidenced in the Polish parts of the Vistula and Szczecin Lagoons (Fig. 1) during the vegetation studies performed on 6 July 2011 in the Vistula Lagoon and on 31 July and 14 August 2012 in the Szczecin Lagoon. The distribution and species composition of aquatic vegetation was determined by direct observations and from samples collected with an anchor. Additionally, in the Vistula

* Corresponding author. Email: paulina.brzeska@im.gda.pl

Handling Editor: Andrzej Bodył

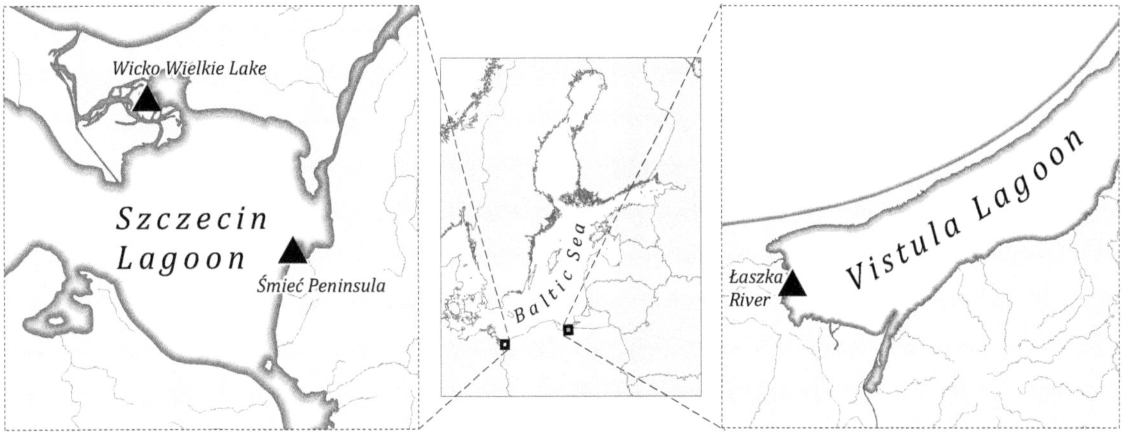

Fig. 1 Sites of *C. connivens* occurrence in the Szczecin and Vistula Lagoons.

Lagoon the floristic data collection was supplemented with measurements of water conductivity and temperature taken with a portable CTD probe.

Results and discussion

Characteristics of individuals

Chara connivens [syn. *Chara globularis* f. *connivens* (P. Salzmann ex A. Braun) R.D. Wood 1962] is a small alga, usually up to 15 cm long, rarely longer (25–50 cm) [4,12,13,15,19]. The slightly encrusted and slender thalli are delicate green and lustrous [4]. Sterile specimens of *C. connivens* are very similar to those of *C. globularis* Thuill. [15] due to presence of triplostichous stem cortex (usually isostichous or prime cells slightly larger than the secondaries) with lacking or rudimentary spine cells [4] and two rows of rudimentary stipulodes (uppers slightly larger than lowers or all imperceptible).

The features important in distinguishing the two species include the branchlets (6–10 in a whorl composed of 6–13 segments each, out of which the terminal 1–2 celled segment is ecorticate) which in *C. connivens* are strongly incurved in fertile male specimens, often also in fertile female individuals [4,6,12,15]. This feature is marked on the photographs of individuals from the Szczecin (Fig. 2) and Vistula (Fig. 3) Lagoons. Additionally, the axial internodes are very long, up to 6 times as long as the branchlets [12]; this was observed in our collections (Fig. 3). The most important distinguishing trait is that *C. connivens* is dioecious, while *C. globularis* is monoecious. In *C. connivens*, antheridia and oogonia develop at the lowermost 1–4 and 1–3 branchlet nodes, respectively [4]. The red solitary antheridia of up to 1 mm in diameter in *C. connivens* are larger than those in *C. globularis* [12,16]. The oogonia are up to 1.1 mm long and the oospores are dark. The specimens collected in the two lagoons showed the characters described above, typical of the vegetative thallus parts of *C. connivens*. In addition, besides sterile specimens, both lagoons yielded fertile male specimens which were more abundant in the Szczecin Lagoon (Fig. 2), where fertile female individuals were found as well.

Characteristics of *Chara connivens* sites

On 6 July 2011, *C. connivens* was found in the western part of the Vistula Lagoon (Fig. 1), on sandy mud at a depth of 1.1 m. The area was partially isolated by abundant monospecific assemblages of emergent vegetation, particularly *Phragmites australis* (Cav.) Steud. and *Schoenoplectus lacustris* (L.) Palla. As the site is shallow and the water dynamics is low, the water column temperature was high (19.5°C). The site is located 2 km away from the nearest river mouth; the riverine inflow resulted in low salinity (0.8 psu). The water was transparent down to the bottom; however, it was not possible to examine the submerged vegetation distribution from the shipboard as the bottom was densely covered by the filamentous green algae *Cladophora* sp., *Oedogonium* sp. and *Rhizoclonium riparium* (Roth) Harvey (those accounted for 90% of the biological material collected). The site supported a high diversity of macrophyte species occurring as single and small-sized specimens. They were dominated by *Potamogeton pusillus* L. (scoring 3 in the cover-abundance scale), young specimens of *Sparganium emersum* Rehmann (3) and *Stuckenia pectinata* (L.) Børner (2). Stoneworts such as *Chara tomentosa* L., *C. contraria* A. Braun ex Kützing, and *C. globularis* co-occurred in small quantities (1). *Chara connivens* was assigned to the category of the least numerous species (r), along with *C. aspera* Detharding ex Willdenow, *Ceratophyllum demersum* L., *Myriophyllum spicatum* L., *Zannichellia palustris* L. and *Nitellopsis obtusa* (N.A. Desvaux) J. Groves. In the 1970s, single specimens of *C. connivens* were reported from the same area by Pliński et al. [17].

In 2012, two sites supporting *C. connivens* were identified in the Szczecin Lagoon. The first site (visited on 31 July) was located in a small cove in the western part of Wicko Wielkie Lake, the other site (visited on 14 August) was situated off the western shore of the Śmieć Peninsula (Fig. 1). The first site, with muddy and sandy-muddy bottom and an average depth lower than 1 m was virtually entirely vegetated by a dense patch of elodeids represented by *Potamogeton perfoliatus* L., *Stuckenia pectinata*, *Potamogeton crispus* L., *Elodea canadensis* Michx., *Ceratophyllum demersum* and *Myriophyllum spicatum*. *Chara connivens* formed dense patches in the southern and south-western parts of the cove,

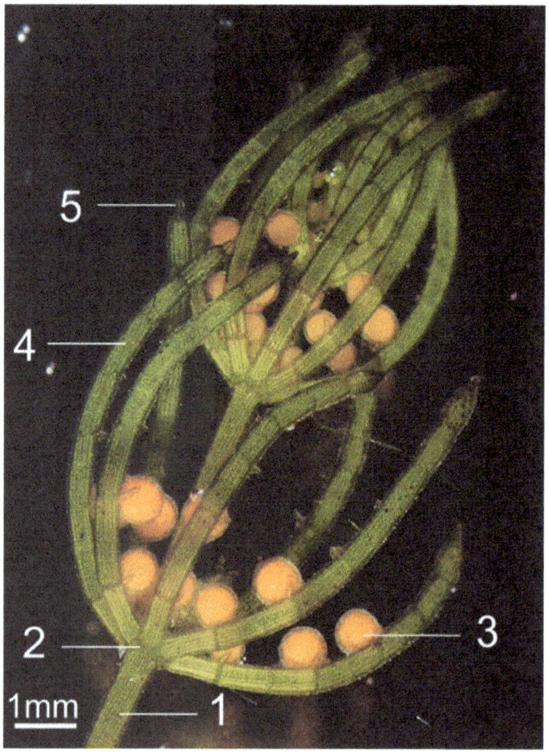

Fig. 2 A male *C. connivens* collected in the Szczecin Lagoon. 1 – main axis (stem) with triplostichous isostichous cortex; 2 – a node with a double whorl of rudimentary stipulodes; 3 – a male gametangium (antheridium) only, showing the dioecious character; 4 – an incurved branchlet; 5 – an ecorticate terminal segment of the branchlet.

Fig. 3 A *C. connivens* specimen from the Vistula Lagoon. 1 – an internode of the main axis (stem), usually longer than branchlets; 2 – an incurved sterile branchlet. The specimen's thalli are delicately green and lustrous as described by Dąmbska [4].

at a depth of 0.5–1.2 m, frequently in a close proximity to the shore and the reed belt. The area covered by *C. connivens* was about 750 m long and up to a few tens of meters wide. Single specimens were encountered as admixtures to other elodeids throughout the cove. The site supported both male and female individuals, the first being distinctly more abundant (at a 3:1 ratio). The male specimens occurred throughout the depth range, the females were present at larger depths (about 1 m) only. When growing abundantly, *C. connivens* forms a community that, in terms of phytosociological attempt to vegetation study [20], is classified as *Charetum conniventis* Velayos et al. 1989 association, with *C. connivens* to be both diagnostic and dominant species [19]. In the cited work, *Charetum conniventis* was classified within I category of threat – vague threat, no threat [19].

The other site was a shallow (depth of about 1 m), muddy embayment, distinctly set off from the Szczecin Lagoon by the course of the shoreline and a dense, wide belt of *P. australis*. The shores were lined by patches of nympheids (*Nuphar lutea* L., *Nymphaea alba* L., *Hydrocharis morsus-ranae* L.) and pleuston species (*Lemna minor* L., *L. gibba* L., *L. trisulca* L.) The central part of the embayment supported an assemblage consisting of *Stratiotes aloides* L. accompanied by *Elodea canadensis*, *Myriophyllum spicatum*, and covered densely by *Lemna trisulca*. The easternmost part of the embayment was found to host a few scattered patches of *C. connivens* – exclusively female individuals.

This paper provides information on new stands of *C. connivens*, an extremely rare and protected stonewort species which has been, for many years, considered to be extinct. Interestingly, following the investigations reported here, the species was found in other localities in both lagoons in 2012 [19]. The cited authors suggested that the species' distribution in Poland can be wider (not only coastal but also saline inland waters) and proposed to change the category of threat of *C. connivens*, species formerly considered as extinct, to LC category – the least concern species. In our opinion such a significant shift in category of threat is disputable and premature due to the very few records. Nevertheless, similar conclusions on a possibly wider distribution of the species have been recently drawn for the Estonian and Swedish coastal waters [21,22].

The fact that *C. connivens* was rediscovered in the Vistula Lagoon (eutrophicated water body), might have resulted from the natural ability of charophytes to survive in unfavorable environmental conditions, probably via their oospore bank in the sediments. Under favorable conditions, oospores can germinate even after decades [23]. During the last few years water column phosphorus concentrations in the Vistula Lagoon have declined [24]. It is known that under lower phosphorus concentrations and improved light conditions charophythes recover their meadows [25]. New sites of charophytes, like in the case of Szczecin Lagoon, can be established not only through vast floristic studies but, interestingly, also by animals, particularly by birds distributing oospores over considerable distances in their intestinal tracks [7].

Considering the findings presented in this paper, it can be concluded that *C. connivens* requires further field studies. Since both Vistula and Szczecin Lagoons still experience a strong anthropogenic pressure (eutrophication), it is essential to identify the species' distribution for its preservation and potential conservation actions.

Acknowledgments

The authors are grateful to Lena Szymanek for producing the map, to Sven Dahlke and Aleksandra Pełechata for valuable taxonomic consultations, and to colleagues who helped with the field work. The three anonymous reviewers are acknowledged for precious comments and suggestions that improved the manuscript. Research in the lagoons was supported by the Polish Ministry of Science and Higher Education throughout the statutory funds.

Authors' contributions

The following declarations about authors' contributions to the research have been made: designing and performing field studies: PB, AW; species identification: PB, AW, MP, IB; photographs: MP, AW; writing of the manuscript: PB, MP, AW, IB.

Competing interests

No competing interests have been declared.

References

1. Wood RD, Imahori K. Geographical distribution of Characeae. Bull Torrey Bot Club. 1959;86:172–183.
2. Guiry MD, Guiry GM. AlgaeBase. World-wide electronic publication, National University of Ireland, Galway [Internet]. 2012 [cited 2012 Dec 8]; Available from:
3. Hutchinson GE. A treatise on Limnology. 3: limnological botany. New York, NY: John Willey and Sons, Inc.; 1975.
4. Dąmbska I. Charophyta - Ramienice. Warsaw: Polish Scientific Publishers PWN; 1964. (Flora słodkowodna Polski; vol 13).
5. Krause W. Characeen als Bioindikatoren für den Gewässerzustand. Limnologica. 1981;13(2):399–418.
6. Krause W. Charales (Charophycae). Süsswasserflora von Mitteleuropa. Jena: Gustav Fischer Verlag; 1997.
7. Martin G, Torn K, Blindow I, Schubert H, Munsterhjelm R, Henricson C. Introduction to charophytes. In: Schubert H, Blindow I, editors. Charophytes of the Baltic Sea. Ruggell: Baltic Marine Biologists. Gantner Verlag; 2003. p. 3–14. (vol 19).
8. Pełechaty M, Pukacz A. Klucz do oznaczanie gatunków ramienic (Characeae) w rzekach i jeziorach. Warszawa: Inspekcja Ochrony Środowiska. Biblioteka Monitoringu Środowiska; 2008.
9. Apolinarska K, Pełechaty M, Pukacz A. $CaCO_3$ sedimentation by modern charophytes (Characeae): can calcified remains and carbonate $\delta^{13}C$ and $\delta^{18}O$ record the ecological state of lakes? – a review. Studia Limnologica et Telmatologica. 2011;5(2):55–66.
10. Schubert H, Blindow I, editors. Charophytes of the Baltic Sea. Ruggell: Baltic Marine Biologists. Gantner Verlag; 2003. (vol 19).
11. Schubert H, Mathiesen L, Blindow I. The brackish environment. In: Schubert H, Blindow I, editors. Charophytes of the Baltic Sea. Ruggell: The Baltic Marine Biologists Publications No. 19, Gantner Verlag; 2003. p. 15–26.
12. Torn K, Martin G. *Chara connivens* Salzm. ex A. Braun 1835. In: Schubert H, Blindow I, editors. Charophytes of the Baltic Sea. Ruggell: Baltic Marine Biologists. Gantner Verlag; 2003. p. 82–88. (vol 19).
13. Guiry MD, Guiry GM. AlgaeBase. World-wide electronic publication, National University of Ireland, Galway [Internet]. 2015 [cited 2012 Dec 21]; Available from:
14. Groves J, Bullock-Webster GR. The British Charophyta. Volume II. Charae with plater, concluding articles, geological sketch, bibliography and index. London: Ray Society; 1924.
15. Wood RD, Imahori K. A revision of the Characeae. First part: monograph of the Characeae. Weinheim: J. Cramer Verlag; 1965.
16. Luther H. *Chara connivens* in the Baltic Sea area. Ann Bot Fennici. 1979;16:141–150.
17. Pliński M, Kreńska B, Wnorowski T. Stosunki florystyczne i biomasa roślinności naczyniowej Zalewu Wiślanego. Studia i Materiały Oceanologiczne. 1978;21:161–196.
18. Siemińska J, Bąk M, Dziedzic J, Gąbka M, Gregorowicz P, Mrozińska T et al. Red list of the algae in Poland. In: Mirek Z, Zarzycki K, Wojewoda W, Szeląg Z., editors. Red list of plants and fungi in Poland. Kraków: W. Szafer Institute of Botany, Polish Academy of Sciences; 2006. p. 37–52.
19. Urbaniak J, Gąbka M. Polish Charophytes. An illustrated guide to identification. Wydawnictwo Uniwersytetu Przyrodniczego we Wrocławiu, Monograph CLXXIII, Wrocław; 2014.
20. Braun-Blanquet J. Pflanzensoziologie, Grundzüge der Vegetationskunde 3. Aufl. Vienna: Springer; 1964.
21. Torn K, Kovtun-Kante A, Herkül K, Martin G, Mäemets H. Distribution and predictive occurrence model of charophytes in Estonian waters, Aquat Bot. 2015;120(A):142–149.
22. Blindow I. Åtgärdsprogram för hotade kransalger: tuvsträfse och barklöst sträfse 2008-2011. Tuvsträfse (*Chara connivens*), Barklöst sträfse (*Chara braunii*). Rapport 5851, April 2009, Naturvårdsverket; 2009.
23. Rodrigo MA, Alonso-Guillén JL, Soulié-Märsche I. Reconstruction of the former charophyte community out of the fructifications identified in Albufera de València lagoon sediments. Aquat Bot. 2010;92:14–22. doi:10.1016/j.aquabot.2009.09.002
24. Zalewski T. Raport o stanie środowiska województwa warmińsko-mazurskiego w 2013 roku. Olsztyn: Inspekcja Ochrony Środowiska, Wojewódzki Inspektorat Ochrony Środowiska w Olsztynie, Biblioteka Monitoringu Środowiska; 2013.
25. Richter D, Gross, E. *Chara* can outcompete *Myriophyllum* under low phosphorus supply. Aquat Sci. 2013;75(3):457–467.

The number of cell types, information content, and the evolution of complex multicellularity

Karl J. Niklas[1]*, Edward D. Cobb[1], A. Keith Dunker[2]
[1] Section of Plant Biology, School of Integrative Plant Science, Cornell University, Ithaca, NY 14853, USA
[2] Center for Computational Biology and Bioinformatics, School of Medicine, Indiana University, Indianapolis, IN 46202, USA

Abstract

The number of different cell types (NCT) characterizing an organism is often used to quantify organismic complexity. This method results in the tautology that more complex organisms have a larger number of different kinds of cells, and that organisms with more different kinds of cells are more complex. This circular reasoning can be avoided (and simultaneously tested) when NCT is plotted against different measures of organismic information content (e.g., genome or proteome size). This approach is illustrated by plotting the NCT of representative diatoms, green and brown algae, land plants, invertebrates, and vertebrates against data for genome size (number of base-pairs), proteome size (number of amino acids), and proteome functional versatility (number of intrinsically disordered protein domains or residues). Statistical analyses of these data indicate that increases in NCT fail to keep pace with increases in genome size, but exceed a one-to-one scaling relationship with increasing proteome size and with increasing numbers of intrinsically disordered protein residues. We interpret these trends to indicate that comparatively small increases in proteome (and not genome size) are associated with disproportionate increases in NCT, and that proteins with intrinsically disordered domains enhance cell type diversity and thus contribute to the evolution of complex multicellularity.

Keywords: algae; alternative splicing; embryophytes; genome size; G paradox; intrinsically disordered proteins; land plants

Introduction

The number of different cell types characterizing an organism has been used to measure organismic complexity because it quantifies the extent to which cellular structure is differentiated into phenotypically different entities (e.g., [1–5]), because this approach is indifferent to whether an organism is a fungus, plant, or animal, and because it is insensitive to most methods of categorizing grade or clade levels of organization. It can even be used to quantify the complexity of unicellular organisms because the vast majority of species with this body plan achieve different cell functionalities and morphologies at different stages in their life cycles (e.g., resting cysts versus actively motile cells). Nevertheless, the use of the number of cell types as a measurement of complexity has some drawbacks. For example, some workers may disagree about whether two cells differ sufficiently to be called different cell types (e.g., xylary versus extra-xylary fibers, and parenchyma versus prosenchyma cells). Gauging complexity by counting the number of different cell types an organism can make also becomes problematic when dealing with coenocytic organisms that can achieve considerable morphological complexity without benefit of cellularization (e.g., the green alga *Caulerpa*), or when dealing with comparatively simple morphologies that nevertheless have highly organized patterns of nuclear or cellular placement (e.g., angiosperm megagametophytes).

There are two other important concerns. The first is that using the number of cell types to quantify complexity results in an untested tautology – "complexity increases as the number of different types of cells increases, and this number increases as an organism's complexity increases" – which requires collateral evidence to show that this reasoning is more than a self-sustaining description of a phenotypic property. The second concern is that cellular simplicity need not reflect an absence of complexity. There is no reason a priori to argue that an organism capable of producing two different types of cells is more complex than an organism capable of producing only one. And, if there is justification, is an organism composed of two cell types twice as complex as an organism composed of one cell type? Further, the supposition that cellular simplicity denotes low

* Corresponding author. Email: kjn2@cornell.edu
Handling Editor: Beata Zagorska-Marek

levels of organismic complexity hinges on the notion that "complexity" must be defined at the level of the diversity of cell phenotypes, a notion that is challenged when considering the physiological (as opposed to the structural) complexity of parasitic organisms that often undergo morphological and anatomical reduction (e.g., the Indian Pipe *Monotropa* and the tapeworm *Taenia*).

Nevertheless, although the concept of biological complexity is philosophically slippery, intuition encourages the notion that more "information" of some sort is required to produce or sustain more different kinds of cells, and that "information content" is a measure of "complexity". If we accept this premise, the number of different kinds of cells is not a direct measure of complexity per se but rather an indirect reflection of information content and thus complexity. This logic fosters a research agenda requiring the quantification of "information context" and a statistical assessment of its correlation with the number of cell types an organism produces. Only in this way can the tautology be tested.

A number of candidates for biological "information" present themselves in this capacity. For example, genome size as measured by the number of base-pairs, or proteome size as measured by the number of amino acids arguably provide reasonable measures of an organism's information content, particularly if attention is paid to the effects of polyploidy on genome size. Another measure of information content is the number of intrinsically disordered protein domains or residues in an organism's proteome, which gauge a proteome's functional versatility. This measure is particularly interesting because intrinsically disordered domains lack an equilibrium 3D structure under normal physiological conditions. Consequently, each domain can assume multiple conformations (and thus multiple developmental regulatory functions) within the same cell without an unnecessary and inefficient expansion of genome or proteome size [6–8]. In this way, proteins with intrinsically disordered domains contain information that is not encoded directly in the genome, a feature that helps in part to explain the G paradox. Finally, two of the advantages of using these three measurements of organismic information content is that each can be applied to unicellular as well as to multicellular organisms, and each can be used to assess whether complexity has increased over any lineage's evolutionary history.

The goal of this paper is to examine whether genome size, proteome size, and, in particular, the number of intrinsically disordered proteins correlate with the number of different cell types diverse organisms produce, and, if they do, to explore the implications of these correlations to understanding the evolution of multicellularity. Toward this goal, we present and analyze data drawn from animal as well as algal and land plant species because there is no reason a priori to expect plant and animal evolution to produce vastly different relationships between information content and cell type diversity, and because, should different patterns emerge among different kinds of organisms, they might shed light on how multicellularity emerged in different lineages. Finally, although it is not conventional, for the purposes of this paper, we define "plants" as photosynthetic eukaryotes so as to include the polyphyletic algae as well as the monophyletic embryophytes because excluding the algae would preclude an examination of unicellular and colonial organisms that are conventionally used to assemble ancestor-descendant transformation series for the evolution of multicellular body plans [9–13].

Cell type numbers do not increase in all lineages

Despite the tautology that results from the supposition that the number of different cell types increases as organismic complexity increases, it is informative to examine whether any trends of this sort are evident in successively evolutionarily divergent taxa within well-documented clades, particularly if these trends flout the conventional wisdom that "plants are invariably simpler than animals". There are however a number of considerations and limitations to this enterprise. For example, the level of taxonomic resolution and the availability of sufficiently resolved phylogenies can present problems. A too finely resolved phylogeny may reveal no pattern because very closely related taxa may share the same or very similar cellular structure and thus the same or very similar number of different cell types. Conversely, too coarsely a resolved phylogeny may fail to reveal a pattern because distantly related taxa may have adapted to very different niches, or undergone anatomical reduction. Two other limitations exist: phylogenies resolved at the optimal level may not be available, and authoritative tabulations of the number of different cell types for species in sufficiently resolved phylogenies may not exist.

These concerns are illustrated by mapping estimates of the number of different cell types reported for representative taxa onto published phylogenies based on molecular data. Specifically, we use the maximum number of cell types (NCT) reported in the primary literature as tabulated by Bell and Mooers [2] and map NCT onto the different plant and animal lineages for which there are published phylogenies using molecular data. The data set assembled by Bell and Mooers [2] undoubtedly reflects divergent philosophies about how to distinguish and classify animal and plant cell types (i.e., the data reflect the opinions of "splitters" and "lumpers"), and thus may be biased in unforeseen ways. However, this data set has the advantage of reflecting the opinions of authorities in diverse fields of research. It also consolidates a large literature scattered in diverse journals.

The first example of this approach employs a multi-locus, time calibrated phylogeny for the brown algae (Phaeophyceae) [14] and a broader phylogeny for the heterokont algae [15]. Mapping maximum NCT onto these phylogenies reveals no clearly discernable pattern within the phylogeny of the Phaeophyceae or the heterokont algae (Fig. 1). Within the Phaeophyceae, the early divergent Sphacelariales, Syringodermatales, and Dictyotales are represented by species with a maximum of four to seven different cell types, whereas the late divergent Fucales and Nemodermatales share similar and small cell type numbers (Fig. 1a). The exception is the Laminariales with a maximum of 14 different types of cells. This order is notable because it is characterized by species with large (up to 60 m in length), highly differentiated sporophytes with a holdfast-stipe-blade construction (e.g., *Nereocystis luetkaena* and *Pelagophycus porra*) [16], and because many species have highly specialized tissues (e.g., "trumpet cells" in *Macrocystis pyrifera*) that provide symplastic conduits for nutrient transport [17–19]. These features are associated with the ecological expansion of the kelps into intertidal zones, which provides an ecological explanation for the morphological and anatomical complexity seen in the kelps.

When the Phaeophyceae is embedded within the larger phylogeny of the heterokont algae based on a concatenated analysis of SSU rRNA and *rbcL* reported by Andersen [15], we see that this group as a whole contains species with larger cell type numbers compared to groups preceding the divergence of the Phaeophyceae within the clade and that these groups contain predominantly unicellular or colonial organisms with two to four cell types. For example, the species in the sister group of the Phaeophyceae, the raphidophytes, characteristically have two cell types (Fig. 1b). Unfortunately, comparisons of the various molecular phylogenies for the heterokont algae identify few consensus relationships among many lineages, with the exception perhaps of the diatoms and bolidophytes [14]. Consequently, until future studies accurately ascertain the evolutionary relationships within this large clade, little more can be said about the evolution of NCT within this important group of organisms.

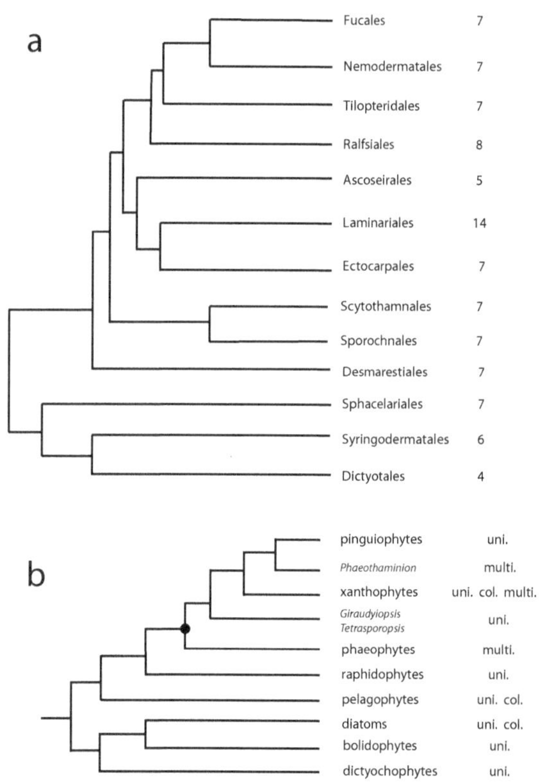

Fig. 1 Redacted phylogenetic relationships among (**a**) Phaeophyceae orders and the maximum number of cell types reported for any species within each order, and (**b**) redacted phylogenetic relationships among the heterokont algae showing the characteristic body plans reported for each group (uni. – unicellular; col. – colonial; multi. – multicellular). Data for number of cell types taken from Bell and Mooers [2]; Phaeophyceae phylogeny based on Silberfield et al. [14]; heterokont phylogeny based on Andersen [15].

In contrast, the consensus phylogeny of the viridiplantae (the chlorophycean and charophycean algae, and the embryophytes) indicates that NCT has increased, albeit not monotonically within this important clade (Fig. 2a). The early divergent chlorophytes are unicellular and have two cell types (e.g., the prasinophyte *Ostreococcus*), whereas later divergent green algae have a maximum of five different types of cells (e.g., *Fritschiella*). Among the streptophytes (the charophycean algae and the embryophytes), the early divergent charophycean alga *Mesostigma* has two cell types, whereas the more derived charophycean alga *Chara* has ten. Curiously, a phylogenetic analysis using 360 plastid genes [20] identifies the Zygnematophyceae as the closest related group to the land plants. This relationship contrasts with the hypothesis that the charophycean algae, such as *Chara* and *Coleochaete*, are more closely related to the land plants (e.g., [15]). This hypothesis is consistent with a large number of what have been traditionally considered shared derived

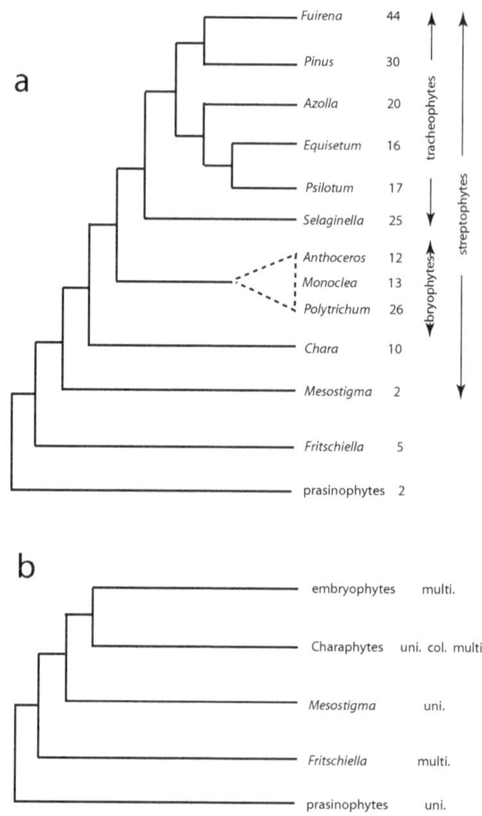

Fig. 2 Redacted phylogenetic relationships among representative taxa within the viridiplantae, the maximum number of cell types reported for species within each taxon (**a**), and the body plans represented within each taxon (uni. – unicellular; col. – colonial; multi. – multicellular; **b**). With the exception of the early divergent chlorophytes (the prasinophytes), each major taxon is represented by the genus reported to have the greatest diversity of cell types in its lineage (e.g., *Chara* and *Polytrichum* are reported to have the largest number of cell types among the charophycean algae and the mosses, respectively). The uncertainty in the phylogenetic relationships among the nonvascular land plants (the bryophytes) is depicted by dashed lines. Data for number of cell types taken from Bell and Mooers [2]. Phylogenetic relationships based in part on Ruhfel [20].

features such as oogamy and plasmodesmata [21], and IAA polar transport [22]. The perplexing placement of the Zygnematophyceae in the streptophytes based on molecular data is explicable if considerable evolutionary reduction and loss of numerous synapomorphies (such as oogamy and flagellated sperm cells) occurred. For example, the zygnematacean algae do not produce flagella and reproduce sexually by conjugation. They also produce a maximum number of only three or four cell types, which is significantly less than that observed for *Chara*, *Nitella*, or *Coleochaete*. Perhaps, like the complex Laminariales, which adapted to a structurally demanding intertidal environment, the comparatively simple Zygnematophyceae, may have radiated into ecologically less demanding environments.

Turning to the land plants, we see that the maximum number of cell types among the nonvascular embryophytes is reported for the mosses (i.e., *Polytrichum*; NCT = 26), which is approximately twice that reported for the hornworts and liverworts (e.g., *Anthoceros* and *Monoclea*; NCT = 12 and 13, respectively). Among the lycopods, the maximum NCT is 25 for *Selaginella*, whereas among the ferns and horsetails, the number of different cell types ranges between 17 and 20 (*Psilotum* and *Azolla*, respectively). Finally, *Pinus monophylla* is reported to have the maximum number of different cell types among conifers, whereas *Fuirena ciliaris* is described as having 44 different kinds of cells, which is the maximum reported for flowering plants ([2]; see [23] and [24]).

Turning to metazoans, it is worth noting that the maximum number of different cell types reported for the flowering plants exceeds that reported for Placozoa (NCT = 4) and early divergent animal groups such as the cnidarians and ctenophorans, and is comparable to those reported for the protostomes (i.e., ecdysozoans and lophotrochozoans; Fig. 3, Fig. 4). Although the resolution of the phylogeny of animals is not complete [25], the available data indicate that maximum cell type numbers range between 16 and 22 for the early divergent poriferans and cniderians, and between 9 and 69 for the mollusks and arthropods. The greatest number of different cell types among vertebrates is 240, which is 5.5 times that of the maximum number of angiosperm cell types (Fig. 4). The diversity of cell types across plant and animals is perhaps best illustrated when we compare the mean number of cell types reported for plants and metazoans (Fig. 3).

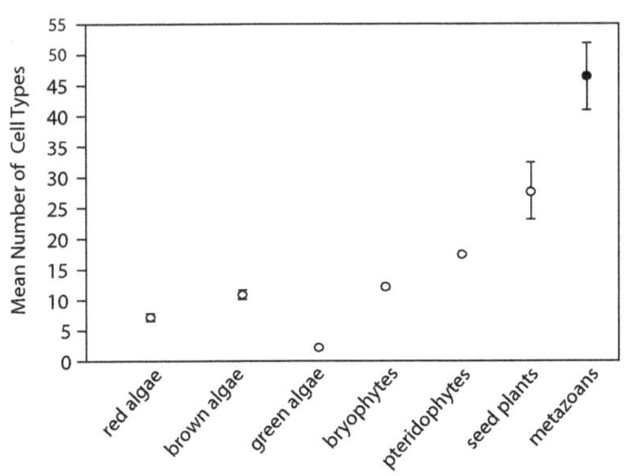

Fig. 3 Mean number of cell types (±*SE*) reported for extant red, brown, and green algae, bryophytes, pteridophytes, seed plants, and metazoans. Data taken from Bell and Mooers [2].

Among the plants, the red and brown algae produce a larger number of different kinds of cells compared to the green algae (chlorophytes and charophycean algae), whereas the NCT of brown algae is statistically indistinguishable from the NCT of the bryophytes. Within the viridiplantae, mean NCT increases progressively across the bryophytes, pteridophytes, and vascular plants (tracheophytes). The metazoans, on average, manifest the greatest diversity of cell types and the greatest range among all of the multicellular lineages.

The preceding provides very limited insights into how the numbers of cell types are distributed in the major plant lineages. It does, however, illustrate that the ability to detect patterns in NCT depends on the degree to which a lineage or clade is phylogenetically resolved. Nevertheless, it is clear that the number of different cell types does not invariably increase over the course of evolutionary history. Consequently, if the generalization that "cell type numbers gauge complexity" is accepted as true, it follows that evolution has not always resulted in more complex organisms in all lineages or clades (even in the absence of the evolution of adopting a parasitic life style). They also dispel the myth that plants are invariably less complex than animals, since many land plants have as many or more different cell types as annelids or arthropods (see Fig. 2, Fig. 4).

Cell type numbers increase with increasing proteome size

If the maximum number of cell types does not invariably increase over the course of a lineage's evolutionary history, does it correlate across taxa with organismic information content as gauged by genome or proteome size?

The expectation is that statistically significant and positive correlations will exist between NCT and proteome size based on a consideration of how alternative splicing and intrinsically disordered protein domains (which occur widely in all eukaryotic lineages) amplify protein functionalities without inefficiently increasing genome size. These molecular processes result in "moonlighting" proteins (i.e., proteins that

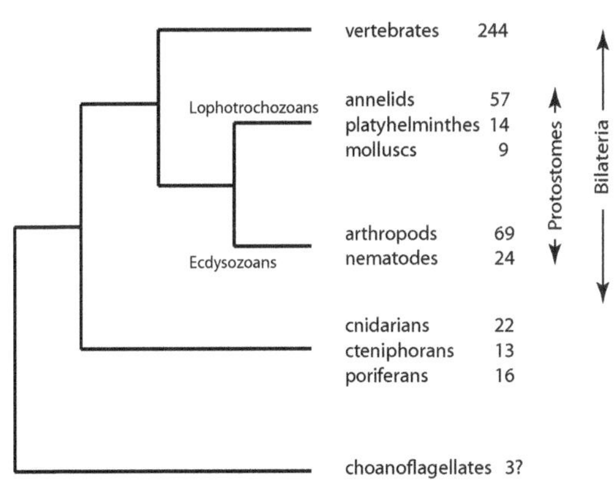

Fig. 4 Redacted phylogenetic relationships among the major groups of metazoans and the maximum number of cell types reported for each group. Data for number of cell types (NCT) taken from Bell and Mooers [2]. Phylogenetic relationships based on Adoutte et al. [25].

can take on a variety of functions, some of which increase the ability to diversify cell types and thus increase morphological and anatomical complexity).

Alternative splicing is an adaptive and highly conserved mechanism because it increases the number of proteins encoded by pre-mRNA in a context-dependent manner in both plants and animals. For example, Chang et al. [26] report a conserved alternative splicing pattern for heat shock transcription factors in the moss *Physcomitrella patens* and the flowering plant *Arabidopsis thaliana*, and show that the alternative splicing mechanism for heat regulation among the land plants is an ancestral condition. Using mRNA sequence data, Pan et al. report that transcripts from ≈95% of human multi-exon genes undergo alternative splicing and that ≈100 000 intermediate to high abundance alternative splicing events occur in a tissue-specific manner [27]. These results are similar to those reported by Johnson et al. using microarray analyses of human tissues [28]. In turn, alternative splicing produces a disproportionate number of intrinsically disordered proteins that lack an equilibrium 3D structure and thus can take on multiple functionalities under normal physiological conditions [6–8]. The majority of transcription factors, which are key players in cell fate specification, have intrinsically disordered protein domains affected by alternative splicing. This enables functional and regulatory diversity while avoiding structural complications [29]. It is reasonable to surmise therefore that the diversity of cell types will increase as a function of the size of a proteome and (in particular) as the number of intrinsically disordered domains increases.

This speculation is consistent with an examination of the data gathered from the primary literature reporting NCT, genome and proteome size, and the number of intrinsically disordered protein domains as gauged by the number of intrinsically disordered residues (IDResidues). Across 19 species of algae (n = 8), nonvascular and vascular land plants (n = 6), and invertebrates and vertebrates (n = 5) whose genomes have been completely sequenced, we find that NCT correlates positively and significantly ($P < 0.0001$) with each of the three metrics of information content (Fig. 5) as gauged by ordinary least squares protocols (which stipulate Y and X as the dependent and independent variables, respectively) using $\log_{10}$-transformed data (Tab. 1). Equally informative is the numerical values of the slopes of each of the bivariate regression curves, because these slopes are scaling exponents (denoted here by α) that indicate numerically the extent to which NCT proportionally increases or decreases with respect to increasing values of each of the other variables on a log scale [30]. Inspection of these slopes indicates that NCT scales roughly as the 2.18 power of the number of intrinsically disordered residues and as the 2.36 power of proteome size. Thus, the degree of cellular specialization increases dramatically as each of these two measures of information content increases. Although the number of intrinsically disordered residues scales nearly one-to-one (isometrically) with respect to proteome size (i.e., $\alpha = 0.97$), NCT fails to increase one-to-one with increasing genome size (i.e., $\alpha = 0.88$), which is a reflection of the so-called G paradox (for a discussion in the context of alternative splicing, see [31]). Collectively, these scaling relationships indicate that the number of intrinsically disordered residues is a highly significant measure of information content if we accept the hypothesis that NCT reflects organismic information content) and that the information contained in these residues helps to explain where some of the "missing" genetic information exists.

Doubts and admonitions

A number of caveats with regard to the preceding analyses exist. The first is that correlations among variables of interest do not constitute proof for cause and effect relationships. This can be illustrated by noting that each of the organisms used in the preceeding analyses is determinate in its growth in overall size. Using data for the maximum body length of the species examined (reported by Lang et al. [32]), regression of the $\log_{10}$-transformed data obtains a statistically significant and positive correlation between maximum body size and NCT ($r^2 = 0.861$, $P < 0.0001$) and a numerically large scaling exponent (i.e., $\alpha = 2.84$; Fig. 6). However, there is no biologically valid reason to assume that body size is dependent on the number of different cell types an organism can produce. The vascular plants produce a maximum of 44 different cell types [24], yet plants are the largest organisms on Earth. In contrast, *Homo sapiens* produce approximately 240 different cell types, yet humans reach a maximum body length of approximately 2.5 m. This example shows that strong correlations can be spurious, if not irrelevant.

Another concern is the nearly flat-line relationship between the number of different cell types and the information content of unicellular plants. For example, once again using $\log_{10}$-transformed data, regression of NCT against the number of intrinsically disordered residues in the six unicellular plants yields a scaling exponent of 0.41 ($r^2 = 0.282$, $P = 0.278$), as opposed to a scaling exponent of 1.84 for the 13 multicellular plants and animals ($r^2 = 0.536$, $P = 0.004$). Taken at face value, these relationships indicate that increases in the information content of unicellular organisms have less effect on NCT compared to multicellular organisms. A third concern is that we know comparatively little about the ancestors of many of the species used in these analyses. Drawing inferences about ancestral morphologies or life styles based on the anatomy or ecology of extant species is highly problematic because we cannot discount the possibility that a seemingly simple species is the descendant of a morphologically or reproductively more complex species.

Clearly, much more research is required to draw definitive conclusions. All that can be said with any certainty is that NCT increases with proteome size and with the number of intrinsically disordered protein residues for the species used in these analyses.

Mechanisms for cell fate specification existed before the evolution of multicellularity

Much has been written about the evolution of multicellularity from the perspective of the mechanisms giving rise to it and the selective advantages, if any, it confers. The traditional

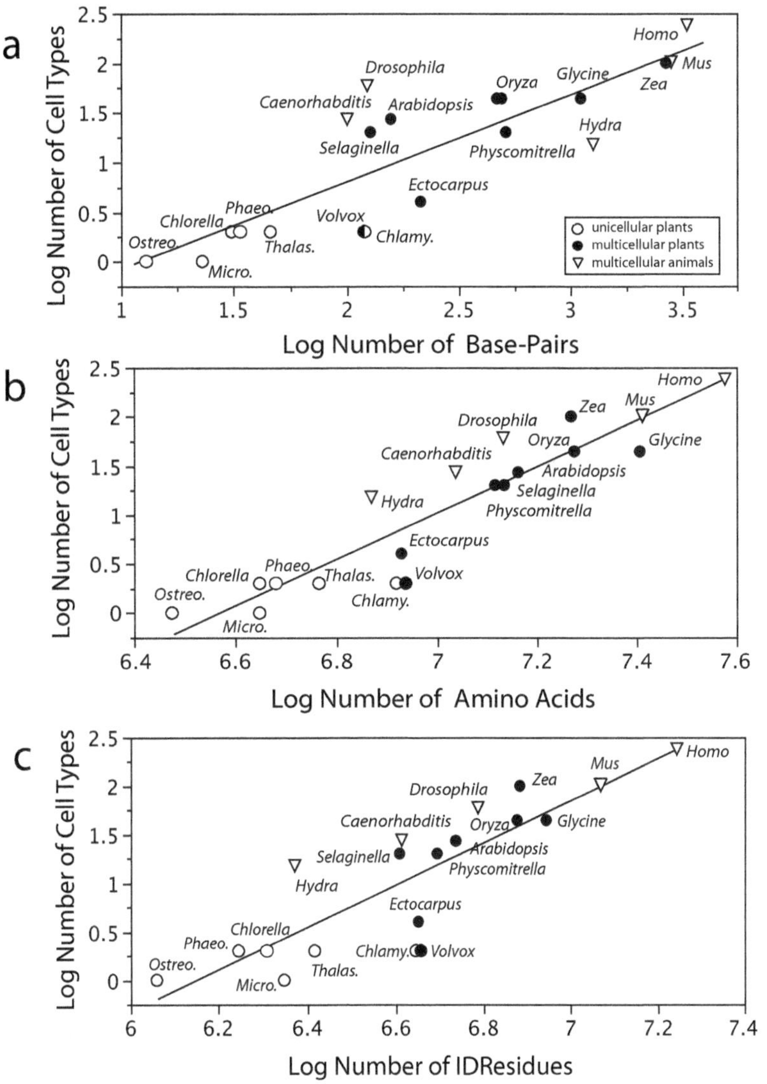

Fig. 5 Log_{10}-transformed data for different cell types (NCT) plotted against log_{10}-transformed data for genome size (number of base-pairs; **a**), proteome size (number of amino acids; **b**), and protein functional diversity (number of intrinsically disordered protein residues, IDResidues; **c**) reported for unicellular algae, multicellular algae and land plants, invertebrates, and vertebrates. Straight lines are ordinary least squares regression curves. *Chlamy.* - *Chlamydomonas reinhardii*; *Micro.* - *Micromonas pusilla* NOUM 17; *Ostreo.* - *Ostreococcus tauri*; *Phaeo.* - *Phaeodactylum tricornutum*; *Thalas.* - *Thalassiosira pseudonana*. For statistical regression parameters, see Tab. 1. Data for NCT taken from Bell and Mooers [2]. Data for IDResidues and AA taken from Oates et al. [46]. Data for IDResidues are based in predictions made by PONDR® VLS2b (see [45]), which ranked as the best for predicting segments of disorder and gave the highest per-residue accuracy for long regions of disorder [52]. Data for genome size taken from Internet sources.

and widely accepted scenario for the evolution of multicellularity postulates a transformation series of body plans starting with a unicellular ancestor, passing through an intermediate organism with a colonial body plan, and culminating with the evolution of simple multicellularity, i.e., a unicellular → colonial → (simple) multicellular body plan transformation series (e.g., [9,10]). This scenario complies reasonably well with phylogenetic trends in body plans within plant and animal groups (see Fig. 2b, Fig. 4), although instances of coenocytic → (simple) multicellular transformation series are known (see [33]). In some lineages, the unicellular → colonial → (simple) multicellular transformation series led to the acquisition of complex multicellularity (for reviews, see [12,33]) and, in some cases, the segregation of germ and soma cell lines [34–36]. Simple multicellularity occurs when every cell in a body plan makes direct contact with the external environment (e.g., filaments of cells), whereas complex multicellularity occurs when some cells make contact only with neighboring cells (e.g., solid blastula-like cellular structures) [37] (see also [38]). The internalization of cells is often associated with the presence of super-cellular transport systems (e.g., vascular plant tissues and mammalian

Tab. 1 Bivariate regression parameters (see Fig. 5) for $\log_{10}$-transformed data of the number of different cell types (NCT), genome size (G, in mbp), number of amino acids (AA), the number of intrinsically disordered residues (IDResidues), and maximum body length (BL) reported for 19 species of algae, land plants, and animal species[a].

Log Y vs. Log X	slope (α-value)	r^2	*P*	*F*
NCT vs. IDResidues	2.18	0.721	<0.0001	44
NCT vs. AA	2.36	0.894	<0.0001	110
IDResidues vs. AA	0.97	0.94	<0.0001	46.7
NCT vs. G	0.88	0.709	<0.0001	43.9
BL vs. NCT	2.84	0.861	<0.0001	112

[a] Species composition: brown algae (n = 1 species; *Ectocarpus siliculosus*), green algae (n = 3 species: *Chlorella* sp., *Chlamydomonas reinhardtii*, and *Volvox carteri*), diatoms (n = 4 species: *Micromonas pusilla* NOUM 17, *Ostreococcus tauri*, *Phaeodactylum tricornutum*, and *Thalassiosira pseudonana*), mosses (n = 1: *Physcomitrella patens*), lycophytes (n = 1 species: *Selaginella moellendorfii*), flowering plants (n = 4 species, *Arabidopsis thaliana*, *Glycine max*, *Oryza sativa*, and *Zea mays*), and metazoans (n = 5 species: *Hydra attenuate*, *Caenorhabditis elegans*, *Drosophila melanogaster*, *Mus musculus*, and *Homo sapiens*). Data for NCT taken from Bell and Mooers [2]. Data for IDResidues and AA taken from Oates et al. [46]. IDResidues values based on PONDR® VLS2b (see [45]) predictions, which is ranked as the best protocol for predicting segments of disorder and gives the highest per-residue accuracy for long regions of disorder [52]. Data for G taken from Internet sources.

pulmonary system), because it makes reliance on passive diffusion for the acquisition of nutrients insufficient to keep pace with metabolic demands (see [34]).

However, it is apparent that the ability to produce more than one cell type is not confined to complex or even simple multicellular life forms. Numerous unicellular, colonial, and filamentous organisms produce different cell types, e.g., the unicellular cyanobacterium *Nostoc* and the colonial volvocine alga *Pleodorina* produce two cell types, the filamentous cyanobacterium *Rivularia* produces three cell types, and the filamentous brown alga *Ectocarpus* produces four cell types. The ability of unicellular organisms to produce two or more different cell types reflects the existence of alternative stable states of gene expression patterns. It is reasonable therefore to suppose that the mechanism responsible for cell fate specification is ancient and prokaryotic in origin, and that its initial adaptive value had nothing to do with multicellularity per se but rather with adapting to different internal or external conditions.

The traditionally accepted mechanism for cell fate specification relies on the inherent multi-stability of gene regulatory networks (e.g., [39–41]). Each stable state provides the conditions that specify a cell type that can be "fixed" in a developmental repertoire ad hoc by natural selection as an organism adapts to its different environmental conditions. Mathematical models show that cellular differentiation can emerge among genetically identical cells as a response to poor compatibility among competing physiological processes [42], whereas, in more derived lineages, an alignment of fitness among adhering cells in a colony might compensate for conflicts of interest among cellular components such that a division of cellular labor becomes possible and even necessary (for a review, see [12]).

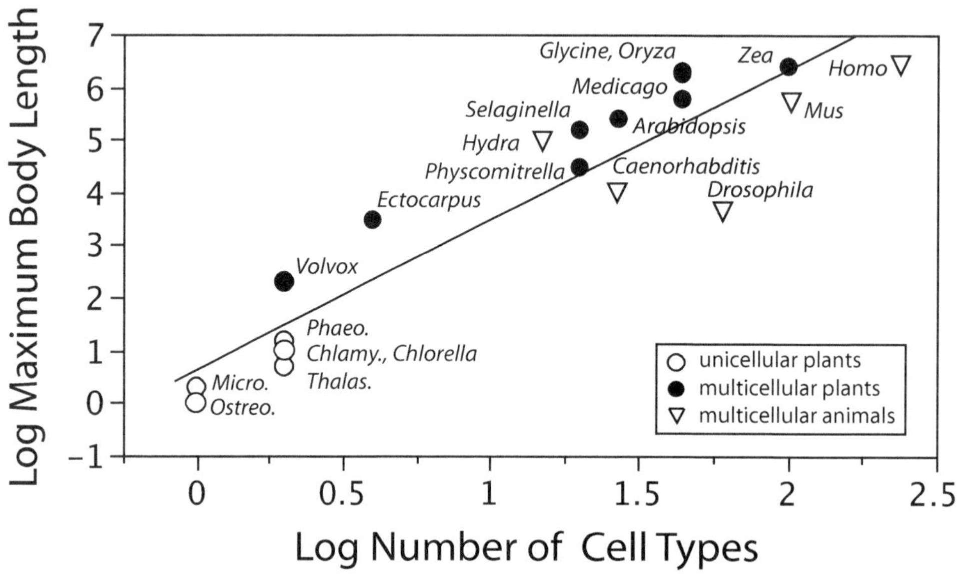

Fig. 6 Log_{10}-transformed data for maximum body length (original units in µm) plotted against $\log_{10}$-transformed data for number of different cell types (NCT) reported for unicellular and multicellular algae, land plants, invertebrates, and vertebrates. Straight lines are ordinary least squares log-log regression curves. *Chlamy.* – *Chlamydomonas reinhardii*; *Micro.* – *Micromonas pusilla* NOUM 17; *Ostreo.* – *Ostreococcus tauri*; *Phaeo.* – *Phaeodactylum tricornutum*; *Thalas.* – *Thalassiosira pseudonana*. For regression parameters, see Tab. 1. Data for NCT taken from Bell and Mooers [2]. Data for body length taken from Lang et al. [32] (supplementary Tab. 7 therein) and supplemented with data reported on the Internet (e.g., tallest human).

A critical assumption shared by most attempts to explain cell fate specification is that the multi-stable states achieved by gene regulatory networks are largely deterministic, i.e., cell fate specification is a direct and by and large invariant result of each state of multi-stable gene regulatory networks. However, there is sufficient evidence that the joint effects of alternative splicing (AS) and intrinsically disordered protein (IDP) domains on cell type specification and differentiation, and post-translational modification (PTM) are critical to the operations of gene regulatory networks. As noted, AS provides a mechanism that increases protein functional diversity without increasing proteome or genome size, whereas IDP domains further amplify protein functionalities. Further, both AS and IDP are dependent on intra- and extracellular conditions, which makes their operation context-dependent and therefore potentially highly adaptive. These properties engender the hypothesis that AS and IDP operate as a very ancient motif that functions in an interactive and context-dependent manner, and that this motif has been elaborated evolutionarily to yield more diverse cell types in successively divergent eukaryotic lineages.

This hypothesis is consistent with the data presented here and by other workers. For example, Schad et al. [5] observe a significant and positive correlation between organismic complexity (as gauged by the number of different cell types) and proteome size (as gauged by the total number of amino acids), and show that the fraction of intrinsically disordered proteins (IDPs) increases significantly between prokaryotes and eukaryotes. However, these authors could find no evidence for further increases in the fraction of IDPs "over the course of evolution" and concluded that "complexity is ... determined by interaction potential, alternative splicing capacity, tissue-specific protein disorder and, above all, proteome size" [5]. These trends are consistent with most, but not all of those reported in Tab. 1, because the number of different cell types produced by the 19 different organisms examined here increases in a log-log monotonic manner as the number of amino acids and the number of intrinsically disordered residues increases. Further, as observed by Schad et al. [5], these trends have scaling exponents that exceed unity, which indicates a disproportionate increase in cell types with increasing proteome size. In contrast to the findings of Schad et al. [5], the trends shown in Fig. 5 indicate that the fraction of IDPs has likely increased "over the course of evolution" from unicellular diatoms and green algae (e.g., *Thalassiosira* and *Chlorella*) to simple multicellular organisms (e.g., *Volvox*, and *Hydra*), and culminating in highly derived plant and animal multicellular life forms (e.g., *Zea* and *Homo*).

The manner by which IDP within the AS–IDP motif can expand cell type diversity is illustrated by a model based on the appearance of five isoforms of a hypothetical regulatory (transcription factor) IDP and a simple Turing reaction-diffusion system (Fig. 7a). Beginning with an undifferentiated (meristematic or stem) cell, each successive cycle of cell division is accompanied by the context-dependent appearance of new isoforms. Cells acquire their "context", which in this case is their position with respect to neighboring cells, by means of the reaction-diffusion system, i.e., cells to the left of each neighbor produce an even-numbered isoform, whereas cells to the right produce an odd-numbered isoform. This scenario does not invoke alternative splicing (which theoretically would expand the repertoire of protein functionality well beyond what is illustrated by this scenario) and depends on only two parameters: (***i***) a chronometer (the cell cycle) that defines when different IDP isoforms are produced, and (***ii***) a spatial discriminator ("left" vs. "right") that defines which different IDP isoforms are produced. Yet, theoretically, even in the absence of AS, within three cell division cycles, eight different cell types are produced as defined by their IDP isoform compositions. Extension of this scenario to eight cell division cycles (and a total of 8 IDP isoforms) results in 258 different cell types, which exceeds the maximum number of cell types reported for any metazoan.

Clearly, this scenario is a naive contrivance if for no other reason than that it presupposes cell type specification depends on the presence or absence of a few protein isoforms, rather than on determinant gene regulatory dynamics. However, it can be expanded easily at the level of an individual cell by considering the synergistic effects of alternative splicing (AS), proteins with intrinsically disordered domains (IDP), and post-translational modifications (PTMs) as shown in Fig. 7b. In this more elaborate model for cell fate specification, a single pre-mRNA undergoes alternative splicing to produce three protein isoforms, each of which has an intrinsically disordered domain that can take on one of three functional variations. Alternatively, each of the three isoforms with disordered domains can be post-translationally modified in one of three ways to yield nine proteins, each of which can assume one of three functional domains. Regardless of the exact numbers, it is clear that a single pre-mRNA can be modified by an AS–IDP–PTMs motif to produce a large number of functionally different proteins.

Concluding remarks

We have tested the supposition that the number of different cell types is a measure of organismic complexity, and we conclude that it is a useful gauge for this purpose, albeit not invariably so. We have also made two assertions based on the relationships between the number of different cell types and proteome size: (***i***) the mechanism for cell fate specification depends on a context-dependent motif consisting of alternative splicing (AS) and intrinsically disordered residues (IDResidues) as well as post-translational modifications (PTMs), and (***ii***) this motif seriously challenges the notion that complex gene regulatory networks regulate development in a deterministic fashion.

Evidence for the operation of the AS–IDR–PTM motif in both plants and metazoans is extensive. As in animals, the pre-mRNAs of model plant organisms, such as *Arabidopsis thaliana*, undergo extensive AS, contain substantial fractions of IDR, and undergo PTM, typically in the form of phosphorylation (e.g., [4,43,44]). Plants and animals use small molecules for inter-cellular signaling that are transduced by binding to disorder-containing receptor proteins, and PONDR VSL2b [45] and other disorder predictors indicate that the various plant homeodomain-containing transcription factors contain significant amounts of disorder, and

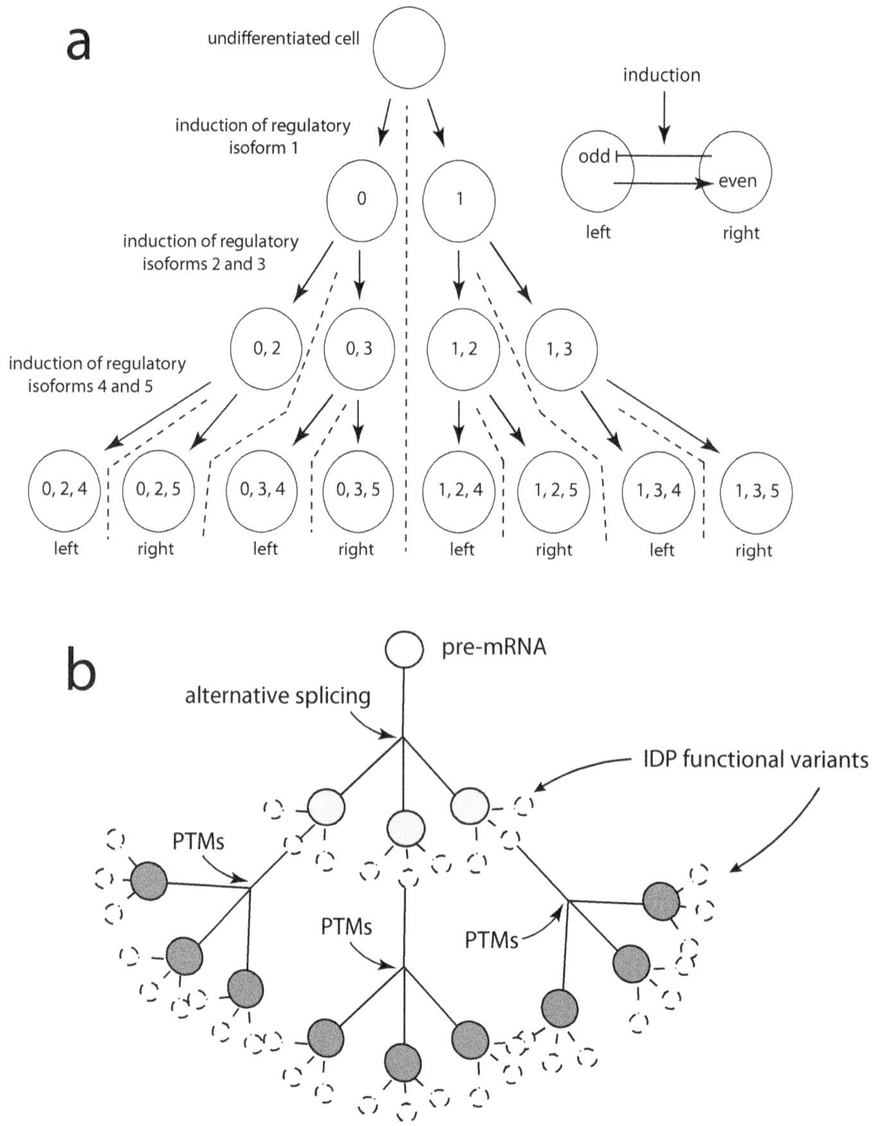

Fig. 7 Scenarios for cell fate specification involving the effects of alternative splicing, intrinsically disordered proteins (IDP), and post-translational modification (PTM). **a** A two-parameter scenario for the specification of eight different cell types using five isoforms (1–5) of an intrinsically disordered regulatory protein (transcription factor). The two parameters are (***i***) a chronometer (using successive cell division cycles) and (***ii***) a reaction-diffusion induction system (upper right diagram) in which paired derivative cells to the right of their neighbor inhibit the formation of odd numbered isoforms and cells to the left of their derivative neighbor activate the formation of even numbered isoforms. Each of three successive divisions (top to bottom) produces successively different isoform compositions, each of which specifies a different cell fate. **b** A single precursor mRNA (pre-mRNA) undergoes alternative splicing to yield three protein isoforms (shown in light gray), each of which has an intrinsically disordered domain that can take on three functions (circles with dashed outlines). Alternatively, each isoform is subjected to PTMs in one of three ways to yield nine proteins (shown in darker gray), each of which has an intrinsically disordered domain that can take on three functions (circles with dashed outlines), to yield a total of 27 IDP functional variants.

they undergo PTM [46]. For example, rice KNOX undergoes tissue-specific AS [47], whereas DELLA proteins are phosphorylated for degradation [48].

The operation of this motif casts doubt on the notion that cell fate specification is genetically deterministic. Across diverse unicellular and multicellular plants and animals, the number of different cell types does not increase one-to-one with increasing genome size (the G paradox), but it does increase (and in a log-log disproportionate manner) with increasing proteome size, particularly with increases in the faction of intrinsically disordered residues, which is a reflection of protein functional versatility. We are sensitive to the fact that the number of species examined in this study is small, and that even very significant and positive correlations between variables of interest do not provide proof of cause and effect. Yet, the majority of transcription factors contain intrinsically disordered protein domains whose functionalities are context-dependent [29,49], and post-translational modification further diversifies protein functionalities. These features, taken even in isolation, challenge the conventional perspective on the deterministic control of gene regulatory networks. They also help us to understand how evolutionarily conserved transcription factors can evolve new functionalities because the AS–IDR–PTM motif can buffer and even isolate a genome from the immediate consequences of initially maladaptive gene mutations and provide a genome sufficient time to evolve adaptively as transcription factors assume new roles. The origin of cellular differentiation may reside therefore in the ability of the AS–IDP–PTM motif to permit somatic or reproductive functional roles for different cell types to become established ad hoc by natural selection. This speculation is consistent with the hypothesis that the evolution of complex genomic architecture was driven by non-adaptive stochastic events, rather than by adaptive evolution by means of natural selection [50], and it is consistent with the observation that metazoan embryonic cell lineages are significantly simpler than would be expected by chance, which suggests that selection for decreased complexity plays a significant role in shaping cell lineages [51]. Under any circumstances, our perspective indicates that sequence homologies do not invariably indicate protein functional homologies, which makes BLAST analyses potentially misleading.

Acknowledgments

The authors thank Dr. Beata Zagorska-Marek for inviting this submission, and Drs. Dominick Paolillo Jr. and Randy Wayne (Cornell University), and two anonymous reviewers for constructive suggestions. Funding from the College of Agriculture and Life Science is also gratefully acknowledged.

Authors' contributions

The following declarations about authors' contributions to the research have been made: conceived of and wrote the paper: KJN, EDC, AKD; prepared the figures: KJN.

Competing interests

No competing interests have been declared.

References

1. Valentine JW, Collins AG, Meyer CP. Morphological complexity increase in metazoans. Paleobiology. 1994;20(2):131–142.
2. Bell G. Size and complexity among multicellular organisms. Biol J Linn Soc. 1997;60(3):345–363.
3. Erwin DH. Early origin of the bilaterian developmental toolkit. Phil Trans R Soc B. 2009;364(1527):2253–2261.
4. Chen L, Bush SJ, Tovar-Corona JM, Castillo-Morales A, Urrutia AO. Correcting for differential transcript coverage reveals a strong relationship between alternative splicing and organism complexity. Mol Biol Evol. 2014;31:1402–1413.
5. Schad E, Tompa P, Hegyi H. The relationship between proteome size, structural disorder and organism complexity. Genome Biol. 2011;12:R120.
6. Liu J, Perumal NB, Oldfield CJ, Su EW, Uversky VN, Dunker AK. Intrinsic disorder in transcription factors. Biochemistry (Mosc). 2006;45(22):6873–6888.
7. Oldfield CJ, Meng J, Yang JY, Yang MQ, Uversky VN, Dunker AK. Flexible nets: disorder and induced fit in the associations of p53 and 14-3-3 with their partners. BMC Genomics. 2008;9(1 suppl):S1.
8. Tompa P, Fuxreiter M. Fuzzy complexes: polymorphism and structural disorder in protein-protein interactions. Trends Biochem Sci. 2008;33(1):2–8.
9. Bonner JT. First signals: the evolution of multicellular development. Princeton, NJ: Princeton University Press; 2000.
10. Kirk DL. A twelve-step program for evolving multicellularity and a division of labor. Bioessays. 2005;27(3):299–310.
11. Herron MD, Michod RE. Evolution of complexity in the volvocine algae: transitions in individuality through Darwin's eye. Evolution. 2008;62(2):436–451.
12. Folse HJ, Roughgarden J. What is an individual organism? A multilevel selection perspective. Q Rev Biol. 2010;85(4):447–472.
13. Niklas KJ. The evolutionary-developmental origins of multicellularity. Am J Bot. 2014;101(1):6–25.
14. Silberfeld T, Leigh JW, Verbruggen H, Cruaud C, de Reviers B, Rousseau F. A multi-locus time-calibrated phylogeny of the brown algae (Heterokonta, Ochrophyta, Phaeophyceae): investigating the evolutionary nature of the "brown algal crown radiation". Mol Phylogenet Evol. 2010;56(2):659–674.
15. Andersen RA. Biology and systematics of heterokont and haptophyte algae. Am J Bot. 2004;91(10):1508–1522.
16. Graham LE. Algae. 2nd ed. San Francisco, CA: Pearson/Benjamin Cummings; 2009.
17. Parker BC. Translocation in the giant kelp *Macrocystis*. I. Rates, direction, quantity of C^{14}-labeled products and fluorescein. J Phycol. 1965;1(2):41–46.
18. Parker BC. Translocation in *Macrocystis*. III. Composition of sieve tube exudate and identification of the major C^{14}-labeled products. J Phycol. 1966;2(1):38–41.
19. Buggeln RG, Fensom DS, Emerson CJ. Translocation of ^{14}C-photoassimilate in the blade of *Macrocystis pyrifera* (Phaeophyceae). J Phycol. 1985;21(1):35–40.
20. Ruhfel BR, Gitzendanner MA, Soltis PS, Soltis DE, Burleigh JG. From algae to angiosperms-inferring the phylogeny of green plants (*Viridiplantae*) from 360 plastid genomes. BMC Evol Biol. 2014;14:23.
21. Cook M, Graham L, Botha C, Lavin C. Comparative ultrastructure of plasmodesmata of *Chara* and selected bryophytes: toward an elucidation of the evolutionary origin of plant plasmodesmata. Am J Bot. 1997;84(9):1169–1178.
22. Boot KJM, Libbenga KR, Hille SC, Offringa R, van Duijn B. Polar auxin transport: an early invention. J Exp Bot. 2012;63(11):4213–4218.
23. Foster AS. Comparative morphology of vascular plants. 2nd ed. San Francisco, CA: W.H. Freeman; 1974.
24. Govindarajalu E. The systematic anatomy of south Indian Cyperaceae. Bot J Linn Soc. 1969;62:27–40.
25. Adoutte A, Balavoine G, Lartillot N, Lespinet O, Prud'homme B, de Rosa R. The new animal phylogeny: reliability and implications. Proc Natl Acad Sci USA. 2000;97(9):4453–4456.
26. Chang CY, Lin WD, Tu SL. Genome-wide analysis of heat-sensitive alternative splicing in *Physcomitrella patens*. Plant Physiol. 2014;165:826–840.
27. Pan Q, Shai O, Lee LJ, Frey BJ, Blencowe BJ. Deep surveying of alternative splicing complexity in the human transcriptome by high-throughput sequencing. Nat Genet. 2008;40(12):1413–1415.
28. Johnson JM, Castle J, Garrett-Engele P, Kan Z, Loerch PM, Armour CD, et al. Genome-wide survey of human alternative pre-mRNA splicing

with exon junction microarrays. Science. 2003;302(5653):2141–2144.

29. Liu J, Perumal NB, Oldfield CJ, Su EW, Uversky VN, Dunker AK. Intrinsic disorder in transcription factors. Biochemistry (Mosc). 2006;45(22):6873–6888.
30. Niklas KJ. Plant allometry: the scaling of form and process. Chicago, IL: University of Chicago Press; 1994.
31. Hahn MW, Wray GA. The g-value paradox. Evol Dev. 2002;4(2):73–75.
32. Lang D, Weiche B, Timmerhaus G, Richardt S, Riaño-Pachón DM, Corrêa LGG, et al. Genome-wide phylogenetic comparative analysis of plant transcriptional regulation: a timeline of loss, gain, expansion, and correlation with complexity. Genome Biol Evol. 2010;2:488–503.
33. Niklas KJ, Cobb ED, Crawford DR. The evo-devo of multinucleate cells, tissues, and organisms, and an alternative route to multicellularity. Evol Dev. 2013;15(6):466–474.
34. Niklas KJ, Newman SA. The origins of multicellular organisms. Evol Dev. 2013;15(1):41–52.
35. Buss LW. The evolution of individuality. Princeton, NJ: Princeton University Press; 1987.
36. Michod RE. Evolution of the individual. Am Nat. 1997;150 suppl 1:S5–S21.
37. Schlichting CD. Origins of differentiation via phenotypic plasticity. Evol Dev. 2003;5(1):98–105.
38. Knoll AH. The multiple origins of complex multicellularity. Annu Rev Earth Planet Sci. 2011;39(1):217–239.
39. Britten RJ, Davidson EH. Gene regulation for higher cells: a theory. Science. 1969;165(3891):349–357.
40. Laurent M, Kellershohn N. Multistability: a major means of differentiation and evolution in biological systems. Trends Biochem Sci. 1999;24(11):418–422.
41. Jaeger J, Monk N. Bioattractors: dynamical systems theory and the evolution of regulatory processes. J Physiol. 2014;592(11):2267–2281.
42. Ispolatov I, Ackermann M, Doebeli M. Division of labour and the evolution of multicellularity. Proc R Soc B. 2012;274:1768–1776.
43. Yao Q, Gao J, Bollinger C, Thelen JJ, Xu D. Predicting and analyzing protein phosphorylation sites in plants using musite. Front Plant Sci. 2012;3:186.
44. Yruela I, Contreras-Moreira B. Protein disorder in plants: a view from the chloroplast. BMC Plant Biol. 2012;12(1):165.
45. Peng K, Radivojac P, Vucetic S, Dunker AK, Obradovic Z. Length-dependent prediction of protein intrinsic disorder. BMC Bioinformatics. 2006;7(1):208.
46. Oates ME, Romero P, Ishida T, Ghalwash M, Mizianty MJ, Xue B, et al. D2P2: database of disordered protein predictions. Nucl Acids Res. 2012;41:D508–516.
47. Ito Y, Hirochika H, Kurata N. Organ-specific alternative transcripts of KNOX family class 2 homeobox genes of rice. Gene. 2002;288(1–2):41–47.
48. Qin Q, Wang W, Guo X, Yue J, Huang Y, Xu X, et al. *Arabidopsis* DELLA protein degradation is controlled by a type-one protein phosphatase, TOPP4. PLoS Genet. 2014;10(7):e1004464.
49. Minezaki Y, Homma K, Nishikawa K. Genome-wide survey of transcription factors in prokaryotes reveals many bacteria-specific families not found in archaea. DNA Res. 2005;12(5):269–280.
50. Lynch M. The origins of eukaryotic gene structure. Mol Biol Evol. 2006;23(2):450–468.
51. Azevedo RBR, Lohaus R, Braun V, Gumbel M, Umamaheshwar M, Agapow PM, et al. The simplicity of metazoan cell lineages. Nature. 2005;433(7022):152–156.
52. Peng ZL, Kurgan L. Comprehensive comparative assessment of in-silico predictors of disordered regions. Curr Protein Pept Sci. 2012;13(1):6–18.

25

Involvement of plastid, mitochondrial and nuclear genomes in plant-to-plant horizontal gene transfer

Maria Virginia Sanchez-Puerta*

IBAM-CONICET & FCA, FCEN, Universidad Nacional de Cuyo, Chacras de Coria, Mendoza 5505, Argentina

Abstract

This review focuses on plant-to-plant horizontal gene transfer (HGT) involving the three DNA-containing cellular compartments. It highlights the great incidence of HGT in the mitochondrial genome (mtDNA) of angiosperms, the increasing number of examples in plant nuclear genomes, and the lack of any convincing evidence for HGT in the well-studied plastid genome of land plants. Most of the foreign mitochondrial genes are non-functional, generally found as pseudogenes in the recipient plant mtDNA that maintains its functional native genes. The few exceptions involve chimeric HGT, in which foreign and native copies recombine leading to a functional and single copy of the gene. Maintenance of foreign genes in plant mitochondria is probably the result of genetic drift, but a possible evolutionary advantage may be conferred through the generation of genetic diversity by gene conversion between native and foreign copies. Conversely, a few cases of nuclear HGT in plants involve functional transfers of novel genes that resulted in adaptive evolution. Direct cell-to-cell contact between plants (e.g. host-parasite relationships or natural grafting) facilitate the exchange of genetic material, in which HGT has been reported for both nuclear and mitochondrial genomes, and in the form of genomic DNA, instead of RNA. A thorough review of the literature indicates that HGT in mitochondrial and nuclear genomes of angiosperms is much more frequent than previously expected and that the evolutionary impact and mechanisms underlying plant-to-plant HGT remain to be uncovered.

Keywords: angiosperm; horizontal gene transfer; chimeric; gene conversion; parasitic

Introduction

Horizontal gene transfer (HGT), the exchange of genetic material between non-mating organisms, is widespread in prokaryotes and, to a lesser extent, in unicellular eukaryotes [1–3]. Most of the evidence for HGT in eukaryotes comes from genes derived from bacteria, describing transfers between phagotrophic, unicellular eukaryotes and their bacterial prey [1,4]. In the last few years, horizontal transfer in multicellular eukaryotes has been increasingly reported [5–8]. Plants in particular have been recognized as donors and targets of HGT involving the three domains of life [5,9]. Of the genetic exchanges reported between multicellular eukaryotes, the frequency of HGT between flowering plants is surprisingly elevated [10–16]. Most cases of plant-to-plant HGT comprise mitochondrial sequences (introns or genes) that were transferred to the mitochondrial genome of another angiosperm [15–20].

This review focuses on plant-to-plant HGT events involving the three DNA-containing compartments in plant cells, with emphasis on transfers between angiosperm mitochondria, especially those involving host-parasite relationships.

HGT among plants: impact in each of the three genomes

Horizontal transfers among angiosperms has been increasingly recognized and slowly accepted by the scientific community. HGT between two plants may involve any of the three DNA-containing compartments, though certain genomes are more prone to genetic exchanges than others (Fig. 1).

Mitochondria-to-mitochondria HGT

Angiosperm mitochondrial genomes (mtDNAs) engage in HGT remarkably often compared to all other organellar genomes and most nuclear genomes (Fig. 2) [10–12,15–18,20–27]. Several properties of plant mitochondria and their genomes contribute to this propensity for HGT. Plant mitochondria are capable of importing DNA or RNA [28,29] and readily undergo fusion [30,31]. Furthermore, plant mtDNAs have an active homologous recombination system [32,33] that is reflected in their structurally dynamic

* Email: mvsanchezpuerta@fca.uncu.edu.ar

Handling Editor: Andrzej Bodył

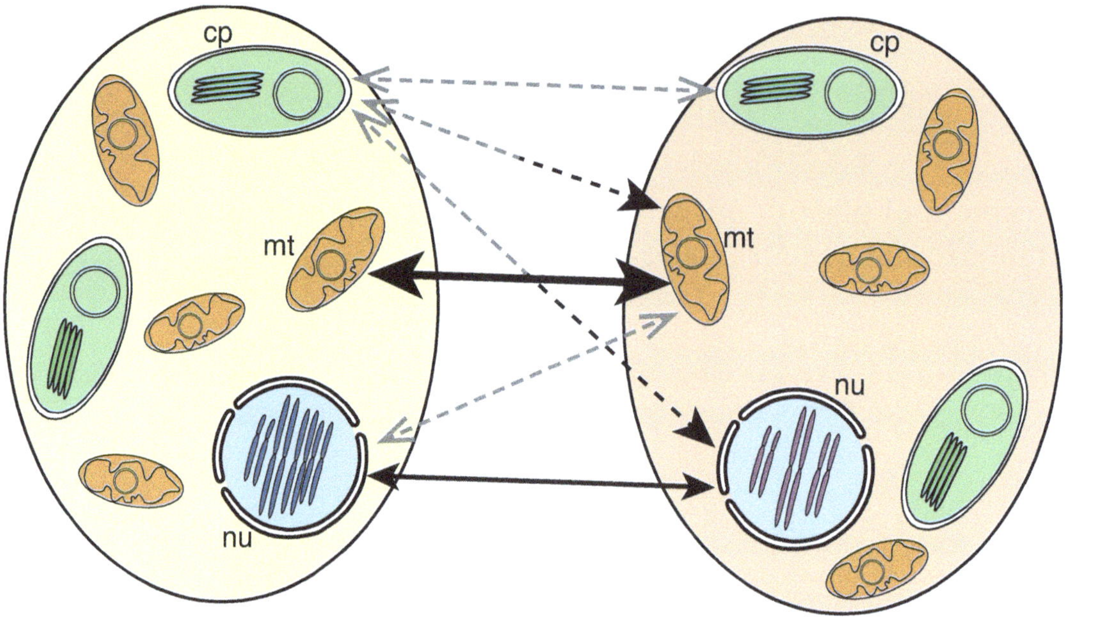

Fig. 1 Horizontal gene transfers (HGT) between two plant species. Solid black lines depict the number of reported HGT events between those cell compartments. Thicker lines indicate more frequent events than thinner lines. Dashed black and grey lines depict putative transfers or lack of evidence for such transfers, respectively. mt: mitochondria; cp: chloroplast; nu: nucleus.

and rearranged genomes [34–36], allowing for a high frequency of intracellular gene transfer (IGT) of chloroplast and nuclear DNA into the mitochondria [37,38]. Plant mitochondrial genomes have generally very low substitution rates [39,40] and relatively large and variable sizes (from 222 kb to 11.3 Mb) [41–43]. Regardless of the genome length, plant mtDNAs encode only 37–67 genes surrounded by long intergenic regions of, mostly, unknown origin with no similarity to any other sequenced DNA [41]. These large intergenic regions facilitate the acquisition of foreign DNA without disruption of the expression of native genes.

In most cases, foreign genes acquired by plant mitochondria are mitochondrial genes donated by another angiosperm species via mitochondria-to-mitochondria horizontal transfer (Fig. 1). Only rarely foreign genes from a non-land plant [44] or from prokaryotes [15] were found in plant mtDNAs.

The most frequently documented case of HGT in multicellular eukaryotes, and the first one to be reported for plants, involves an intron present in the mitochondrial gene *cox1* of flowering plants, which had spread among many diverse angiosperms via an estimated 73 separate plant-to-plant transfer events (Fig. 2) [16,45]. This constitutes the most rampant case of horizontal transmission known among multicellular, if not all eukaryotes, with multiple lines of evidence strongly supporting the highly frequent horizontal transfer of this intron [16,19,45,46].

In the past two decades, more than 24 different angiosperm species were found to contain foreign genes in their mitochondrial genomes (not counting *cox1* intron transfers). From those, two well-studied angiosperm lineages stand out for their extraordinarily large numbers of foreign mitochondrial genes (Fig. 2). The complete sequence of the ~3.9 Mb mtDNA of *Amborella trichopoda* revealed the presence of two mitochondrial genome equivalents from other

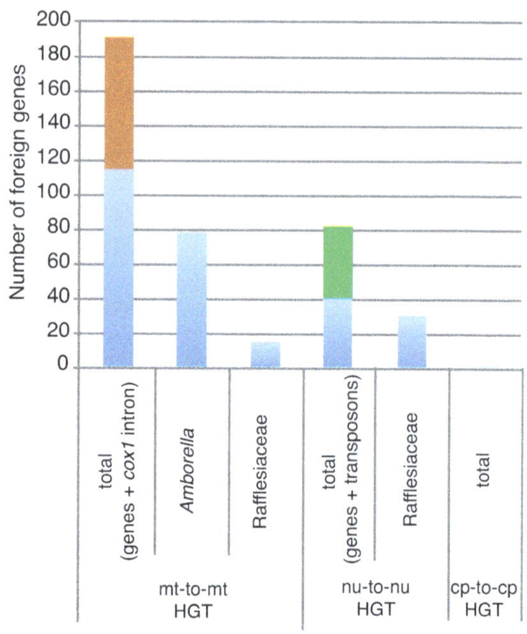

Fig. 2 Number of foreign genes transferred by HGT between nuclear (nu), chloroplast (cp) and mitochondrial (mt) genomes of angiosperms, respectively. It is possible that several foreign genes were transferred together in a single HGT event. Blue, orange and green bars represent coding genes, *cox1* introns and transposable elements, respectively.

angiosperms (~80 mitochondrial genes; Fig. 2), one from mosses and three from green algae – all acquired by HGT [15]. Foreign genes in *Amborella* mtDNA are predominantly pseudogenes, while the native homologs remain intact and presumably functional [15]. A next-generation sequencing study on three endophytic holoparasites of the angiosperm family Rafflesiaceae uncovered a minimum of 16 foreign mitochondrial genes (Fig. 2) presumably obtained from their host plants in a number of, both ancient and recent, independent horizontal transfers [47]. In contrast to *Amborella*, a fraction of the foreign genes appear to be functional in the parasite and may have replaced the native genes via homologous recombination [47].

Most foreign mitochondrial transfers, including those in *Amborella* and holoparasites within the Rafflesiaceae, were inferred to occur as genomic DNA, with no evidence of RNA-transfer followed by retrotranscription [15,23,24,47]. Evidence for this comes from the presence of introns, long tracts of foreign mitochondrial sequences encompassing genes and intergenic regions, or unedited cytidine nucleotides at sites known to undergo RNA editing in plant mitochondria.

Nuclear genes

An increasing number of reports on plant nuclear genes horizontally transferred to the nuclear genome of a foreign plant have been published (Fig. 1). Nuclear-to-nuclear transfers involve both genes [22,48–52] and transposable elements [14,53,54] exchanged between angiosperms (Fig. 2) and also between bryophytes and ferns [55]. The most striking example of nucleus-to-nucleus HGT involves the holoparasitic plant *Rafflesia cantleyi* (Rafflesiaceae), whose nuclear genome contains a minimum of 31 genes obtained from its plant host, *Tetrastigma rafflesiae* (Vitaceae) [22]. The foreign genes, transferred as DNA, are expressed and many of them are probably functional [22].

Given their high abundance in nuclear genomes and their ability to move and become functional in a new host genome [56], transposable elements are often maintained in the recipient genome after being horizontally transferred [57]. Therefore, transposable elements, and other forms of selfish DNA, are expected to be the predominant type of DNA transferred among eukaryotes once enough comparative data is available across this domain of life. Nuclear-to-nuclear plant HGT likely occurs in a higher (even extremely higher) frequency than we are able to detect today given the paucity of plant nuclear genome sequences available for comparison. It is likely that most nuclear transfers are non-functional and therefore transient, disappearing before detection.

Plastid genes

Chloroplast genomes (cpDNAs) from photosynthetic eukaryotes are only rarely involved in HGT. This is likely not an underestimation because cpDNAs have been thoroughly sequenced across the diversity of the eukaryotic tree of life and phylogenetic analyses of plastid genes intensively performed [5,58]. Even though it is an unusual event, a few confirmed cases of foreign plastid genes in algal cpDNAs exist [59–62], whereas no convincing evidence is available for HGT between angiosperm plastid genomes and only few examples are available for horizontal transfers of plastid genes involving different compartments in two plant species (Fig. 1).

Angiosperm plastid genomes are highly resistant to acquiring genes from other cellular compartments, as well as from other organisms. Two examples of IGT into to the plastid genome of angiosperms include a mitochondrial fragment and possible retrotransposon in the cpDNA of *Daucus carota* [63,64] and a 2.4 kb mitochondrial segment transferred to an intergenic spacer in the cpDNA of the milkweed *Asclepias syriaca* [65]. The limited acquisition of sequences from other compartments or other organisms by angiosperm chloroplasts relates to their particular properties: the absence of plastid fusion or uptake mechanisms, and the reduced plastid genome with minimal intergenic regions. However, once a foreign DNA fragment enters the plastid, this DNA can be incorporated into the cpDNA by homologous recombination [35,66].

There are a few examples of horizontally transferred plastid genes found in angiosperms, but it is not always clear in which cellular compartment they reside or whether the transfer occurred directly from the chloroplast genome of the donor (Fig. 1). A cp-to-cp HGT of the plastid gene *rps2* identified in parasitic angiosperms of the genus *Phelipanche* [13] cannot be ruled out (Fig. 2). However, the recently sequenced plastid genome of *P. ramosa* [67] showed that the foreign copy of *rps2* is not present in its cpDNA. In other cases, foreign plastid genes were located in the mitochondrial genomes of the fabid *Phaseolus vulgaris* [68], the early-diverging angiosperm *Amborella trichopoda* [15], the holoparasite *Sapria himalayana* [47], and in the Solanaceae *Hyoscyamus niger* [69]. The route by which these plastid genes entered the recipient mitochondria remains unknown, but given the frequency of mt-to-mt HGT, it has been proposed that plastid genes were first transferred to the mitochondrial genome within the donor plant by IGT and then horizontally transferred via mitochondrial HGT. Likewise, the presence of foreign plastid genes in the nuclear genome of the holoparasite *Rafflesia lagascae* (Rafflesiaceae) has been described [70]. IGT from the chloroplast to the nucleus of the host followed by nucleus-to-nucleus HGT could have occurred.

A thorough study of the three genomes of plant parasites in the holoparasitic plant family Rafflesiaceae indicated that mitochondrial HGT is more frequent than nuclear HGT in these host-parasite relationships, with no strong evidence for plastid-to-plastid transfers [47].

Mechanisms of HGT among plants

A long-standing question in plant evolutionary biology relates to the mechanisms of gene transfer among angiosperms. Postulated natural mechanisms that facilitate HGT include direct transmission involving parasitic or epiphytic plants, tissue grafts, illegitimate pollination, or indirect transmission via a vector intermediate, such as virus, bacteria, insects, and fungi [5,11,12,16,18,21,25,71]. These proposed mechanisms can explain the horizontal transfer across the three cellular compartments (Fig. 1).

The frequency of HGT events between plant mitochondria is biased towards host-parasite systems [51], even though more than 55 mitochondrial genomes from free-living angiosperms have been sequenced and only a draft assembly of the mtDNA of the parasitic plant *Rafflesia lagascae* is available [70]. This bias suggests the importance of direct physical contact as a mechanism for plant-to-plant HGT. Parasitic plants form vascular connections with the host plant that allow transfer of water, nutrients, proteins, mRNAs, and pathogens [72–74]. Even though the exchange of genomic DNA in host-parasite relationships has not been shown, it is not hard to imagine that DNA fragments can occasionally travel through the haustorium.

A few studies focused on an experimental model of HGT through grafting. Natural grafting may enable HGT between two plants [9,15]. In a recent study, three HGT events were observed on axillary shoots of grafted cotton plants, but the genetic variations were not stably inherited across three generations [75]. Two research groups recreated HGT by grafting two lines or species of tobacco and found that nuclear genomes, entire chloroplasts and organellar DNA travelled across the graft junction through plasmodesmata [71,76–78]. These observations demonstrate that plant grafting can result in the exchange of DNA from the three DNA-containing compartments.

Regardless of how mitochondrial genes physically reach a recipient plant, a mitochondrial-fusion model for HGT in plants has been proposed [15]. According to this model, plant mitochondrial HGT occurs by capture of entire mitochondria from foreign, donor plants, followed by fusion of native and foreign mitochondria, and then recombination of their genomes. The fusion of foreign and native mitochondria seems to be restricted to phylogenetically related taxa, such as those that belong to the Viridiplantae, due to putative incompatibilities in the mechanism of mitochondrial fusion with other lineages [15]. This restriction would explain the almost exclusive occurrence of foreign mitochondrial genes from plants integrated in angiosperm mitochondrial genomes.

To study the cellular and molecular processes of plant-to-plant transfers, experimental assays recreating HGT events have been undertaken [69,71,76–78]. In one study, the effects of mitochondrial fusion followed by mtDNA recombination were assessed in a cybrid (cytoplasmic hybrid) plant between two Solanaceae [69]. It has been known for decades that mitochondrial genomes recombine after protoplast fusions between two plant species [79]. Recently, the complete sequence of the mtDNA of a cybrid plant indicated that the two parental genomes recombined intensely after protoplast fusion to form an intensely chimeric mtDNA that encodes a single allele of most of its genes, including nine chimeric genes [69].

Outcomes of HGT in plants

The evolutionary implications of HGT in land plants are still largely unknown. Examples of putative adaptive roles of HGT in plants have been recently reviewed [7,8]. Most cases involve genes obtained from bacteria [80–83], viruses [81] and fungi [6,81], and reside in the plant nuclear genome. Only a handful of nuclear genes were acquired through plant-to-plant HGT [22,48,55], and a few of those represent novel, functional genes in the recipient genome that may lead to ecological adaptations [48,55].

Horizontal transfers between flowering plant mitochondria generally results in the initial duplication of a mitochondrial gene, in which two different copies (the foreign and the native) co-exist for some time in the recipient mtDNA. Over time, this initial situation may have different outcomes (Fig. 3):

(*i*) deletion of the foreign gene copy: in this case, the process of HGT goes unseen (silent HGT);

(*ii*) deletion of the native gene copy once the foreign copy becomes functional in the recipient genome (replacement HGT);

(*iii*) both copies are maintained in the recipient mtDNA (duplicative HGT);

(*iv*) homologous recombination and gene conversion between the two copies lead to a single functional chimeric copy of the gene (chimeric HGT).

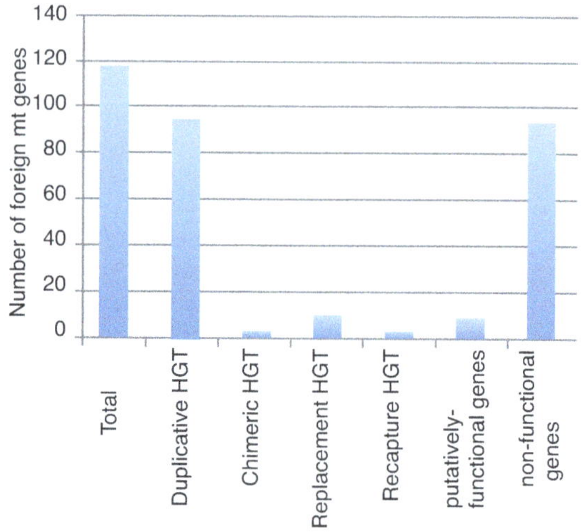

Fig. 3 Number of foreign genes acquired by HGT between angiosperm mitochondria. The total number of foreign mitochondrial (mt) genes are subdivided by HGT type and whether they are putatively functional or not (genes with undetermined functional status are not shown). Non-functional genes include those foreign genes that are pseudogenes, not significantly expressed, or not fully edited.

Foreign mitochondrial genes are commonly found as extra genes, coexisting with native mitochondrial copies of the same gene (duplicative HGT). In those cases, transferred genes usually become pseudogenes while the native alleles remain functional. Occasionally, the foreign and native copies may undergo continuous or discontinuous gene conversion, leading to one or two chimeric gene copies, but often with a single functional allele (chimeric HGT). The functional chimeric gene may translate into a protein with novel residues in the recipient genome. This increase in genetic diversity through HGT could impact the evolution of plant mitochondria [20,23]. A duplicative HGT and differential gene conversion (DH-DC) model has been postulated [23] in which gene conversion of foreign (from other organisms or organelles) and native genes generates mitochondrial genetic variability. A striking case of chimeric HGT has been recently discovered in an extensive survey of the mitochondrial *cox1* gene across almost 900 angiosperms (Sanchez-Puerta et al., unpublished). A divergent short region of the gene *cox1* has most likely been horizontally acquired and gene converted about 20 times during the evolution of the angiosperms examined. The chimeric *cox1* genes are the only *cox1* alleles present in the mtDNAs analyzed and are therefore most likely functional (Sanchez-Puerta et al., unpublished).

In a few cases, a gene that has been functionally transferred to the nuclear genome of the recipient angiosperm is reacquired by its mtDNA via HGT (recapture HGT). On the other hand, introducing a novel gene into angiosperm mitochondria is uncommon because plant mitochondria contain a reduced gene repertoire, unless it brings a gene that was, in turn, acquired from another cell compartment, such as a plastid or nuclear gene (as discussed in the previous section).

In the plant mtHGTs described to date, no clear evidence for an adaptive gain of function was found. It is likely that genetic drift rather than selection has fixed mitochondrial HGT events. Most foreign genes are non-functional pseudogenes due to truncation of the gene or presence of premature stop codons, while those that have a complete open reading frame may not be significantly expressed or efficiently edited (Fig. 3). About nine cases have been reported where chimeric [10,23] or foreign [47] genes are full-length, transcribed and edited in the recipient mtDNA suggesting they are functional (plus the 20 chimeric *cox1* genes described above; unpublished results). The chimeric or foreign functional genes do not show any apparent increase in fitness due to the genetic diversity introduced by HGT. On the contrary, HGT could impair overall gene expression through gene conversion with a foreign allele or when two alleles (a foreign and a native) are simultaneously present and functional in a single mitochondrion [69].

Conclusions

Little is known about the functional implications of horizontal gene transfer among plants, in addition to the real incidence of nuclear and mitochondrial HGT in the plant kingdom. This phenomenon raises a number of questions regarding the extent and mechanism of the exchange of genetic material between plants, its consequences for transgene containment, and the co-evolution of organellar DNA with the nuclear genome. The current accelerated generation of comparative genomic data will contribute to quickly uncover evolutionary aspects of plant-to-plant HGT, such as the expected prevalence of nuclear HGT.

Acknowledgments
This work is supported by the National Institutes of Health (R03-TW-008353) and by Universidad Nacional de Cuyo (Sectyp 06/M053). M.V.S.P. is a researcher of CONICET.

Competing interests
No competing interests have been declared.

References
1. Andersson JO. Lateral gene transfer in eukaryotes. Cell Mol Life Sci. 2005;62(11):1182–1197.
2. Ochman H, Lawrence JG, Groisman EA. Lateral gene transfer and the nature of bacterial innovation. Nature. 2000;405(6784):299–304.
3. Koonin EV, Makarova KS, Aravind L. Horizontal gene transfer in prokaryotes: quantification and classification. Annu Rev Microbiol. 2001;55(1):709–742.
4. Ford Doolittle W. You are what you eat: a gene transfer ratchet could account for bacterial genes in eukaryotic nuclear genomes. Trends Genet. 1998;14(8):307–311.
5. Keeling PJ, Palmer JD. Horizontal gene transfer in eukaryotic evolution. Nat Rev Genet. 2008;9(8):605–618.
6. Richards TA, Soanes DM, Foster PG, Leonard G, Thornton CR, Talbot NJ. Phylogenomic analysis demonstrates a pattern of rare and ancient horizontal gene transfer between plants and fungi. Plant Cell. 2009;21(7):1897–1911.
7. Huang J. Horizontal gene transfer in eukaryotes: the weak-link model. Bioessays. 2013;35(10):868–875.
8. Schönknecht G, Weber APM, Lercher MJ. Horizontal gene acquisitions by eukaryotes as drivers of adaptive evolution. Bioessays. 2014;36(1):9–20.
9. Bock R. The give-and-take of DNA: horizontal gene transfer in plants. Trends Plant Sci. 2010;15(1):11–22.
10. Bergthorsson U, Adams KL, Thomason B, Palmer JD. Widespread horizontal transfer of mitochondrial genes in flowering plants. Nature. 2003;424(6945):197–201.
11. Davis CC, Wurdack KJ. Host-to-parasite gene transfer in flowering plants: phylogenetic evidence from Malpighiales. Science. 2004;305(5684):676–678.
12. Mower JP, Stefanović S, Young GJ, Palmer JD. Plant genetics: gene transfer from parasitic to host plants. Nature. 2004;432(7014):165–166.
13. Park JM, Manen JF, Schneeweiss GM. Horizontal gene transfer of a plastid gene in the non-photosynthetic flowering plants *Orobanche* and *Phelipanche* (Orobanchaceae). Mol Phylogenet Evol. 2007;43(3):974–985.
14. Baidouri ME, Carpentier MC, Cooke R, Gao D, Lasserre E, Llauro C, et al. Widespread and frequent horizontal transfers of transposable elements in plants. Genome Res. 2014;24:831–838.
15. Rice DW, Alverson AJ, Richardson AO, Young GJ, Sanchez-Puerta MV, Munzinger J, et al. Horizontal transfer of entire genomes via mitochondrial fusion in the angiosperm *Amborella*. Science. 2013;342(6165):1468–1473.
16. Sanchez-Puerta MV, Cho Y, Mower JP, Alverson AJ, Palmer JD. Frequent, phylogenetically local horizontal transfer of the *cox1* group I intron in flowering plant mitochondria. Mol Biol Evol. 2008;25(8):1762–1777.
17. Bergthorsson U, Richardson AO, Young GJ, Goertzen LR, Palmer JD. Massive horizontal transfer of mitochondrial genes from diverse land plant donors to the basal angiosperm *Amborella*. Proc Natl Acad Sci USA. 2004;101(51):17747–17752.
18. Barkman TJ, McNeal JR, Lim SH, Coat G, Croom HB, Young ND, et

al. Mitochondrial DNA suggests at least 11 origins of parasitism in angiosperms and reveals genomic chimerism in parasitic plants. BMC Evol Biol. 2007;7(1):248.

19. Cho Y, Qiu YL, Kuhlman P, Palmer JD. Explosive invasion of plant mitochondria by a group I intron. Proc Natl Acad Sci USA. 1998;95(24):14244–14249.
20. Mower JP, Stefanović S, Hao W, Gummow JS, Jain K, Ahmed D, et al. Horizontal acquisition of multiple mitochondrial genes from a parasitic plant followed by gene conversion with host mitochondrial genes. BMC Biol. 2010;8(1):150.
21. Davis CC, Anderson WR, Wurdack KJ. Gene transfer from a parasitic flowering plant to a fern. Proc Biol Sci. 2005;272(1578):2237–2242.
22. Xi Z, Bradley RK, Wurdack KJ, Wong K, Sugumaran M, Bomblies K, et al. Horizontal transfer of expressed genes in a parasitic flowering plant. BMC Genomics. 2012;13:227.
23. Hao W, Richardson AO, Zheng Y, Palmer JD. Gorgeous mosaic of mitochondrial genes created by horizontal transfer and gene conversion. Proc Natl Acad Sci USA. 2010;107(50):21576–21581.
24. Hepburn NJ, Schmidt DW, Mower JP. Loss of two introns from the Magnolia tripetala mitochondrial *cox2* gene implicates horizontal gene transfer and gene conversion as a novel mechanism of intron loss. Mol Biol Evol. 2012;29(10):3111–3120.
25. Nickrent DL, Blarer A, Qiu YL, Vidal-Russell R, Anderson FE. Phylogenetic inference in Rafflesiales: the influence of rate heterogeneity and horizontal gene transfer. BMC Evol Biol. 2004;4(1):40.
26. Schönenberger J, Anderberg AA, Sytsma KJ. Molecular phylogenetics and patterns of floral evolution in the Ericales. Int J Plant Sci. 2005;166(2):265–288.
27. Cho Y, Adams KL, Qiu YL, Kuhlman P, Vaughn JC, Palmer JD. A highly invasive group I intron in the mitochondrial *cox1* gene. In: Moller K, Gardestrom P, Glimelius K, Glaser E, editors. Plant mitochondria: from gene to function. Leiden: Backhuys Publishers; 1998. p. 19–23.
28. Koulintchenko M, Konstantinov Y, Dietrich A. Plant mitochondria actively import DNA via the permeability transition pore complex. EMBO J. 2003;22(6):1245–1254.
29. Duchêne AM, Pujol C, Maréchal-Drouard L. Import of tRNAs and aminoacyl-tRNA synthetases into mitochondria. Curr Genet. 2009;55(1):1–18.
30. Arimura S, Yamamoto J, Aida GP, Nakazono M, Tsutsumi N. Frequent fusion and fission of plant mitochondria with unequal nucleoid distribution. Proc Natl Acad Sci USA. 2004;101(20):7805–7808.
31. Sheahan MB, McCurdy DW, Rose RJ. Mitochondria as a connected population: ensuring continuity of the mitochondrial genome during plant cell dedifferentiation through massive mitochondrial fusion. Plant J. 2005;44(5):744–755.
32. Manchekar M, Scissum-Gunn K, Song D, Khazi F, McLean SL, Nielsen BL. DNA recombination activity in soybean mitochondria. J Mol Biol. 2006;356(2):288–299.
33. Shedge V, Arrieta-Montiel M, Christensen AC, Mackenzie SA. Plant mitochondrial recombination surveillance requires unusual *RecA* and *MutS* homologs. Plant Cell. 2007;19(4):1251–1264.
34. Backert S, Lynn Nielsen B, Börner T. The mystery of the rings: structure and replication of mitochondrial genomes from higher plants. Trends Plant Sci. 1997;2(12):477–483.
35. Maréchal A, Brisson N. Recombination and the maintenance of plant organelle genome stability. New Phytol. 2010;186(2):299–317.
36. Palmer JD, Shields CR. Tripartite structure of the *Brassica campestris* mitochondrial genome. Nature. 1984;307(5950):437–440.
37. Stern DB, Lonsdale DM. Mitochondrial and chloroplast genomes of maize have a 12-kilobase DNA sequence in common. Nature. 1982;299(5885):698–702.
38. Unseld M, Marienfeld JR, Brandt P, Brennicke A. The mitochondrial genome of *Arabidopsis thaliana* contains 57 genes in 366,924 nucleotides. Nat Genet. 1997;15(1):57–61.
39. Palmer JD, Herbon LA. Plant mitochondrial DNA evolves rapidly in structure, but slowly in sequence. J Mol Evol. 1988;28(1–2):87–97.
40. Wolfe KH, Li WH, Sharp PM. Rates of nucleotide substitution vary greatly among plant mitochondrial, chloroplast, and nuclear DNAs. Proc Natl Acad Sci USA. 1987;84(24):9054–9058.
41. Mower JP, Sloan DB, Alverson AJ. Plant mitochondrial genome diversity: the genomics revolution. In: Wendel JF, Greilhuber J, Dolezel J, Leitch IJ, editors. Plant genome diversity. Vienna: Springer; 2012. p. 123–144. (vol 1).
42. Sloan DB, Alverson AJ, Chuckalovcak JP, Wu M, McCauley DE, Palmer JD, et al. Rapid evolution of enormous, multichromosomal genomes in flowering plant mitochondria with exceptionally high mutation rates. PLoS Biol. 2012;10(1):e1001241.
43. Ward BL, Anderson RS, Bendich AJ. The mitochondrial genome is large and variable in a family of plants (Cucurbitaceae). Cell. 1981;25(3):793–803.
44. Vaughn JC, Mason MT, Sperwhitis GL, Kuhlman P, Palmer JD. Fungal origin by horizontal transfer of a plant mitochondrial group I intron in the chimeric *cox1* gene of *Peperomia*. J Mol Evol. 1995;41:563–572.
45. Sanchez-Puerta MV, Abbona CC, Zhuo S, Tepe EJ, Bohs L, Olmstead RG, et al. Multiple recent horizontal transfers of the *cox1* intron in Solanaceae and extended co-conversion of flanking exons. BMC Evol Biol. 2011;11(1):277.
46. Cho Y, Palmer JD. Multiple acquisitions via horizontal transfer of a group I intron in the mitochondrial *cox1* gene during evolution of the Araceae family. Mol Biol Evol. 1999;16(9):1155–1165.
47. Xi Z, Wang Y, Bradley RK, Sugumaran M, Marx CJ, Rest JS, et al. Massive mitochondrial gene transfer in a parasitic flowering plant clade. PLoS Genet. 2013;9(2):e1003265.
48. Christin PA, Edwards EJ, Besnard G, Boxall SF, Gregory R, Kellogg EA, et al. Adaptive evolution of C_4 photosynthesis through recurrent lateral gene transfer. Curr Biol. 2012;22(5):445–449.
49. Vallenback P, Jaarola M, Ghatnekar L, Bengtsson BO. Origin and timing of the horizontal transfer of a *PgiC* gene from *Poa* to *Festuca* ovina. Mol Phylogenet Evol. 2008;46(3):890–896.
50. Yoshida S, Maruyama S, Nozaki H, Shirasu K. Horizontal gene transfer by the parasitic plant *Striga hermonthica*. Science. 2010;328(5982):1128–1128.
51. Zhang D, Qi J, Yue J, Huang J, Sun T, Li S, et al. Root parasitic plant *Orobanche aegyptiaca* and shoot parasitic plant *Cuscuta australis* obtained Brassicaceae-specific strictosidine synthase-like genes by horizontal gene transfer. BMC Plant Biol. 2014;14(1):19.
52. Zhang Y, Fernandez-Aparicio M, Wafula EK, Das M, Jiao Y, Wickett NJ, et al. Evolution of a horizontally acquired legume gene, albumin 1, in the parasitic plant *Phelipanche aegyptiaca* and related species. BMC Evol Biol. 2013;13(1):48.
53. Diao X, Freeling M, Lisch D. Horizontal transfer of a plant transposon. PLoS Biol. 2006;4(1):e5.
54. Roulin A, Piegu B, Wing RA, Panaud O. Evidence of multiple horizontal transfers of the long terminal repeat retrotransposon RIRE1 within the genus *Oryza*. Plant J. 2008;53(6):950–959.
55. Li FW, Villarreal JC, Kelly S, Rothfels CJ, Melkonian M, Frangedakis E, et al. Horizontal transfer of an adaptive chimeric photoreceptor from

bryophytes to ferns. Proc Natl Acad Sci USA. 2014;111(18):6672–6677.

56. Feschotte C, Pritham EJ. DNA transposons and the evolution of eukaryotic genomes. Annu Rev Genet. 2007;41:331–368.

57. Schaack S, Gilbert C, Feschotte C. Promiscuous DNA: horizontal transfer of transposable elements and why it matters for eukaryotic evolution. Trends Ecol Evol. 2010;25(9):537–546.

58. Huang J, Yue J. Horizontal gene transfer in the evolution of photosynthetic eukaryotes. J Syst Evol. 2013;51(1):13–29.

59. Delwiche CF, Palmer JD. Rampant horizontal transfer and duplication of rubisco genes in eubacteria and plastids. Mol Biol Evol. 1996;13(6):873–882.

60. Khan H, Parks N, Kozera C, Curtis BA, Parsons BJ, Bowman S, et al. Plastid genome sequence of the cryptophyte alga *Rhodomonas salina* CCMP1319: lateral transfer of putative DNA replication machinery and a test of chromist plastid phylogeny. Mol Biol Evol. 2007;24(8):1832–1842.

61. Rice DW, Palmer JD. An exceptional horizontal gene transfer in plastids: gene replacement by a distant bacterial paralog and evidence that haptophyte and cryptophyte plastids are sisters. BMC Biol. 2006;4(1):31.

62. Moszczynski K, Mackiewicz P, Bodyl A. Evidence for horizontal gene transfer from bacteroidetes bacteria to dinoflagellate minicircles. Mol Biol Evol. 2012;29(3):887–892.

63. Goremykin VV, Salamini F, Velasco R, Viola R. Mitochondrial DNA of *Vitis vinifera* and the issue of rampant horizontal gene transfer. Mol Biol Evol. 2009;26(1):99–110.

64. Iorizzo M, Senalik D, Szklarczyk M, Grzebelus D, Spooner D, Simon P. De novo assembly of the carrot mitochondrial genome using next generation sequencing of whole genomic DNA provides first evidence of DNA transfer into an angiosperm plastid genome. BMC Plant Biol. 2012;12(1):61.

65. Straub SCK, Cronn RC, Edwards C, Fishbein M, Liston A. Horizontal transfer of DNA from the mitochondrial to the plastid genome and its subsequent evolution in milkweeds (Apocynaceae). Genome Biol Evol. 2013;5(10):1872–1885.

66. Maliga P. Plastid transformation in higher plants. Annu Rev Plant Biol. 2004;55:289–313.

67. Wicke S, Müller KF, de Pamphilis CW, Quandt D, Wickett NJ, Zhang Y, et al. Mechanisms of functional and physical genome reduction in photosynthetic and nonphotosynthetic parasitic plants of the broomrape family. Plant Cell. 2013;25(10):3711–3725.

68. Woloszynska M, Bocer T, Mackiewicz P, Janska H. A fragment of chloroplast DNA was transferred horizontally, probably from non-eudicots, to mitochondrial genome of *Phaseolus*. Plant Mol Biol. 2004;56(5):811–820.

69. Sanchez-Puerta MV, Zubko MK, Palmer JD. Homologous recombination and retention of a single form of most genes shape the highly chimeric mitochondrial genome of a cybrid plant. New Phytol. 2014 (in press).

70. Molina J, Hazzouri KM, Nickrent D, Geisler M, Meyer RS, Pentony MM, et al. Possible loss of the chloroplast genome in the parasitic flowering plant *Rafflesia lagascae* (Rafflesiaceae). Mol Biol Evol. 2014;31:793–803.

71. Stegemann S, Bock R. Exchange of genetic material between cells in plant tissue grafts. Science. 2009;324(5927):649–651.

72. Kim G, LeBlanc ML, Wafula EK, de Pamphilis CW, Westwood JH. Genomic-scale exchange of mRNA between a parasitic plant and its hosts. Science. 2014;345(6198):808–811.

73. Roney JK, Khatibi PA, Westwood JH. Cross-species translocation of mRNA from host plants into the parasitic plant dodder. Plant Physiol. 2007;143(2):1037–1043.

74. Westwood JH, Roney JK, Khatibi PA, Stromberg VK. RNA translocation between parasitic plants and their hosts. Pest Manag Sci. 2009;65(5):533–539.

75. Hao J, Jia X, Yu J, Deng S. Direct visualization of horizontal gene transfer in cotton plants. J Hered. 2014;105(6):834–836.

76. Fuentes I, Stegemann S, Golczyk H, Karcher D, Bock R. Horizontal genome transfer as an asexual path to the formation of new species. Nature. 2014;511(7508):232–235.

77. Stegemann S, Keuthe M, Greiner S, Bock R. Horizontal transfer of chloroplast genomes between plant species. Proc Natl Acad Sci USA. 2012;109(7):2434–2438.

78. Thyssen G, Svab Z, Maliga P. Cell-to-cell movement of plastids in plants. Proc Natl Acad Sci USA. 2012;109(7):2439–2443.

79. Earle ED. Mitochondrial DNA in somatic hybrids and cybrids. In: Levings C, Vasil IK, editors. The molecular biology of plant mitochondria. Dordrecht: Kluwer Academic Publishers; 1995. p. 557–584.

80. Emiliani G, Fondi M, Fani R, Gribaldo S. A horizontal gene transfer at the origin of phenylpropanoid metabolism: a key adaptation of plants to land. Biol Direct. 2009;4:7.

81. Yue J, Hu X, Sun H, Yang Y, Huang J. Widespread impact of horizontal gene transfer on plant colonization of land. Nat Commun. 2012;3:1152.

82. Zardoya R, Ding X, Kitagawa Y, Chrispeels MJ. Origin of plant glycerol transporters by horizontal gene transfer and functional recruitment. Proc Natl Acad Sci USA. 2002;99(23):14893–14896.

83. Knie N, Polsakiewicz M, Knoop V. Horizontal gene transfer of chlamydial-like tRNA genes into early vascular plant mitochondria. Mol Biol Evol. 2014 (in press).

26

The Old and New RNA World

Zofia Szweykowska-Kulińska*
Institute of Molecular Biology and Biotechnology, Faculty of Biology, Adam Mickiewicz University, Umultowska 89, 61-614 Poznań, Poland

Abstract

Among the numerous hypotheses offering a scenario for the origin of life on Earth, the one called "The RNA World" has gained the most attention. According to this hypothesis RNA acted as a genetic information storage material, as a catalyst of all metabolic reactions, and as a regulator of all processes in the primordial world. Various experiments show that RNA molecules could have been synthesized abiotically, with the potential to mediate a whole repertoire of metabolic reactions. Ribozymes carrying out aminoacyl-tRNA reactions have been found in SELEX (systematic evolution of ligands by exponential enrichment) approaches and the development of a ribosome from a RNA-built protoribosome is easy to imagine. Transfer RNA aminoacylation, protoribosome origin, and the availability of amino acids on early Earth allowed the genetic code to evolve. Encoded proteins most likely stabilized RNA molecules and were able to create channels across membranes. In the modern cell, DNA replaced RNA as the main depositor of genetic information and proteins carry out almost all metabolic reactions. However, RNA is still playing versatile, crucial roles in the cell. Apart from its classical functions in the cell, a huge small RNA world is controlling gene expression, chromatin condensation, response to environmental cues, and protecting the cell against the invasion of various nucleic acids forms. Long non-coding RNAs act as crucial gene expression regulators. Riboswitches act at the level of transcription, splicing or translation and mediate feedback regulation on biosynthesis and transport of the ligand they sense. Alternative splicing generates genetic variability and increases the protein repertoire in response to developmental or environmental changes. All these regulatory functions are essential in shaping cell plasticity in the changing milieu. Recent discoveries of new, unexpected and important functions of RNA molecules support the hypothesis that we live in a New RNA World.

Keywords: RNA World hypothesis; ribozyme; small RNAs; long noncoding RNAs; riboswitches; alternative splicing; New RNA World

The origin of life

Among numerous hypotheses offering a scenario for the origin of life on Earth the one called "The RNA World" has gained the most attention. The break-through for this hypothesis, originally described independently by Leslie Orgel (working then at the Salk Institute for Biological Studies in San Diego) [1], Francis Crick (then at the Medical Research Council in England) [2], and Charles Woese (of the University of Illinois) [3] in the late 1960s of the XXth century suggested a world in which RNA catalyzed all the reactions necessary to create an ancestor of a living cell that was able to survive and replicate. In 1977, Manfred Eigen and Peter Schuster described the key features of a molecule with the potential to start life: ability to replicate, in other words to have molecular memory; to carry out all metabolic reactions, to be a catalyst; and to exhibit genetic variability, i.e. to be able to evolve [4]. The only key feature of the RNA molecules that raised problems those days as an originator of Life was the possibility to catalyze biochemical reactions. In 1983, Thomas R. Cech (University of Colorado at Boulder) and Sidney Altman (Yale University) independently discovered the first known enzymes made of RNA [5,6]. The term ribozyme was coined for RNA molecules that exhibit catalytic activities. These discoveries revealed that ancient RNA might have been catalytic and also a progenitor molecule that gave rise to life.

Abiotic synthesis of RNA components and RNA

If RNA molecules started life on Earth they must have been synthesized abiotically. RNA is built from nucleotides, which in turn are composed of a ribose, nitrogen base, and phosphate group(s). The experiments that were carried out during the last six decades show that nitrogen bases can be formed non-enzymatically in a range of experiments that simulate conditions scientists believe to be present on early

* Email: zofszwey@amu.edu.pl
Handling Editor: Beata Zagórska-Marek

Earth. However, for a long time there was no evidence that ribose molecule could be formed selectively. The addition of nucleobases to ribose is inefficient in the case of purines and does not occur at all in the case of pyrimidines. These difficulties have been recently circumvented through a non-enzymatic pathway in which arabinose amino-oxazline and anhydronucleoside intermediates play a key role leading to the synthesis of activated pyrimidine ribonucleotides. These intermediates are formed from plausible prebiotic feedstock molecules and phosphate. Phosphate, apart from its other functions in the pathway, is added to the ribonucleoside at the late stage of the synthesis [7,8]. In another course of experiments, an alternative, efficient and selective synthesis of pyrimidine ribonucleoside-2′,3′-cyclic phosphates has been demonstrated in conditions simulating the prebiotic environment of the early Earth [9].

Assuming the ribonucleotides could be formed on early Earth, there is still a need to show these molecules can polymerize and form RNA. Recently, it has been shown that the 2′-hydroxyl group of oligoribonucleotide-3′-phosphates can be chemoselectively acetylated in water under prebiotically credible conditions, which allows rapid and efficient template-directed 3′-5′ ligation. Moreover, the 2′-O-acetyl group at the ligation junctions of the product RNA strand can be removed under conditions that leave the newly formed 3′-5′ phosphodiester bonds intact. These results suggest a prebiotic pathway from ribonucleoside-2′,3′-cyclic phosphates to predominantly 3′,5′-linked RNA via partially 2′-O-acetylated RNA [10].

In addition, a recent paper by Adamala and Szostak provides strong evidence for nonenzymatic template-directed RNA synthesis inside model protocells representing fatty acid vesicles [10]. All these data concerning the possibility to abiotically generate ribonucleotides and to synthesize non-enzymatically template-directed RNA support the RNA World hypothesis.

RNA can promote various metabolic activities

Tom Cech and Sidney Altman provided the first evidence of RNA being a catalyst [5,6]. Since then naturally occurring ribozymes have been found in prokaryotes, eukaryotes, and viruses. Their functions can be grouped as follows: self-cleaving ribozymes [e.g. hammerhead, hairpin, HDV, VS ribozymes, autocatalytic RNA cleavage in the human β-globin pre-mRNA – so called CoTC element (co-transcriptional cleavage)], self-splicing ribozymes (e.g. self-splicing introns of group I and II, and GRI1 ribozyme), ribonucleases acting in trans (RNase P) and ribozymes displaying peptidyl-transferase activity (ribosome) [11].

However, the existence of the RNA World would require a much larger number of pathways and catalysts that would resemble modern metabolism. Invented over two decades ago, the experimental approach termed SELEX (systematic evolution of ligands by exponential enrichment) represents a powerful tool to select unnatural ribozymes selected in vitro. The overwhelming variety of RNA catalysts that were obtained with the help of the SELEX approach shed light on the potential of catalytic activities in the RNA World [12]. Ribozymes with the ligase activity that utilize a RNA with a 5′-triphosphate as a substrate and form 3′-5′ phosphodiester bonds at the ligation junctions have been isolated [13]. Moreover, the group of Bartel selected a ribozyme capable of extending a primer for more than one helical turn of RNA in a template-dependent polymerization reaction [14]. The activity of this ribozyme resembles the reaction carried out by the RNA dependent RNA polymerase and strengthens the possibility for the existence of a ribozyme acting as an RNA polymerase in the primordial, RNA World. A ribozyme able to conduct specific tRNA aminoacylation has been obtained supporting the idea of RNA-controlled protein synthesis in the RNA World [15]. Redox-ribozymes, resembling the activity of alcohol dehydrogenase have been isolated [16]. Secondary metabolites are thought to be synthesized via Diels–Alder reaction (e.g. enantiospecific carbon-carbon bond reaction). Diels–Alderase ribozyme has been obtained and its structure and mechanism of the reaction was solved [17].

Coenzymes represent small compounds that provide reactive groups for various protein enzymes. Many coenzymes contain ribonucleotide moieties suggesting the existence of coenzyme-binding RNAs in the RNA World and probably represent the present-day cell surviving vestiges of that time. Jadhav and Yarus managed to isolate in frames of the SELEX approach trans-capping ribozymes able to add coenzyme A (CoA) to the 5′ end of RNA containing only a short terminal specific sequence followed by a random internal sequence. From the trans-capped RNA pool they obtained a ribozyme able to transfer CoA via its SH group to biotinylated AMP. These ribozymes also produce crucial metabolic intermediates like acetyl-CoA and butyryl-CoA [18]. From this short overview it is shown that RNA molecules can offer a broad spectrum of catalytic activities necessary to support various metabolic pathways.

First RNA-encoded protein syntheses

Modern cell metabolism is based mainly on protein enzymes carrying out almost the entire repertoire of cellular catalytic reactions. Assuming the existence of the RNA World, how did proteins come into existence in the world controlled by RNA? Firstly, almost all amino acids were present in the primordial world as proved by the experiments simulating the conditions present on the early Earth [19]. Secondly, every step of biological peptide synthesis can be catalyzed by ribozymes [20]: (***i***) ribozymes can aminoacylate the 2′(3′) tRNA termini via 5′ aminoacyl-AMP; (***ii***) a small ribozyme can catalyze peptidyl-RNA synthesis [21], and (***iii***) peptide bond formation is observed in the ribosome, which is a ribozyme [22]. On the other hand, a protoribosome was probably built from RNA and evolved from a duplication of a ribozyme center able to bind aminoacyl-tRNA-like molecules. Having binding sites for two aminoacyl-tRNA-like molecules, one can speculate that the duplicated ribozyme center became capable of peptide bond formation. Assuming the specificity of ribozymes carrying out aminoacyl-tRNA reactions and the development of a ribosome one can easily imagine the origin of the genetic code.

Why proteins substituted RNA catalysts?

There are 20 amino acids that can be incorporated into the peptide chain during ribosomal protein synthesis in modern cells. These amino acids contain more chemical groups than the four nucleotides present in RNA molecules, and thereby offer a broader potential for different functions. Since the majority of amino acids were already present in the primordial soup, one cannot neglect their existence. What was the role of amino acids in the RNA World? One attractive possibility comes from the observations of the modern cell nucleotide de novo synthesis: purine rings are synthesized from aspartic acid, glutamine, and glycine, while the formation of a pyrimidine ring involves aspartic acid. Thus, amino acids could serve in the RNA World as substrates for the synthesis of RNA components. What could be the attractiveness of primitive first RNA-encoded peptides to the RNA world? A synthesis of the first peptides could not lead to the production of efficient catalysts. However, even short peptides composed of serine/leucine motifs are able to form ion-channels across membranes that allows the exchange of metabolites between the membrane-bound riboorganism and the environment [23]. The possibility to exchange the information between the proto-cell and the environment must have preceded the closure of RNA ribozyme metabolic centers within the fatty acid vesicles. Other attractive functions of simple peptides are the stabilization of RNA molecules. In a modern cell practically all RNA molecules form complexes with proteins, referred to as ribonucleoporteins (RNPs). RNase P, snRNPs, hnRNPs, snoRNPs, ribosomes, and telomerases are the best known examples of ribonucleoproteins in which proteins play structural and protective roles (RNase P, snRNP, snoRNP, hnRNP), structural, protective and catalytic roles (ribosome) and a catalytic role (telomerase). The role of proteins within RNPs mentioned above may reflect the evolutionary progress of protein functions. Finally enzymes that bind and use coenzymes as donors of reactive groups could have acted as ribonucleoproteins earlier in the course of evolution. In these RNPs, RNA was systematically shortened and coenzymes having only a ribonucleotide represent a final stage of RNA moiety atrophy.

DNA represents relatively young evolutionary achievement

Apart of several viruses genetic information is stored in DNA molecules. However, in the RNA World era, RNA can mediate both the storage of genetic material and catalytic activity. The comparison of DNA and RNA as genetic information storage material clearly shows DNA more suitable to fulfill these functions. DNA is a double stranded molecule in which antiparallel DNA strands contain complementary nucleotide sequences. Watson–Crick base pairing rule secures an additional control over the nucleotide sequence in both DNA strands. In addition, the lack of a 2′OH group in the deoxyribose renders DNA molecules more stable and resistant to chemical degradation. Thus DNA is more efficient for safe storage of genetic information. Are there any arguments supporting the relatively young age of DNA molecules? Firstly, in the modern cell deoxyribonucleotides are synthesized from ribonucleotides. Assuming RNA building blocks serve as substrates to synthesize deoxyribonucleotides one can speculate that ribonucleotides had to exist earlier than their deoxy- counterparts. Secondly, cellular metabolism uses ATP, CTP, UTP, and GTP but not their deoxy- counterparts in all biochemical reactions. This in turn suggests the existence of cellular metabolism long before DNA was created.

New RNA World

Is RNA playing a minor, secondary role in modern cell metabolism? Did proteins really replace RNA? The essential role of tRNAs, rRNAs, mRNAs, snRNAs, and snoRNAs in RNA metabolism or protein synthesis are known. Moreover, DNA replication is controlled by RNA in two ways: deoxyribonucleotides are made from ribonucleotides and main replicating enzymes, DNA-dependent DNA polymerases require an RNA primer to start DNA replication. Recent discoveries show that RNA molecules also play a crucial role in the regulation of gene expression (small RNAs and long non-coding RNAs, riboswitches) as well as in the boosting of cellular genetic diversity (alternative splicing). This chapter will focus on newly discovered roles of RNA molecules in plant cells.

Small RNA realm

The discovery of small RNAs (sRNAs) acting in the transcriptional and posttrancriptional gene silencing (PTGS) pathway was first reported by Hamilton and Baulcombe in plants and by Fire et al. in *Caenorhabditis elegans* [24,25]. Today there are two main classes of small RNAs operating in the plant cell: microRNAs and siRNAs (small interfering RNAs). The latter can be further divided into multiple categories of which the best-known examples are siRNAs (small interfering RNAs), ta-siRNAs (trans-acting siRNAs), ra-siRNA (repeat-associated siRNAs), and nat-siRNAs (natural-antisense siRNAs).

Small RNAs control gene expression at the posttranscriptional level

MicroRNAs

MicroRNAs represent a class of small RNAs of 18–24 nt in length that are encoded in eukaryotic genomes and posttranscriptionally downregulate the expression of target mRNAs. Plant miRNA genes (*MIRs*) are transcribed by RNA polymerase II and contain a cap structure at the 5′ end and polyA tail at the 3′ end [26]. A specific feature of primary microRNA transcripts is the formation of a stem and loop structure; within the stem microRNA and its partially complementary microRNA* are embedded. A duplex of microRNA and microRNA* is excised in a two-step reaction from the primary transcript by a protein complex referred to as the microprocessor in which DCL1 (Dicer Like 1) enzyme plays a crucial role. miRNA : miRNA* is exported from the nucleus to the cytoplasm where the microRNA is incorporated into the RISC (RNA induced silencing complex) containing Argonaute 1 protein (Ago1, slicer) while miRNA* is in the majority of cases degraded. RISC complex guided by miRNA binds to a complementary mRNA and promotes mRNA cleavage (for review, see [27]). In some cases mRNA is not cleaved but its translation is inhibited [28]. The main targets of miRNAs are genes regulating key developmental processes such as flowering, flower, leaf and root morphogenesis (these are genes encoding mainly transcription factors) and providing plant response to stresses [stress-related genes, e.g. nucleotide binding site–leucine-rich repeat (NBS-LRR)]. MicroRNAs can be regarded as a rapid reaction force that enables fast developmental transitions and responses to various stresses.

Trans-acting siRNAs (ta-siRNAs)

Trans-acting siRNAs are a plant-specific class of siRNAs that were discovered several years ago [29]. Ta-siRNAs are produced from long non-coding RNAs that are derived from *TAS* loci. *TAS* loci-derived transcripts are initially cleaved by a microRNA the product of which is mainly recognized by RNA dependent RNA polymerase 6 (RdRP 6) and converted into the dsRNA. This class of dsRNA is recognized by DCL4 enzyme, which produces phased, 21 nt long ta-siRNAs proceeding in a step-wise manner along the dsRNA [30]. Unlike other siRNAs, ta-siRNAs silence the expression of genes that have little overall resemblance to the genes from which they originate. In this respect they resemble microRNAs. Similarly to microRNAs, ta-siRNAs downregulate posttranscriptionally cellular mRNAs by targeting RISC complex to a complementary mRNA. Trans-acting siRNAs are involved in the regulation of crucial plant developmental processes (e.g. the juvenile-to-adult transition in leaf development, lateral organ development) [31,32]. Increasing experimental evidence suggests that the originally triggered slicing of TAS transcripts leading to the ta-siRNA generation may induce a cascade of secondary siRNA production that targets other mRNAs [33,34]. Multiple steps in ta-siRNA-derived siRNA production enable the cell to precisely and quantitatively control metabolic processes. In this respect, the siRNA cascade resembles well-known cascades of kinases (e.g. MAP kinases) and transcription factors [35].

Small RNAs control plant response to abiotic stress

Natural antisense siRNAs (nat-siRNAs)

Nat-siRNAs (22–41 nt long) are produced from double stranded RNA regions formed by antisense transcripts derived from overlapping genes transcribed in opposite directions. They have been discovered in plants subjected to stress conditions. Generally one of the partially antisense transcripts is expressed constitutively while the other is stress-induced. dsRNA regions are cleaved by DCL2, DCL1/DCL4 thereby producing nat-siRNAs. Their production also requires RdRP6 activity, HYL1 (HYPONASTIC LEAVES 1) as well RNA Pol IV. Their biogenesis pathway is still not fully understood. It is known, that nat-siRNAs target the constitutively expressed mRNA [36–38]. Further studies on nat-siRNAs have revealed the existence of nat-siRNAs derived from partially antisense transcripts encoded in different parts of the genome [39]. Both types of nat-siRNAs are involved in the regulation of plant response to abiotic or biotic stresses.

Small RNAs control chromatin condensation

Repeat-associated siRNAs (ra-siRNAs)

Repeat-associated siRNAs [called also heterochromatic siRNAs (hc-siRNAs)] are a class of 23–24 nt long siRNAs generated from transposons, heterochromatic and repetitive genomic regions, representing the most abundant group of small RNAs in the plant cell [40,41]. They lead to de novo synthesis or maintain already existed DNA methylation of cognate genomic loci keeping them transcriptionally inactive. Biogenesis of ra-siRNA is nucleolar and requires the activity of atypical, plant specific RNA polmerases, RNA Pol IV and RNA Pol V. The requirement for the activity of RNA Pol II has also been shown in at least some cases [42]. RNA Pol IV is thought to transcribe methylated loci and transcripts are converted into double stranded RNA by RNA-dependent RNA polymerase 2 (RdRP 2). DsRNA is cleaved by DCL3 and ra-siRNAs associate with the Ago4 protein [43–45]. Ra-siRNA-guided Ago4 promotes DNA methylation in the cognate DNA regions leading to transposon, heterochromatin and repeat associated regional silencing. Surprisingly, ra-siRNAs can also move over long distances via phloem and guide DNA-target methylation in recipient cells [46]. Ra-siRNAs regulate gene expression via controlling chromatin condensation and control the accessibility of DNA to transcriptional machinery [47].

Small RNAs protect the eukaryotic cell against viral invasions

Short interfering RNAs (siRNAs)

Post-transcriptional gene silencing (i.e. RNAi in plants) represents mainly the viral RNA-targeting host innate immunity pathway. These siRNAs (also called viRNAs) represent a class of 21–24 nt long short RNAs that are produced mainly from duplexes of long RNA molecules having perfect complementarity. These can be found within the plant cell as a result of viral infection and viral RNA replication. dsRNA is recognized mainly by DCL2 and DCL4 family members which cleave the dsRNA producing short duplexes called siRNAs [48]. These are loaded into the RISC complex containing mainly Ago1, a family member of Argoanute proteins. One strand of the siRNA duplex is removed while the other one guides the RISC complex to execute posttranscriptional gene silencing. siRNAs can also be produced as a result of the activity of one of the cellular RNA-dependent RNA polymerases activity (RdRPs) which converts a single stranded RNA (ssRNA) into a double stranded RNA (dsRNA) in the case of transgenes, aberrant transcripts or single stranded long RNA containing hairpin structures (inverted repeats,

IR) expression. Primary siRNAs derived from viral- or transgene-dsRNA may also recruit RNA-dependent RNA polymerases to homologous ssRNA, and after dsRNA generation, to produce secondary siRNAs. This process leads to RNA silencing amplification. In plants, the silencing signal that requires the production of secondary siRNAs moves from cell-to-cell through plasmodestmata (short range signaling – up to 15 cells) or over long distances through the phloem (long-range silencing). Both processes require the activity of cellular RNA-dependent RNA polymerases [49]. The first one likely impacts virus movement while the other one helps immunizing the recipient tissues [50]. Taken together, viRNAs as well as transgene or IR-derived siRNAs, function to combat viral infection or to degrade transgene-derived or aberrant transcripts. Fig. 1 summarizes functions of different small RNA classes in the plant cell.

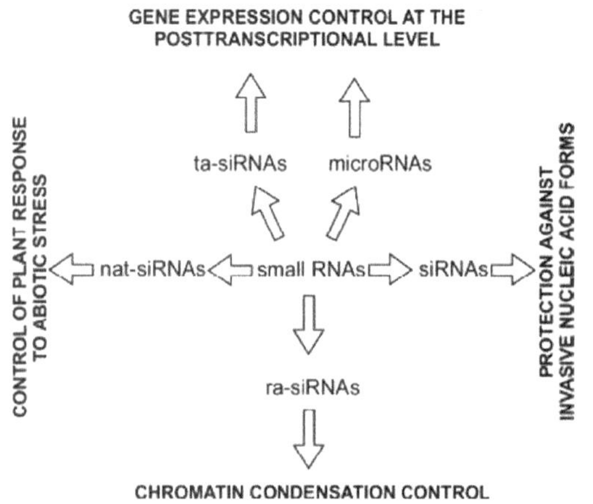

Fig. 1 Small RNA realm. A diagram showing different classes of plant sRNAs and their functions.

Long non-coding RNAs represent crucial regulators of gene expression

In recent years long intergenic non-coding RNAs (lincRNAs) have been identified as crucial regulators of gene expression. In *A. thaliana*, more than 2700 lincRNAs have been detected [51]. Plant lincRNAs are mainly transcribed by RNA Pol II and contain 5′ cap and polyA tail. Some lincRNAs are transcribed by RNA Pol IV and RNA Pol V and are connected with DNA methylation guided by 24 nt siRNAs produced from these transcripts [45]. Recently, Ariel et al have shown that auxin-driven transcription of lincRNA called APOLLO by RNA polymerases II and V controls the formation of a chromatin loop. Within this loop a promoter of a neighbor gene is included. Altering the expression of APOLLO regulates the chromatin loop formation. This control occurs via RNA-dependent DNA methylation, DNA demethylation, and Polycomb complexes. The dynamics of this loop is responsible for oscillatory gene expression pattern in response to developmental cues. This implies the existence of a general type of dynamic regulatory mechanism that enables rapid plant adaptation when exposed to environmental and developmental signals [52].

Riboswitches control cellular metabolism

Riboswitches represent metabolite-sensing domains of RNA [53,54]. They act at the level of transcription, splicing or translation and mediate feedback regulation on biosynthesis and transport of the ligand they sense. They are composed of an aptamer binding domain and a so-called expression platform. The first is responsible for metabolite binding while the other one executes gene expression regulation functions. The variety of ligands that riboswitches are able to bind includes metal ions, nucleobases, nucleosides, coenzymes, and amino acids. Riboswitches also sense physical parameters such as temperature [55]. More than 20 structurally different riboswitches have been identified so far [56]. The majority of riboswitches were found in prokaryotes, however there is one type discovered in filamentous fungi, green algae and land plants. This is a thiamin pyrophosphate (TPP)-binding riboswitch [57]. In all of these TPP-sensing riboswitches studied so far, the control of splicing reactions seems to play a key role in the gene regulatory mechanism. In land plants, the TPP riboswitch is found in the 3′UTR of thiamine metabolic genes. The best-studied example is the TPP riboswitch of *THIC* (*THIAMIN C*) gene. This gene encodes a key regulatory enzyme in the vitamin B1 (thiamin) biosynthesis. The riboswitch regulation involves splice site control by formation of mutually exclusive RNA structures. In the absence of TPP, pre-mRNA 3′UTR forms a structure in which access to the 5′ splice site for splicing machinery is blocked. These results in the presence of an unspliced 3′UTR in the mature *THIC* mRNA yielding stable mRNA and expression of THIC protein. However, in the elevated level of TPP *THIC* pre-mRNA binds this metabolite and 5′ slice site in the 3′UTR becomes accessible for splicing machinery and splicing takes place. As a result, a less-stable THIC mRNA is produced resulting in the downregulation of THIC protein levels (for review, see [58]). The plant TPP-sensing riboswitch as well as other riboswitch examples elucidated in prokaryotes show that a mechanism in which RNA structure is sensitive to an aptamer binding that results in the fine-tuning of cell metabolism. One can easily imagine the existence of such regulatory devices that allowed riboorganisms to sense and respond to environmental cues in the Old RNA World. Thus today operating riboswitches can be also regarded as putative descendants of RNA molecular regulators present already in the early forms of life.

Alternative splicing as a source of genetic variety

Alternative splicing (AS) generates multiple mRNA isoforms from a single primary transcript. There are several types of AS events: exon skipping/inclusion, intron retention, and alternative 5′/3′ splice site selection [59]. The AS mRNA isoforms pattern is regulated by tissue-specific or developmentally/environmentally induced expression of multiple SR proteins and hnRNPs that regulate quantitatively and qualitatively splicing events. In plants, over 60% of intron-containing pre-mRNAs undergo AS to produce a vast repertoire of mRNA isoforms [60]. In humans, more than 95% of protein mRNA coding genes undergo AS, generating multiple mRNA isoforms and proteins [61]. Thus the value of over 60% of protein-coding transcripts undergoing AS events calculated for plants can still be underestimated since not all tissues, developmental stages and life conditions of plants have been studied in this respect. Is AS of functional importance? More and more studies indicate that the functionality of mRNA isoforms, and in many cases consequently isoproteins, is crucial for plant development and response to environmental changes. A transcript of *Arabidopsis ZINC-INDUCED FACILITATOR-LIKE 1* (*ZIFL1*) gene belonging to the major facilitator superfamily (MFS) transporter genes undergoes AS via alternative 3′ splice site selection in one of the multiple introns generating two mRNAs encoding longer and truncated protein forms. AS is tissue-specific and the full-length protein (ZIFL1.1) is detected in the tonoplasts of root cells while the truncated version (ZIFL1.3) – in the plasma membrane of leaf stomatal guard cells. ZIFL1.1 transporter regulates various root auxin-related processes, while ZIFL1.3 isoform mediates drought tolerance by regulating stomatal closure. This example shows that AS is responsible for tissue-specific expression of protein isoforms exhibiting different functions in plant development and response to environmental stimuli [62]. Another example shown in the work of the same group proved that intron retention in the 5′UTR of the novel zinc induced facilitator 2 (ZIF2) transporter enhances translation to promote zinc tolerance in *Arabidopsis*. Two mRNA isoforms with shorter (ZIF2.1, spliced version) or longer (ZIF2.2 intron retaining version) 5′UTR encoding the same protein are obtained. The higher amount of Zn in the soil induces the generation of a longer mRNA isoform, which confers greater zinc tolerance due to enhanced protein synthesis. Moreover, a stem and loop structure that is formed in the case of a longer 5′UTR immediately upstream of the start codon is largely responsible for the zinc-dependent efficient translation [63]. This example shows that AS may generate heavy metal-responsive mRNA isoforms with different susceptibility for translation. Recently published examples by the group of Paula Duque underscore the importance of alternative splicing in the regulatory network of a eukaryotic cell.

Conclusions

The Old and the New RNA Worlds seem to be metabolically connected. In the hypothetical RNA World, RNA molecules played roles of genomes, catalysts, and regulators. Some of these functions in modern cells have been overtaken by other types of molecules, which were most likely created by RNA in the Old RNA World: DNA maintains and protects genetic information, proteins are the main cell catalysts. However, RNA is still playing versatile and crucial roles in the cell. For example, RNA is significant in DNA and protein synthesis and in transcriptome and protein biodiversity. The newly discovered realm of small RNA molecules and riboswitches also show that RNA is a powerful tool in gene expression regulation and in shaping cell metabolism. Fig. 2 summarizes functions played by RNA in the Old and New RNA Worlds.

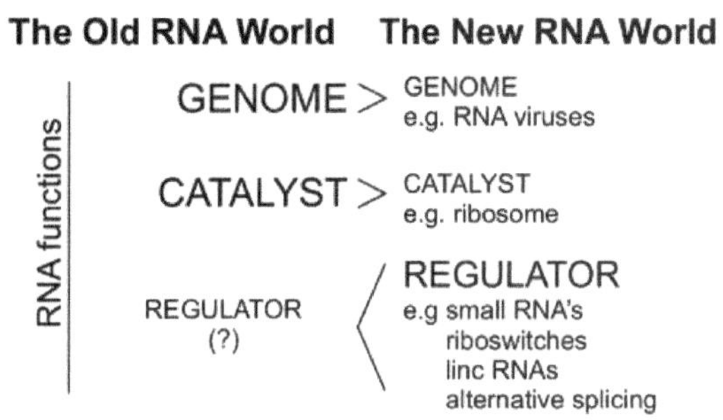

Fig. 2 A scheme presenting RNA functions in the Old RNA World and in the New RNA World.

Acknowledgments
This work was supported by a grant UMO-2013/11/B/NZ1/02099 and KNOW RNA Research Centre in Poznań, 01/KNOW2/2014.

Competing interests
No competing interests have been declared.

References
1. Orgel LE. Evolution of the genetic apparatus. J Mol Biol. 1968;38(3):381–393.
2. Crick FH. The origin of the genetic code. J Mol Biol. 1968;38(3):367–379.
3. Woese CR. The genetic code: the molecular basis for genetic expression. New York, NY: Harper & Row; 1967.
4. Eigen M, Schuster P. A principle of natural self-organization. Naturwissenschaften. 1977;64(11):541–565.
5. Kruger K, Grabowski PJ, Zaug AJ, Sands J, Gottschling DE, Cech TR. Self-splicing RNA: autoexcision and autocyclization of the ribosomal RNA intervening sequence of Tetrahymena. Cell. 1982;31(1):147–157.
6. Guerrier-Takada C, Gardiner K, Marsh T, Pace N, Altman S. The RNA moiety of ribonuclease P is the catalytic subunit of the enzyme. Cell. 1983;35(3 pt 2):849–857.
7. Leslie E. O. Prebiotic chemistry and the origin of the RNA world. Crit Rev Biochem Mol Biol. 2004;39(2):99–123.
8. Powner MW, Gerland B, Sutherland JD. Synthesis of activated pyrimidine ribonucleotides in prebiotically plausible conditions. Nature. 2009;459(7244):239–242.
9. Bowler FR, Chan CKW, Duffy CD, Gerland B, Islam S, Powner MW, et al. Prebiotically plausible oligoribonucleotide ligation facilitated by chemoselective acetylation. Nat Chem. 2013;5(5):383–389.
10. Adamala K, Szostak JW. Nonenzymatic template-directed RNA synthesis inside model protocells. Science. 2013;342(6162):1098–1100.
11. Talini G, Gallori E, Maurel MC. Natural and unnatural ribozymes: back to the primordial RNA world. Res Microbiol. 2009;160(7):457–465.
12. Gold L, Janjic N, Jarvis T, Schneider D, Walker JJ, Wilcox SK, et al. Aptamers and the RNA world, past and present. Cold Spring Harb Perspect Biol. 2012;4(3):a003582.
13. Bartel DP, Szostak JW. Isolation of new ribozymes from a large pool of random sequences. Science. 1993;261(5127):1411–1418.

14. Johnston WK, Unrau PJ, Lawrence MS, Glasner ME, Bartel DP. RNA-catalyzed RNA polymerization: accurate and general RNA-templated primer extension. Science. 2001;292(5520):1319–1325.

15. Lee N, Bessho Y, Wei K, Szostak JW, Suga H. Ribozyme-catalyzed tRNA aminoacylation. Nat Struct Mol Biol. 2000;7(1):28–33.

16. Tsukiji S, Pattnaik SB, Suga H. An alcohol dehydrogenase ribozyme. Nat Struct Mol Biol. 2003;10(9):713–717.

17. Serganov A, Keiper S, Malinina L, Tereshko V, Skripkin E, Höbartner C, et al. Structural basis for Diels-Alder ribozyme-catalyzed carbon-carbon bond formation. Nat Struct Mol Biol. 2005;12(3):218–224.

18. Jadhav VR, Yarus M. Acyl-CoAs from coenzyme ribozymes. Biochemistry (Mosc). 2002;41(3):723–729.

19. Parker ET, Cleaves HJ, Dworkin JP, Glavin DP, Callahan M, Aubrey A, et al. Primordial synthesis of amines and amino acids in a 1958 Miller H2S-rich spark discharge experiment. Proc Natl Acad Sci USA. 2011;108(14):5526–5531.

20. Joyce GF. The antiquity of RNA-based evolution. Nature. 2002;418(6894):214–221.

21. Illangasekare M, Yarus M. A tiny RNA that catalyzes both aninoacyl-tRNA and peptidyl-RNA synthesis. RNA. 1999;5:1482–1489.

22. Steitz TA, Moore PB. RNA, the first macromolecular catalyst: the ribosome is a ribozyme. Trends Biochem Sci. 2003;28(8):411–418.

23. Lear JD, Wasserman ZR, DeGrado WF. Synthetic amphiphilic peptide models for protein ion channels. Science. 1988;240(4856):1177–1181.

24. Hamilton AJ, Baulcombe DC. A species of small antisense RNA in post-transcriptional gene silencing in plants. Science. 1999;286(5441):950–952.

25. Fire A, Xu S, Montgomery MK, Kostas SA, Driver SE, Mello CC. Potent and specific genetic interference by double-stranded RNA in *Caenorhabditis elegans*. Nature. 1998;391(6669):806–811.

26. Xie Z, Allen E, Fahlgren N, Calamar A, Givan SA, Carrington JC. Expression of *Arabidopsis MIRNA* genes. Plant Physiol. 2005;138(4):2145–2154.

27. Szweykowska-Kulińska Z, Jarmolowski A, Vazquez F. The crosstalk between plant microRNA biogenesis factors and the spliceosome. Plant Signal Behav. 2013;8(11):e26955.

28. Brodersen P, Sakvarelidze-Achard L, Bruun-Rasmussen M, Dunoyer P, Yamamoto YY, Sieburth L, et al. Widespread translational inhibition by plant miRNAs and siRNAs. Science. 2008;320(5880):1185–1190.

29. Vazquez F, Vaucheret H, Rajagopalan R, Lepers C, Gasciolli V, Mallory AC, et al. Endogenous trans-acting siRNAs regulate the accumulation of *Arabidopsis* mRNAs. Mol Cell. 2004;16(1):69–79.

30. Allen E, Xie Z, Gustafson AM, Carrington JC. microRNA-directed phasing during trans-acting siRNA biogenesis in plants. Cell. 2005;121(2):207–221.

31. Fahlgren N, Montgomery TA, Howell MD, Allen E, Dvorak SK, Alexander AL, et al. Regulation of *AUXIN RESPONSE FACTOR3* by *TAS3* ta-siRNA affects developmental timing and patterning in *Arabidopsis*. Curr Biol. 2006;16(9):939–944.

32. Zhou C, Han L, Fu C, Wen J, Cheng X, Nakashima J, et al. The trans-acting short interfering RNA3 pathway and NO APICAL MERISTEM antagonistically regulate leaf margin development and lateral organ separation, as revealed by analysis of an *argonaute7/lobed leaflet1* mutant in *Medicago truncatula*. Plant Cell. 2013;4845–4862.

33. Rajagopalan R, Vaucheret H, Trejo J, Bartel DP. A diverse and evolutionarily fluid set of microRNAs in *Arabidopsis thaliana*. Genes Dev. 2006;20(24):3407–3425.

34. Chen HM, Li YH, Wu SH. Bioinformatic prediction and experimental validation of a microRNA-directed tandem trans-acting siRNA cascade in *Arabidopsis*. Proc Natl Acad Sci USA. 2007;104(9):3318–3323.

35. Vazquez F, Hohn T. Biogenesis and biological activity of secondary siRNAs in plants. Scientifica. 2013;2013:1–12.

36. Borsani O, Zhu J, Verslues PE, Sunkar R, Zhu JK. Endogenous siRNAs derived from a pair of natural cis-antisense transcripts regulate salt tolerance in *Arabidopsis*. Cell. 2005;123(7):1279–1291.

37. Katiyar-Agarwal S, Gao S, Vivian-Smith A, Jin H. A novel class of bacteria-induced small RNAs in *Arabidopsis*. Genes Dev. 2007;21(23):3123–3134.

38. Katiyar-Agarwal S, Morgan R, Dahlbeck D, Borsani O, Villegas A, Zhu JK, et al. A pathogen-inducible endogenous siRNA in plant immunity. Proc Natl Acad Sci USA. 2006;103(47):18002–18007.

39. Zhou X, Sunkar R, Jin H, Zhu JK, Zhang W. Genome-wide identification and analysis of small RNAs originated from natural antisense transcripts in *Oryza sativa*. Genome Res. 2009;19(1):70–78.

40. Pikaard CS, Haag JR, Ream T, Wierzbicki AT. Roles of RNA polymerase IV in gene silencing. Trends Plant Sci. 2008;13(7):390–397.

41. Matzke M, Kanno T, Daxinger L, Huettel B, Matzke AJ. RNA-mediated chromatin-based silencing in plants. Curr Opin Cell Biol. 2009;21(3):367–376.

42. Zheng B, Wang Z, Li S, Yu B, Liu JY, Chen X. Intergenic transcription by RNA polymerase II coordinates Pol IV and Pol V in siRNA-directed transcriptional gene silencing in *Arabidopsis*. Genes Dev. 2009;23(24):2850–2860.

43. Smith LM, Pontes O, Searle I, Yelina N, Yousafzai FK, Herr AJ, et al. An SNF2 protein associated with nuclear RNA silencing and the spread of a silencing signal between cells in *Arabidopsis*. Plant Cell. 2007;19(5):1507–1521.

44. Wierzbicki AT, Ream TS, Haag JR, Pikaard CS. RNA polymerase V transcription guides ARGONAUTE4 to chromatin. Nat Genet. 2009;41(5):630–634.

45. Wierzbicki AT, Haag JR, Pikaard CS. Noncoding transcription by RNA polymerase Pol IVb/Pol V mediates transcriptional silencing of overlapping and adjacent genes. Cell. 2008;135(4):635–648.

46. Molnar A, Melnyk CW, Bassett A, Hardcastle TJ, Dunn R, Baulcombe DC. Small silencing RNAs in plants are mobile and direct epigenetic modification in recipient cells. Science. 2010;328(5980):872–875.

47. Vazquez F, Legrand S, Windels D. The biosynthetic pathways and biological scopes of plant small RNAs. Trends Plant Sci. 2010;15(6):337–345.

48. Bouche N, Lauressergues D, Gasciolli V, Vaucheret H. An antagonistic function for *Arabidopsis* DCL2 in development and a new function for DCL4 in generating viral siRNAs. EMBO J. 2006;25(14):3347–3356.

49. Voinnet O. Non-cell autonomous RNA silencing. FEBS Lett. 2005;579(26):5858–5871.

50. Ding S-W, Voinnet O. Antiviral immunity directed by small RNAs. Cell. 2007;130(3):413–426.

51. Liu J, Jung C, Xu J, Wang H, Deng S, Bernad L, et al. Genome-wide analysis uncovers regulation of long intergenic noncoding RNAs in *Arabidopsis*. Plant Cell. 2012;24(11):4333–4345.

52. Ariel F, Jegu T, Latrasse D, Romero-Barrios N, Christ A, Benhamed M, et al. Noncoding transcription by alternative RNA polymerases dynamically regulates an auxin-driven chromatin loop. Mol Cell. 2014;55(3):383–396.

53. Mironov AS, Gusarov I, Rafikov R, Lopez LE, Shatalin K, Kreneva RA, et al. Sensing small molecules by nascent RNA: a mechanism to

control transcription in bacteria. Cell. 2002;111(5):747–756.

54. Winkler W, Nahvi A, Breaker RR. Thiamine derivatives bind messenger RNAs directly to regulate bacterial gene expression. Nature. 2002;419(6910):952–956.

55. Reining A, Nozinovic S, Schlepckow K, Buhr F, Fürtig B, Schwalbe H. Three-state mechanism couples ligand and temperature sensing in riboswitches. Nature. 2013;499(7458):355–359.

56. Roth A, Breaker RR. The structural and functional diversity of metabolite-binding riboswitches. Annu Rev Biochem. 2009;78(1):305–334.

57. Cheah MT, Wachter A, Sudarsan N, Breaker RR. Control of alternative RNA splicing and gene expression by eukaryotic riboswitches. Nature. 2007;447(7143):497–500.

58. Wachter A. Riboswitch-mediated control of gene expression in eukaryotes. RNA Biol. 2010;7(1):67–76.

59. Simpson CG, Manthri S, Raczynska KD, Kalyna M, Lewandowska D, Kusenda B, et al. Regulation of plant gene expression by alternative splicing. Biochem Soc Trans. 2010;38(2):667.

60. Reddy ASN, Marquez Y, Kalyna M, Barta A. Complexity of the alternative splicing landscape in plants. Plant Cell. 2013;25(10):3657–3683.

61. Pan Q, Shai O, Lee LJ, Frey BJ, Blencowe BJ. Deep surveying of alternative splicing complexity in the human transcriptome by high-throughput sequencing. Nat Genet. 2008;40(12):1413–1415.

62. Remy E, Cabrito TR, Baster P, Batista RA, Teixeira MC, Friml J, et al. A major facilitator superfamily transporter plays a dual role in polar auxin transport and drought stress tolerance in *Arabidopsis*. Plant Cell. 2013;25(3):901–926.

63. Remy E, Cabrito TR, Batista RA, Hussein MAM, Teixeira MC, Athanasiadis A, et al. Intron retention in the 5'UTR of the novel ZIF2 transporter enhances translation to promote zinc tolerance in *Arabidopsis*. PLoS Genet. 2014;10(5):e1004375. http://dx.doi.org/10.1371/journal.pgen.1004375

27

Coronilla vaginalis – a species new for the flora of Poland

Zbigniew Mirek, Agnieszka Nikel, Łukasz Wilk*
W. Szafer Institute of Botany, Polish Academy of Sciences, Lubicz 46, 31-512 Kraków, Poland

Abstract

Coronilla vaginalis Lam., a Central European montane species, is reported for the first time from Poland, from the Tatra Mountains. A general map of the taxon's distribution is given, and the location of the newly recorded stand is shown on a map of Poland and the Tatra National Park. Its habitat, calcareous grassland of the order *Seslerietalia*, is characterized in detail.

Keywords: *Coronilla vaginalis*; Poland; Tatra Mountains; calcareous grassland; new locality

Introduction

The vascular plant flora of Poland is so well known that except for native microspecies and alien newcomers we do not expect to discover previously unknown indigenous representatives of this group. All the more interesting, then, to find a new montane species, in such a well-researched area as the Tatra Mountains. The wrong season or bad weather is enough to make one overlook a new species; another visit to the same spot might produce such a record. That was the case with the present find, in Tatra National Park, of *Coronilla vaginalis*, a species not reported from there or anywhere else in Poland. It is a Central European mountain species known from the Western Carpathians but has not been previously reported from its northern section, including the Polish part of the range.

Area of the Tatra Mountains, in particular some calcareous ranges of its western part, it is a place of exceptional accumulation of rare, threatened and unique elements of flora in the scale both of Tatra Mountains, the Carpathians as well as Poland. Apart from Czerwone Wierchy range, undoubtedly the part of Tatra National Park most abounding with occurrence of such kinds of vascular plants [1] also an area of the widely understood massif of the Kominiarski Wierch is significant. This is a place where another species of vascular plants, new for the flora of Poland were found, what is still enhancing this special position of listed area.

New discoveries of this type being all at the same time discoveries of some geographical element are affecting phytogeographical discussions both in regional and general scale, getting in the process out of the side phytogeographical identity of the flora and are supplementing our knowledge about history. This is why such aspects are discussed in the paper.

* Corresponding author. Email: l.wilk@botany.pl

Handling Editor: Zygmunt Kącki

Material and methods

The stand is characterized based on data collected during the 2013 field season. The species was determined from material collected at the locality (fragment of fruiting stem), which is deposited as voucher material in the herbarium of the W. Szafer Institute of Botany, Polish Academy of Sciences (KRAM).

Results

Locality

The newly discovered locality is near the trail leading to the Polana na Stołach glade (ca. 1125 m a.s.l., 49°15′06.6″ N, 19°51′46.6″ E, Datum: WGS-84; Fig. 1).

Habitats

At the recently discovered locality, *Coronilla vaginalis* was found on a rocky calcareous ridge ca. 150 m above the bottom of the Dolina Kościeliska valley (Fig. 2).

The habitat of the plant is calcareous grassland related to *Carici sempervirentis-Festucetum tatrae* with admixture of species characteristic of the alliance *Pulsatillo slavicae-Pinion* (*Erico-Pinion*) [2,3], developed on shallow rendzina soil. The locality is on a very steep slope (50–70°) with a SSE exposure (Fig. 3).

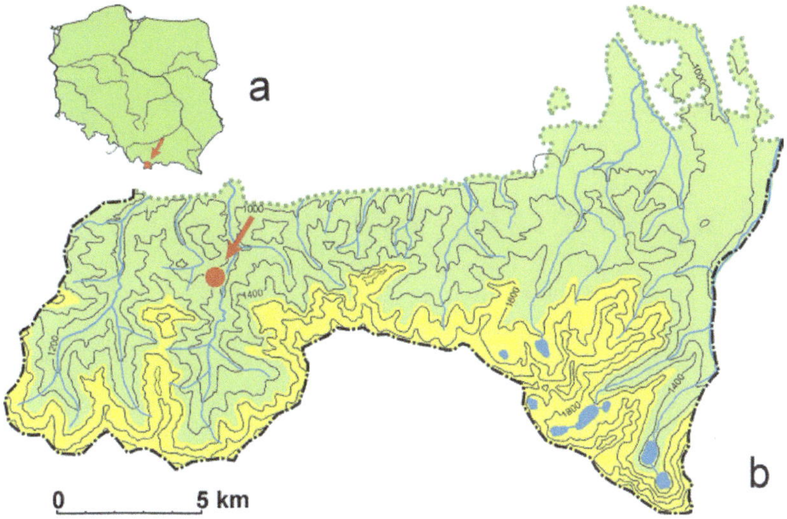

Fig. 1 Locality of the newly discovered stand of *Coronilla vaginalis* on a map of Poland and Tatra National Park.

Fig. 2 View of the locality of *Coronilla vaginalis* (arrow) from the bottom of the Dolina Kościeliska valley.

Fig. 3 Closer view of the habitat of *Coronilla vaginalis*.

Species characteristics

The description provided is based on specimens found in the locality; supplemented with the information from flora of Poland [4], flora of Slovakia [5] and flora of Europe [6].

Coronilla vaginalis Lam., Small Scorpion vetch, is a perennial plant often having a lignified stem base (semi-shrub, chamaephyte), blue-green in color (Fig. 4, Fig. 5).

Stem prostrate to ascending, lignified in lower part, 5–25 cm tall, usually unbranched, leaves petiolate, imparipinnate with 2–6 leaflet pairs. Leaflets relatively thick, somewhat fleshy, with weakly marked venation and scarious margin, obovate, broadly elliptic or roundish, often with asymmetrical wedge-shaped base, rounded at top, more rarely emarginate and mucronulate, 3–12.5 mm long and 2–9 (11) mm wide. Stipules large, membranous, connate, forming a sheath opposite the peduncle base. Head inflorescence, 4–7-flowers, apparently terminally standing on peduncles up to three times (and more) longer than the leaves supporting them. Flowers ±nodding, 8–9 mm long, on pedicels 2–2.5 (3) mm long (±equal to length of calyx). Bracts very small. Calyx campanulate, membranous; sepals very short, broadly triangular, upper ones ±fused. Corolla bright yellow (greenish when dried); petals on ungues slightly longer than tubes of calyx; vexillum (standard) in upper part ±circular, with red-brown venation, cut at top, with small ears at base; wings as long as vexillum or slightly longer, with edges of lamina often folded pouch-like; keel tapering, almost half the length of other petals. Pods hanging or protruding, ±bent, light-brown (often reddish) or blackish purple, 9–35 mm long and 2–3 mm broad, lomentaceous, consisting of (1)3–7(8) segments, 6-angled, 4 of the angles winged. Seeds cylindrical, brown. Flowering June–July.

Coronilla vaginalis differs from its only relative in the flora of Poland, *Coronilla varia* L. [also known as *Securigera varia* (L.) Lassen], mostly by corolla color, which is white, pink or purple in the latter. *Coronilla varia* is a perennial herbaceous plant (hemicryptophyte) with lesser, free and herbaceous stipules; also unlike *Coronilla vaginalis* its pods are erect. Another representative of the genus not occurring in Poland but known from localities close to the country's southern and eastern borders is *Coronilla coronata* L., differing from *Coronilla vaginalis* in having more flowers (12–24) in the heads, longer pedicels (4–6.5 mm), an obovate vexillum (standard) and unwinged, 4-angled pods consisting of 1–4(5) segments.

General distribution

Coronilla vaginalis is a Central European-submediterranean oreophyte (Fig. 6). Its center of distribution is in the Alps and Dinarides, where it is common; the Western Carpathian sites form the northeast section of the range. To the north its distribution reaches central Germany (isolated enclaves in Thuringia and the Franconian Jura). Only a few sites have been recorded in the central and southern parts of the Apennine Peninsula, and a single one (the easternmost) in the Southeastern Carpathians in Romania [7]. Its

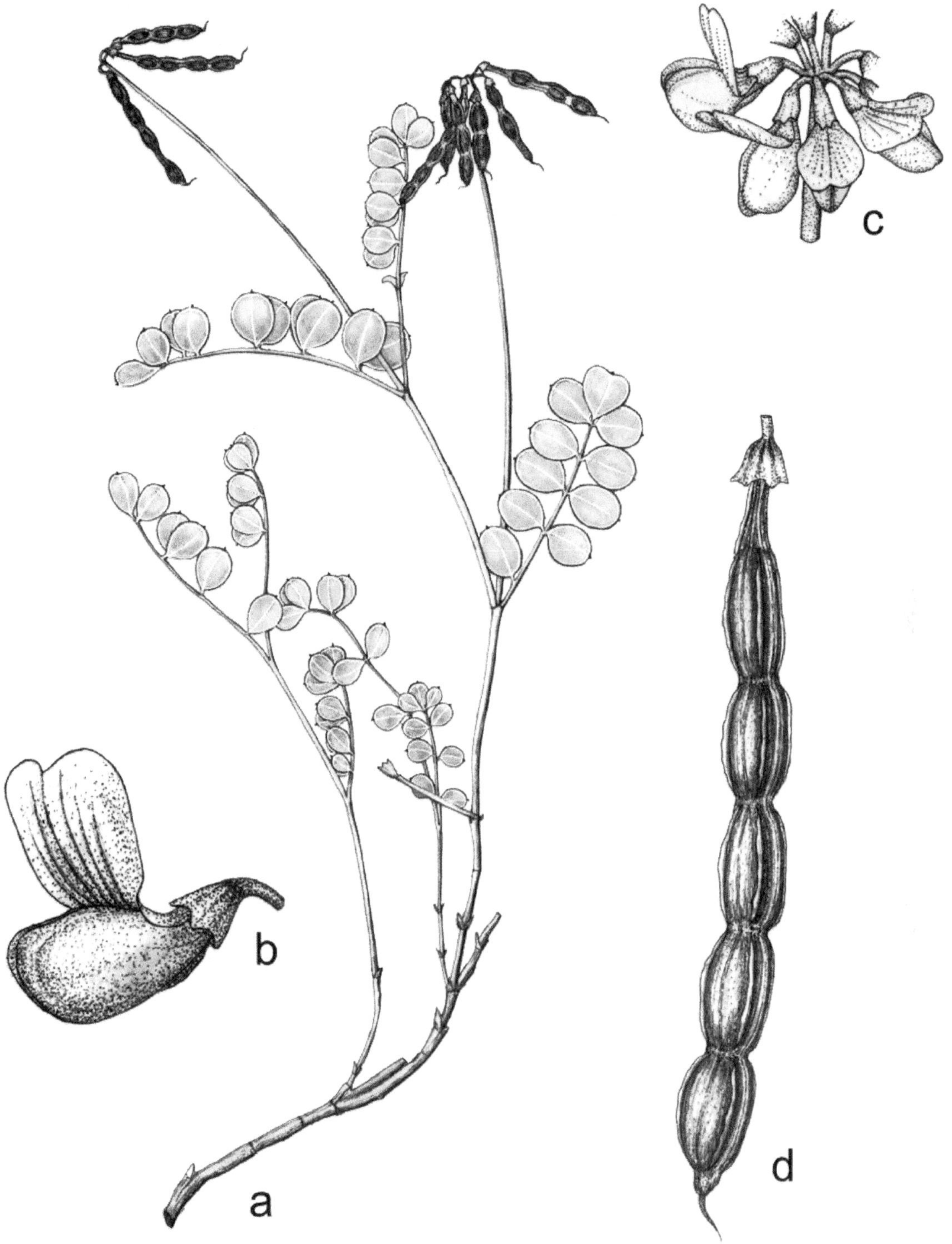

Fig. 4 *Coronilla vaginalis*: habit (**a**), flower (**b**), inflorescence (**c**), pod (**d**). Original drawing based on herbarium sheet (**a,d**) and photos of living individuals (**b,c**) by Jolanta Urbanik.

distribution known thus far, characterized above, includes France, Germany, Switzerland, Austria, the Czech Republic, Slovakia, Hungary, Italy, Slovenia, Croatia, Bosnia and Herzegovina, Montenegro, Albania, Serbia and Romania [6,8], and now also Poland.

Closest sites

The Tatra site is the northeasternmost locality of Coronilla vaginalis across the species range and is not very distant from the relatively compact Western Carpathian section of its range (Fig. 6).

Fig. 5 Living *Coronilla vaginalis* plants at the described locality above the Dolina Kościeliska valley.

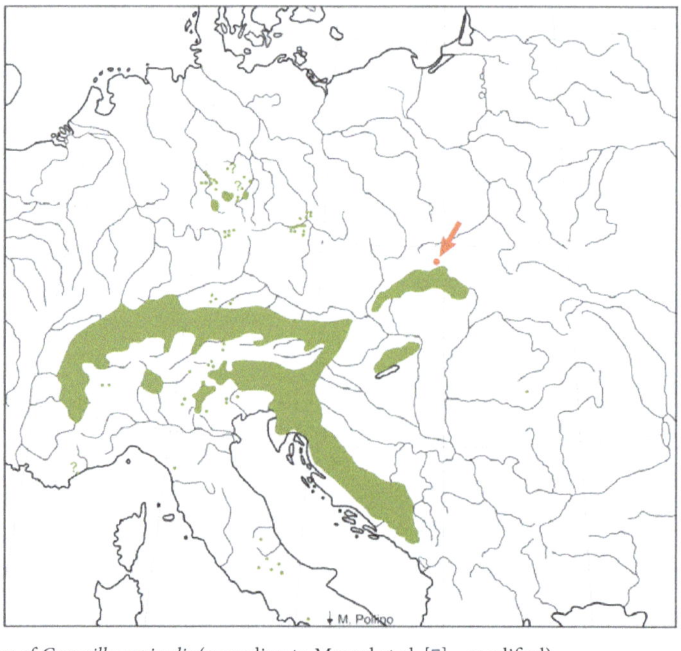

Fig. 6 General distribution of *Coronilla vaginalis* (according to Meusel et al. [7] – modified).

The closest sites in the Liptov Basin, western part of Choč and the Sivý Vrch range [5] lie ca. 20–27 km (straight-line distance) to the south and southwest of the site in the Polish Tatras.

Discussion

Possible origins of the site in the Tatras. Three scenarios might explain the origin of the site above the Dolina Kościeliska valley.

The species is a relict of the warmest postglacial period and moved into the Tatras 7500–7000 years ago [9,10].

The species is a newcomer and has reached the Tatras very recently, perhaps as late as the last quarter-century, in a period of considerable warming and "Atlantisation" of the climate. Some examples argue for that scenario; *Galium hercynicum* Weig. (subatlantic species) is one of them [11,12].

The species was deliberately or accidentally introduced.

It is difficult to resolve the case. One can only assign a higher or lower likelihood to each of these scenarios.The third is least likely. Nearly a century ago, alpine species were planted on experimental plots in the Polana na Stołach glade [13,14], and some of the species planted at that time spread out in the immediate vicinity of the plots [15,16], with some, such as *Crepis aurea*, surviving on that clearing even today. Hence one might wonder whether *Coronilla* could have got here that way; however, the cited sources give detailed lists of the taxa planted, which do not include *Coronilla vaginalis*. Also, the new locality of *Coronilla* is relatively well isolated and distant from the sites of the experimental plots mentioned above. It is true that when the area was under intensive grazing the entire area was penetrated by sheep. That would have mattered to our considerations if there were any mentions of the species having been planted, but there are none. The site where *Coronilla* grows is off the trail and can be visited only sporadically by someone wandering away from it. Lastly, there is no evidence of active reforestation on previously grazed mountain meadows, nor any evidence of the use of seeds originating from outside of Poland. So the third scenario is highly unlikely.

The circumstances presented above, as well as the general phytogeography of the species (Central European-submediterranean montane) indicate that it may have appeared at the newly discovered site spontaneously. This leaves only the first two scenarios, but without evidence for one or the other. The fact that very rare species of similar nature are long known to be present both in the part of the Kominy Tylkowe mountain pasture and in other sections of the West Tatras points to the first scenario positing the locality's older age and relict nature; but as the population of *Coronilla vaginalis* there is extremely sparse – only two or three plants at the discovered site – the locality might be very young indeed. All that can be said with much certainty is that the site most likely is a natural one and that its anthropogenic origin is very unlikely.

Acknowledgments

We thank Jolanta Urbanik for making the drawings. This research was supported by the Polish Ministry of Science and Higher Education, project grant No. N N304 276640 to ZM.

Authors' contributions

The following declarations about authors' contributions to the research have been made: field research, determination of species, writing the manuscript, preparation of distribution maps and illustrations: ZM, AN, ŁW.

Competing interests

No competing interests have been declared.

References

1. Piękoś-Mirkowa H. Rare taxa of vascular plants of the territory of the Tatra National Park – their endangerment by tourism and problems of their conservation. Stud Nat Ser A 1982;22:79–132.
2. Matuszkiewicz W. Przewodnik do oznaczania zbiorowisk roślinnych Polski. Warszawa: Wydawnictwo Naukowe PWN; 2008.
3. Kącki Z, Czarniecka M, Swacha G. Statistical determination of diagnostic, constant and dominant species of the higher vegetation units of Poland. Łódź: Polish Boanical Society; 2013. (Monographiae Botanicae; vol 103).
4. Szafer W, Pawłowski B, editors. Flora polska – rośliny naczyniowe Polski i ziem ościennych. Warszawa: Państwowe Wydawnictwo Naukowe; 1959. (vol 8).
5. Chrtková A, Jasičová M. *Coronilla vaginalis* L a m. In: L. Bertová, editor. Flóra Slovenska. Bratislava: Veda, Vyd. Slov. Akad. Vied; 1988. p. 361–363. (vol 4).
6. Ball PW. Genus *Coronilla* L. In: Tutin TG, Heywood VH, Burges NA, Moore DM, Valentine DH, Walters SM, et al. Flora Europaea. Cambridge: Cambridge University Press; 1968. p. 182–184. (vol 2).
7. Meusel H, Jäger E, Weinert E. Vergleichende Chorologie der Zentraleuropäischen Flora. Jena: Veb Gustav Fischer Verlag; 1965. (vol I).
8. Josifović M, editor. Flora SR Srbije (Flora of the Republic of Serbia). Belgrade: Serbian Academy of Science and Art; 1972. (vol 4).
9. Ralska-Jasiewiczowa M, Latałowa M, Wasylikowa K, Tobolski K, Madeyska E, Wright HE, et al., editors. Late Glacial and Holocene history of vegetation in Poland based on isopollen maps. Kraków: W. Szafer Institute of Botany, Polish Academy of Sciences; 2004.
10. Obidowicz A, Madeyska E, Turner C. Postglacial history of vegetation in the Polish part of the Western Carpathians based on isopollen maps. Kraków: W. Szafer Institute of Botany, Polish Academy of Sciences; 2013.
11. Piękoś-Mirkowa H, Mirek Z. O rzadkich lub dotychczas z obszaru Tatr nieznanych gatunkach roślin naczyniowych. Fragm Flor Geobot. 1978;24(3):363–368.
12. Mirek Z, Piękoś-Mirkowa H. Distribution and habitats of *Galium saxatile* L. in the Carpathians. Acta Soc Bot Pol. 1984;53(3):419–427.
13. Krzemieniewski S. Próby podniesienia uprawy łąk i pastwisk w Tatrach. Rocz Nauk Roln. 1907;3:145–221.
14. Żmuda A. Łąki i hale w Tatrach i próby podniesienia ich uprawy. Pam Tow Tatrz. 1912;33:6–20.
15. Zwolińska Z. Historia pierwszych prób podniesienia wartości pastewnej łąk i hal tatrzańskich oraz ich wpływ na florę Tatr. In: Antoniewicz W, editor. Pasterstwo Tatr Polskich i Podhala. Wrocław: Zakład Narodowy im. Ossolińskich, PAN; 1960. p. 121–135. (vol 2).
16. Mirek Z. Relikty doświadczeń łąkarskich na Polanie Stoły w Tatrzańskim Parku Narodowym. Chrońmy Przyr Ojcz. 1995;51:72–77.

28

Downregulation of chloroplast protease AtDeg5 leads to changes in chronological progression of ontogenetic stages, leaf morphology and chloroplast ultrastructure in *Arabidopsis*

Małgorzata Baranek[1]*, Tomasz P. Wyka[2], Grzegorz Jackowski[1]

[1] Department of Plant Physiology, Institute of Experimental Biology, Faculty of Biology, Adam Mickiewicz University, Umultowska 89, 61-614 Poznań, Poland

[2] Department of General Botany, Institute of Experimental Biology, Faculty of Biology, Adam Mickiewicz University, Umultowska 89, 61-614 Poznań, Poland

Abstract

The chloroplast protein AtDeg5 is a serine-type protease peripherally attached to thylakoid membrane at its lumenal side. Since reliable data regarding the role of AtDeg5 in controlling the course of growth and developmental processes are extremely limited, two independent T-DNA insertional lines with different extent of AtDeg5 reduction were prepared and ontogenesis stage-based analysis performed. Both mutant lines displayed a compensatory overaccumulation of AtDeg8. The repression of AtDeg5 protease altered a range of phenotypic features in at least one of the mutants, with the most prominent being changes in chronological progression of development and growth of individual rosette leaves, flower production and silique ripening as well as in the area of fully expanded leaves and chloroplast ultrastructure. By analyzing the results of parallel-mutant screening we conclude that AtDeg8 overdose may rescue 23% of AtDeg5 deficiency with regard to some AtDeg5-controlled traits; alternatively AtDeg5 may have catalytic sites in excess so that these traits might remain unaltered when AtDeg5 pool is reduced by 23%. For some other AtDeg5-dependent traits the absence of excessive amount of AtDeg5 catalytic sites, lack of AtDeg5 dosage effect and inability of AtDeg8 to compensate deficiency or absence of AtDeg5 occurred.

Keywords: AtDeg5 protease; chronological progression; leaf morphology; chloroplast ultrastructure; starch grains

Introduction

The results of studies performed in recent years show that regulatory protein hydrolysis, catalyzed by proteolytic enzymes is directly or indirectly involved in a majority of cellular processes of all living organisms including plants and it is thus not surprising that the MEROPS database (9.10 release) lists as many as 783 known and putative proteases in *Arabidopsis thaliana*, representing about 3% of all proteins identified in this taxon. Proteolysis is considered a vital factor influencing whole-plant maintenance, structure and functions under changing environmental conditions and in changing ontogenetical context [1]. It is widely believed that proteases are involved in protein quality control and protein turnover processes. The protein quality control comprises a hydrolysis of proteins which have been damaged due to mutations or exposition of plants to stressing environmental conditions, synthesized at redundant quantity or sorted to incorrect cell compartment. Protein turnover, in its turn comprises a hydrolysis of proteins which become unnecessary in a defined spatio-temporal context under non-stressing conditions.

Deg proteases were first discovered and described in late 1980s by studying *Escherichia coli* mutants unable to grow at elevated temperatures (>37°C) due to their inability to dispose of periplasmic proteins damaged under heat shock conditions [2,3]. In fact the quality of *E. coli* periplasmic proteins is controlled by three Deg proteins namely DegP, DegQ and DegS whose structure and function have been solved in much detail [4–6]. Later Deg proteases were found to occur ubiquitously in cells of all domains of life, from Bacteria through Archea to Eukarya [7]. Sixteen genes coding for proteins orthologous to DegP, Q and S have been identified in the *A. thaliana* nuclear genome and designated *AtDEG1-16* [8]. It has been experimentally proven that five of the genes (*AtDEG1, 2, 5, 7* and *8*) encode proteins targeted to chloroplasts, one (*AtDEG10*) encodes a mitochondrion-targeted protein and another one (*AtDEG15*) a peroxisomal protein. The remaining nine *AtDEG* genes code for proteins which have been predicted in silico to be targeted either to

* Corresponding author. Email: gosiab@amu.edu.pl

Handling Editor: Beata Zagórska-Marek

mitochondria or chloroplasts or their localization has not been predicted in silico yet (for a review see [9]).

It was demonstrated that a linear structure of chloroplast-targeted AtDeg proteases involves one or two protease domains (if the second protease domain is present – it is a catalytically inactive version [10]) and 0 to 4 of PDZ domains which are thought to mediate formation of proteolytically active oligomeric forms [11] and select proper substrates [12]. In fact, several lines of evidence show that chloroplast-targeted AtDeg proteases exist as both homo- and heterooligomeric assemblies with trimer as a basic structural unit. For instance, a catalytically active AtDeg1 particle was demonstrated to be a hexamer of a "sealed-cage" type, consisting of two trimeric rings stacked upon each other in such a way that they form an inner, catalytic chamber. A hexamer is stabilized through interactions of protease and PDZ domains originating from adjacent monomers of opposing trimers [11]. In turn AtDeg5 forms trimers in which individual monomers are associated through hydrophobic interactions, hydrogen bonds and π-cation interaction [13]. AtDeg5 and AtDeg8 have been suggested to make assemblies mediated by association of equimolar amounts of homooligomers of both molecules [14].

In spite of a significant progress which has been made with regard to understanding structural organization of chloroplast-targeted AtDeg oligomers the knowledge of their functions is extremely limited both in terms of identity of chloroplast proteins representing native substrates for AtDeg proteases under stressing or non-stressing conditions and in terms of their role in controlling the course of growth and developmental processes at various levels of plant body organization. Specifically, it was demonstrated for AtDeg1, 7 and 8 that recombinant versions of these enzymes catalyze in vitro degradation of photodamaged PSII reaction center protein PsbA (D1) which in vivo is inherent to PSII repair cycle. The results have suggested that the photodamaged PsbA protein is cleaved in vivo by introduction of multiple cuts by cooperating AtDegs 1, 5/8, 7 and probably AtDeg2 as well [12,14–17]. Similarly, recombinant version of AtDeg1 and AtDeg7 has been demonstrated to be responsible for in vitro degradation of PsbD, PsbS, Lhcb4 and cytb6 as well as PsbB-D, respectively under photoinhibitory conditions [12,18]. Later it was demonstrated in a more direct way that Lhcb6 and PsbF apoproteins are degraded in a stress-related manner by AtDeg2 and AtDeg5, respectively [16,19]. Reliable data on the role of chloroplast targeted-AtDegs in controlling the course of growth and developmental processes at various levels of the plant body plan are very scarce – in two separate studies using 4-weeks-old long day grown wild type (WT) plants and mutants devoid of AtDeg2 and AtDeg5 we have suggested recently that AtDeg2 and AtDeg5 have some functions in controlling morphology and anatomy of leaves as well as the chloroplast life cycle under non-stressing conditions [16,19].

To gain a deeper insight into the functional significance of AtDeg chloroplast proteases, and AtDeg5 in particular, we investigated AtDeg5 involvement in plant growth and development by a large-scale, comparative phenotypic screen of WT plants and two *deg5* mutants deficient in AtDeg5 protease when growing under non-stressing conditions. As the extent of the AtDeg5 reduction varied between 23% (*deg5-2*) and 100% (*deg5-1*) in this way we could study the importance of AtDeg5 protein dosage effect on phenotypic traits. We demonstrate here that AtDeg5 is needed for normal plant ontogenesis since AtDeg5 repression produces significant changes in chronological progression of some growth and development stages, leaf morphology, chloroplast ultrastructure and diurnal transitory starch metabolism. Our results show that AtDeg5 deficiency/absence leads to a compensatory overaccumulation of AtDeg8 – this protease may be functionally interchangeable with AtDeg5, possibly with regard to expression of some phenotypic alterations or AtDeg5 has enough catalytic sites in excess to function normally even when its pool is lowered to 77% of the WT plants value. In the case of other phenotypic traits which are altered in *deg5* mutants AtDeg8 is unable to compensate AtDeg5 deficiency or absence.

Material and methods

Plant material and growth conditions

Seeds of WT and mutant plants were sown on ½ Murashige and Skoog [20] medium with agar (for early analyses) or 42 mm Jiffy peat pellets on sphagnum peat moss and wood pulp (AgroWit, Przylep, Poland; for soil-based analysis) and grown in a growth chamber (NEMA, Netzschkau, Germany) under long day conditions (16 h : 8 h, light : dark), with the irradiance set at 110 μmol quanta m^{-2} s^{-1}, at a 22°C (day)/18°C (night) temperature cycle and 65% relative humidity. Seeds of *Arabidopsis thaliana* insertion lines SAIL_645_D07 (*deg5-2* mutant) and SALK_099162 (*deg5-1* mutant) were obtained from Nottingham Arabidopsis Stock Center (NASC). T-DNA insertions in *DEG5* gene (At4g18370) were confirmed and hetero or homozygosity was analyzed by PCR using the following primers for *DEG5* sequence: forward 5′- GAGGATGAAGAAGAGAGAAATG-3′ and reverse 5′-CCCAACCCACTTACTACCTAC-3′ for *deg5-2* mutant, forward 5′-GCTTTTTCCTCAATCTTCAATAC-3′ and reverse: 5′-AGGATTTAGTTCACGTCCCTC-3′ for *deg5-1* mutant, and LBb 5′-GCGTGGACCGCTTGCTGCAACT-3′ for the insertion.

Preparation of thylakoids samples

Intact chloroplasts were prepared from leaves of WT or mutant plants using centrifugation on Percoll gradient, essentially as described in [21]. Thylakoids were isolated by resuspending the pellet of intact chloroplasts in water, centrifuging the suspension for 10 min at 12 500 *g* and collected as a green pellet, finally suspended in 10% glycerol and stored frozen at −18°C.

SDS-PAGE, immunoblotting and protein quantitation

Thylakoid samples were resolved by SDS-PAGE [22] on the basis of equal amount of chlorophyll, electrotransferred onto PVDF membranes and immunostained with primary antibodies against Deg1, Deg2, Deg5, Deg8 and PsbC apoproteins. The immunostaining and the quantification of immunostained bands were performed essentially as described by us previously [23].

Chronological progression of selected principal and secondary stages of growth and development of WT and mutant plants

Chronological progression of seed germination (principal stage 0, [24]) was analyzed by light microscopy (observations performed two times a day) of seeds growing on agar plates. The secondary stages i.e. the time needed to reach seed imbibition, radicle emergence and hypocotyl/cotyledon emergence were determined.

Chronological progression of leaf development (principal stage 1, [24]) was analyzed by visual inspection (performed two times a day) of seedlings growing on Jiffy peat pellets. The secondary stages i.e. the time when cotyledons were fully opened, leaves 1 and 2 and all remaining leaves (3–9) became longer than 1 mm, were determined.

Chronological progression of leaf growth (principal stage 3 defined in a modified way, see "Results" for details) was evaluated by light microscopy (the data were collected once a day). The secondary stages were determined for each leaf individually as the time needed to reach 10%, 50% and 100% of its final size.

Progression of inflorescence emergence (principal stage 5, [24]) was analyzed by visual inspection (performed two times a day). The only secondary stage, i.e. the time when the first flower bud became longer than 1 mm, had been determined.

Chronological progression of flower production and silique ripening (principal stages 6 and 8, respectively, [24]) were determined by visual inspection (performed two times a day). The secondary stages of the principal stage 6 were determined as the time when the first flower as well as 10%, 50% and 100% of flowers opened and the secondary stages of the principal stage 8 as the time when the first silique as well as 10%, 50% and 100% of the siliques became shattered.

Microscopy

Whole leaves were digitally photographed using a SteReo Lumar V12 stereomicroscope (Carl Zeiss, Jena, Germany) at 6.4× magnification, and the images were analyzed with AxioVision Rel 4.8 (Carl Zeiss) and leaf length, width, area, perimeter and shape coefficient (defined according to [25]) were determined.

Transmission electron microscopy of the leaves (JEM 1200 EXII, JEOL, Tokyo, Japan), the negative scanning and digital image analysis were performed essentially as described by us earlier [16].

Scanning electron microscopy of the seeds (EVO 40, Carl Zeiss Microscopy GMBH, Jena, Germany) was performed according to [26], the negative scanning and digital image analysis was done essentially as described in [16].

Starch quantification

Leaves of WT and mutant plants were harvested at various time points during the day, night and an extended night. The leaves were frozen in liquid nitrogen and soluble carbohydrates were extracted by placing samples into boiling 80% ethanol and incubating in 80% ethanol for 18–20 h at 37°C. The ethanol fraction was removed by centrifugation (10 min at 8000 *g* at 4°C) and the dried pellet was homogenized in 0.2 N KOH and boiled for 1.5 h. Then the mixture was acidified to pH 5.0, incubated with amyloglucosidase (10 U) for 24 h at 55°C, centrifuged (for 3 min at 8000 *g*) and the glucose was measured enzymatically as described in [27]. The amount of glucose was calculated against a glucose standard. To determine starch the glucose amount was multiplied by 0.9 coefficient.

Gas exchange analysis

Determination of A/C_i responses was performed using the Li-6400 gas exchange system equipped with the standard 6 cm^2 chamber with a built-in red-and-blue LED light source (Li-Cor Inc., Lincoln, NE, USA). For about 60 min prior to the measurement plants were exposed to 500 μmol quanta m^{-2} s^{-1} and the measurements were conducted at 1500 μmol quanta m^{-2} s^{-1} while leaf temperature was maintained at 22°C and vapor pressure deficit in the chamber did not exceed 1.4 MPa. To generate A/C_i response curves, photosynthetic rate (A) was measured at 400, 300, 200, 100, 75, 50 μmol mol^{-1} CO_2, followed by restoration of initial photosynthetic rate at 400 μmol mol^{-1} CO_2. Further measurements were done at 500, 750, 1000, 1500 and 1800 μmol mol^{-1} CO_2. Values of A were corrected for diffusional leaks as determined with an empty chamber using the same range of CO_2 concentrations.

Leaf areas were determined using a flatbed scanner and an image processing software. Since leaves were smaller than the chamber area and varied in size and shape, leaf boundary layer conductances were determined individually for each leaf. Leaves were photocopied, leaf silhouettes were cut out of filter paper, moistened and steady-state evaporation was measured. Boundary layer conductance was computed using the Li-6400 routine. Internal concentrations of CO_2 (C_i) were calculated using equations provided in the Li-6400 manual (Li-Cor 2005).

Equations of Farquhar et al. [28] two-phase model were fitted to the A/C_i data, with measurements at $C_i < 400$ μmol mol^{-1} CO_2 considered to be limited by Rubisco amount, kinetics and activity whereas those at $C_i > 400$ μmol mol^{-1} CO_2 to be limited by electron transport used in the regeneration of ribulose 1,5-bisphosphate (RuBP) [29]. Fitting was done with the maximum-likelihood based curve fitting routine of SigmaPlot 11 (Systat Software, San Jose, CA, USA) using kinetic parameters of Rubisco given in [30]. Maximal carboxylation rate Vc,max and maximal electron transport rate Jmax were obtained as parameters in the best fit models. Both $V_{c,max}$ and J_{max} were normalized to 25°C values using equations given in [31].

Chlorophyll determination

Chlorophyll concentration was assayed spectrophotometrically according to [32].

Statistical analysis

Statistical significance of differences in measured parameters were analyzed according to one-way analysis of variance (ANOVA) and Tukey–Kramer multiple comparison test. Means were regarded as significantly different at $P < 0.01$.

Results

T-DNA insertion mutants for *AtDEG5*

To study AtDeg5 functions seeds of the insertion SAIL_645_D07 line (*deg5-2*) were obtained from the Nottingham Arabidopsis Stock Centre. For comparison our study also included SALK_099162 line (*deg5-1*) which was studied previously [14,19]. For *deg5-2* a confirmed position of the T-DNA insertion is in the 4th intron 872 bp downstream of ATG, whereas the *deg5-1* line contains an insertion in the 2nd exon of *At4g18370* gene 429 bp downstream of ATG codon (Fig. 1a). Homozygous mutants of both lines were identified in F2 progeny by PCR-based genotyping (Fig. 1b). In accordance with earlier studies [19] the *deg5-1* mutant was found to be entirely depleted of AtDeg5 protein, whereas *deg5-2* plants showed a modest (but significant) reduction of AtDeg5 level, to 77% of the value found in WT plants (Fig. 2). No changes in the amount of AtDeg1 and AtDeg2 proteins have been detected in both mutants vs. WT plants. Interestingly, the amount of AtDeg8 increased about 2-fold in the mutant lines (Fig. 2).

Chronological progression of ontogenetic stages in WT and *deg5* mutants

As a first step in performing comparative phenotypic screens a chronological progression of selected stages covering the plant ontogenesis from seed imbibition to fruit ripening has been studied for homozygous *deg5-2* and *deg5-1* mutants as well as WT plants growing under a 16/8-h photoperiod. The study involved five principal stages defined essentially according to BBCH scale [33] as adapted for *Arabidopsis* phenotypes [24], namely: 0 – seed germination, 1 – leaf development, 5 – inflorescence emergence, 6 – flower production and 8 – silique ripening and their secondary stages. In the case of the principal stage 3 an altered BBCH-based definition was developed by us allowing to recognize growth of individual leaves – instead of rosette growth – by introducing numbers indicating accomplishment of various percentages of final sizes by individual leaves (= completion of various secondary stages). Thus the secondary stages of the principal stage 3 are coded by three digits representing the principal stage, the secondary stage and the leaf number, respectively (e.g. 3.45 means the moment when the secondary stage 4 of the principal stage 3 was reached by leaf 5). Fig. 3 shows the time (in days) required to complete individual secondary stages of growth and development by WT plants and the mutants. No significant alterations in chronological progression of secondary ontogenesis stages were observed for *deg5-2*. In contrast, a variety of WT/*deg5-1* significant differences were seen involving progression of secondary stages of principal stages 1, 3, 6 and 8. Rosette leaves 1 and 2 as well as 5 of *deg5-1* mutant appeared (secondary stages 1.2 and 1.5) before and after respective leaves of WT plants, indicating that the mutant produced these leaves at an altered rate. The majority of rosette leaves of *deg5-1* mutant (i.e. the leaves 1, 3, 4, 5, 6 and 7) reached their final size 2–6 days later than WT plants, for instance the leaf 1 accomplished its final size (secondary stage 3.91) five days later than the same leaf of WT plant. The remaining leaves of the mutant (2, 8 and 9) were delayed in comparison to leaves of WT plants

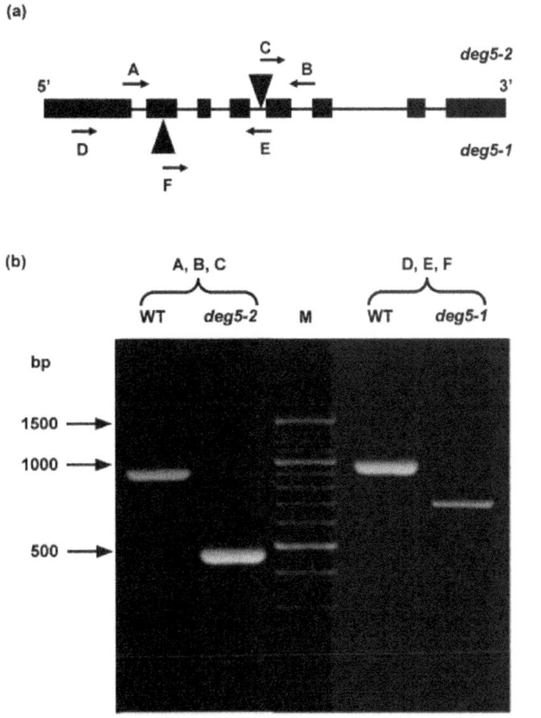

Fig. 1 Identification of *deg5* mutants. **a** Schematic diagram of genomic *AtDEG5* gene. Black boxes represent exons, thin lines – introns and triangle – site of T-DNA insertion. Arrows indicate the annealing sites of primers used for PCR analysis of genomic DNA. **b** Confirmation of homozygosity of *deg5-2* and *deg5-1* mutants by electrophoretic separation of PCR products of genomic DNA of WT and mutant plants. Amplification was performed with primers A, B, C, D, E and F according to the descriptions in Fig. 1a.

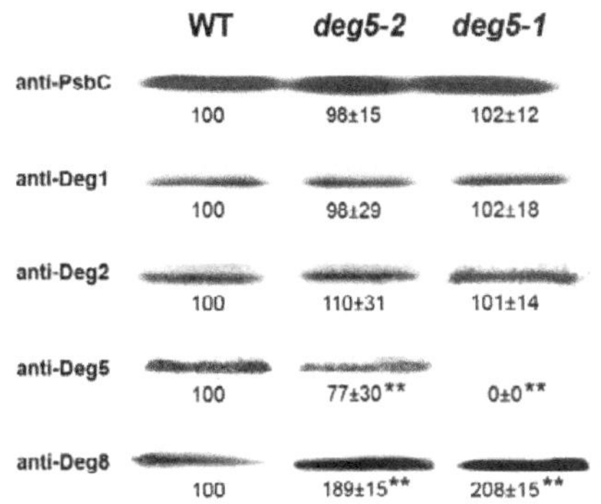

Fig. 2 Immunoblot analysis of levels of chloroplast AtDeg proteins in leaves of WT plants and *deg5* mutants. Samples of the thylakoids isolated from leaves of WT and *deg5-2* and *deg5-1* mutant plants which reached secondary stage 6.0 were subjected to immunoblot analysis with anti-Deg1, Deg2, Deg5, Deg8 and PsbC antibodies. Individual signals of AtDeg proteases or PsbC apoprotein in mutant thylakoids were quantified relative to signals identified for thylakoids of WT plants (100%) using Gelix One software (Biostep, Jahnsdorf, Germany). Signal strength ±*SD* is indicated below the corresponding lanes. The signals were determined by scanning six blots representing two biological replicates. Asterisks indicate the data for which mutant/WT plants differences were significant ($P < 0.01$).

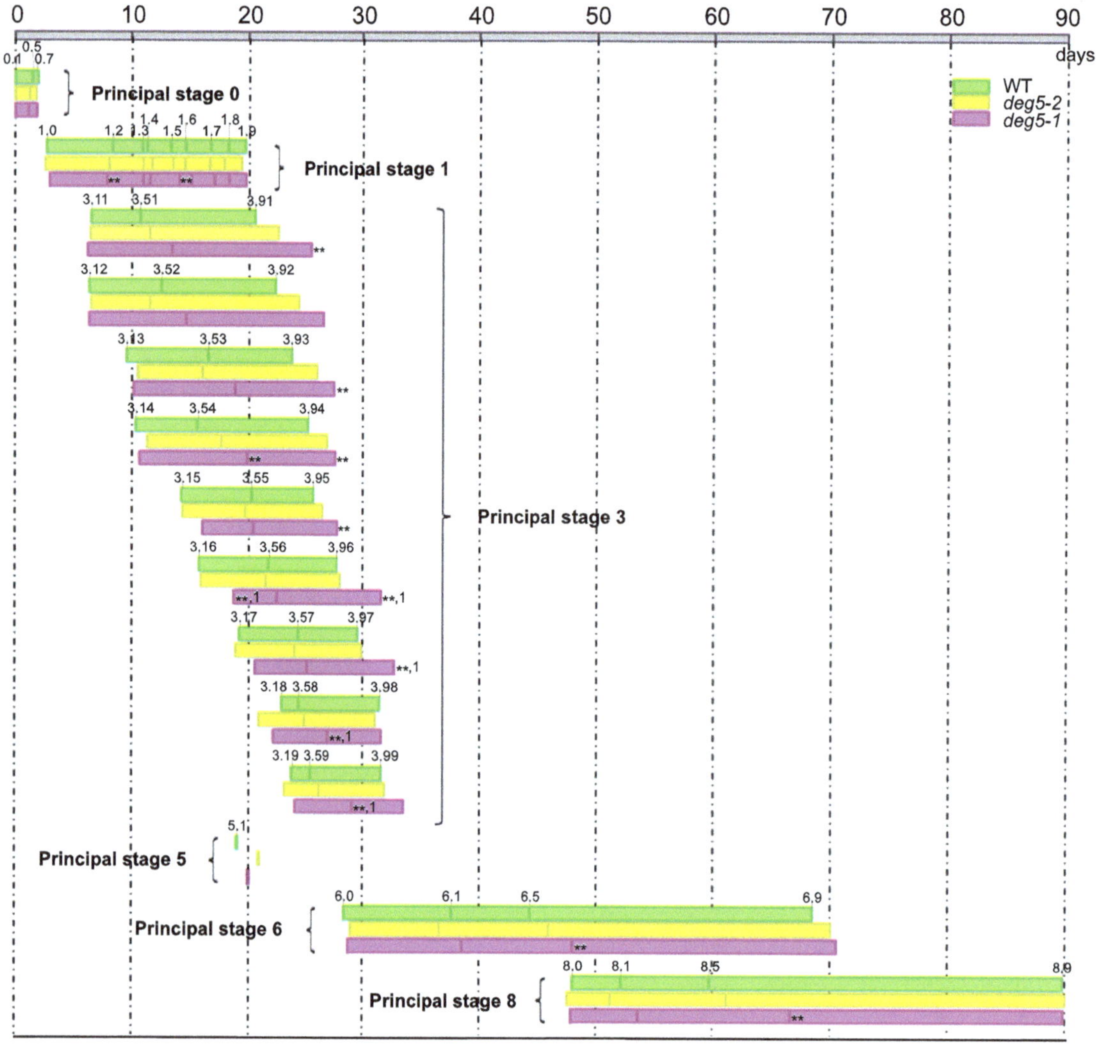

Fig. 3 Analysis of the chronological progression of the principal stages 0, 1, 3, 5, 6 and 8 of plant ontogenesis in WT and *deg5* mutant plants. Horizontal, colored boxes indicate the time (in days) which is necessary for WT and *deg5-2* and *deg5-1* mutant plants to complete individual secondary stages (marked as small vertical bars within colored boxes). Numbers above the colored boxes correspond to secondary stages of ontogenesis. The asterisks indicate the data for which mutant/WT plants differences were significant ($P < 0.01$) and 1 indicate the values for which *deg5-1*/*deg5-2* differences were significant ($P < 0.01$).

as well but the delay was not significant. Furthermore, some leaves of the *deg5-1* mutant (No. 4, 8 and 9) were delayed transiently at the moment when they reached 50% of their final sizes. A delay in growth rate of the majority of leaves of the *deg5-1* mutant was accompanied by a modest delay in flower production and silique ripening – as a result 50% of the flowers opened (secondary stage 6.5) and 50% of the siliques shattered (secondary stage 8.5) significantly later than in WT plants.

Morphology of fully expanded leaves, ultrastructure and diurnal metabolism of transitory starch of chloroplasts of fully expanded leaves of WT and mutant plants

To investigate the effect of the delay in growth exhibited by the majority of leaves of *deg5* mutants on final leaf morphology digital images of the leaves of both mutants and WT plants were collected and analyzed at secondary stage 6.0, i.e. the moment when the first flower was opened [24]. The leaves 3 and 5 were selected for these analyses since they represent juvenile and mature leaf populations, respectively [34] and AtDeg5 deficiency could have different impact on their morphological features (both leaves became fully expanded at or before secondary stage 6.0 both in mutants and WT plants). The morphological properties of *deg5-2* mutant leaves were found to be unchanged relative to those of WT plants whereas *deg5-1* mutant plants exhibited a considerable increase in the area and perimeter of both juvenile and mature leaves relative to WT plants and this was due to relevant increases in the leaf length and width (Fig. 4). Since the increases in length and width of the leaf blades were roughly proportional leaf shape coefficients remained unaffected. No statistically significant *deg5-1*/*deg5-2* differences with respect to morphological features of either juvenile or mature leaves were identified.

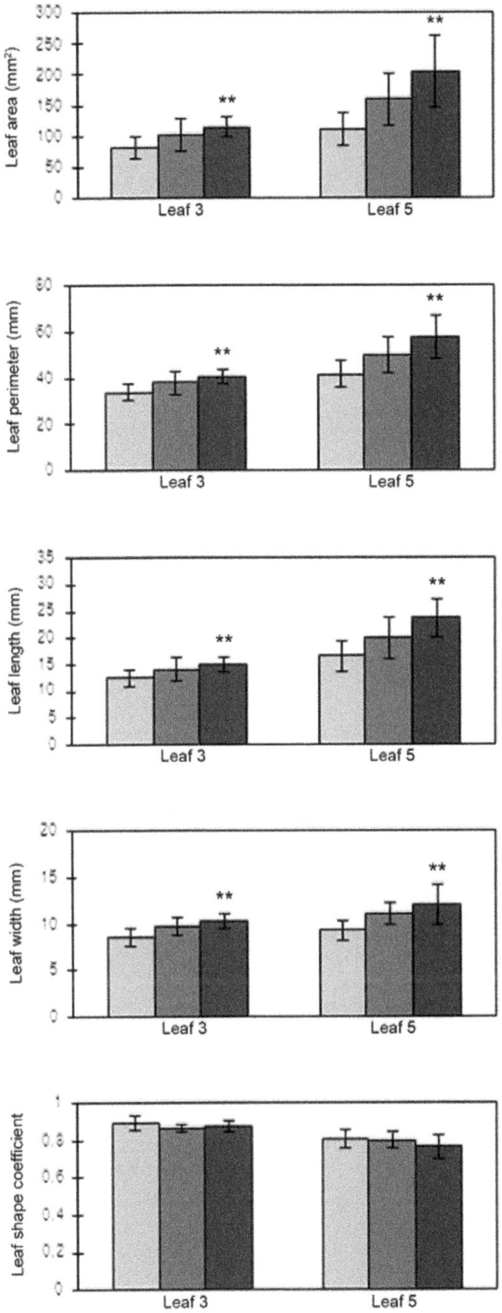

Fig. 4 Analysis of the morphology of leaves of WT (pale grey bars) and *deg5* mutant plants (*deg5-2*, dark grey bars; *deg5-1*, black bars). Leaf area, perimeter, length, width and shape coefficient were measured for juvenile (leaf 3) and mature (leaf 5) leaves of WT as well as *deg5-2* and *deg5-1* mutant plants that had reached secondary stage 6.0. Asterisks indicate the data for which mutant/WT plants differences were significant ($P < 0.01$). Values are mean ±*SD*.

To further understand the consequences of the loss of chloroplast protease AtDeg5 transmission electron microscopy was applied to investigate the ultrastructure of juvenile and mature leaves of both mutants and WT plants which reached secondary stage 6.0. Significant differences were found in the thylakoid system architecture, namely mature leaves (but not juvenile ones) of the mutants had more thylakoids stacked per granum and no intermutant differences were found to exist in this respect (Fig. 5); however, the thylakoid width was not altered in leaves of mutants vs WT plants. A striking difference was that juvenile leaves of *deg5-2* mutant as well as both juvenile and mature leaves of *deg5-1* showed an extensive transitory starch accumulation at the end of the night period while, according to the expectations, no starch grains were observed in chloroplasts of WT plants. To provide more detailed information about the impact of deficiency/loss of AtDeg5 on leaf transitory starch metabolism diurnal changes in contents of starch were measured in leaves of juvenile and mature leaves of both mutants and WT plants. It was found (Fig. 6) that the rate of accumulation of starch in the light in mature leaves of *deg5-1* and juvenile leaves of both *deg5-1* and *deg5-2* mutants was higher than in WT plants and as result starch amount in mutant chloroplast reached significantly higher values at the end of the light period (16 h).

During the night the starch pool declined at approximately equal rates in both types of leaves of all three plant genotypes thus at the end of the dark period more unhydrolized starch was left in chloroplasts of mutant plants although the mutants were shown to be able to hydrolyze starch completely during an extended night. Thus the starch excess phenotype of the *deg5* mutants seems to be caused by their ability to synthesize starch during light period at an elevated rate in particular leaves.

A/C_i responses of fully expanded leaves of WT and mutant plants

To address the question why *deg5* mutants synthesize excessive amounts of starch during the light period, A/C_i responses were recorded at saturating irradiance for mature leaves of deg mutants and WT plants (no reliable measurements of photosynthetic gas exchange was possible with juvenile leaves due to their small area). These data allowed us to calculate the apparent maximal carboxylation rates limited by Rubisco amount, kinetics and activity ($V_{c,max}$) and maximal carboxylation rates limited by electron transport used in the regeneration of RuBP (J_{max}). Mature leaves of WT plants and *deg5* mutants exhibited very similar A/C_i curve patterns and $V_{c,max}$ and Jmax values were not statistically different among all three genotypes (Fig. 7).

Interpretation of mutant vs. WT plants differences in terms of AtDeg5 dosage effect and compensatory accumulation of AtDeg8

To facilitate interpretation of the mutants vs WT differences in terms of importance of AtDeg5 dosage and compensatory AtDeg8 accumulation, 33 phenotypic traits that are firmly established as AtDeg5-related since they are significantly altered significantly in at least *deg5-1* (i.e. at least in the null mutant *deg5-1*) were classified into three groups according to similarity in the profile of the alterations (Tab. 1). The remaining traits, i.e. those not altered in any of

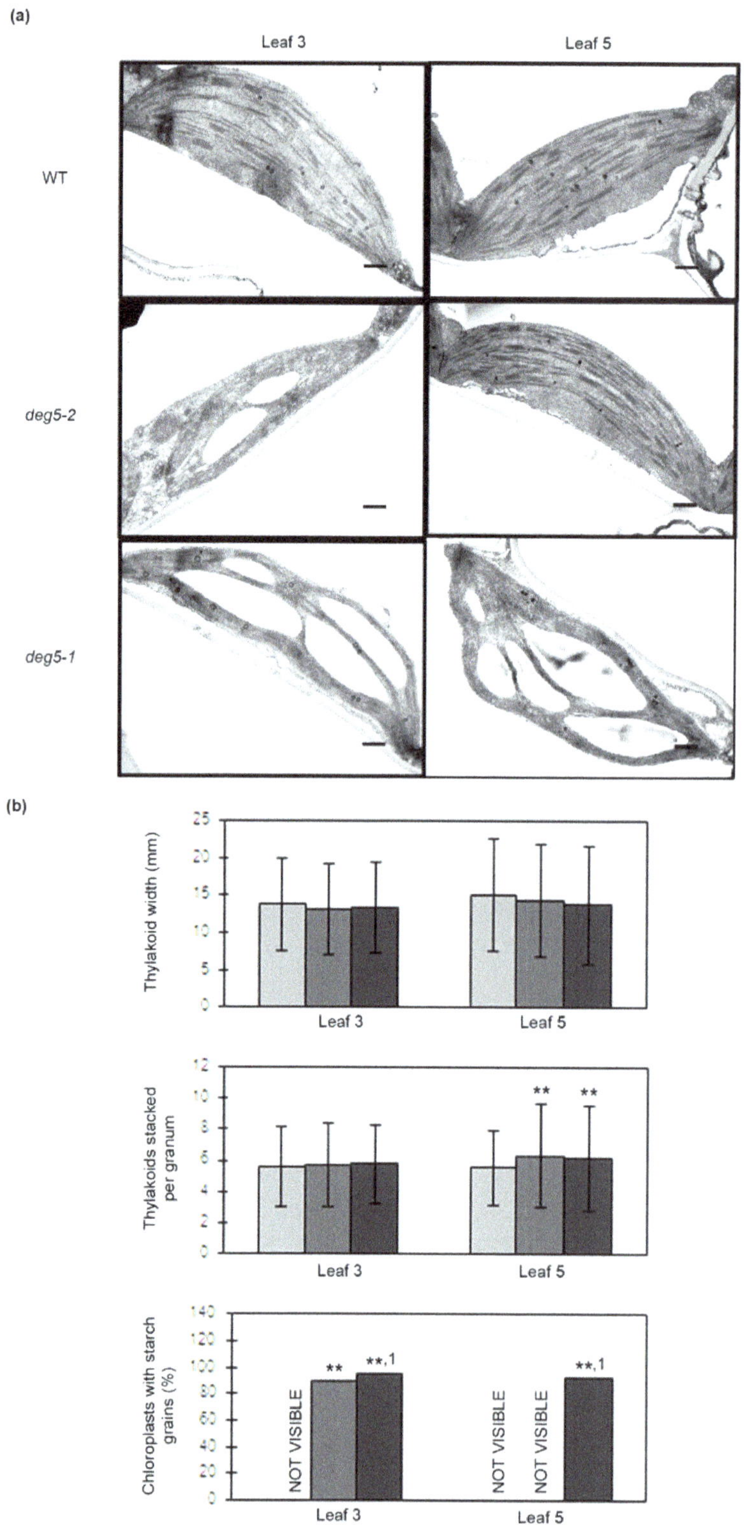

Fig. 5 Transmission electron microscopy analysis of chloroplast of WT (pale grey bars) and *deg5* mutant plants (*deg5-2*, dark grey bars; *deg5-1*, black bars). **a** Electron micrographs of chloroplasts of juvenile (leaf 3) and mature (leaf 5) leaves of WT as well as *deg5-2* and *deg5-1* mutant plants that had reached secondary stage 6.0. Micrographs were taken at the end of the night period. Scale bar: 500 nm. **b** Number of thylakoids stacked per granum, thylakoid width and frequency of chloroplasts with starch grains of juvenile (leaf 3) and mature (leaf 5) leaves of WT, *deg5-1* and *deg5-2* mutant plants which had reached secondary stage 6.0. Thylakoid width was measured as end-to-end distance between the two membranes of the same disk. Asterisks indicate the data for which mutant/WT plants differences were significant ($P < 0.01$) and 1 indicate the values for which *deg5-1*/*deg5-2* differences were significant ($P < 0.01$). Values are mean ±*SD*.

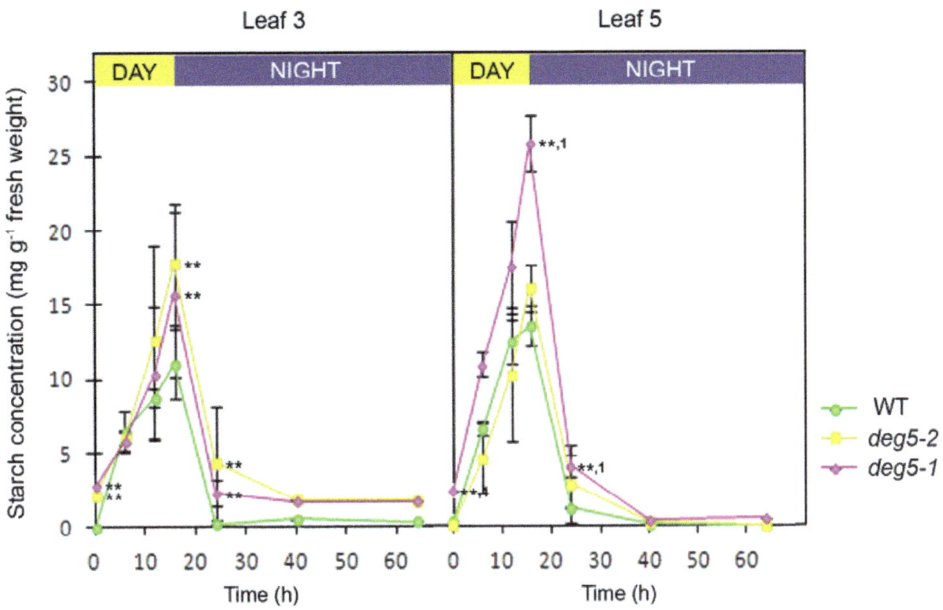

Fig. 6 Diurnal changes in starch content in leaves of WT and *deg5* mutant plants. Starch content was determined in juvenile (leaf 3) and mature (leaf 5) leaves of WT as well as *deg5-2* and *deg5-1* mutant plants that had reached secondary stage 6.0. The measurements were performed over a 24-h photoperiod and an extended night. Asterisks indicate the data for which mutant/WT plants differences were significant ($P < 0.01$) and 1 indicate the values for which *deg5-1*/*deg5-2* differences were significant ($P < 0.01$). Values are mean ±*SD*. The absence of error bars means that they are smaller than the symbols.

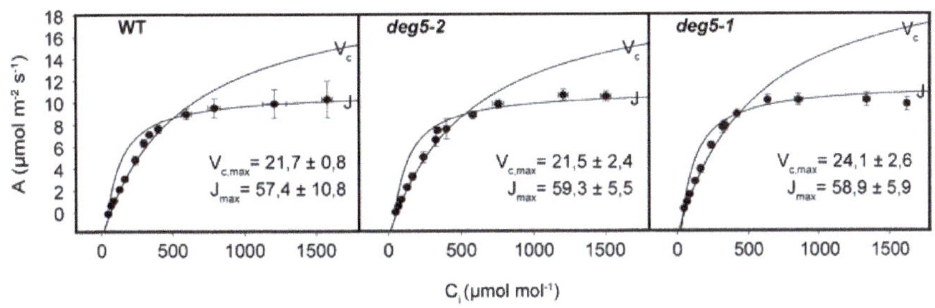

Fig. 7 Average A/C_i curve for leaves of WT and *deg5* mutant plants and Farquhar model fitted to the data. Closed circles represent measured rates of CO_2 assimilation of mature (leaf 5) leaves (*A*) whereas solid curves were obtained by fitting equations of the photosyntheis model [28] to experimental data. Datapoints corresponding to $C_i < 400$ μmol mol^{-1} were used to generate the curve for photosynthetis limited by Rubisco amount, kinetics and activity, whereas points corresponding to $C_i > 400$ μmol mol^{-1} were used to generate the curve for photosynthetis limited by rate of electron transport used for the regeneration of RuBP. Estimated apparent maximal Rubisco carboxylation rate $V_{c,max}$ and maximal photosynthetic electron transport rate Jmax are shown in each panel. Values are mean ±*SD*.

the two mutants are regarded either as unrelated to AtDeg5 or as AtDeg5-related but fully rescuable by AtDeg8 overdose.

For the prevailing majority of AtDeg5-dependent traits (26/33) belonged to groups 1 and 2 involving those for which no WT/*deg5-2* alterations and significant WT/*deg5-1* alterations were detected, with (group 2) or without (group 1) significant between-mutants differences. It may be suspected that AtDeg8 is functionally interchangeable with AtDeg5 with regard to expression of these traits and AtDeg8 overaccumulation may rescue 23% of AtDeg5 deficiency, but not AtDeg5 absence. As a consequence AtDeg5 dosage determines whether intermutant differences are visible (group 2) or are not visible (group 1). The other possibility cannot be ruled out, however, that AtDeg8 is not able to rescue the loss of 23% of AtDeg5 pool but AtDeg5 has enough catalytic sites in excess for AtDeg5/AtDeg8 heterocomplex to function normally even when AtDeg5 pool is reduced by 23% with regard to WT plants level. Depending on whether the number of AtDeg5 catalytic sites is substantially or only modestly excessive in relation to functional needs a phenotypic trait may fall into group 1 or 2, respectively.

Tab. 1 Phenotypic traits which are altered in *deg5* mutants with regard to WT plants.

Group of traits	Phenotypic trait	Significance of WT/*deg5-2*, WT/*deg5-1* and *deg5-1*/*deg5-2* differences			Interpretation
		WT	*deg5-2*	*deg5-1*	
1	Secondary stage 1.2 Secondary stage 1.5 Secondary stage 3.54 Secondary stage 3.91 Secondary stage 3.93 Secondary stage 3.94 Secondary stage 3.95 Secondary stage 6.5 Secondary stage 8.5 Area of leaf 3 Area of leaf 5 Perimeter of leaf 3 Perimeter of leaf 5 Length of leaf 3 Length of leaf 5 Width of leaf 3 Width of leaf 5	100	100	**	Chloroplast has a substantial excess of AtDeg5 catalytic sites with regard to functional requirements or AtDeg8 overdose is able to compensate for 23% reduction of AtDeg5 pool
2	Secondary stage 3.16 Secondary stage 3.58 Secondary stage 3.59 Secondary stage 3.96 Secondary stage 3.97 Frequency of chloroplasts containing starch grains in leaf 5 Starch accumulation at 0 h in leaf 5 Starch accumulation at 16 h in leaf 5 Starch accumulation at 24 h in leaf 5	100	100	**, 1	Chloroplast has a modest excess of AtDeg5 catalytic sites with regard to functional requirements and AtDeg5 dosage effect is important or AtDeg8 overdose is able to compensate for 23% reduction of AtDeg5 pool
3	Frequency of chloroplasts containing starch grains in leaf 3 Thylakoids number per granum in leaf 5 Starch accumulation at 0 h in leaf 3 Starch accumulation at 16 h in leaf 3 Starch accumulation at 24 h in leaf 3 Seed length Seed width	100	**	**	Chloroplast does not have excess of AtDeg5 catalytic sites with regard to functional requirements, no AtDeg5 dosage effect exists, AtDeg8 overdose is not able to compensate for AtDeg5 pool reduction/absence

The traits have been selected so that they were altered significantly ($P < 0.01$) relative to WT plants (marked as 100) at least in *deg5-1* mutant. The traits for which no significant *deg5-2*/WT differences were detected are marked as 100. The asterisks indicate the data for which mutant/WT plants differences were significant ($P < 0.01$) and 1 indicates the values for which *deg5-1*/*deg5-2* differences were significant ($P < 0.01$).

Markedly fewer phenotypic traits (7 out of 33) belonged to the group 3 including those for which significant WT vs. *deg5-2* and WT vs. *deg5-1* differences were detected, without significant *deg5-2*/*deg5-1* differences. Such pattern was due to the absence of excessive amount of AtDeg5 catalytic sites, lack of AtDeg5 dosage effect and inability of AtDeg8 to compensate AtDeg5 deficiency or absence.

Discussion

In this study we have shown that the presence of T-DNA insertion in the 4th intron 872 bp downstream of ATG of *At4g18370* gene (*deg5-2* mutant, Fig. 1) results in lowering of AtDeg5 protein level to 77% of WT value. The deficiency or absence of AtDeg5 (*deg5-1* null mutant) did not lead to changes in abundances of AtDeg1 and AtDeg2 proteins but the amount of AtDeg8 increased about twofold in both mutants (Fig. 2). Since AtDeg5 and AtDeg8 are thought to form a functional heterocomplex [14] a remarkable increase in AtDeg8 in both studied mutants may be regarded as a compensatory response and it could be inferred that AtDeg8 is functionally interchangeable with AtDeg5. The overdose of AtDeg8 could be substituting for the absent AtDeg5 in AtDeg5/AtDeg8 oligomeric heterocomplex. Consistent with this suggestion are the observations that subunits of other chloroplast proteolytic complexes are structurally interchangeable and functionally redundant, i.e. AtFtsH2/8 and AtFtsH1/5 protease pairs of AtFstH heterocomplex [35,36] and ClpR1 and ClpR3 subunits of Clp complex [37]. Alternatively, the excessive AtDeg8 pool could form functional oligomeric homocomplexes. AtDeg8 overdose could in fact substitute AtDeg5 in its catalytic functions as judged by the fact that for as many as 26 phenotypic traits that were altered in *deg5-1* mutant no alteration was observable for the weak mutant (*deg5-2*) presumably due to ability of AtDeg8 to partially substitute for AtDeg5. Furthermore, the traits that were not altered in any of the mutants may be considered as AtDeg5-dependent/AtDeg8-rescuable although these traits may be just unrelated to both AtDeg5 and AtDeg8 as well. In any case, our data do not preclude the possibility that AtDeg8 may be functionally interchangeable with AtDeg5, at least with regard to regulation of some phenotypic traits (groups 1 and 2, Tab. 1).

Our results show that AtDeg5 is involved in leaf development, leaf growth, flower production and silique ripening as judged by a large array of examples of significant WT/*deg5-1* delays in chronological progression of individual ontogenesis secondary stages (Fig. 3, Tab. 1). In contrast, *deg5-2* mutant was not changed with regard to WT plants in this respect although intermutant differences have been identified with respect to five traits demonstrating the importance of AtDeg5 dosage in these events (Tab. 1). Leaf production (development) and growth were shown to be controlled by genetic factors [38,39] linked to epidermal cell expansion and cell division with a key role of QTL at *ERECTA* marker in interaction with SNP295 and SNP21. Since functional relationships between these factors are complicated and far from clear it is not easy to imagine how AtDeg5 may influence these network. It is conceivable that the alterations in leaf growth identified in *deg5-1* may be linked in a causal way to alterations in flower production related to AtDeg5 absence (Fig. 3) since epidermal cell area is controlled by mechanisms associated with the onset of flowering [40] with two highly conserved miRNA – miR156 and miR172 and their targets regarded as key components in reproductive phase change [41]. The final leaf area was demonstrated to be strongly correlated with epidermal cell number and epidermal cell area [38,40] and AtDeg5 must be engaged in respective regulatory events since fully expanded leaves, both juvenile and mature ones were significantly affected in their morphological features as a result of AtDeg5 repression (Fig. 4). It was shown earlier by us that a mixed population of mature leaves (7 and 8) of *deg5-1* mutant had significantly increased area with regard to WT plants [19] thus our data strongly indicate that AtDeg5 is engaged in determining final leaf area of all types of rosette leaves.

Interestingly enough, mature leaves of mutant plants had more thylakoids stacked per granum than WT ones (6.2 vs. 5.6; Fig. 5), the effect described earlier for chloroplasts of juvenile leaves of *deg2* mutants [16]. It may be suggested that early symptoms of senescence-related disintegration of thylakoid system occur in chloroplasts of WT plants which reach secondary stage 6.0 whereas in both *deg5* mutants the symptoms are retarded. This notion is consistent with the appearance of other ultrastructural signs of AtDeg5-related retardation of early chloroplast senescence e.g. a reduction in cross-sectional area of plastoglobules with regard to WT plants [19]. It was suggested that senescence-related disintegration of thylakoid membranes may be controlled by chlorophyll *b* reductase which paves the way for LHCII destabilization by starting off chlorophyll *b* degradation [42]. However we did not find any differences in the steady-state level of LHCII apoproteins in fully expanded leaves of both *deg5* mutants vs WT plants (data not shown) thus the reason why the mutants have more thylakoids per granum remains unclear.

Transmission electron microscopy showed an extensive transitory starch accumulation at the end of the night period in fully expanded juvenile leaves of *deg5-2* mutant as well as both juvenile and mature leaves of *deg5-1* (Fig. 6). It was demonstrated by enzymatic assays that during the light period the mutants accumulated more starch than WT plants (Fig. 6) but the three genotypes hydrolyzed starch during the night at approximately equal rates. Since carboxylation rates as well as electron transport rates were similar in mutants and WT plants (Fig. 7) it appears that AtDeg5 absence presumably results in an increase in the amount/activity of the enzyme(s) which operate(s) in starch biosynthesis pathway downstream of fructose-6-phosphate.

Although our results show that AtDeg5 is needed for normal plant growth and development, leaf morphology and chloroplast ultrastructure this protease does not seem to have any impact on the fitness of *Arabidopsis*, defined as seed production per plant, since both *deg5* mutants did not differ from WT plants in this respect (data not shown).

One of two ways to globally interpret the heterogeneity of AtDeg5-regulated phenotypic traits (besides the involvement of AtDeg8 compensatory effect) is that the most numerous groups of traits which are altered in at least *deg5-1* mutant (groups 1 and 2, Tab. 1) involve those for which AtDeg5 has a substantial (group 1) or a modest (group 2) excess of catalytic sites to function normally even when its pool is reduced by 23% with regard to WT plants whereas the group 3 involve the traits for which no AtDeg5 excess or dosage effect exists. The heterogeneity of AtDeg5-controlled phenotypic traits with regard to the degree of their dependence on AtDeg5 dosage/excess of catalytic sites would not be unexpected altogether since similar effects were described for other chloroplast proteases. It was demonstrated in this respect that *var1-1* mutants – which are deficient in the chloroplast proteolytic AtFtsH heterocomplex – retain sufficient amount of AtFtsH catalytic sites for some processes to proceed normally (chloroplast biogenesis) but this amount is not sufficient for other processes to function normally (stress-related degradation of Lhcb1-3 apoproteins) [23].

Acknowledgments

This work was supported by a Polish Ministry of Science and Higher Education in the form of statutory subsidy for the Department of Plant Physiology, Institute of Experimental Biology, Faculty of Biology, Adam Mickiewicz University, Poznań, Poland. We thank professor Jacek Oleksyn (Institute of Dendrology, Polish Academy of Sciences, Kórnik, Poland) for permission to use gas exchange system.

Authors' contributions

The following declarations about authors' contributions to the research have been made: designed the experiments: GJ, MB; performed the experiments: MB, TW; analyzed the experimental data: MB, TW, GJ; wrote the paper: GJ

Competing interests

No competing interests have been declared.

References

1. Pesquet E. Plant proteases – from detection to function. Physiol Plant. 2012;145(1):1–4.
2. Lipińska B, Ang D, Georgopoulos C. Sequence analysis and transcriptional regulation of the *Escherichia coli grpE* gene, encoding a heat shock protein. Nucl Acids Res. 1988;16(15):7545–7562.
3. Strauch KL, Beckwitt J. An *Escherichia coli* mutation preventing degradation of abnormal periplasmic proteins. Proc Natl Acad Sci USA. 1988;85(5):1576–1580.
4. Wilken C, Kitzing K, Kurzbauer R, Ehrmann M, Clausen T. Crystal structure of the DegS stress sensor: how a PDZ domain recognizes misfolded protein and activates a protease. Cell. 2004;117(4):483–494.
5. Jiang J, Zhang X, Chen Y, Wu Y, Zhou ZH, Chang Z, Sui SF. Activation of DegP chaperone-protease via formation of large cage-like oligomers upon binding to substrate proteins. Proc Natl Acad Sci USA. 2008;105(33):11939–11944.
6. Bai XC, Pan XI, Wang XJ, Ye YY, Chang LF, Leng D, et al. Characterization of the structure and function of *Escherichia coli* DegQ as a representative of the DegQ-like proteases of bacterial HtrA family proteins. Structure. 2011;19(9):1328–1337.
7. Ortega J, Iwańczyk J, Jomaa A. *Escherichia coli* DegP: a structure-driven functional model. J Bacteriol. 2009;191(15):4705–4713.
8. Schuhmann H, Huesgen PF, Adamska I. The family of Deg/HtrA proteases in plants. BMC Plant Biol. 2012;12(1):1–14.
9. Schuhmann H, Adamska I. Deg proteases and their role in protein quality control and processing in different subcellular compartments of the plant cell. Physiol Plant. 2012;145(1):224–234.
10. Schuhmann H, Mogg U, Adamska I. A new principle of oligomerization of plant DEG7 protease based on interactions of degenerated protease domains. Biochem J. 2011;435(1):167–174.
11. Kley J, Schmidt B, Boyanov B, Stolt-Bergner PC, Kirk R, Ehrmann M, et al. Structural adaptation of the plant protease Deg1 to repair photosystem II during light exposure. Nat Struct Mol Biol. 2011;18(6):728–731.
12. Sun XW, Fu TJ, Chen N, Guo JK, Ma JF, Zou MJ, et al. The stromal chloroplast Deg7 protease participates in the repair of photosystem II after photoinhibition in *Arabidopsis*. Plant Physiol. 2010;152(3):1263–1273.
13. Sun W, Gao F, Fan H, Shan X, Sun R, Liu L, et al. The structures of *Arabidopsis* Deg5 and Deg8 reveal new insights into HtrA proteases. Acta Crystallogr D Biol Crystallogr. 2013;69(5):830–837.
14. Sun XW, Peng L, Guo J, Chi W, Ma J, Lu C, et al. Formation of DEG5 and DEG8 and their involvement in the degradation of photodamaged photosystem II reaction center D1 protein in *Arabidopsis*. Plant Cell. 2007;19(4):1347–1361.
15. Kapri-Pardes E, Naveh L, Adam Z. The thylakoid lumen protease Deg1 is involved in the repair of photosystem II from photoinhibition in *Arabidopsis*. Plant Cell. 2007;19(3):1039–1047.
16. Luciński R, Misztal L, Samardakiewicz S, Jackowski G. The thylakoid protease Deg2 is involved in stress-related degradation of the photosystem II light-harvesting protein Lhcb6 in *Arabidopsis thaliana*. New Phytol. 2011;192(1):174–186.
17. Kato Y, Sun X, Zhang L, Sakamoto W. Cooperative D1 degradation in the photosystem II repair mediated by chloroplastic proteases in *Arabidopsis*. Plant Physiol. 2012;159(4):1428–1439.
18. Zienkiewicz M, Ferenc A, Wasilewska W, Romanowska E. High light stimulates Deg1-dependent cleavage of the minor LHCII antenna proteins CP26 and CP29 and the PsbS protein in *Arabidopsis thaliana*. Planta. 2012235(2):279–288.
19. Luciński R, Misztal L, Samardakiewicz S, Jackowski G. Involvement of Deg5 protease in wounding-related disposal of PsbF apoprotein. Plant Physiol Biochem. 2011;49(3):311–320.
20. Murashige T, Skoog F. A revised medium for rapid growth and bio assays with tobacco tissue cultures. Physiol Plant. 1962;15(3):473–497.
21. Grabsztunowicz M, Jackowski G. Isolation of intact and pure chloroplasts from leaves of *Arabidopsis thaliana* plants acclimated to low irradiance for studies on Rubisco regulation. Acta Soc Bot Pol. 2013;82(1):91–95.
22. Laemmli UK. Cleavage of structural proteins during the assembly of the head of bacteriophage T4. Nature. 1970;227(5259):680–685.
23. Luciński R, Jackowski G. AtFtsH heterocomplex-mediated degradation of apoproteins of the major light harvesting complex of photosystem II (LHCII) in response to stresses. J Plant Physiol. 2013;170(12):1082–1089.
24. Boyes DC, Zayed AM, Ascenzi R, McCaskill AJ, Hoffman NE, Davis KR, et al. Growth stage-based phenotypic analysis of *Arabidopsis*: a model for high throughput functional genomics in plants. Plant Cell. 2001;13(7):1499–1510.
25. Kincaid DT, Schneider RB. Quantification of leaf shape with a microcomputer and Fourier transform. Can J Bot. 1983;61(9):2333–2342.
26. Western TL, Skinner DJ, Haughn GW. Differentiation of mucilage secretory cells of the *Arabidopsis* seed coat. Plant Physiol. 2000;122(2):345–355.
27. Chow PS, Landhausser SM. A method for routine measurements of total sugar and starch content in woody plant tissues. Tree Physiol. 2004;24(10):1129–1136.
28. Farquhar GD, von Caemmerer S, Berry JA. A biochemical-model of photosynthetic CO_2 assimilation in leaves of C3 species. Planta. 1980;149(1):78–90.
29. Tanaka Y, Sugano SS, Shimada T, Nishimura I. Enhancement of leaf photosynthetic capacity through increased stomatal density in *Arabidopsis*. New Phytol. 2013;198(3):757–764.
30. Long SP, Bernacchi CJ. Gas exchange measurements, what can they tell us about the underlying limitations to photosynthesis? Procedures and sources of error. J Exp Bot. 2003;54(392):2393–2401.
31. Sharkey TD, Bernacchi CJ, Farquhar GD, Singsaas EL. Fitting photosynthetic carbon dioxide response curves for C3 leaves. Plant Cell Environ. 2007;30(9):1035–1040.
32. Arnon D. Copper enzymes in isolated chloroplasts. Polyphenol oxidase in *Beta vulgaris*. Plant Physiol. 1949;24(1):1–15.
33. Lancashire PD, Bleiholder H, van der Boom T, Langeluddeke P, Stauss R, Weber E, et al. A uniform decimal code for growth stages of crops and weeds. Ann Appl Biol. 1991;119(3):561–601.
34. Telfer A, Bollman KM, Poethig RS. Phase change and the regulation of trichome distribution in *Arabidopsis thaliana*. Development. 1997;124(3):645–654.
35. Yu F, Park S, Rodermel SR. The *Arabidopsis* FtsH metalloprotease gene family: interchangeability of subunits in chloroplast oligomeric complexes. Plant J. 2004;37(6):864–876.
36. Yu F, Park S, Rodermel SR. Functional redundancy of AtFtsH

metalloproteases in thylakoid membrane complexes. Plant Physiol. 2005;138(4):1957–1966.

37. Stanne TM, Sjögren LL, Koussevitzky S, Clarke AK. Identification of new protein substrates for the chloroplast ATP-dependent Clp protease supports its constitutive role in *Arabidopsis*. Biochem J. 2009;417(1):257–268.

38. Tisné S, Reymond M, Vile D, Fabre J, Dauzat M, Koornneef M, et al. Combined genetic and modeling approaches reveal that epidermal cell area and number in leaves are controlled by leaf and plant developmental processes in *Arabidopsis*. Plant Physiol. 2008;148(2):1117–1127.

39. Massonnet C, Vile D, Fabre J, Hannah MA, Caldana C, Lisec J, et al. Probing the reproducibility of leaf growth and molecular phenotypes: a comparison of three *Arabidopsis* accessions cultivated in ten laboratories. Plant Physiol. 2010;152(4):2142–2157.

40. Cookson SJ, Chenu K, Granier C. Day length affects the dynamics of leaf expansion and cellular development in *Arabidopsis thaliana* partially through floral transition timing. Ann Bot. 2007;99(4):703–711.

41. Huijser P, Schmid M. The control of developmental phase transitions in plants. Development. 2011;138(19):4117–4129.

42. Kusaba M, Ito H, Morita R, Iida S, Sato Y, Fujimoto M, et al. Rice NON-YELLOW COLORING1 is involved in light-harvesting complex II and grana degradation during leaf senescence. Plant Cell. 2007;19(4):1362–1375.

Are all red algal parasites cut from the same cloth?

Eric D. Salomaki*, Christopher E. Lane
Department of Biological Sciences, University of Rhode Island, Kingston, RI 02881, USA

Abstract

Parasitism is a common life strategy throughout the eukaryotic tree of life. Many devastating human pathogens, including the causative agents of malaria and toxoplasmosis, have evolved from a photosynthetic ancestor. However, how an organism transitions from a photosynthetic to a parasitic life history strategy remains mostly unknown. This is largely because few systems present the opportunity to make meaningful comparisons between a parasite and a close free-living relative. Parasites have independently evolved dozens of times throughout the Florideophyceae (Rhodophyta), and often infect close relatives. The accepted evolutionary paradigm proposes that red algal parasites arise by first infecting a close relative and over time diversify and infect more distantly related species. This provides a natural evolutionary gradient of relationships between hosts and parasites that share a photosynthetic common ancestor. Elegant microscopic work in the late 20th century provided detailed insight into the infection cycle of red algal parasites and the cellular interactions between parasites and their hosts. Those studies led to the use of molecular work to further investigate the origins of the parasite organelles and reveal the evolutionary relationships between hosts and their parasites. Here we synthesize the research detailing the infection methods and cellular interactions between red algal parasites and their hosts. We offer an alternative hypothesis to the current dogma of red algal parasite evolution and propose that red algae can adopt a parasitic life strategy through multiple evolutionary pathways, including direct infection of distant relatives. Furthermore, we highlight potential directions for future research to further evaluate parasite evolution in red algae.

Keywords: Rhodophyta; parasite evolution; pit connections; Florideophyceae, organelles

Introduction

Parasitism has evolved innumerable times throughout the eukaryotic tree of life [1]. Some of the more virulent parasites have transitioned from a once photosynthetic ancestor, including the causative agents of malaria and related mammalian diseases [2,3]. Therefore, understanding the evolutionary trajectory between photosynthesis and abandoning autotrophy for a parasitic strategy, is of particular importance. Red algal parasites are uniquely valuable to study this path because they have independently evolved many times, providing literally dozens of discrete events to compare [4–7]. This system may provide novel insights into the evolution of parasitism, especially with regard to the early stages of transitioning from a photosynthetic past.

Red algal parasites exclusively infect other red algae, typically ones with which they share a recent common ancestor [5–8]. The relationship between host and parasite was first recognized using morphological similarities in the life-cycles of parasites and their hosts [9]. More recently, molecular data have confirmed this hypothesis [4–6,10]. Traditionally red algal parasites have been placed into two different groups, based on their phylogenetic relationships with their hosts [6]. Adelphoparasites (adelpho is Greek for "kin") are closely related to their host and often infect only one host, whereas alloparasites are more divergent from their host(s) [4,6]. Currently, adelphoparasites are believed to make up roughly 90% of all red algal parasites [6].

Among the Florideophyceae, parasites belonging to at least 66 different red algal genera have evolved independently over 100 times (Tab. 1) [11]. The accepted evolutionary paradigm proposes that adelphoparasitism is the initial state, followed by parasite diversification, which leads to the development of alloparasites (Fig. 1). These "older" parasites can infect more distantly related taxa, and make up roughly 10% of red algal parasites [5,7]. Due to their rarity, alloparasites are relatively unstudied, with the exceptions of *Choreocolax polysiphoniae* and *Harveyella mirabilis*.

The importance of red algal pit connections

One of the defining characteristics of the florideophycean red algae is the ability of cells to form connections with their adjacent cells [12]. There are two distinct forms of these "pit connections" formed by red algae. Primary pit connections

* Corresponding author. Email: eric_salomaki@my.uri.edu
Handling Editor: Andrzej Bodył

Tab. 1 Taxonomic summary of red algal parasites including number of described species and type of parasite by family. The shaded boxes are adelphoparasite numbers, highlighted to indicate their abundance, relative to alloparasites.

Order	Family	Type and number of parasites
Ceramiales	Ceramiaceae	Adelphoparasites = 1
		Alloparasites = 2
	Delesseriaceae	Adelphoparasites = 14
	Rhodomelaceae	Adelphoparasites = 37
		Alloparasites = 8
	Spyridiaceae	Adelphoparasites = 1
	Wrangeliaceae	Adelphoparasites = 1
Corallinales	Corallinaceae	Adelphoparasites = 2
		Alloparasites = 1
	Hapalidiaceae	Adelphoparasites = 5
Gigartinales	Cystocloniaceae	Adelphoparasites = 2
		Alloparasites = 1
	Kallymeniaceae	Adelphoparasites = 3
	Phyllophoraceae	Adelphoparasites = 1
	Solieriaceae	Adelphoparasites = 1
Gracilariales	Gracilariaceae	Adelphoparasites = 2
	Pterocladiophilaceae	Adelphoparasites = 1
		Alloparasites = 12
Halymeniales	Halymeniaceae	Adelphoparasites = 1
Palmariales	Palmariaceae	Adelphoparasites = 1
	Rhodophysemataceae	Adelphoparasites = 1
Plocamiales	Plocamiaceae	Adelphoparasites = 2
Rhodymeniales	Faucheaceae	Adelphoparasites = 2
	Rhodymeniaceae	Adelphoparasites = 2
Incertae sedis	Incertae sedis	Adelphoparasites = 1
		Alloparasites = 6

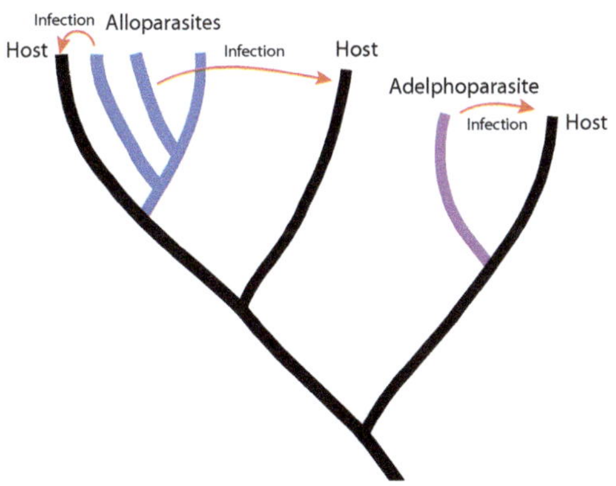

Fig. 1 Red algal parasites often evolve from a recent common ancestor with their host. These "adelphoparasites" (purple branch) are usually host-specific. However, parasites that have diversified into more than one species and/or infect distantly-related hosts, are known as "alloparasites" (blue branches).

arise between a mother and daughter cell during apical growth [13]. These connections result from a seemingly incomplete cell division where the septum begins to develop from the cell walls growing inward to separate the daughter nuclei [13]. However, cytokinesis is incomplete and the septum does not fuse, leaving an opening that connects the two cells [13]. A pit plug composed of a polysaccharide-protein complex then forms sealing the pit connection and separating the two cells [13,14]. Though pit connections result in an aperture that is not entirely sealed by a septum, the pit plug prevents the transfer of cellular contents and photosynthate between adjacent cells [15].

In addition to the primary pit connections, florideophytes also form secondary pit connections between adjacent non-daughter cells. These secondary pit connections are known to occur in a wide range of Florideophyceae and form between two genetically similar red algal cells [8,16]. Red algal parasite spores utilize secondary pit connections as a way to enter the cells of the host [8,17,18]. As evidence for the importance of secondary pit connections in parasitic infections, parasites are not known from red algal orders where secondary pit connections do not occur [19]. An advantage of this strategy is that the similarity between host and parasite at the genetic level allows parasite spores to simply deposit their organelles into the host cell and take over, spreading through primary or secondary connections [8,20]. The widespread existence of secondary pit connections among red algae is undoubtedly a primary factor in the promiscuous nature of parasitism as a life history strategy in the lineage.

Differences in adelphoparasite and alloparasite infection cycles

Spore germination and host infection

Rhodophytes lack flagella in all stages of their life cycle, making the initial stages of locating a host a passive process. Once a parasite spore lands upon a susceptible host, the parasite carpospore (2N) or tetraspore (1N) will germinate and undergo an initial cell division [21]. Adelphoparasite cells will divide between 1 and 3 more times before one of the cells forms a rhizoid that penetrates the surface of the host, growing into the wall of a host epidermal cell [21]. The tip of the rhizoid swells isolating a single parasite nucleus along with its organelles into a conjunctor cell which divides from the infection rhizoid [21,22]. This conjunctor cell then fuses, via a secondary pit connection, with the adjacent host cell (Fig. 2a). The contents of the conjunctor cell, which include the parasite nucleus and organelles, are deposited into the host cell, thus forming a heterokaryotic cell (containing both parasite and host nuclei) [8,17,21]. This connection between the parasite infection rhizoid and the transformed host cell is sealed by a pit plug that was formed previously during the initial fusion between the conjunctor cell and the parasite infection rhizoid [8,21].

The differences between adelpho- and alloparasites become evident at the initial infection of the host cell. In alloparasites, rather than going through a few cell divisions before penetrating the surface, the parasite spore attaches to a suitable host, penetrating and forming a hyphae-like

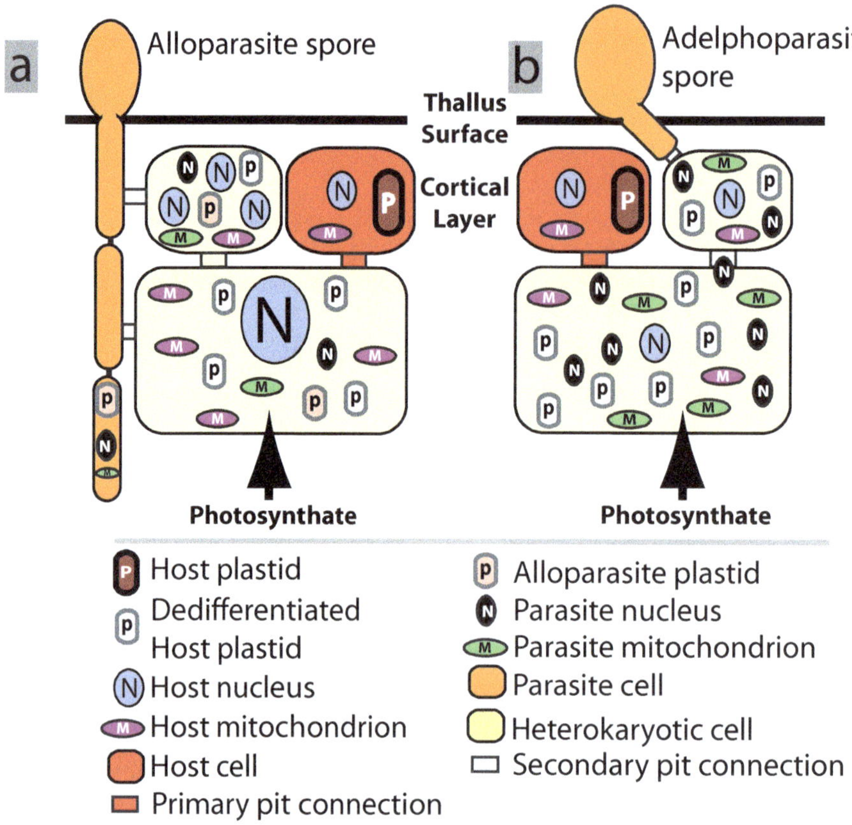

Fig. 2 Infection strategies of red algal alloparasites and adelphoparasites. **a** Alloparasites penetrate the host thallus and grow a network of filamentous cells into the host. Each cell is capable of fusing to a host cell via secondary pit connection and depositing its contents. Once inside the cell, the alloparasite nucleus does not divide, but causes the host nucleus to enlarge, or in the case of cortical cells, multiply. It is currently unclear whether the alloparasite plastid is derived from the host or parasite. **b** Adelphoparasite spores fuse with a cortical cell and inject their contents. These parasite nuclei can multiply within the host, however no nuclear DNA synthesis has been observed by the host after infection. Parasite nuclei and organelles spread via the host primary pit connections. The host organelles also multiply in response to infection.

network of multicellular filaments between the host cells [8,23,24]. These filaments enable the parasite to spread numerous cells deep into the host away from the initial site of infection (Fig. 2b). Each alloparasite cell in these filaments contains a single nuclei and can form a conjunctor cell and secondary pit connection through which the parasite deposits its cellular contents into the host, creating a heterokaryotic cell [8,23,24].

Inside the heterokaryon

After the host cell becomes heterokaryotic, both adelpho- and alloparasites take control of the host cellular machinery [8,21]. Almost immediately upon infection, the hosts' central vacuole tonoplast is lysed, allowing cytoplasm to spread throughout the space previously occupied by the vacuole [4,25]. Subsequently, the number of organelles including plastids, mitochondria, and ribosomes increases throughout the cytoplasm causing it to appear denser [8,21,23,25]. Along with the increase in cytoplasmic organelles comes an increase in cell size (hypertrophy), a process in which the cell can grow to 40 times its original size [8 22,26]. In addition to the increased organelles, either the parasite or host nuclei also increase in size and/or number [8,23,24].

Goff and Coleman used 4',6-diamidino-2-phenylindole (DAPI) staining and microspectrofluorometry to examine the interactions between the alloparasite, *Choreocolax polysiphoniae* and its host *Vertebrata lanosa* [8]. Their study showed that infected *V. lanosa* central cells will become enlarged and the nuclei will undergo DNA synthesis but not nuclear division resulting in polyploid host nuclei [8]. Alternatively, in infected *V. lanosa* pericentral cells, the host nuclei will either increase in size, increase in number, or some combination of both [8]. An increase in the number of host nuclei is the most common response [8]. Host cells that are adjacent to infected cells and connected by pit connections will not show any cytological transformation [8,23].

Conversely, no DNA synthesis or nuclear division has been observed from the host nucleus after infection by the adelphoparasite *Janczewskia gardneri* [24]. Instead, the adelphoparasite nucleus rapidly undergoes DNA synthesis, generating numerous parasite nuclei inside a single host cell (Fig. 2) [24]. The adelphoparasite subsequently spreads to additional host cells through the formation of conjunctor cells that can infect adjacent host cells [21]. Adelphoparasites can also form rhizoidal infection cells, which are multinucleate and contain large numbers of mitochondria, ribosomes, and dedifferentiated host-derived proplastids and can fuse with more distant host cells [21].

Alloparasites are capable of mitotic divisions to create the multicellular filaments that spread between host cells [23,24]. However, the alloparasite nucleus does not undergo DNA synthesis inside the host cell [8,23,24]. Therefore, parasite nuclei remain at a 1:1 ratio with the number of secondary pit connections between parasite and host cells [8,23,24]. Once inside a heterokaryon, the adelphoparasite continues spreading to adjacent host cells via pit connections, while utilizing the host to progress through its lifecycle and reproduce [21].

Formation of reproductive structures

As the adelphoparasite spreads throughout the host, a gall or "erumpent pustule" begins to form as host cells continually expand upon infection by the parasite. Eventually the adelphoparasite will start to form reproductive structures [21]. If the original infecting spore was haploid the parasite will form caprosporangia that can be fertilized by a spermatia from another parasite forming a diploid carposporophyte that will eventually release carpospores [21,27]. If the original infection was from a diploid carpospores, the parasite will undergo meiosis forming haploid tetraspores that will be released from the erumpent pustule [21,27].

An alloparasite does not spread through the host like the adelphoparasite. Rather than forming a gall from invaded cells, a host pericentral cell containing a parasite nuclei and many host nuclei will form a protuberance [8]. This protuberance will become isolated from the original host cell and undergo mitotic divisions, which produces the mature parasite pustule containing reproductive cells similar to that of adelphoparasites [8]. After the reproductive cells are released, the former host tissue that made the erumpent pustule becomes necrotic [8]. While the fate of future adelphoparasites lies in the released spores, alloparasites are able to continue their infection of the same host as parasite filaments continue to grow into uninfected areas [8].

Organelles

Studies of cellular organelles have yielded particularly interesting findings during investigations of red algal parasite biology. Early studies established the role of secondary pit connections between the parasite and host cells and demonstrated their role in transferring the parasite nucleus to the host cell [8,17,21]. However, it was unclear whether the parasite maintained its own mitochondrion and plastid or if it utilized the host organelles once the parasite nucleus was transferred into the host cell. Cytoplasmic organelles support many major metabolic pathways as well as play major roles in cellular energy and carbohydrate production. It has been well established that purifying selection is relaxed on parasite organellar genes that become unnecessary, leading to genome reduction in parasites as they increasingly rely on a host for energy and carbohydrates [28–32]. Therefore, it seems likely that some red algal parasite mitochondrion and plastid genes would be truncated or even lost over time. The origin and roles of mitochondria and plastids in the parasite-host interaction may reveal key information regarding red algal parasite biology and their ability to infect and control the host cellular machinery.

First using the alloparasite *Choreocolax polysiphoniae*, and later the adelphoparasites *Gracilariophila oryzoides* and *Gardneriella tuberifera*, researchers observed that in addition to the parasite nuclei, organelles are also transferred to the host cell via the conjunctor cell upon infection [21,24]. Once molecular tools became more widely available, Goff and Coleman utilized restriction fragment length polymorphisms (RFLPs) to investigate the origin of mitochondria in the adelphoparasites *Plocamiocolax pulvinata*, *Gracilariophila oryzoides*, and *Gardneriella tuberifera* and their respective hosts, *Plocamium cartilagineum*, *Gracilariopsis lemaneiformis*, and *Sarcodiotheca gaudichaudii* [22]. This work revealed that *P. pulvinata* and *G. oryzoides* maintain a genetically unique mitochondrion and that both the parasite and host mitochondria are present within the heterokaryotic cells [22]. The study was unable to conclusively demonstrate that the mitochondrion from *Gardneriella tuberifera* is unique from that of *Sarcodiotheca gaudichaudii*, due to the extremely close relationship between the two species [22].

The mitochondrion genomes of the adelphoparasites *Plocamiocolax pulvinata* and *Gracilariophila oryzoides*, as well as its host *Gracilariopsis andersonii*, were recently sequenced. These data paved the way for a new level of fine-scaled investigations of red algal parasites, elucidating details of the organellar genome architecture that was previously unattainable [33]. When comparing the mitochondrion genome sequences of the parasites with the free-living host, the *atp8* and *sdhC* genes from *G. oryzoides* were determined to be pseudogenes [33]. Furthermore, *atp8* was determined to be absent from *P. pulvinata*. However, the authors noted that according to sequenced cDNA libraries, the genes were still transcribed [33]. Recent second generation sequencing of additional samples of these species has revealed that the "missing" genes are all present and lack frameshift mutations (Salomaki and Lane, unpublished). These data suggest that purifying selection is maintained on red algal parasite mitochondria and that even though much of the parasite life cycle exists inside a host cell, red algal parasites still require their own mitochondrion for their survival.

While evident that parasites maintain their native mitochondrion, microscopy and molecular studies have demonstrated that red algal parasites do not maintain their own plastid [21,22]. Microscopy shows that the spores of adelphoparasites *Gracilariophila oryzoides*, and *Gardneriella tuberifera* contain proplastids lacking photosynthetic pigments, phycobilisomes, and thylakoids [21]. Once the parasite injects its nuclei and organelles into a host cell, the host plastids transform and their light harvesting pycobilisomes disappear from the thylakoids [21]. After rapid dedifferentiation, simple proplastids that are similar to the infecting parasite plastid bud off the host plastid [21]. As the parasite nuclei, mitochondrion, and host derived proplastids spread to adjacent cells through pit connections, plastids from the newly infected host cells also rapidly dedifferentiate into proplastids [21]. Eventually cells emerge from the heterokaryotic host cell containing only parasite nuclei, parasite mitochondrion and a host-derived proplastid [21]. RFLP analysis was utilized to investigate whether this plastid was a genetically unique parasite plastid, or instead, the parasite was incorporating a host-derived proplastid [22]. This study revealed that adelphoparasites *Plocamiocolax pulvinata*, *Gracilariophila oryzoides*, and *Gardneriella tuberifera* and their hosts had identical banding patterns [22]. Subsequent DNA sequencing of the variable plastid *rbc*L-*rbc*S spacer region revealed that the plastid from both hosts and parasites were genetically identical, confirming that the parasite plastid is a dedifferentiated host plastid [22].

Interestingly, two studies suggest that the proplastid may be capable of differentiating again into a photosynthetic plastid. While investigating the photosynthetic rates and C^{14} transfer from *Polysiphonia* (=*Vertebrata*) *lanosa* to the alloparasite *Choreocolax polysiphoniae*, Callow et al. examined the alloparasite pustules after dissection from their host [34]. The authors note that many of the parasite pustules had a pinkish hue and incorporated radioactively labeled C^{14} into their thallus [34]. Furthermore, the carbon fixation rate increased over time (up to 66 hours) leading the authors to conclude that *C. polysiphoniae* is capable of photosynthesis on its own. However, the source of photosynthetic activity in dissected *C. polysiphoniae* pustules may be from host cells that have been incorporated into the pustule as observed in that study, and independently, by Kugrens and West, and Goff examining *Janczewskia gardneri* [21,26,34]. Additionally, the status of *J. gardneri* as a parasite or obligate epiphyte has been debated due to its pigmentation [24,35,36]. Most recently, it was noted that during the early stages of the interaction between *J. gardneri* and its host, *Laurencia spectabilis*, *J. gardneri* exists as colorless cells and "infects" host cells in the same manner as other adelphoparasites. As *J. gardneri* cells erupt from the host they remain colorless but the cells become pigmented once the adelphoparasite becomes reproductively mature [24]. Whether this pigmentation originates from host cells in the pustule matrix, or if the proplastid differentiates back into a photosynthetic plastid remains unknown.

Nutrient transfer

With the exception of a few adelphoparasites that gain pigmentation upon reproductive maturity, red algal parasites are not capable of photosynthesis on their own and must obtain carbohydrates and other nutrients from a host. After parasite infection, the host (now heterokaryotic) cell loses the ability to photosynthesize as a result of plastid dedifferentiation [21]. This leads to a differential gradient of carbon between the heterokaryotic cell and the adjacent normally functioning host cells [21,22,25]. To account for the loss of carbon fixation, uninfected host cells direct photosynthate to heterokaryon and parasite cells that they are connected to via pit connections [24,37].

The first studies investigating carbon transfer between a red algal host and its parasite found three products of photosynthesis (floridoside, isofloridoside, and manitol) were transferred from the host to its parasite via a concentration gradient [34,38]. Later five different sugar species were identified to be assimilated by the host *Rhodomela confervoides* and translocated to its parasite *Harveyella mirabilis* [39]. Investigations into carbon translocation in *H. mirabilis* demonstrated the localization of carbon, from being fixed by the photosynthetic host through its movement into the parasite cells and revealed that heterokaryon cells incorporated more C^{14} than neighboring uninfected host cells [37]. Furthermore, it was determined that starch was not distributed evenly throughout the parasite cells as might be expected, but instead was being directed preferentially to parasite reproductive cells [8,37]. Given the capabilities of parasites for obtaining carbon from the host, the role of the maintained proplastid in parasite cells remains in question.

Host specificity and parasite resistance

Red algal parasites are known to be extremely host specific, usually infecting one to a few, closely related host species [7,40,41]. A study using the adelphoparasite *Janczewskia morimotoi* tested its ability to infect 15 other species including close relatives of its natural host, *Laurencia nipponica*, as well as members of different genera [42]. While *J. morimotoi* was capable of infecting two close relatives of its natural host, the more distantly related potential hosts prevented parasite infections [42]. Additionally, the host specificity of *Leachiella pacifica* was assessed through culture studies attempting to use parasites isolated from *Polysiphonia paniculata* to infect *Pterocladia bipinnata* and vice-versa [41]. Parasites isolated from *P. paniculata* could infect other populations of the same species as well as some other *Polysiphonia* species, however they could not infect *Pt. bipinnata* populations that were susceptible to parasites isolated from other *Pt. bipinnata* specimens [41]. These *L. pacifica* isolates showed strong genus-level host specificity. However, due to the greatly reduced morphology of red algal parasites, it cannot be ruled out that parasites isolated from different genera are, in fact, different host-specific species. Revisiting this study with molecular data would strengthen our understanding of host specificity and potentially reveal cryptic parasite species.

Dawsoniocolax bostrychiae and *Bostrychiocolax australis* are parasites that infect *Bostrychia radicans* [40]. A study on host range and specificity of these parasites on a variety of potential hosts, yielded similar results to the *J. morimotoi* study: the genetic distance between parasite and host has a strong negative correlation with susceptibility to parasite penetration and infection [40]. The authors note that they encountered hosts that are resistant to parasite infection, including some host populations that contained resistant and susceptible specimens [40]. In several cases the parasite was capable of forming an initial infection in a resistant host, however the host cell or cells adjacent to the infected cell died off, preventing the parasite from spreading further into the host [40]. However, subsequent molecular studies revealed phenotypic plasticity and cryptic diversity in *B. radicans* [43]. Therefore, the possibility remains that resistant and susceptible hosts from the host resistance study were actually different species. These findings emphasize the need for ongoing taxonomic evaluation of red algal parasites and their hosts. Without the taxonomic framework, questions about whether or not the host is actually resisting parasite infection cannot be answered conclusively.

Many questions remain

Why does the parasite maintain a copy of the host plastid as it is forming its own reproductive cells and spores? Other parasites that have evolved from a plastid bearing ancestor, including the apicomplexans *Eimeria tenella* and *Plasmodium falciparum*, the parasitic plant *Epifagus virginiana* and many others, maintain a reduced plastid for cellular functions other than photosynthesis, such as fatty acid biosynthesis [2,30,31]. However, none of these plastid-bearing parasites steal a plastid from their host like the red algal adelphoparasites. Are adelphoparasites genetically similar enough to their hosts that they can target nuclear-encoded proteins to the host-derived proplastid and utilize those products for fatty acid biosynthesis? Genomic analyses of signaling and targeting peptides for plastid targeted nuclear genes in red algal parasites, combined with transcriptomic and proteomic approaches, will provide valuable insight into the role of the plastids in the infection mechanism and parasite life cycle. The use of additional molecular tools including in-situ hybridization would enable researchers to localize parasite nuclear-encoded proteins in the heterokaryotic cell.

Furthermore, the taxonomic range and multiple independent origins of red algal parasites makes it difficult to make generalizations based on a few observations. Thus far, the origin of red algal parasite plastids has only been investigated in adelphoparasites and the origin of the alloparasite plastid remains unknown. The assumption is that alloparasites first progress through an adelphoparasite stage and also maintain a co-opted host plastid. However, there are distinct developmental differences between adelpho- and alloparasites, including the initial steps of infection and alloparasites inability to synthesize DNA in heterokaryon cells (Fig. 2). Therefore, it seems plausible to propose an alternative hypothesis that, rather than passing through an adelphoparasite stage, alloparasites are capable of directly evolving infection mechanisms to parasitize distantly related hosts. In the proposed scenario, alloparasites would presumably maintain their own plastid, as they are likely incapable of utilizing such a genetically distant host plastid. Preliminary genomic data from *C. polysiphoniae* indicates this may, in fact, be the case (Salomaki and Lane, unpublished)

Future research investigating red algal parasite evolution will provide unique insight into the effects of transitioning from a free-living to a parasitic life strategy. Molecular data has supported morphological observations that red algal parasites share a recent common ancestor with their hosts [4–6,10]. However, further use of molecular tools is essential to provide a robust taxonomic framework of red algal parasites and their hosts. Only then can meaningful observations be made about host specificity and parasite resistance. With the technological advances of the past few decades and continually decreasing costs of DNA sequencing, information about the relationships between parasites and their hosts, unraveling the roles of parasite and host interactions, and the origins and function of organelles is within our grasp.

Acknowledgments
The authors express their thanks to two anonymous reviewers who provided helpful feedback in improving the manuscript. Funding for this work was provided by the National Science Foundation (grant #1257472).

Authors' contributions
The following declarations about authors' contributions to the research have been made: conceived of the topic for this review and wrote the initial draft: EDS; provided editorial feedback, conceptual guidance and figures for the final manuscript: CEL.

Competing interests
No competing interests have been declared.

References
1. Lafferty KD, Allesina S, Arim M, Briggs CJ, de Leo G, Dobson AP, et al. Parasites in food webs: the ultimate missing links: parasites in food webs. Ecol Lett. 2008;11(6):533–546.
2. Wilson RJMI, Denny PW, Preiser PR, Rangachari K, Roberts K, Roy A, et al. Complete gene map of the plastid-like DNA of the malaria parasite *Plasmodium falciparum*. J Mol Biol. 1996;261(2):155–172.
3. Köhler S, Delwiche CF, Denny PW, Tilney LG, Webster P, Wilson RJM, et al. A plastid of probable green algal origin in apicomplexan parasites. Science. 1997;275(5305):1485–1489.
4. Goff LJ, Moon DA, Nyvall P, Stache B, Mangin K, Zuccarello G. The evolution of parasitism in the red algae: molecular comparisons of adelphoparasites and their hosts. J Phycol. 1996;32(2):297–312.
5. Zuccarello GC, Moon D, Goff LJ. A phylogenetic study of parasitic genera placed in the family Choreocolacaceae (Rhodophyta). J Phycol. 2004;40(5):937–945.
6. Goff LJ, Ashen J, Moon D. The evolution of parasites from their hosts: a case study in the parasitic red algae. Evolution. 1997;51(4):1068.
7. Goff LJ. The biology of parasitic red algae. In: Round F, Chapman D, editors. Progress in phycological research. Amsterdam: Elsevier Biomedical Press; 1982. p. 289–369. (vol 1).
8. Goff LJ, Coleman AW. The role of secondary pit connections in red algal parasitism. J Phycol. 2004;21(3):483–508.
9. Setchell WA. Parasitism among red algae. Proc Am Philos Soc. 1918;57(2):155–172.
10. Kurihara A, Abe T, Tani M, Sherwood AR. Molecular phylogeny and evolution of red algal parasites: a case study of *Benzaitenia*, *Janczewskia*, and *Ululania* (Ceramiales). J Phycol. 2010;46(3):580–590.
11. Verbruggen H, Maggs CA, Saunders GW, Le Gall L, Yoon H, de Clerck O. Data mining approach identifies research priorities and data requirements for resolving the red algal tree of life. BMC Evol Biol. 2010;10(1):16.
12. Pueschel CM, Cole KM. Rhodophycean pit plugs: an ultrastructural survey with taxonomic implications. Am J Bot. 1982;69(5):703.
13. Ramus J. Pit connection formation in the red alga *Pseudogloiophloea*.

J Phycol. 1969;5(1):57–63.

14. Ramus J. Properties of septal plugs from the red alga *Griffithsia pacifica*. Phycologia. 1971;10(1):99–103.

15. Turner CHC, Evans LV. Translocation of photoassimilated ^{14}C in the red alga *Polysiphonia lanosa*. Br Phycol J. 1978;13(1):51–55.

16. Hawkins EK. Observations on the developmental morphology and fine structure of pit connections in red algae. Cytologia. 1972;37(4):759–768.

17. Wetherbee R, Quirk H. The fine structure of secondary pit connection formation between the red algal alloparasite *Holmsella australis* and its red algal host *Gracilaria furcellata*. Protoplasma. 1982;176:166–176.

18. Wetherbee R, Quirk HM, Mallett JE, Ricker RW. The structure and formation of host-parasite pit connections between the red algal alloparasite *Harveyella mirabilis* and its red algal host *Odonthalia floccosa*. Protoplasma. 1984;119(1–2):62–73.

19. Blouin NA, Lane CE. Red algal parasites: models for a life history evolution that leaves photosynthesis behind again and again. Bioessays. 2012;34(3):226–235.

20. Wetherbee R, Quirk HM. The fine structure and cytology of the association between the parasitic red alga *Holmsella australis* and its red algal host *Gracilaria furcellata*. Protoplasma. 1982;110(3):153–165.

21. Goff LJ, Zuccarello G. The evolution of parasitism in red algae: cellular interactions of adelphoparasites and their hosts. J Phycol. 1994;30(4):695–720.

22. Goff LJ, Coleman AW. Fate of parasite and host organelle DNA during cellular transformation of red algae by their parasites. Plant Cell. 1995;7(11):1899.

23. Goff LJ, Coleman AW. Transfer of nuclei from a parasite to its host. Proc Natl Acad Sci USA. 1984;81(17):5420–5424.

24. Goff LJ, Coleman AW. Nuclear transfer from parasite to host: a new regulatory mechanism of parasitism. Ann NY Acad Sci. 1987;503(1):402–423.

25. Goff LJ. The biology of *Harveyella mirabilis* (Cryptonemiales, Rhodophyceae). V. Host responses to parasite infection. J Phycol. 1976;12(3):313–328.

26. Kugrens P, West JA. The ultrastructure of an alloparasitic red alga *Choreocolax polysiphoniae*. Phycologia. 1973;12(3–4):175–186.

27. Goff LJ, Coleman AW. Elucidation of fertilization and development in a red alga by quantitative DNA microspectrofluorometry. Dev Biol. 1984;102(1):173–194.

28. Borza T, Popescu CE, Lee RW. Multiple metabolic roles for the nonphotosynthetic plastid of the green alga *Prototheca wickerhamii*. Eukaryot Cell. 2005;4(2):253–261.

29. de Koning AP, Keeling PJ. The complete plastid genome sequence of the parasitic green alga *Helicosporidium* sp. is highly reduced and structured. BMC Biol. 2006;4(1):12.

30. Wolfe KH, Morden CW, Palmer JD. Function and evolution of a minimal plastid genome from a nonphotosynthetic parasitic plant. Proc Natl Acad Sci USA. 1992;89(22):10648–10652.

31. Cai X, Fuller AL, McDougald LR, Zhu G. Apicoplast genome of the coccidian *Eimeria tenella*. Gene. 2003;321:39–46.

32. Vaidya AB, Mather MW. Mitochondrial evolution and functions in malaria parasites. Annu Rev Microbiol. 2009;63(1):249–267.

33. Hancock L, Goff L, Lane C. Red algae lose key mitochondrial genes in response to becoming parasitic. Genome Biol Evol. 2010;2:897–910.

34. Callow JA, Callow ME, Evans LV. Nutritional studies on the parasitic red alga *Choreocolax polysiphoniae*. New Phytol. 1979;83(2):451–462.

35. Setchell WA. Parasitic florideae. I. Berkeley, CA: University of California Press; 1914. (University of California publications in botany; vol 6).

36. Court GJ. Photosynthesis and translocation studies of *Laurencia spectabilis* and its symbiont *Janczewskia gardneri* (Rhodophyceae). J Phycol. 1980;16(2):270–279.

37. Goff LJ. The biology of *Harveyella mirabilis* (Cryptonemiales, Rhodophyceae). VI. Translocation of photoassimilated ^{14}C. J Phycol. 1979;15(1):82–87.

38. Evans LV, Callow JA, Callow ME. Structural and physiological studies on the parasitic red alga *Holmsella*. New Phytol. 1973;72(2):393–402.

39. Kremer BP. Carbon economy and nutrition of the alloparasitic red alga *Harveyella mirabilis*. Mar Biol. 1983;76(3):231–239.

40. Zuccarello GC, West JA. Host specificity in the red algal parasites *Bostrychiocolax australis* and *Dawsoniocolax bostrychiae* (Choreocolacaceae, Rhodophyta). J Phycol. 1994;30(3):462–473.

41. Zuccarello GC, West JA. Genus and race specificity in the red algal parasite *Leachiella pacifica* (Choreocolacaceae, Rhodophyta). Phycologia. 1994;33(3):213–218.

42. Nonomura AM, West JA. Host specificity of *Janczewskia* (Ceramiales, Rhodophyta). Phycologia. 1981;20(3):251–258.

43. Zuccarello GC, West JA. Multiple cryptic species: molecular diversity and reproductive isolation in the *Bostrychia radicans*/*B. moritziana* complex (Rhodomelaceae, Rhodophyta) with focus on North American isolates. J Phycol. 2003;39(5):948–959.

30

A sea slug's guide to plastid symbiosis

Jan de Vries[1], Cessa Rauch[1], Gregor Christa[1,2], Sven B. Gould[1*]

[1] Institute for Molecular Evolution, Heinrich Heine-University, Düsseldorf 40225, Germany

[2] CESAM, University of Aveiro, 3810-193 Aveiro, Portugal

Abstract

Some 140 years ago sea slugs that contained chlorophyll-pigmented granules similar to those of plants were described. While we now understand that these "green granules" are plastids the slugs sequester from siphonaceous algae upon which they feed, surprisingly little is really known about the molecular details that underlie this one of a kind animal-plastid symbiosis. Kleptoplasts are stored in the cytosol of epithelial cells that form the slug's digestive tubules, and one would guess that the stolen organelles are acquired for their ability to fix carbon, but studies have never really been able to prove that. We also do not know how the organelles are distinguished from the remaining food particles the slugs incorporate with their meal and that include algal mitochondria and nuclei. We know that the ability to store kleptoplasts long-term has evolved only a few times independently among hundreds of sacoglossan species, but we have no idea on what basis. Here we take a closer look at the history of sacoglossan research and discuss recent developments. We argue that, in order to understand what makes this symbiosis work, we will need to focus on the animal's physiology just as much as we need to commence a detailed analysis of the plastids' photobiology. Understanding kleptoplasty in sacoglossan slugs requires an unbiased multidisciplinary approach.

Keywords: kleptoplasty; sacoglossan slugs; photosynthesis; plastid biology; photosynthetic slugs; evolution

Of leaves that crawl

Plastids are the main difference that distinguishes a plant or algal cell from animal cells. However, in 1876 de Negri and de Negri [1] described some green sea slugs to harbor granules that appeared to be stained green through chlorophyll pigments, similar to those of plant and algae plastids. It took almost another century before Kawaguti and Yamasu [2] could demonstrate that the globular chlorophyll bodies were identical to the plastids of the slug's algal food source (Fig. 1). Due to the nature by which the slugs acquire the plastids from their algal food source, these stolen organelles were termed kleptoplasts: stolen plastids. In his work on *Elysia crispata* (at that time known as *Tridachia crispata*), Trench [3] was one of the first to suggest that the slugs might specifically sequester the organelles for their ability to photosynthesize. Trench [3] not only demonstrated the incorporation of $^{14}CO_2$ through the plastids that are embedded within the epithelial cells that form the digestive glandular tubules, but also analyzed the perpetuation of the kleptoplast-slug relationship by separating the slugs from their algal prey. Ever since, starving the slugs has been a common approach to determine the slug's capacity to maintain functional kleptoplasts [4]. Along these lines the presence of photosynthesizing kleptoplasts was generally associated with the ability of some sacoglossans to survive starvation periods that can last many months [5]. This led researchers to coin the term of "leaves that crawl" [6].

In contrast to plants and algae, plastids of slugs are not vertically inherited; kleptoplasts have to be acquired by each new slug generation. Sacoglossan sea slugs have a highly specialized radula that consists of individual, serially organized teeth [7]. Only one tooth is used at a time and, when idle, stored in an autapomorphic structure called "saccus" [8], eponymous for the sacoglossan group. Some slug species can feed only on a single algal species and this might be associated with a specialized radula of mature animals [7]. In other cases, such as *Elysia viridis* that can feed for instance on *Codium* and *Bryopsis*, animals seem to have a more generally adapted radula allowing them to feed on a variety of different species [7]. However, this remains an observed correlation and it is difficult to imagine how one could provide empirical evidence at this point. Feeding experiments in the laboratory alone do not do the trick and radula mutation is far from feasible. Currently we do not know where food source selection actually begins. From what we know it could very well be that the "selective" animals can penetrate different siphonaceous algae and selection occurs downstream, not mechanically (hard and soft or small and large), but biochemically (sweet and sour or fresh and putrid).

* Corresponding author. Email: gould@hhu.de

Handling Editor: Andrzej Bodył

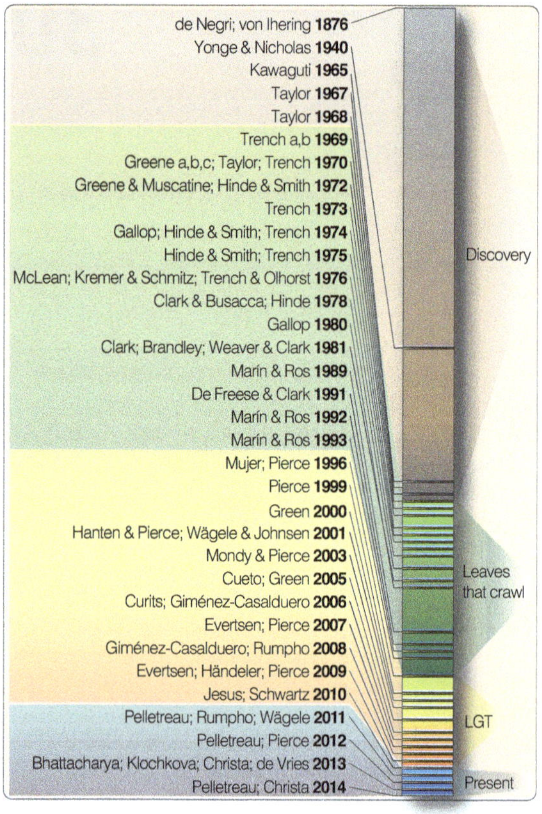

Fig. 1 Overview of about 140 years of research on green sacoglossan slugs. The timeline highlights key publications [1–6,8,10–17,20–23,27,33,40–43,46–49,52,53,56,61,63–94] on sacoglossan slugs since 1876. Four main periods (on the right) can be distinguished: the discovery phase, in which slugs and "chlorophyll-pigmented granules" were morphologically described (**a**), evidence for the incorporation of CO_2 suggested that slugs are "leaves that crawl" (**b**), and at a time where in general more and more gene transfers from one genome to another were identified, the concept was born that lateral gene transfer (LGT) from algae to slugs could support kleptoplasty in sacoglossans (**c**). The LGT concept dominated the field for 15 years until it was challenged for the first time in 2011 [12], which changed the way kleptoplasty in sacoglossan slugs is now viewed and studied. The listed manuscripts all refer to primary data manuscripts except the 1975 review by Trench, in which the term 'leaves that crawl' was coined. If an article has more than two authors, only the first author is listed.

The slugs do not feed on the entire alga, but rather use the radula's tooth to penetrate the cell wall of siphonaceous algae. They then suck out the entire cytosolic content of the algae including the organelles and all other compartments. This is not yet special, but in a few sacoglossan species only the plastids are selectively sequestered from the phagocytosed material. Individual food vacuoles can initially contain several kleptoplasts [9], but these are subsequently released into the cytosol after the vacuole is degraded. Considering that the vast majority of sacoglossans appear to treat incorporated plastids just like any other food particle, we can assume that kleptoplast retention happens on purpose. Still, the molecular mechanism of how plastids are initially recognized and subsequently released into the cytosol remains entirely unknown.

The ability of some kleptoplasts to remain functional inside the animal cells was long spearheaded by the idea that the slugs express genes they obtained through lateral gene transfer (LGT) from the algal nuclei, and which encode proteins that maintain plastid functionality [10,11]. When the first slug transcriptomes emerged that concept was challenged [12,13]. It has been discussed elsewhere in detail why LGT cannot support the stolen plastids [12,14], but, in brief, the reasons are (***i***) the meagre amount of photosynthesis-related transcript identified among slug messenger RNA – "one in a million reads" – and (***ii***) that algal genes have never been identified within in the genomic context of slug nuclear DNA. It now seems that intrinsic properties the stolen plastid bring along render some algal plastids more robust (that is "longer-living" in a foreign environment outside the algal/plant cytosol) than others [15–19]. How exactly is not known, but it is important to note that plastid transcription and translation can continue for months in some species and that their genomic coding capacity varies from that of land plastids [11,17].

The topsy-turvy of functional plastid retention among sacoglossan slugs

Based on pulse amplitude modulation (PAM)-fluorometry, and the determined maximum quantum yield (F_v/F_m) of photosystem II (which is commonly used to determine the photosynthetic capacity of slugs [20]), the majority of sacoglossans do not retain any functional plastids (non-retention species, NR) or for only a few days and up to two weeks (short-term retention species, StR). Seven non-monophyletic species are currently known to retain kleptoplasts with F_v/F_m values that remain on a level that is generally considered to account for the presence of a functional PSII for several months: *Elysia chlorotica*, *E. timida*, *E. crispata*, *E. clarki*, *E. viridis*, *Plakobranchus ocellatus* and *Costasiella ocellifera* [3,4,19,21–25]. All latter species are referred to as long-term retention (LtR) slugs.

But what really makes an LtR species? Genotyping incorporated algal plastids shows that the algal food sources of NR species sometimes matches to those of StR and LtR species [26]. The amount of plastids sequestered and stored in the digestive tubules also does not seemingly differ between long- and short-term retainers (Fig. 2; [19]). To make the matter even more complicated, some LtR species feed on several algae simultaneously [27–29]. What that means is that the food source alone cannot explain the long-term retention of functional kleptoplasts in LtR species. Interestingly though, plastid genome barcoding throughout starvation in the two polyphagous LtR species *E. clarki* and *P. ocellatus* suggest that the speed with which kleptoplasts are digested differs and depends on the plastid source [27,30,31]. It demonstrates that it takes two to tango: the right slug and the right plastid source.

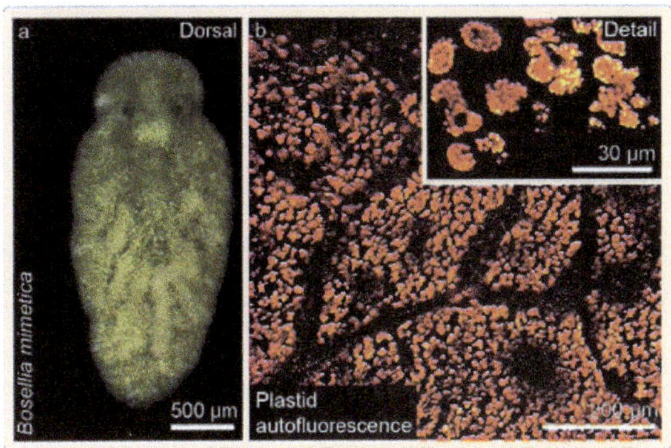

Fig. 2 Quantity does not equal quality. **a** The sacoglossan slug *Bosellia mimetica* harbours numerous kleptoplasts in its digestive tubular network, which (**b**) appear much denser in comparison to some LtR species (see [19,91]). Yet, *B. mimetica* is classified as a StR species; within a few days of starvation F_v/F_m values drop below those that are considered to represent functional photosynthesis [4].

Phylogenetic and photosynthetic analyses recently provided evidence that functional kleptoplasty evolved multiple times within the Plakobranchacea [25,32]. In turn, all basal shelled species are NR forms, but these species do not form a monophyletic group [4,19]. Virtually no molecular study has yet addressed the issue of uncovering the animals' mechanisms that determine the mode of retention. Comparative analyses of different species, in particular the molecular differences between NR, StR and LtR species, represent a promising tool to do so. For example, the trend that NR slugs engulf plastids in phagosomal membranes was noticed before [33,34]. NR species seem to keep the incorporated plastids inside phagosomes for immediate degradation all the time; they do not appear to ever release them into the cytosol [33,34]. StR and LtR forms on the contrary are known to retain kleptoplasts – that are released from phagosomes – in a similar fashion throughout starvation [19]. That is, StR do not appear to degrade kleptoplasts quicker than LtR species, but empirical evidence (for example based on $^{14}CO_2$ incorporation data) is currently lacking. Both, StR and LtR species, should have, in theory, the same "point of departure" when starvation commences and should have the same potential to make use of the kleptoplasts sequestered. Yet they do not. Notably, all LtR species have in common that cytosolic kleptoplast are left surrounded by two membranes only; they are those that constitute the canonical two membranes we are familiar with from land plant plastids [35–37]. This is even true for *E. chlorotica* that feeds on the stramenopile *Vaucheria*, but which houses complex plastids that in the alga are surrounded by four membranes [36,38,39]. That always these two membranes remain – the two that trace back to outer membrane and plasma membrane of the original cyanobacterial endosymbiont [37] – might hint at how, and in fact that, substrate and metabolites are actively exchanged between animal and plastid.

Darkness is more than the absence of photosynthesis

Although it is commonly accepted that the slugs benefit from photosynthesis, direct evidence is surprisingly scarce. Various studies on kleptoplastic slugs have analyzed survival rates or demonstrated that $^{14}CO_2$ is fixed by the acquired plastids [40–45]. To what degree the quantity of such carbon compounds is then physiologically relevant has, however, not yet been satisfactorily shown. Photosynthates sustaining the slugs might not be the sole necessity for sacoglossans to survive (long lasting) starvation periods. Note that for example the NR species *Costasiella nonatoi* survives starvation for about a month without showing any measurable PSII activity [25]. Similar observations have been made for *Elysia nigrocapitata* [46]. The question remains to what degree carbon fixation in sacoglossan slugs is needed to endure starvation (Fig. 3) and whether there is maybe another primary reason for starvation survival. If so, benefiting from functional kleptoplasts comes second.

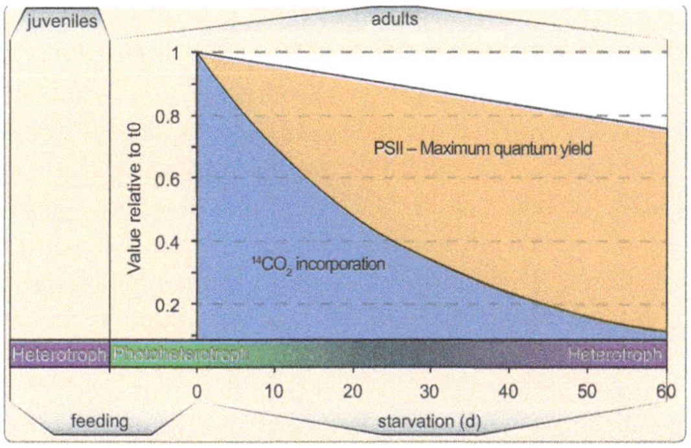

Fig. 3 Schematic overview of kleptoplast performance in LtR species. When the slugs hatch, they immediately need to feed on algal cytosol and begin to sequester their first plastids. During this phase, called transient kleptoplasty, juveniles are basically heterotrophic. Adult slugs are photoheterotrophic animals: they graze upon algae as long as these are available, but at the same time house CO_2-fixing kleptoplasts. When deprived of their food LtR species likely benefit from photosynthesis during the early phases of starvation, but the amount of incorporated carbon cannot sustain the animal. They switch back to heterotrophy through efficiently digesting their own tissue and degrading kleptoplasts. The latter is the reason why all starving LtR species are observed to shrink. Published CO_2 fixation rates for *C. ocellifera*, *E. viridis*, *P. ocellatus* and *E. timida* (blue curve; [14,22,40,43,69]) and photosystem II maximum quantum yield (F_v/F_m) values for *C. ocellifera*, *P. ocellatus* and *E. timida* (orange curve; [14,25]) of various studies on LtR species were pooled and plotted in relation to the values measured for freshly fed animals. The contrast between the two curves highlights that caution is warranted when kleptoplast productivity is evaluated on F_v/F_m values alone.

Previous studies reported the incorporation of photosynthetically fixed carbon (stemming from [^{14}C] bicarbonate) in a variety of slug metabolites [47,48]. Neglecting for now that we do not know enough about what photosynthate supports the animal (and to what degree), it is important to ask: how do the slugs acquire products of photosynthesis? Kleptoplast-synthesized substrate will inevitably end up being metabolized, no matter whether that substrate was actively provided by an intact kleptoplast or whether it stems from an organelle degraded (e.g. through an autophagosome). It is important to determine which initial route the labeled CO_2 took to be incorporated into slug-specific metabolites. Microscopic analyses of *E. viridis* suggest that kleptoplasts that originate from *Codium fragile* accumulate substrate such as starch during starvation [49,50], which raises the question if any is actively secreted by the stolen organelles. Apart from any energy support the plastids might provide for the slugs, the animals might benefit from various other biochemical pathways the stolen plastids bring along [19,51]. It is also possible that photosynthates and kleptoplast-derived metabolites play a more crucial role for the proper development of juveniles than for the maintenance of mature adult slugs as recently suggested [14]. This is supported by recent observations on juveniles of *E. chlorotica* [52]. The data provided evidence with regard to the need of kleptoplast-derived lipid production for the proper development of the animals.

Animals are often kept in the dark, as a control for the slugs' dependency on photosynthates [52–54]. This is a good time to remember that the absence of light translates into more than just the absence of plastid photosynthesis. A few essential biochemical processes, such as the synthesis of vitamin D [55], occur in a light-dependent manner. Sacoglossans can synthesize a series of unique secondary metabolites including elysiapyrones [56] and tridachione [44] and it has been suggested that their synthesis is light dependent. Their synthesis could still be linked to the presence of kleptoplasts, but not photosynthesis. It has been hypothesized that an essential function of these polypropionates is linked to the quenching of ROS [56,57], hereby protecting the kleptoplasts' photosystems and/or the slugs' own tissue from oxidative stress.

Morphology does not equal function

The lateral foot expansions – the parapodia – are the morphological foundation from which the term "leaves that crawl" originates. In many plastid-bearing slugs the parapodia are intervened with numerous digestive tubules that harbor the green kleptoplasts; the wing-shaped structures are reminiscent of a leaf (Fig. 2). Some studies claim that the opening and closing of the parapodia is a controlled response to different light intensities, to either expose or shield the kleptoplasts to and from sunlight, respectively [58,59]. But these observations are to be interpreted cautiously [60]. There is no question that the position of the parapodia affects the amount of light reaching the kleptoplasts [61], but the response of some species is slow and it takes many minutes for the slugs to close the parapodia. Moreover, the LtR species *P. ocellatus* does not appear to alter the position of its parapodia at all. Even species that completely lack parapodia can retain functional plastids for a long time, as the case of the LtR species *C. ocellifera* demonstrates [25]. Parapodia are furthermore not limited to sacoglossan slugs. A wide range of heterobranchs have evolved parapodia that are used for various purposes including swimming and digging. In a nutshell, there is currently no evidence, or reason to assume, that sacoglossan parapodia have evolved as a consequence of housing kleptoplasts.

Two steps forward, one step back

Sacoglossan sea slugs puzzle researchers, as much as they fascinate. Ed Yong said it best when he commented on a recent analysis, which provided evidence that some adult slug species survive in the dark and do not loose weight faster than those kept in the light [14]: "Good science is about resisting the pull of easy conclusions. It's about testing stories that seem like they should be right to see if they actually are right. This is no easy task. Consider the case of the 'solar-powered' slugs" [62]. Several concepts have been put forward trying to explain how an organelle, adapted to plant cells, remains functional inside the cytosol of a eumetazoan cell. Two key concepts, slugs become photoautotrophs through kleptoplasty and kleptoplasts are supported by laterally transferred genes, are currently challenged [12,14,46], in particular the latter [12,13]. Research on the animal-plastid symbiosis in sacoglossan sea slugs is at a turning point. We know that some slugs sequester plastids through a sophisticated phagocytic mechanism whose details remain currently unknown. The kleptoplasts can continue to photosynthesize in the cytosol of the slugs' epithelial cells, but adults do not strictly depend on ongoing photosynthesis to survive starvation periods long-term. Laterally transferred algal genes do not support stolen plastids: the kleptoplasts' very own

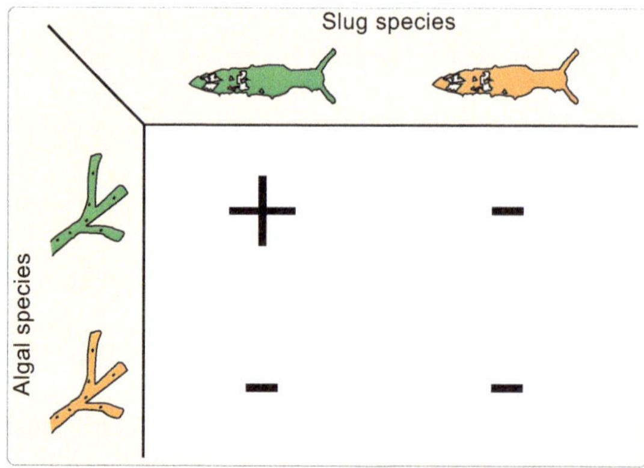

Fig. 4 Animal-plastid compatibility is determined by a two component system. In order to profit from functional kleptoplasts long-term (+), a slug species – component A – with an adapted physiology and digestive mechanism must acquire plastids from an algal species – component B – that naturally has robust plastids (both in green). The multifaceted factors that determine the compatibility of both partners in the plastid-slug system are yet to be discovered, but they could include dealing with plastid-derived toxins in case of the slugs and an adapted photobiology in case of the plastids. If any of the two partners lacks these requirements (orange), only short- or non-functional kleptoplast retentions are established (–).

biochemistry continues to function more independently than previously anticipated. From what we can tell, animal-plastid symbiosis depends on a combination of a robust plastid and a slug species whose physiology has evolved to tolerate (and likely service through substrate and metabolite exchange) an alien organelle (Fig. 4).

From the animals' perspective the following questions are hence key to better understand this unique symbiosis and how functional kleptoplasty has evolved multiple times in different slug species: (*i*) What are the benefits for the slugs next to carbon fixation? (*ii*) What underpins kleptoplast compatibility in the few species so far identified that house functional plastids for months? (*iii*) What are the molecular details of how plastids are recognized by the animals and released into the cytosol? A multidisciplinary approach will contribute to our understanding of how an organelle functions in a cytosol for which it did not evolve, and how some algae plastids have evolved to become as robust as they are. Through recent developments we now have the opportunity to make progress on this unique biological system of general evolutionary interest, for the purpose of better understanding plant animal symbioses.

Acknowledgments

Funding through the Deutsche Forschungsgemeinschaft to SBG (DFG; GO1825/4-1) and the DAAD to GC (P.R.I.M.E. program) is gratefully acknowledged. We thank Steffen Köhler (CAi, HHU) for photography and Bill Martin for suggesting Fig. 4.

Authors' contributions

The following declarations about authors' contributions to the research have been made: all authors have contributed equally to the review of literature. JV and SBG drafted the manuscript, whose final version was written and approved by all four authors.

Competing interests

No competing interests have been declared.

References

1. de Negri A, de Negri G. Farbstoff aus *Elysia viridis*. Ber Deut Chem Gesell. 1876;9:84.
2. Kawaguti S, Yamasu T. Electron microscopy on the symbiosis between an elysioid gastropod and chloroplasts of a green alga. Biol J Okayama Univ. 1965;11(3-4):57–65.
3. Trench RK. Chloroplasts as functional endosymbionts in the mollusc *Tridachia crispata* (Bërgh), (Opisthobranchia, Sacoglossa). Nature. 1969;222(5198):1071–1072.
4. Händeler K, Grzymbowski YP, Krug PJ, Wägele H. Functional chloroplasts in metazoan cells – a unique evolutionary strategy in animal life. Front Zool. 2009;6(1):28.
5. Trench RK, Boyle JE, Smith DC. The association between chloroplasts of *Codium fragile* and the mollusc *Elysia viridis*. III. Movement of photosynthetically fixed ^{14}C in tissues of intact living *E. viridis* and in *Tridachia crispata*. Proc R Soc B. 1974;185(1081):453–464.
6. Trench RK. Of "leaves that crawl": functional chloroplasts in animal cells. Symp Soc Exp Biol. 1975;(29):229–265.
7. Jensen KR. Morphological adaptations and plasticity of radular teeth of the Sacoglossa (=Ascoglossa) (Mollusca: Opisthobranchia) in relation to their food plants. Biol J Linn Soc. 1993;48(2):135–155.
8. von Ihering H. Versuch eines natürlichen Systems der Mollusken. Jahrb Dtsch Malakozool Ges. 1876;3:97–174.
9. Schmitt V, Händeler K, Gunkel S, Escande ML, Menzel D, Gould SB, et al. Chloroplast incorporation and long-term photosynthetic performance through the life cycle in laboratory cultures of *Elysia timida* (Sacoglossa, Heterobranchia). Front Zool. 2014;11(1):5.
10. Pierce S, Biron R, Rumpho M. Endosymbiotic chloroplasts in molluscan cells contain proteins synthesized after plastid capture. J Exp Biol. 1996;199(10):2323–2330.
11. Rumpho ME, Worful JM, Lee J, Kannan K, Tyler MS, Bhattacharya D, et al. Horizontal gene transfer of the algal nuclear gene *psbO* to the photosynthetic sea slug *Elysia chlorotica*. Proc Natl Acad Sci USA. 2008;105(46):17867–17871.
12. Wägele H, Deusch O, Handeler K, Martin R, Schmitt V, Christa G, et al. Transcriptomic evidence that longevity of acquired plastids in the photosynthetic slugs *Elysia timida* and *Plakobranchus ocellatus* does not entail lateral transfer of algal nuclear genes. Mol Biol Evol. 2011;28(1):699–706.
13. Bhattacharya D, Pelletreau KN, Price DC, Sarver KE, Rumpho ME. Genome analysis of *Elysia chlorotica* egg DNA provides no evidence for horizontal gene transfer into the germ line of this kleptoplastic mollusc. Mol Biol Evol. 2013;30(8):1843–1852.
14. Christa G, Zimorski V, Woehle C, Tielens AGM, Wägele H, Martin WF, et al. Plastid-bearing sea slugs fix CO_2 in the light but do not require photosynthesis to survive. Proc Biol Sci. 2014;281(1774):20132493.
15. Trench RK, Boyle JE, Smith DC. The association between chloroplasts of *Codium fragile* and the mollusc *Elysia viridis*. II. Chloroplast ultrastructure and photosynthetic carbon fixation in *E. viridis*. Proc R Soc B. 1973;184(1074):63–81.
16. Trench RK, Ohlhorst S. The stability of chloroplasts from siphonaceous algae in symbiosis with sacoglossan molluscs. New Phytol. 1976;76(1):99–109.
17. de Vries J, Habicht J, Woehle C, Huang C, Christa G, Wägele H, et al. Is *ftsH* the key to plastid longevity in sacoglossan slugs? Genome Biol Evol. 2013;5(12):2540–2548.
18. Giles KL, Sarafis V. Chloroplast survival and division in vitro. Nat New Biol. 1972;236(63):56–58.
19. de Vries J, Christa G, Gould SB. Plastid survival in the cytosol of animal cells. Trends Plant Sci. 2014;19(6):347–350.
20. Wägele M, Johnsen G. Observations on the histology and photosynthetic performance of "solar-powered" opisthobranchs (Mollusca, Gastropoda, Opisthobranchia) containing symbiotic chloroplasts or zooxanthellae. Org Divers Evol. 2001;1(3):193–210.
21. Taylor DL. Photosynthesis of symbiotic chloroplasts in *Tridachia crispata* (Bërgh). Comp Biochem Physiol Physiol. 1971;38(1):233–236.
22. Marín A, Ros JD. The chloroplast-animal association in four Iberian sacoglossan opisthobranchs: *Elysia timida*, *Elysia translucens*, *Thuridilla hopei* and *Bosellia mimetica*. Sci Mar. 1989;53(2–3):429–440.
23. Mujer CV, Andrews DL, Manhart JR, Pierce SK, Rumpho ME. Chloroplast genes are expressed during intracellular symbiotic association of *Vaucheria litorea* plastids with the sea slug *Elysia chlorotica*. Proc Natl Acad Sci USA. 1996;93(22):12333–12338.
24. Pierce SK, Curtis NE, Massey SE, Bass AL, Karl SA, Finney CM. A morphological and molecular comparison between *Elysia crispata* and a new species of kleptoplastic sacoglossan sea slug (Gastropoda: Opisthobranchia) from the Florida Keys, USA. Mollus Res. 2006;26(1):23–38.
25. Christa G, Gould SB, Franken J, Vleugels M, Karmeinski D, Handeler K, et al. Functional kleptoplasty in a limapontioidean genus: phylogeny, food preferences and photosynthesis in *Costasiella* with a focus on

C. ocellifera (Gastropoda: Sacoglossa). J Mollus Stud. 2014.

26. Christa G, Händeler K, Schäberle TF, König GM, Wägele H. Identification of sequestered chloroplasts in photosynthetic and nonphotosynthetic sacoglossan sea slugs (Mollusca, Gastropoda). Front Zool. 2014;11(1):15.
27. Christa G, Wescott L, Schäberle TF, König GM, Wägele H. What remains after 2 months of starvation? Analysis of sequestered algae in a photosynthetic slug, *Plakobranchus ocellatus* (Sacoglossa, Opisthobranchia), by barcoding. Planta. 2013;237(2):559–572.
28. Curtis NE, Schwartz JA, Pierce SK. Ultrastructure of sequestered chloroplasts in sacoglossan gastropods with differing abilities for plastid uptake and maintenance: chloroplast degradation in four sacoglossans. Invertebr Biol. 2010;129(4):297–308.
29. Maeda T, Hirose E, Chikaraishi Y, Kawato M, Takishita K, Yoshida T, et al. Algivore or phototroph? *Plakobranchus ocellatus* (Gastropoda) continuously acquires kleptoplasts and nutrition from multiple algal species in nature. PLoS ONE. 2012;7(7):e42024.
30. Christa G, de Vries J, Jahns P, Gould SB. Switching off photosynthesis: the dark side of sacoglossan slugs. Commun Integr Biol. 2014;7(1):e28029.
31. Christa G, Händeler K, Kück P, Vleugels M, Franken J, Karmeinski D, et al. Phylogenetic evidence for multiple independent origins of functional kleptoplasty in Sacoglossa (Heterobranchia, Gastropoda). Org Divers Evol. 2014 (in press).
32. Maeda T, Kajita T, Maruyama T, Hirano Y. Molecular phylogeny of the Sacoglossa, with a discussion of gain and loss of kleptoplasty in the evolution of the group. Biol Bull. 2010;219(1):17–26.
33. McLean N. Phagocytosis of chloroplasts in *Placida dendritica* (Gastropoda: Sacoglossa). J Exp Zool. 1976;197(3):321–329.
34. Wägele H, Martin WF. Endosymbioses in sacoglossan seaslugs: plastid-bearing animals that keep photosynthetic organelles without borrowing genes. In: Löffelhardt W, editor. Endosymbiosis. Vienna: Springer; 2014. p. 291–324.
35. Jensen PE, Leister D. Chloroplast evolution, structure and functions. F1000Prime Rep. 2014;6(40).
36. Zimorski V, Ku C, Martin WF, Gould SB. Endosymbiotic theory for organelle origins. Curr Opin Microbiol. 2014;22:38–48.
37. Cavalier-Smith T. Membrane heredity and early chloroplast evolution. Trends Plant Sci. 2000;5(4):174–182.
38. Archibald JM. The puzzle of plastid evolution. Curr Biol. 2009;19(2):R81–R88.
39. Keeling PJ. The endosymbiotic origin, diversification and fate of plastids. Phil Trans R Soc B. 2010;365(1541):729–748.
40. Greene RW. Symbiosis in sacoglossan opisthobranchs: functional capacity of symbiotic chloroplasts. Mar Biol. 1970;7(2):138–142.
41. Kremer BP, Schmitz K. Aspects of $^{14}CO_2$-fixation by endosymbiotic rhodoplasts in the marine opisthobranchiate *Hermaea bifida*. Mar Biol. 1976;34(4):313–316.
42. Hinde R. The metabolism of photosynthetically fixed carbon by isolated chloroplasts from *Codium fragile* (Chlorophyta: Siphonales) and by *Elysia viridis* (Mollusca: Sacoglossa). Biol J Linn Soc. 1978;10(3):329–342.
43. Clark KB, Jensen KR, Stirts HM, Fermin C. Chloroplast symbiosis in a non-elysiid mollusc, *Costasiella lilianae* Marcus [Hermaeidae: Ascoglossa (=Sacoglossa)]: effects of temperature, light intensity, and starvation on carbon fixation rate. Biol Bull. 1981;160(1):43–54.
44. Ireland C, Scheuer PJ. Photosynthetic marine mollusks: in vivo ^{14}C incorporation into metabolites of the sacoglossan *Placobranchus ocellatus*. Science. 1979;205(4409):922–923.
45. Yamamoto S, Hirano YM, Hirano YJ, Trowbridge CD, Akimoto A, Sakai A, et al. Effects of photosynthesis on the survival and weight retention of two kleptoplastic sacoglossan opisthobranchs. J Mar Biol Assoc UK. 2013;93(1):209–215.
46. Klochkova TA, Han JW, Chah KH, Kim RW, Kim JH, Kim KY, et al. Morphology, molecular phylogeny and photosynthetic activity of the sacoglossan mollusc, *Elysia nigrocapitata*, from Korea. Mar Biol. 2013;160(1):155–168.
47. Trench ME, Trench RK, Muscatine L. Utilization of photosynthetic products of symbiotic chloroplasts in mucus synthesis by *Placobranchus ianthobapsus* (Gould), Opisthobranchia, Sacoglossa. Comp Biochem Physiol. 1970;37(1):113–117.
48. Greene RW, Muscatine L. Symbiosis in sacoglossan opisthobranchs: photosynthetic products of animal-chloroplast associations. Mar Biol. 1972;14(3):253–259.
49. Evertsen J, Johnsen G. In vivo and in vitro differences in chloroplast functionality in the two north Atlantic sacoglossans (Gastropoda, Opisthobranchia) *Placida dendritica* and *Elysia viridis*. Mar Biol. 2009;156(5):847–859.
50. Hawes CR, Cobb AH. The effects of starvation on the symbiotic chloroplasts in *Elysia viridis*: a fine structural study. New Phytol. 1980;84(2):375–379.
51. Gould SB, Waller RF, McFadden GI. Plastid evolution. Annu Rev Plant Biol. 2008;59(1):491–517.
52. Pelletreau KN, Weber APM, Weber KL, Rumpho ME. Lipid accumulation during the establishment of kleptoplasty in *Elysia chlorotica*. PLoS ONE. 2014;9(5):e97477.
53. Hinde R, Smith DC. The role of photosynthesis in the nutrition of the mollusc *Elysia viridis*. Biol J Linn Soc. 1975;7(2):161–171.
54. Akimoto A, Hirano YM, Sakai A, Yusa Y. Relative importance and interactive effects of photosynthesis and food in two solar-powered sea slugs. Mar Biol. 2014;161(5):1095–1102.
55. Lips P. Vitamin D physiology. Prog Biophys Mol Biol. 2006;92(1):4–8.
56. Cueto M, D'Croz L, Maté JL, San-Martín A, Darias J. Elysiapyrones from *Elysia diomedea*. Do such metabolites evidence an enzymatically assisted electrocyclization cascade for the biosynthesis of their bicyclo[4.2.0]octane core? Org Lett. 2005;7(3):415–418.
57. Díaz-Marrero AR, Cueto M, D'Croz L, Darias J. Validating an endoperoxide as a key intermediate in the biosynthesis of elysiapyrones. Org Lett. 2008;10(14):3057–3060.
58. Schmitt V, Wägele H. Behavioral adaptations in relation to long-term retention of endosymbiotic chloroplasts in the sea slug *Elysia timida* (Opisthobranchia, Sacoglossa). Thalassas. 2011;27(225):225–238.
59. Rahat M, Monselise EBI. Photobiology of the chloroplast hosting mollusc *Elysia timida* (Opisthobranchia). J Exp Biol. 1979;79(1):225–233.
60. Serôdio J, Cruz S, Cartaxana P, Calado R. Photophysiology of kleptoplasts: photosynthetic use of light by chloroplasts living in animal cells. Phil Trans R Soc B. 2014;369(1640):20130242.
61. Jesus B, Ventura P, Calado G. Behaviour and a functional xanthophyll cycle enhance photo-regulation mechanisms in the solar-powered sea slug *Elysia timida* (Risso, 1818). J Exp Mar Biol Ecol. 2010;395(1–2):98–105.
62. Yong E. Solar-powered slugs are not solar-powered [Internet]. Natl Geogr. 2013 [cited 2014 Oct 28]; Available from:
63. Yonge CM, Nicholas AG. Structure and function of the gut and

symbiosis with zooxanthellae in *Tridachia crispate* (Oerst.) Bgh. Pap Tortugas Lab. 1940;32:287.

64. Taylor DL. The pigments of the zooxanthellae symbiotic with the intertidal *Anemone, Anemonia sulcata*. J Phycol. 1967;3(4):238–240.

65. Taylor DL. Chloroplasts as symbiotic organelles in the digestive gland of *Elysia viridis* (Gastropoda: Opisthobranchia). J Mar Biol Assoc UK. 1968;48(1):1–15.

66. Trench RK. Chloroplasts as functional organelles in animal tissues. J Cell Biol. 1969;42(2):404–417.

67. Greene RW. Symbiosis in sacoglossan opisthobranchs: symbiosis with algal chloroplasts. Malacologia. 1970;10(2):357–368.

68. Greene RW. Symbiosis in sacoglossan opisthobranchs: translocation of photosynthetic products from chloroplast to host tissue. Malacologia. 1970;10(2):369–380.

69. Hinde R, Smith DC. Persistence of functional chloroplasts in *Elysia viridis* (Opisthobranchia, Sacoglossa). Nat New Biol. 1972;239(88):30–31.

70. Gallop A. Evidence for the presence of a "factor" in *Elysia viridis* which stimulates photosynthate release from its symbiotic chloroplasts. New Phytol. 1974;73(6):1111–1117.

71. Hinde R, Smith DC. "Chloroplast symbiosis" and the extent to which it occurs in Sacoglossa (Gastropoda: Mollusca). Biol J Linn Soc. 1974;6(4):349–356.

72. Clark KB, Busacca M. Feeding specificity and chloroplast retention in four tropical Ascoglossa, with a discussion of the extent of chloroplast symbiosis and the evolution of the order. J Mollus Stud. 1978;44(3):272–282.

73. Gallop A, Bartrop J, Smith DC. The biology of chloroplast acquisition by *Elysia viridis*. Proc R Soc B. 1980;207(1168):335–349.

74. Brandley BK. Ultrastructure of the envelope of *Codium australicum* (Silva) chloroplasts in the alga and after acquisition by *Elysia maoria* (Powell). New Phytol. 1981;89(4):679–686.

75. Weaver S, Clark KB. Light intensity and color preferences of five ascoglossan (=sacoglossan) molluscs (Gastropoda: Opisthobranchia): a comparison of chloroplast-symbiotic and aposymbiotic species. Mar Behav Physiol. 1981;7(4):297–306.

76. de Freese DE, Clark KB. Transepidermal uptake of dissolved free amino acids from seawater by three ascoglossan opisthobranchs. J Mollus Stud. 1991;57(suppl, pt 4):65–74.

77. Marín A, Ros JD. Dynamics of a peculiar plant-herbivore relationship: the photosynthetic ascoglossan *Elysia timida* and the chlorophycean *Acetabularia acetabulum*. Mar Biol. 1992;112(4):677–682.

78. Marín A, Ros J. Ultrastructural and ecological aspects of the development of chloroplast retention in the sacoglossan gastropod *Elysia timida*. J Mollus Stud. 1993;59(1):95–104.

79. Pierce SK, Maugel TK, Rumpho ME, Hanten JJ, Mondy WL. Annual viral expression in a sea slug population: life cycle control and symbiotic chloroplast maintenance. Biol Bull. 1999;197(1):1–6.

80. Green BJ. Mollusc-algal chloroplast endosymbiosis. Photosynthesis, thylakoid protein maintenance, and chloroplast gene expression continue for many months in the absence of the algal nucleus. Plant Physiol. 2000;124(1):331–342.

81. Hanten JJ, Pierce SK. Synthesis of several light-harvesting complex I polypeptides is blocked by cycloheximide in symbiotic chloroplasts in the sea slug, *Elysia chlorotica* (Gould): a case for horizontal gene transfer between alga and animal? Biol Bull. 2001;201(1):34–44.

82. Mondy WL, Pierce SK. Apoptotic-like morphology is associated with annual synchronized death in kleptoplastic sea slugs (*Elysia chlorotica*). Invertebr Biol. 2003;122(2):126–137.

83. Green BJ, Fox TC, Rumpho ME. Stability of isolated algal chloroplasts that participate in a unique mollusc/kleptoplast association. Symbiosis. 2005;40(1):31–40.

84. Curtis NE, Massey SE, Pierce SK. The symbiotic chloroplasts in the sacoglossan *Elysia clarki* are from several algal species. Invertebr Biol. 2006;125(4):336–345.

85. Casalduero FG, Muniain C. Photosynthetic activity of the solar-powered lagoon mollusc *Elysia timida* (Risso, 1818) (Opisthobranchia: Sacoglossa). Symbiosis. 2006;41(3):151–158.

86. Evertsen J, Burghardt I, Johnsen G, Wägele H. Retention of functional chloroplasts in some sacoglossans from the Indo-Pacific and Mediterranean. Mar Biol. 2007;151(6):2159–2166.

87. Pierce SK, Curtis NE, Hanten JJ, Boerner SL, Schwartz JA. Transfer, integration and expression of functional nuclear genes between multicellular species. Symbiosis. 2007;43(2):57–64.

88. Casalduero FG, Muniain C. The role of kleptoplasts in the survival rates of *Elysia timida* (Risso, 1818): (Sacoglossa: Opisthobranchia) during periods of food shortage. J Exp Mar Bio Ecol. 2008;357(2):181–187.

89. Pierce SK, Curtis NE, Schwartz JA. Chlorophyll a synthesis by an animal using transferred algal nuclear genes. Symbiosis. 2009;49(3):121–131.

90. Schwartz JA, Curtis NE, Pierce SK. Using algal transcriptome sequences to identify transferred genes in the sea slug, *Elysia chlorotica*. Evol Biol. 2010;37(1):29–37.

91. Pelletreau KN, Bhattacharya D, Price DC, Worful JM, Moustafa A, Rumpho ME. Sea slug kleptoplasty and plastid maintenance in a metazoan. Plant Physiol. 2011;155(4):1561–1565.

92. Rumpho ME, Pelletreau KN, Moustafa A, Bhattacharya D. The making of a photosynthetic animal. J Exp Biol. 2011;214(2):303–311.

93. Pelletreau KN, Worful JM, Sarver KE, Rumpho ME. Laboratory culturing of *Elysia chlorotica* reveals a shift from transient to permanent kleptoplasty. Symbiosis. 2012;58(1–3):221–232.

94. Pierce SK, Fang X, Schwartz JA, Jiang X, Zhao W, Curtis NE, et al. Transcriptomic evidence for the expression of horizontally transferred algal nuclear genes in the photosynthetic sea slug, *Elysia chlorotica*. Mol Biol Evol. 2012;29(6):1545–1556.

31

Monophyly of Archaeplastida supergroup and relationships among its lineages in the light of phylogenetic and phylogenomic studies. Are we close to a consensus?

Paweł Mackiewicz*, Przemysław Gagat

Department of Genomics, Faculty of Biotechnology, University of Wrocław, Fryderyka Joliot-Curie 14a, 50-383 Wrocław, Poland

Abstract

One of the key evolutionary events on the scale of the biosphere was an endosymbiosis between a heterotrophic eukaryote and a cyanobacterium, resulting in a primary plastid. Such an organelle is characteristic of three eukaryotic lineages, glaucophytes, red algae and green plants. The three groups are usually united under the common name Archaeplastida or Plantae in modern taxonomic classifications, which indicates they are considered monophyletic. The methods generally used to verify this monophyly are phylogenetic analyses. In this article we review up-to-date results of such analyses and discussed their inconsistencies. Although phylogenies of plastid genes suggest a single primary endosymbiosis, which is assumed to mean a common origin of the Archaeplastida, different phylogenetic trees based on nuclear markers show monophyly, paraphyly, polyphyly or unresolved topologies of Archaeplastida hosts. The difficulties in reconstructing host cell relationships could result from stochastic and systematic biases in data sets, including different substitution rates and patterns, gene paralogy and horizontal/endosymbiotic gene transfer into eukaryotic lineages, which attract Archaeplastida in phylogenetic trees. Based on results to date, it is neither possible to confirm nor refute alternative evolutionary scenarios to a single primary endosymbiosis. Nevertheless, if trees supporting monophyly are considered, relationships inferred among Archaeplastida lineages can be discussed. Phylogenetic analyses based on nuclear genes clearly show the earlier divergence of glaucophytes from red algae and green plants. Plastid genes suggest a more complicated history, but at least some studies are congruent with this concept. Additional research involving more representatives of glaucophytes and many understudied lineages of Eukaryota can improve inferring phylogenetic relationships related to the Archaeplastida. In addition, alternative approaches not directly dependent on phylogenetic methods should be developed.

Keywords: Archaeplastida; Glaucophyta; monophyly; phylogenomics; phylogenetic analyses; primary plastid; Rhodophyta; Viridiplantae

Introduction

Two-membrane plastids are characteristic of three eukaryotic lineages, glaucophytes (Glaucophyta), red algae (Rhodophyta) and green plants (Viridiplantae or Chloroplastida) [1]. Glaucophytes comprise a small group of freshwater unicellular algae of only 13 species [2]. Rhodophytes constitute a large assemblage of both unicellular and multicellular algae, mainly sea inhabitants, with about 5000–6000 species [3]. The green clade is the largest, with 350 000 widespread species in both aquatic (mainly green algae) and terrestrial environments (land plants) [4]. It has highly diversified into unicellular, colonial and multicellular forms, including land plants with composite tissue organization. These three major lineages are commonly classified together in the Archaeplastida or Plantae, one of several evolutionary supergroups of the domain Eukaryota [5–7].

There is a general consensus that Archaeplastida plastids are of prokaryotic origin, the result of endosymbiosis between a heterotrophic eukaryotic host and a photosynthetic cyanobacterium [8–10]. In support of this, Archaeplastida plastid genomes share similar gene contents and conserved gene arrangement with cyanobacterial genomes [11,12], including an unusual tRNA-Leu group I intron [13]. Glaucophyte plastids (called cyanelles) retain more cyanobacterial characteristics than do red alga and green plant plastids. The most remarkable is the presence of peptidoglycan, a typical component of bacterial cell walls located between the inner and outer membrane [14], as well as carboxysomes with enzymes involved in carbon fixation [15]. In addition, glaucophytes and rhodophytes still possess phycobilisomes, typical cyanobacterial photosystem II light-harvesting antennae [16].

* Corresponding author. Email: pamac@smorfland.uni.wroc.pl

Handling Editor: Andrzej Bodył

There are many features, beside those mentioned above, suggesting a common origin of Archaeplastida plastids. For example, they all are bound by two membranes, corresponding to the outer and inner membranes of their cyanobacterial ancestor. This distinguishes them from complex plastids surrounded by three or four membranes present in many other photosynthetic (and also heterotrophic) eukaryotes. In contrast to the Archaeplastida plastids, complex plastids did not evolve directly from a cyanobacterium but arose by endosymbioses of green or red algae with other eukaryotes [8,9,17,18]. As a result, Archaeplastida plastids are called primary and the latter secondary or tertiary plastids. Moreover, primary plastids from all Archaeplastida representatives as well as their descendants (higher-order plastids) share the same unique *atpA* gene cluster in the genome [11] and translocase supercomplex (Toc/Tic apparatus) responsible for protein import of nuclear-encoded proteins. The complex is composed of a conserved set of cyanobacterium-derived homologs and subunits that are presumed to have arisen de novo in the common host [19–22]. The Archaeplastida also evolved a common mosaic feature of nuclear plastid-targeted genes of Calvin cycle enzymes [23]. At least three of the original cyanobacterial genes were replaced by host homologs in their common ancestor.

Transformation of an endosymbiont into a true organelle is considered to be a very complicated process requiring many modifications and inventions both in the host and endosymbiont [24,25]. Therefore, it generally is assumed that the primary endosymbiosis happened only once in the common ancestor of all Archaeplastida members [10,26,27]. Such a view requires monophyly of both primary plastids and their hosts, which is, however, still controversial because the clear similarity of plastids and their monophyly are also consistent with alternative scenarios for the evolution of the Archaeplastida supergroup [28–33] (Fig. 1). A good way to test the monophyly of Archaeplastida plastids and hosts is to carry out phylogenetic analyses of genes present in plastids, compared to those inferred from nuclear and mitochondrial genomes.

Testing the monophyly of Archaeplastida based on plastid genes

In support of a common origin of primary plastids, the first phylogenetic analyses involving plastid SSU (16S) rRNA genes grouped all representatives of Archaeplastida together [34–36]. In agreement with earlier assumptions based on structural similarities of plastids and cyanobacteria, the three eukaryotic lineages branched from the cyanobacterial clade in phylogenetic trees. These results were also strongly confirmed by analyses of LSU (23S) rRNA, tRNA genes (for example, see [37,38]) as well as numerous studies of single [39–41] and concatenated sets of plastid and/or nuclear-encoded plastid genes (see [42–46] and Tab. 1 for other references).

The most extensive phylogenetic studies of plastid evolution to date involved 191 genes with a total of more than 90 000 sites [47] and 75 genes with almost 40 000 sites [48]. The results agree with those from phylogenetic analyses of 21 complete genomes using correlation of compositional vectors calculated on frequency of amino acid strings [49]. In all these studies, representatives of the Glaucophyta, Rhodophyta and Viridiplantae clustered together with significant support. However, such a grouping does not exclude the possibility that each or at least two of these groups acquired their plastids independently from closely related cyanobacterial lineages that may now be extinct, or were passed from one Archaeplastida lineage to others via a secondary endosymbiosis (Fig. 1). This could lead to the same phylogenetic tree topology as in the case of a single plastid origin [28,32,33]. Finding a cyanobacterial lineage that clearly breaks up the monophyly of Archaeplastida plastids would refute the single plastid origin concept. So far, however, all Archaeplastida plastids cluster exclusive of cyanobacteria with significant support, even in protein [50] and RNA gene [38,51] phylogenies that are most rich in representatives (up to 127) of known cyanobacteria lineages.

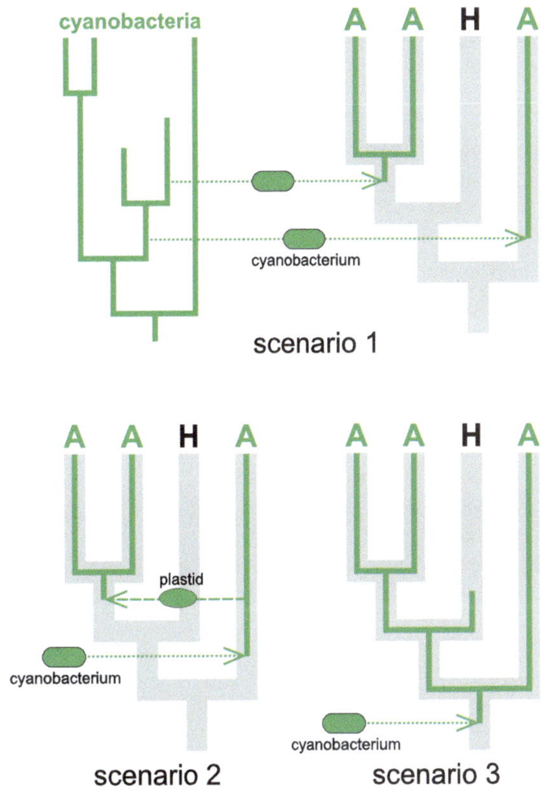

Fig. 1 Three alternative scenarios for primary plastid origin assuming a non-monophyly of Archaeplastida lineages (A) separated by a heterotrophic lineage (H). The first scenario assumes that independent primary endosymbioses involved closely related cyanobacteria and ancestors of separated Archaeplastida lineages. The cyanobacteria finally became extinct and the resulted plastids became similar by convergent evolution. The second hypothesis supposes that a single primary endosymbiosis occurred in one Archaeplastida lineage and next the primary plastid was transferred into the remaining lineages via secondary endosymbioses. The third scenario presumes an ancient primary endosymbiosis to have occurred very early in the eukaryote evolution and losses of primary plastids in lineages that are now heterotrophic.

Tab. 1 Characteristics of phylogenetic studies based on plastid genes and nuclear genes for plastid-realted proteins (with "n" prefix) including three lineages of Archaeplastida. Additional notes were added only at the same references describing a different variant of studies.

Reference	Markers	Taxa	Sites	Data type	Tree topology
Adachi et al. 2000 [43]	45	10	9957	p	[G,[R,V]]
Bachvaroff et al. 2005 [183]	9	13	3452	p	[V,(G,R)]
Bachvaroff et al. 2005 [183]	9	13	6904	g	[V,[G,R]]
Bhattacharya and Medlin 1995 [54][1]	1	40	1403	S	[G,(R,V)]
Bhattacharya and Medlin 1995 [54][2]	1	40	512	S	(V,(G,R))
Criscuolo and Gribaldo 2011 [47][3]	134	74	22 019	np	[G,(R,V)]
Criscuolo and Gribaldo 2011 [47]	134	74	22 019	np	[V,[G,R]]
Criscuolo and Gribaldo 2011 [47][4]	134	74	66 057	ng	[G,(R,V)]
Criscuolo and Gribaldo 2011 [47]	127	83	18 934	p	[R,(G,V)]
Criscuolo and Gribaldo 2011 [47][3]	127	83	18 934	p	[R,[G,V]]
Criscuolo and Gribaldo 2011 [47][3]	191	83	30 149	p+np	[R,[G,V]]
Criscuolo and Gribaldo 2011 [47]	191	83	30 149	p+np	[V,(G,R)]
Criscuolo and Gribaldo 2011 [47][4]	127	83	56 802	g	[R,(G,V)]
Criscuolo and Gribaldo 2011 [47][4]	191	83	90 447	g+ng	[R,(G,V)]
Cuvelier et al. 2010 [184]	22	23	4425	p	[G,(R,V)]
Deschamps and Moreira 2009 [46]	124	23	28 603	np	[V,[G,R]]
Deschamps and Moreira 2009 [46][5]	55	21	9177	p	[G,(R,V)]
Deschamps and Moreira 2009 [46][3]	55	21	9177	p	[V,(G,R)]
Deschamps and Moreira 2009 [46][6]	55	21	9177	p	[V,(G,R)]
Douglas and Turner 1991 [34][7]	1	18	862	S	[V,(G,R)]
Douglas and Turner 1991 [34]	1	18	1118	S	(V,(G,R))
Falcon et al. 2010 [167]	1	65	1255	S	[N,[G,(R,V)]]
Giovannoni et al. 1993 [35]	1	17	1184	S	[G,(R,V)]
Hagopian et al. 2004 [185]	17	25	3347	p	[G,(R,V)]
Hagopian et al. 2004 [185]	41	25	8810	p	[R,(G,V)]
Helmchen et al. 1995 [164]	1	27	1403	S	(G,(R,V))
Helmchen et al. 1995 [164][8]	1	27	1403	S	[V,[G,R]]
Janouskovec et al. 2010 [186]	34	24	7599	p	[R,[G,V]]
Janouskovec et al. 2010 [186]	68	24	15 736	p	[R,[G,V]]
Janouskovec et al. 2010 [186]	23	24	4438	p	[V,[G,R]]
Janouskovec et al. 2012 [187]	1	158	1273	S	[G,[R,V]]
Khan et al. 2007 [188][3]	45	17	9081	p	[G,(R,V)]
Khan et al. 2007 [188]	45	17	9081	p	[R,(G,V)]
Kim et al. 2011 [189]	4	38	4062	t+S+L	[V,(G,R)]
Le Corguille et al. 2009 [190]	44	20	8652	p	[R,[G,V]]
Li et al. 2014 [48][6]	75	42	12 956	p+np	[G,[R,V]]
Li et al. 2014 [48][9]	75	42	12 956	p+np	[G,(R,V)]
Li et al. 2014 [48][5]	75	42	12 956	p+np	[V,[G,R]]
Li et al. 2014 [48][5]	75	42	38 868	g+ng	[V,[G,R]]
Li et al. 2014 [48][9]	75	42	38 868	g+ng	[V,[G,R]]
Li et al. 2014 [48][4]	75	42	38 868	g+ng	[V,(G,R)]
Marin et al. 2005 [37]	4	60	4104	t+S+L	[V,[G,R]]
Marin et al. 2007 [51]	4	101	4126	t+S+L	[V,[G,R]]
Martin et al. 1998 [191]	45	10	11 039	p	G,[R,V]
Martin et al. 2002 [192]	41	16	8303	p	G,[R,V]
Nelissen et al. 1995 [193]	1	36	1277	S	[V,[G,R]]
Ochoa et al. 2014 [38][10]	2	93	935	S+L	(G,(R,V))
Ochoa et al. 2014 [38]	2	99	1029	S+L	[G,[R,V]]
Ochoa et al. 2014 [38][11]	2	69	1029	S+L	[V,(G,R)]
Ochoa et al. 2014 [38]	18	38	2906	p	[V,[G,R]]
Ohta et al. 1997 [42]	7	10	NA	p	(R,(G,V))
Ohta et al. 2003 [194]	41	17	8308	p	(G,[R,V])
Olsen et al. 1994 [36]	1	253	NA	S	(G,(R,V))
Price et al. 2012 [20]	NA	61	17 049	p	[R,(G,V)]
Qui et al. 2012 [163]	30	48	7288	p	[G,[R,V]]
Reyes-Prieto and Bhattacharya 2007 [162]	19	16	5136	np	[G,[R,V]]
Reyes-Prieto et al. 2010 [133]	NA	NA	NA	S	[G,[R,V]]
Rodriguez-Ezpeleta et al. 2005 [45]	50	31	10 334	p	[G,(R,V)]

Tab. 1 *(continued)*

Reference	Markers	Taxa	Sites	Data type	Tree topology
Rogers et al. 2007 [195]	38	28	9108	p	[R,[G,V]]
Sanchez-Puerta et al. 2007 [170]	62	14	11 986	p	[G,[R,V]]
Sanchez-Puerta et al. 2007 [170]	62	14	23 972	g	[V,(G,R)]
Sato 2006 [196]	27	35	NA	p	[G,[R,V]]
Sato 2006 [196]	11	35	NA	p	[R,[G,V]]
Sato 2006 [196]	27	35	NA	g	[V,[G,R]]
Shih et al. 2013 [50]	25	164	NA	p	[G,[R,V]]
Tengs et al. 2000 [197]	1	72	1798	S	[G,(R,V)]
Turner et al. 1999 [161]	1	63	1377	S	[G,[R,V]]
Vogl et al. 2003 [174]	48	7	14 821	p	(R,(G,V))
Vogl et al. 2003 [174][12]	48	7	5404	p	(V,(G,R))
Wang et al. 2008 [198]	20	20	2160	np	[V,[G,R]]
Wang et al. 2008 [198]	10	22	4088	p	[V,[G,R]]
Yoon et al. 2004 [44]	6	46	7111	g+S	[G,[R,V]]
Yoon et al. 2004 [44]	5	46	1934	p	[G,(R,V)]
Yoon et al. 2005 [165]	5	47	1490	p	[R,[G,V]]
Yoon et al. 2009 [166]	2	31	NA	S+L	[R,[G,V]]

[1] Maximum likelihood method was used. [2] Neighbor joining and logDet methods were used. [3] Amino acids were recoded. [4] Codons were degenerated. [5] Stationary composition model was used. [6] CAT model was used. [7] Alignment gaps were excluded. [8] Neighbor joining and maximum parsimony methods were used. [9] Nonstationary (tree-heterogeneous) composition model was used. [10] Data complexity was reduced. [11] Compostional bias was reduced. [12] Fast evolving sites were excluded. g – genes; n – nuclear; p – proteins; S – SSU rRNA gene; L – LSU rRNA gene; t- Ile-tRNA + Ala-tRNA genes; G – Glaucophyta; N – *Nephroselmis olivace*; R – Rhodophyta; V – Viridiplantae; Square brackets indicate a significant grouping with a posterior probabilty >0.9 or a bootstrapp support >90%.

Testing the monophyly of Archaeplastida based on nuclear genes

In contrast to the general concordance of phylogenetic trees based on plastid genetic markers, different nuclear gene data sets produced monophyly, paraphyly, polyphyly or unresolved topologies of Archaeplastida host lineages (Tab. 2). In total, 26 data sets we gathered favor glaucophyte, red alga and green plant monophyly, whereas 50 do not recover the monophyly of these groups. Because their sequences were available and useful as a marker across comparable studies [52,53], initial analyses of representatives of all three Archaeplastida lineages were based on nuclear SSU (18S) rRNA gene [54–56]. However, the trees obtained were inconclusive and had very poorly resolved deeper branches. The three Archaeplastida lineages did not group together and, surprisingly, glaucophytes clustered with cryptophytes, a group of algae bearing a secondary plastid of rhodophyte origin. The lack of Archaeplastida monophyly was also confirmed in subsequent studies based on the same marker, and joined with LSU (28S) rRNA gene, but including many more taxa, even 2551 species ([57,58]; see Tab. 2 for other examples). Interestingly, glaucophytes still clustered with cryptophytes in these trees [59,60] and other recently discovered related lineages, such as heterotrophic katablepharids and picozoans [61–63]; the latter were formerly called picobiliphytes and considered photosynthetic but available data indicate that they are heterotrophic [64,65]. Several later analyses recovered a Rhodophyta-Viridiplantae monophyly [58,62,66,67] and even a clade containing all three Archaeplastida lineages, but still mixed with cryptophytes [67], katablepharids, picozoans and the newly discovered heterotrophic flagellate *Palpitomonas bilix* [68]. However, the relationships were poorly supported or, at best, had moderate support. A relatively high value (Bayesian PP = 1) was obtained only by Kim et al. [69] for the clade including glaucophytes with cryptophytes and katablepharids in a tree based on genes encoding large and small subunit rRNA.

The monophyly of Archaeplastida was recovered by initial global phylogenetic studies of eukaryotes based on several nuclear-encoded proteins [58,70,71], but not with very strong support (Tab. 2). The inclusion of 143 protein sequences with a total length of more than 30 000 sites [45] and the application of a more advanced site-heterogeneous mixed-model (CAT) [72], as well as other approaches, recovered monophyly of primary-plastid host lineage with significant support [46,73]. However, the analyses did not consider representatives of cryptophytes or their relatives, which grouped with Archaeplastida in rRNA gene trees. In fact, when these taxa and other eukaryotic lineages were taken into account, different results were obtained depending on the data set and method used (Tab. 2).

In some cases almost all deep branches, including Archaeplastida lineages, remained completely unresolved [74–76]. Interestingly, in analyses involving the largest number of taxa (451), major eukaryotic supergroups, including the Opisthokonta, Amoebozoa, Excavata and SAR (Stramenopiles, Alveolates and Rhizaria) obtained significant or some support, but not the Archaeplastida. As in rRNA gene phylogenies, they were mixed with cryptophytes, haptophytes and katablepharids [77]. Many other trees inconsistent with the monophyly of Archaeplastida hosts were also obtained

Tab. 2 Characteristics of phylogenetic studies based on nuclear genetic markers including three lineages of Archaeplastida. Additional notes were added only at the same references describing a different variant of studies.

Reference	Markers	Taxa	Sites	Data type	Tree topology
Baldauf et al. 2000 [71]*	4	61	1528	p	(**G**,(**R**,**V**))
Baldauf et al. 2000 [71]*	6	61	NA	p+S+L	(**G**,[**R**,**V**])
Baurain et al. 2010 [199]	108	57	15 392	p	[**G**,(**R**,(C,**V**)),(H,SA)]
Bhattacharya and Medlin 1995 [54]	1	47	1612	S	((C,**G**),(**V**,(**R**,(H+SAR))))
Bhattacharya and Weber 1997 [70]*	1	48	748	actin gene	(**G**,(**R**,**V**))
Bhattacharya et al. 1995 [55]	1	47	1621	S	((C,**G**),(**V**,(**R**,(H+SAR))))
Brown et al. 2013 [93]	159	68	43 615	p	[**G**,[H,T],(C,(P,(**R**,**V**)))]
Burki et al. 2007 [160]	123	49	29 908	p	(**R**,[**G**,**V**])
Burki et al. 2008 [85]	135	65	31 921	p	[**G**,(**R**,**V**)]
Burki et al. 2009 [86]	127	72	29 235	p	[**R**,(**G**,**V**)]
Burki et al. 2012 [81]	258	68	55 881	p	[[P,**G**],[**R**,**V**]]
Burki et al. 2012 [81][1]	258	65	55 881	p	[**G**,[**R**,**V**]]
Cavalier-Smith 2002 [59]	1	53	NA	S	(((C,**G**),**V**),(**R**,(H,(E+SAR))))
Cavalier-Smith 2004 [60]	1	193	1044	S	(C,**G**),others,**R**,others,**V**
Cavalier-Smith et al. 2014 [80][2]	173	73	45 194	p	[**G**,[**R**,**V**]]
Cavalier-Smith et al. 2014 [80][3]	173	98	45 194	p	(**R**,(**G**,**V**))
Cavalier-Smith et al. 2014 [80][4]	178	122	47 510	p	(**R**,(**G**,**V**))
Cavalier-Smith et al. 2014 [80]	192	75	52 824	p	[**G**,[[C,K],[**R**,**V**]]
Cuvelier et al. 2008 [62]	1	124	NA	S	((SA,(**R**,**V**)),(**G**,(P,[C,K])))
Deschamps and Moreira 2009 [46]*	143	28	23 238	p	[**G**,[**R**,**V**]]
Hackett et al. 2007 [84][4,5]	16	44	NA	p	[**G**,[**R**,**V**]]
Hackett et al. 2007 [84][4]	16	46	NA	p	[**G**,[**R**,**V**]]
Hackett et al. 2007 [84]	16	46	6735	p	[**R**,(**G**,**V**)]
Hampl et al. 2009 [79][5]	143	34	35 584	p	((**G**,**R**),(H,**V**))
Hampl et al. 2009 [79]	143	48	35 584	p	[**G**,[**V**,(H,**R**)]]
Hampl et al. 2009 [79][6]	143	49	35 584	p	[(**G**,**R**),(**V**,(C,H))]
Hampl et al. 2009 [79][7]	NA	47	NA	p	(**G**,(**V**,(H,**R**)))
Ishida et al. 2010 [63]	1	270	1020	S	((**R**,((**G**,(P,(C,K))),(H,(Ce+E)))),(SAR,(E,**V**)))
Kim et al. 2006 [69]	2	48	3287	S+L	[[**G**,[C,K]],[**R**,**V**]]
Kim and Graham 2008 [78]	1	40	736	EF2 p	((**G**,(Am,E)),(Am,(**Ri**,(E[(C,(H,K)),[**R**,**V**]]))))
Kim and Graham 2008 [78]	6	37	2797	p	(**G**,[E,[**R**,(**V**,[C,(H,K)])]])
Minge et al. 2009 [200]	78	37	17 283	p	([**R**,**V**],(**G**,(C,H)))
Moreira et al. 2000 [201]*	6	12	1938	p	(**G**,(**R**,**V**))
Moreira et al. 2007 [66]	2	91	2574	S+L	((**R**,**V**),(H,(Ce,(C,**G**))))
Nikolaev et al. 2004 [75]	2	121	1405	actin p+S	(**R**,**V**),others,(C,(Ce,**G**))
Nozaki et al. 2003 [83]*	4	53	1525	p	(**R**,(**G**,(SA,(E,**V**))))
Nozaki et al. 2003 [83]*,[2]	4	34	1525	p	(**R**,[**G**,[SA,(E,**V**)]])
Nozaki 2005 [29]*	4	36	1525	p	(**R**,[**G**,(SA,(E,**V**))])
Nozaki et al. 2007 [98][5]	19	31	5216	p	[**R**,(**G**,(**V**,(H,SA))]
Nozaki et al. 2007 [98]	19	33	5216	p	[(E,**R**),((**G**,**V**),(H,SA))]
Nozaki et al. 2009 [31][5]	19	29	5216	p	[**R**,((**G**,[**V**,[H,SA]]))]
Nozaki et al. 2009 [31]	19	35	5216	p	[(E,**R**),((**G**,(**V**,(H,SA))]
Nozaki et al. 2009 [31][5]	21	29	6048	p	[**R**,[**G**,[**V**,[H,SA]]]]
Nozaki et al. 2009 [31]	21	35	6048	p	[(E,**R**),(**G**,[**V**,(H,SA)])]
Okamoto and Inouye 2005 [61]	1	69	1109	S	((**R**,(Ce,(**G**,(C,K)))),(**V**,(H,SAR)))
Okamoto et al. 2009 [159]	1	67	NA	Hsp90 p	(**V**,(**G**,**R**))
Parfrey et al. 2010 [77][2]	16	88	6578	p+S	(**G**,(**R**,(**V**,(C,H))))
Parfrey et al. 2010 [77]	16	451	6578	p+S	((**G**,**V**),(**R**,(H,(C,K))))
Patron et al. 2007 [202]	102	38	18 425	p	(**R**,(**G**,**V**))
Reeb et al. 2009 [74]	1	64	NA	EF2 p	[[**G**,Am],[SAR,[E,[(C,[H,K]),[**R**,**V**]]]]]
Reeb et al. 2009 [74]	3	108	1088	p	[**R**,(C,[H,K]),(**G**,(**V**,(E+SAR+T))))]
Reeb et al. 2009 [74]	6	75	2302	p	((**G**,**R**),(**V**,(E,SAR)))
Reeb et al. 2009 [74]	7	75	NA	p (+ EF2)	(**G**,(SAR,(E,((**V**,T),[**R**,(C,[H,K])]))))
Reeb et al. 2009 [74]	5	112	4197	p+S+L	[[**G**,[K,C]],([**R**,**V**],[E+SAR+T])]
Rodriguez-Ezpeleta et al. 2005 [45]*	143	34	30 113	p	[**R**,(**G**,**V**)]
Rodriguez-Ezpeleta et al. 2007 [73]*	143	37	30 244	p	(**R**,(**G**,**V**))
Rodriguez-Ezpeleta et al. 2007 [172]*,[5]	143	51	31 604	p	(**G**,(**R**,**V**))
Rodriguez-Ezpeleta et al. 2007 [172]*	143	64	31 604	p	(**R**,(**G**,**V**))

Tab. 2 *(continued)*

Reference	Markers	Taxa	Sites	Data type	Tree topology
Tekle et al. 2007 [203]	4	74	2120	p+S	(**G**,[**R**,**V**])
Tekle et al. 2008 [204]	4	100	2291	p+S	[**R**,[**G**,[**V**,([C,[H,E]],[E,SAR])]]]
van de Peer and de Wachter 1997 [57]	1	500	NA	S	((**R**,(Am,(C,**G**))),(SAR,(H,**V**)))
van de Peer et al. 1996 [56]	1	107	NA	S	((SR,(**R**,(A,Am))),(H,(**V**,(C,**G**))))
van de Peer et al. 2000 [58]*	4	61	1528	p	(**G**,(**R**,**V**))
van de Peer et al. 2000 [58]	1	2551	1574	S	((**R**,**V**),((Am,(C,**G**)),(H,SAR)))
Yabuki et al. 2010 [68]	5	18	4840	p+S+L	[**R**,[SAR,(**G**,(H,(**V**,[Pa,[C,K]])))]]
Yabuki et al. 2010 [68]	6	18	5559	p+S+L	(**G**,[(H,[C,K]),(Pa,(**R**,**V**))])
Yabuki et al. 2010 [68]	1	67	1335	S	(**V**,(**R**,Pa,(P,(**G**,(C,K)))))
Yabuki et al. 2014 [87][8]	157	64	41 372	p	((**G**,**R**),((H,**V**),(Pa,[C,K])))
Yabuki et al. 2014 [87][9]	157	64	41 372	p	[**R**,**G**,**V**]
Yoon et al. 2008 [76][5]	4	92	NA	p+S	[**R**,[**G**,((C,**V**),(E,(H,SAR)))]]
Yoon et al. 2008 [76]	4	105	NA	p+S	[(Ri,**R**),E,(**G**,[C,**V**]),(H,SAR)]
Yoon et al. 2011 [64]	7	78	2594	p	[**R**,[(**V**,(**G**,((H,[P,T]),(C,K))))],[E,SAR]]
Zhao et al. 2012 [67][10]	124	77	27 638	p	[**G**,[**R**,**V**]]
Zhao et al. 2012 [67]	124	79	27 638	p	[**G**,[**R**,**V**]]
Zhao et al. 2012 [67]	2	80	NA	S+L	((**R**,**V**),(C,**G**))
Zhao et al. 2012 [67]	1	142	NA	S	((**R**,**V**),(Ce,(C,**G**)))

* Cryptophyta, Haptophyta or related taxa were not considered. [1] Telonemid, centrohelid and picobiliphyte (picozoan) were excluded. [2] Impact of missing data was decreased. [3] Impact of missing data for more taxa was decreased. [4] Fast evolving sites were excluded. [5] Long-branched taxa were excluded. [6] Cryptophyta was included. [7] Long-branched sequences were excluded. [8] Maximum likelihood method was used. [9] Bayesian method was used. [10] Two *Malawimonas* taxa were excluded. p – proteins; S – SSU rRNA gene; L – LSU rRNA gene; A – Alveolata; Am – Amoebozoa; P – Picozoa; C – Cryptophyta; Ce – Centrohelida; E – Excavata; **G** – Glaucophyta; H – Haptophyta; K – Katablepharida; Pa – *Palpitomonas*; **R** – Rhodophyta; Ri – Rhizaria; SA – Stramenopila + Alveolata; SAR – Stramenopila + Alveolata + Rhizaria; T – *Telonemia*; **V** – Viridiplantae. Square brackets indicate a significant grouping with a posterior probabilty >0.9 or a bootstrapp support >90%.

(indicating paraphyly or polyphyly) but with more significant support (Tab. 2). In one of them, glaucophytes formed a separate clade, whereas red algae and green plants grouped significantly with cryptophytes, haptophytes and katablepharids [78] (see also [74] for another interpretation). In other studies, a clade with the three Archaeplastida lineages was indeed recovered, but they still clustered significantly with haptophytes [79], cryptophytes and katablepharids [80], *Palpitomonas* [68], as well as with picozoans [81]. The latter study was based on the most sequence-rich data set to date, 258 nuclear-encoded proteins with 55 881 sites in total. In addition, other representatives of the main eukaryotic supergroups, such as Excavata and SAR, grouped with the Archaeplastida with high support in other phylogenetic analyses [29,31,64,68,74,76,82,83]. Such relationships were also inferred from many other data sets but with weaker or no support (Tab. 2). It should be emphasized, however, that many other phylogenies based on both smaller [84] and much larger data sets [67,80,81,85–87] have high or very high support values for monophyly of glaucophytes, red algae and green algae/land plants.

Testing the monophyly of Archaeplastida based on mitochondrial genes

Because of the inconsistencies in nuclear multigene phylogenies of main eukaryotic lineages, including glaucophytes, red algae and green plants, mitochondrial genes have been explored as alternative phylogenetic markers to trace host cell history. Successful sequencing of glaucophyte mitochondrial genomes enabled testing the monophyly of Archaeplastida. Studies based on 42 mitochondrion-encoded and nucleus-encoded, mitochondrion-targeted proteins (11 384 sites) and 84 taxa (including two glaucophytes) supported the monophyly of the primary plastid-bearing eukaryotes with 0.99 and 1.0 posterior probabilities [88]. Inclusion of seven glaucophyte sequences confirmed this result for the concatenated amino acid alignment of 14 proteins encoded in mitochondrial genomes (3267 sites) from 49 taxa [89]. However, topology tests of the protein tree showed that placing red algae as a sister branch to haptophytes outside the Archaeplastida, as well as a topology uniting red algae and haptophytes within the Archaeplastida, were not statistically worse than the best tree. In addition to that, the Archaeplastida clade obtained very weak (51%) bootstrap support in maximum-likelihood analysis using the corresponding nucleotide alignment, whereas in the Bayesian tree the clade of Hacrobia (katablepharid, cryptophytes and haptophytes) branched within the Archaeplastida as sister to the red algae with 0.99 posterior probabilities for both the Archaeplastida + Hacrobia and the Hacrobia + red algae clades [89]. Interestingly, this topology corresponds to results obtained from rRNA genes and some nuclear protein data.

The results based on mitochondrial genes should be considered with caution because it was demonstrated that the clustering of mitochondrial sequences from red algae and green plants can be artifactual, resulting from similarly slow evolutionary rates of these sequences compared to many other eukaryotic taxa [90].

Influence of different methodologies on discrepancies in Archaeplastida phylogenies

The striking differences in inferring phylogenetic positions and relationships of Archaeplastida hosts could result from many methodological limitations. On one side there is stochastic error caused by poor phylogenetic signal in the data, which makes it difficult to resolve relationships, especially at deep levels; for example, among very early diverging major eukaryotic lineages [91]. One way to overcome this problem is to increase the number of genes analyzed and use phylogenomic approaches based on huge data sets on a genomic scale [92]. However, even consideration of more than 40 000 sites does not always result in deep robust phylogenies for the Archaeplastida [80,87,93]. Moreover, analyses of many genetic markers are vulnerable to systematic errors, which can lead to false phylogenetic reconstructions, reflecting unreal evolutionary relationships but with high statistical support.

The best-known systematic bias is long-branch attraction (LBA) [94] and the correlated effect that has been called short-branch exclusion (SBE) [95]. These biases cause, respectively, artificial clustering of highly diverged sequences, as well as of sequences with less than average substitution rates. The potential influence of LBA on the Archaeplastida was nicely shown by Rodriguez-Ezpeleta et al. [73]. The supergroup was moderately supported (64%) when two representatives of red algae, *Porphyra* and *Cyanidioschyzon*, were considered. Interestingly, the inclusion of only *Porphyra* increased the support for the whole Archaeplastida clade to 99%, whereas the tree with only *Cyanidioschyzon* moved this taxon to kinetoplastids with 100% support. Both kinetoplastids and *Cyanidioschyzon* are characterized by fast evolving sequences and their clustering is most consistent with LBA. In turn, the SBE effect concerns some sequences of glaucophytes [95] and could be responsible for grouping of short branches leading to glaucophytes and katablepharids in some tested tree topologies [81]. This effect could also explain common clustering of red algal and green plant in phylogenetic trees based on slowly evolving mitochondrial sequences [90].

In addition to that, gene-rich data sets can suffer from inadequate taxon sampling because it is not always possible to collect the full sets of sequences from all species. If, however, including many taxa is required to test their relationships, final alignments can lack many sequences from understudied species. Some of the discrepancies observed in above-described phylogenies likely result from such missing data [96]. For example, the exclusion of missing data and a more distant excavate outgroup changed the polyphyly of Archaeplastida (grouped with a katablepharid and a cryptophyte) to monophyly [80]. However, further removing missing data resulted in weak or lack of support for the clade. It clearly shows that topologies obtained are susceptible to different data selection, and therefore, conclusions based on phylogenomic data should be drawn with caution.

It was shown that increased taxon-sampling could improve phylogenetic accuracy even at the cost of including data that are missing for some organisms [97]. Including new eukaryotic lineages, especially haptophytes, cryptophytes, katablepharids, picozoans and *Palpitomonas* [79–81] broke up the monophyly of Archaeplastida that had been obtained previously using less taxon-rich data sets [45,46,68]. The analyses based on the largest number of proteins (258 with 55 881 sites), represented by 68 species, grouped Archaeplastida lineages, but also included picozoans in this cluster, which significantly affiliated with glaucophytes [81]. As could be expected, exclusion of picozoans from the data recovered Archaeplastida as the monophyletic clade. The influence of taxon-sampling on Archaeplastida relationships with other eukaryotes was also shown by Jackson and Reyes-Prieto [89] using the concatenated alignment of mitochondrial proteins. Archaeplastida monophyly was obtained when all glaucophyte representatives were included. However, after removing glaucophyte taxa other than those previously published (i.e. *Cyanophora paradoxa* and *Glaucocystis nostochinearum*), two phylogenetic methods failed to recover the common origin of Archaeplastida lineages, because the glaucophytes grouped with the katablepharid *Leucocryptos marina* (significantly using the Bayesian approach), whereas red algae clustered with cryptophytes and haptophytes. The whole clade including Archaeplastida and the other eukaryotes obtained significant support by the Bayesian method. Interestingly, the authors recovered the monophyly of Archaeplastida only when excavate taxa were excluded. When they were present, they grouped with glaucophytes and green algae [89]. It indicates that the relationships obtained strongly depend on selection of taxa and sequences, and thus should be treated with caution.

Another problem with the multi-gene data is gene sampling. Its influence on the incongruity of Archaeplastida monophyly between results of Nozaki et al. [98] and Hackett et al. [84] was studied by Inagaki et al. [99], who evaluated the significance of Rhodophyta-Viridiplantae monophyly by analyzing multi-gene data sets of varied sizes [99]. The authors showed that recovery of this relationship depends on gene sampling in phylogenetic inferences with fewer than 10 000 amino acid positions. The sampled data sets consisted of different genes and supported various topologies but the tree based on the full multi-gene data set recovered the monophyly of red and green lineages. However, no representative of glaucophytes was considered in these studies.

Comparing data gathered from nuclear genes (Tab. 2) with more than two markers, we found that the monophyly of the Archaeplastida was, on average, obtained more often with greater numbers of genes (mean 103 vs. 52), longer alignments (26 694 vs. 14 624) but smaller numbers of taxa (59 vs. 69). The first two differences were statistically significant but not the last (Mann–Whitney test, *P*-value: 0.04, 0.04 and 0.75, respectively). These comparisons, however, should be considered with caution because the data compared are not completely independent.

Phylogenomic data sets consist of concatenated alignments of many genes, which most probably evolved subject to different substitution rates and patterns. Therefore, application of improper, usually simple, evolutionary models can also lead to inaccurate phylogenies. Model violation can also result from saturation of mutations, across-site rate variation, heterogeneous substitutions across sites, rate variation across sites through time (heterotachy), site-interdependent substitutions as well as compositional heterogeneity and nonstationarity of substitutions. It was shown in the case of the Archaeplastida that model misspecification could reinforce systematic errors such as LBA of the red alga *Cyanidioschyzon* with non-Archaeplastida taxa, giving an incorrect topology even with a strong support [73].

Several approaches are applied to overcome or at least reduce the impacts of these errors; for example, removing rapidly evolving sites, genes and taxa [100], recoding amino acids into functional categories [101], and using a covarion [102] and site-heterogeneous mixture (CAT) models [72]. Although initial studies involving Archaeplastida were based on simpler phylogenetic models, many later analyses were usually carried out with advanced approaches using more realistic models. However, they still resulted in varied phylogenetic positions and relationships of the Archaeplastida. On one hand, the application of some of the modern methods increased support for the monophyly of Archaeplastida and alleviated the influence of some systematic errors [46,73,84] in comparison to earlier studies. On the other hand, Hampl et al. [79] obtained the non-monophyly of Archaeplastida (specifically placing of haptophytes within this supergroup) even after application of the CAT model and recoding of amino acids by functional categories, as well as progressive exclusion of rapidly evolving gene sequences and long-branch taxa. Therefore, the authors concluded that these relationships do not appear to be due to long-branch attraction artifacts. Similarly, Nozaki et al. used the CAT model and only slowly evolving genes and recovered polyphyly of Archaeplastida lineages [31]. Similarly, the application of a covarion model in rRNA gene-based phylogenies still grouped glaucophytes with cryptophytes but did not improve the resolution of Archaeplastida [67]. There also are differences depending on the phylogenetic methods used. Yabuki et al. [87] obtained significant for monophyly of the Archaeplastida in a Bayesian tree but not using maximum-likelihood. In contrast, Jackson and Reyes-Prieto [89], using mitochondrial genes, obtained some bootstrap support (although very weak) for Archaeplastida monophyly with maximum-likelihood but very strong confidence for separation of Archaeplastida lineages by katablepharid, cryptophytes and haptophytes using Bayesian inference.

To avoid problems with concatenated alignments, Chan et al. [103], used novel genomic data from two mesophilic red algae, *Porphyridium cruentum* and *Calliarthron tuberculosum*, considering each gene phylogeny separately and found that about 50% of the examined protein phylogenies support the monophyly of red and green algae. However, to resolve monophyly of the whole Archaeplastida inclusion of complete glaucophyte genomes is necessary.

It seems that there is no clear interdependence between the tree topology obtained with respect to Archaeplastida monophyly, and the methodological approaches used in phylogenetic analyses.

Inconsistency of Archaeplastida phylogenies in the context of gene and plastid evolution

The disagreement between plastid gene trees supporting the monophyly of Archaeplastida plastids, and many phylogenies based on nuclear (and some mitochondrial) markers that often favor other topologies, is very intriguing. Besides methodological problems with phylogenetic analyses, differences in the inferred trees can result from disparate gene histories that are incongruent with species trees. One of them is gene paralogy [104,105]. When different paralogous gene families are retained across lineages, trees based on these genes will reproduce gene histories and duplication events but not accurate phylogenetic relationships among taxa. Although such genes should be excluded from phylogenetic data sets, it can be difficult to recognize hidden paralogs and separate them from real orthologs. The same problem concerns genes subjected to horizontal gene transfer (HGT), which seems to be a common process, not only in prokaryotic organisms, but also eukaryotes including Archaeplastida lineages [106–111]. Such genes are most likely also present in many alignments analyzed and could cause artificial clustering. One of the genes possibly subjected to this process is *EF2*, which encodes translation elongation factor. Phylogenies based on this marker gave a very strong signal supporting the non-monophyly of Archaeplastida: the significant grouping of rhodophytes and viridiplants with katablepharid, cryptophytes and haptophytes but excluding glaucophytes [78]. Therefore, it was proposed that the gene was likely horizontally transferred [74]. The close relationship of Archaeplastida with katablepharid, cryptophytes and haptophytes was also reproduced by some trees of mitochondrial genes. If these results are reliable (e.g. do not reflect a compositional bias) and we insist on Archaeplastida monophyly, we should also assume HGT at least for some of these genes from Archaeplastida to the other photosynthetic lineages. Widespread horizontal transfer of mitochondrial genes has been reported in plants [112–115].

Another process that can complicate inferred phylogenies of the Archaeplastida is endosymbiotic gene transfer (EGT). During this process, genes from an endosymbiont, including those from both plastid and nuclear genomes, are moved and integrated into the host nuclear genome [107,116–118]. The presence of these genes in an alignment would mask phylogenetic signal from true (endogenous) orthologs (Fig. 2). Interestingly, in many trees based on nuclear genes the primary plastid-containing lineages intermix with other eukaryotes, such as cryptophytes, haptophytes and picozoans, all of which contain genes most likely from EGT during secondary (or tertiary) endosymbioses involving the red algal lineage [8,9,17,18]. In agreement with that, Leigh et al. [119] found an influence of EGT in phylogenetic analyses on attraction of the red alga Porphyra to stramenopiles, which possess plastids derived from an ancient secondary endosymbiosis with a red alga (Fig. 2). The interfering effects of EGT could also explain the poor resolution of deep branches and a failure to recover the monophyly of Archaeplastida [77]. However, the supergroup remained polyphyletic in phylogenetic analyses after exclusion of cryptophyte and haptophyte taxa [77]. It should be noted that Archaeplastida

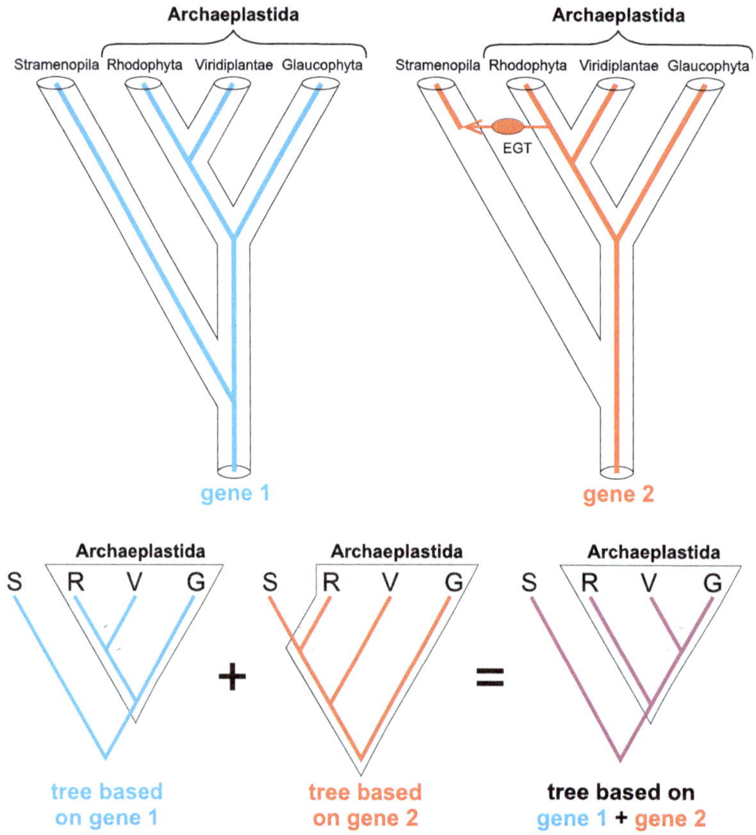

Fig. 2 Influence of endosymbiotic gene transfer (EGT) associated with a secondary endosymbiosis between a rhodophyte and a stramenopile on the inference of phylogenetic relationships between Archaeplastida and other eukaryotic lineages. The gene 1 is a true ortholog (vertically inherited) therefore the tree based on this gene reflects real host relationships, i.e. the monophyly of Archaeplastida. However, the tree based on the gene 2, which was acquired by the stramenopile from the red algae during the secondary endosymbiosis, shows a close affiliation of Stramenopila to Rhodophyta, thereby indicates the paraphyly of Archaeplastida. The simultaneous use of these genes in a tree reconstruction could mask the phylogenetic signal from the true ortholog leading to the attraction of rhodophytes to stramenopiles and its basal placement in respect to other Archaeplastida.

lineages also join other photosynthetic lineages in trees based on conserved nuclear rRNA genes, which are not considered to be involved in the EGT process. Inferences about EGT, especially in aplastidic eukaryotes, should be performed with caution and use appropriate controls; otherwise invoking it could lead to overestimation of EGT events and its impact on eukaryotic phylogenies [120].

It is remarkable that Archaeplastida members show affiliation to cryptophytes, haptophytes and related taxa in rRNA, nuclear and some mitochondrial gene trees. If these groupings are not a result of some compositional bias or another systematic error, but rather reflect true host cell history, it is worthwhile considering alternative views [77,79]. Three scenarios have been proposed (Fig. 1) that are compatible with the unquestionable monophyly of plastid genes, and explain both similar and different plastid features of Archaeplastida members (for wide discussion see [27–33,41,83,121,122]). One of them assumes that multiple independent primary endosymbioses occurred involving closely related cyanobacteria and separate Archaeplastida lineages. The resulting plastids were next subjected to convergent evolution, whereas the intervening cyanobacterial lineages became extinct. The second possibility supposes that a single primary endosymbiosis occurred in one Archaeplastida lineage and the primary plastid was next transferred via secondary endosymbioses into the remaining lineages. The third hypothesis claims that an ancient primary endosymbiosis occurred very early in eukaryote evolution, before divergence of some major lineages, and was followed by subsequent losses of the primary plastids in ancestors of groups that now contain no primary plastid.

To accept or reject these scenarios we should determine if they are consistent with available data on key processes involved in plastid evolution: (*i*) multiple endosymbiotic events followed by transformation of the endosymbionts into true organelles and (*ii*) multiple plastid losses. The problem is still hotly debated [9,17,18,123]. Plastids in the reduced or vestigial form are still present in some eukaryotes that have lost photosynthesis, and are involved in vital non-photosynthetic functions (e.g. amino acid, heme, isoprenoid, and fatty acid biosynthesis); therefore, their loss seems unfavorable [124–127]. The transformation of endosymbiont to organelle is usually considered a complicated process but successful endosymbiotic events might not be as rare as once assumed [128]. In addition to examples of higher-order endosymbiosis [8,9,17,18,129], there is also at least one case of a successful primary endosymbiosis independent of Archaeplastida, namely between a freshwater thecate amoeba, *Paulinella chromatophora*, and a cyanobacterium from a phylogenetic lineage different from the ancestor of Archaeplastida plastids [37,51,130]. Two photosynthetically active chromatophores retained by *Paulinella* fulfill all criteria to be considered true organelles [131]. Their deep integration involved a significant reduction of the endosymbiont's genome [132,133], transfer of many endosymbiont genes to the host nucleus [134,135] and evolution of a machinery for import of host-encoded proteins into the chromatophores [136–139]. Moreover, apart from *Paulinella*, there are many other eukaryotes in endosymbiotic relationships with cyanobacteria [140–143]. However, the first scenario assuming independent primary plastid endosymbioses in Archaeplastida lineages encounters a problem with an unparsimonious convergent evolution of several specific and complex plastidal features, e.g. Toc/Tic import apparatus [20] and composition of Calvin cycle enzymes [23].

At present, a single factor influencing inferred relationship of Archaeplastida with other eukaryotic supergroups in phylogenetic studies cannot be pinpointed and, therefore, be used to remove doubts about the monophyly or non-monophyly of glaucophytes, red algae and green algae/plants.

Testing the relationships among Archaeplastida lineages

If we assume the monophyly of Archaeplastida despite its controversies, we should consider relationships among its three lineages. Glaucophytes are usually considered to be the earliest branch of the Archaeplastida because their plastids retain more ancestral features typical of cyanobacteria than are present in red algae or green plants; for example, peptidoglycan and carboxysomes [14,15]. Other characteristics that distinguish the Glaucophyta from the Rhodophyta and Viridiplantae include (***i***) the presence of a cyanobacterial-like fructose-1,6-bisphosphate aldolase, which was replaced by a duplicated cytosolic copy in the case of the other Archaeplastida lineages [144], (***ii***) lack of triple-helix chlorophyll-binding, light harvesting antenna complexes [145,146] and (***iii***) lack of plastidial phosphate-translocator [20]. All these differences support the hypothesis that glaucophytes are sister to red algae and green plants.

On the other hand, glaucophytes and rhodophytes share the presence of unstacked thylakoid membranes with phycobilisomes also characteristic of cyanobacteria [16]. Moreover, these two Archaeplastida groups lack chlorophyll b present in Viridiplantae and both synthesize starch in the cytoplasm contrary to green plants, which relocated the pathway to plastids[147–150]. The close affiliation of these two groups was noticed by Cavalier-Smith who united them under the subkingdom Biliphyta, non-phagotrophic and phycobilisome-containing algae [26,151]. However, the features shared by Glaucophyta and Rhodophyta represent characteristics of the cyanobacterial ancestor, and therefore should not be used as evidence for the monophyly of the Glaucophyta-Rhodophyta clade.

There is also evidence for the third possibility, i.e. a closer affiliation of glaucophytes and green plants. The Glaucophyta and Viridiplantae retain cyanobacterial RuBisCO genes; in contrast, red algae acquired RuBisCO from a proteobacterium via horizontal gene transfer [152]. Moreover, N-terminal sequences of two plastid proteins of photosystems I and II in the glaucophyte *Cyanophora* show significant similarity to those in green plants but not in red algae [153,154]. It was also found that the moss *Physcomitrella*, representing the "green" lineage, has nine homologous genes related to peptidoglycan biosynthesis that are essential for plastid division but are absent from red algae [155,156].

The first phylogenetic studies based on single genes that included all three representatives of the Archaeplastida were inconclusive and contradictory (see for example [39,40,157,158]), but subsequent analyses significantly resolved relationships within Archaeplastida (Tab. 1 and Tab. 2). In fact, all three possible topologies were proposed; however, phylogenetic analyses based on nuclear-encoded proteins (sometimes concatenated with rRNA genes) more often indicated the early-branching of the Glaucophyta (in 15 data sets) than the early divergence of the Rhodophyta (in 9 data sets) – Tab. 2. A basal Viridiplantae was proposed only from an Hsp90 tree, but without significant support [159]. What is more, the glaucophyte-first hypothesis was significantly supported by the larger number of studies, e.g. [46,67,84], whereas the rhodophyte-first concept obtained significant support only in one investigation, by Burki et al. [160], and with a simultaneous weak support for the whole Archaeplastida. In analyses by Rodriguez-Ezpeleta et al. [73], the glaucophytes + green clade received 90% support with 64% support for the whole Archaeplastida; however, when the highly diverged red alga *Cyanidioschyzon* was removed, and only *Porphyra* retained, these support values changed to 46% and 99%, respectively. It should be noticed that the earlier separation of glaucophytes was obtained using the data set with the largest number of alignment sites (55 881) among those tested [81]. Although, the earlier branching of the Rhodophyta was recovered using data sets with, on average, greater numbers of markers (128 vs. 83), alignment sites (29 885 vs. 22 748) and taxa (62 vs. 57), the differences were not statistically significant (Mann–Whitney or *t*-Student tests, $P > 0.12$).

Analyses of gains and losses of nuclear genes encoding transcription-associated proteins (TAPs, comprising transcription factors and other transcriptional regulators) were also used to assess support for the alternative branching orders in Archaeplastida [20]. The Viridiplantae-first scenario required 48 such events, the Rhodophyta-first scenario 46 events, whereas the Glaucophyta-first scenario 47 changes. However, when the pattern of TAP evolution typical of Viridiplantae (a greater probability of gains than losses) was taken into account, an earlier divergence of glaucophytes was favored. All these results based on nuclear markers tend to support the earlier separation of glaucophytes from the "red-green" lineage.

In contrast, phylogenies based on plastid genes and/or nuclear genes derived from a cyanobacterial endosymbiont are much less conclusive. Thirty data sets (15 significantly) supported the earlier divergence of glaucophytes (e.g. [44,161–163]), 27 (15 significantly) – green plants (e.g. [46,51,164]) and 17 (6 significantly) – rhodophytes (e.g. [47,165,166]) – Tab. 1. In one case involving SSU rRNA gene data, a fourth strange topology was obtained: the prasinophyte *Nephroselmis olivacea* took a significant basal position to glaucophytes and a clade of rhodophytes with the rest of green algae and land plants [167]. Considering cumulative results, support for the three possible topologies does not seem to depend on the type of data sets used, that is, plastid-encoded genes vs. nuclear-encoded genes or rRNA vs. protein genes. As for nuclear gene data, the rhodophyte-first concept was recovered by data sets with larger on average numbers of markers (74 vs. 52 and 48), alignment sites (21 972 vs. 14 715 and 12 301) and taxa (45 vs. 37 and 39) compared to topologies favoring the earliest separation of green plants and glaucophytes, respectively. The differences were, however, not statistically significant in Kruskal–Wallis test ($P > 0.28$). Accordingly, the most site-rich data set (more than 90 000 nucleotide sites) indicated the Rhodophyta-first hypothesis [47].

Other studies based on plastid genomes also produced inconsistent results. Comparisons of conserved gene arrangements showed either the earlier divergence of red algae [42] or glaucophytes [11], using 6 and 10 representatives of Archaeplastida, respectively. On the other hand, general comparisons of 50 plastid genomic sequences using BLAST results in distance phylogenies [168], as well as phylogenetic analysis of 21 complete plastid genomes using correlation analysis of compositional vectors calculated based on frequency of amino acid strings [49], recovered the Viridiplantae as the earliest-diverging lineage. In turn, studies of the presence and absence of 261–277 genes in 17–20 plastid genomes indicated the early-branching of Rhodophyta [169,170] or Viridiplantae [171]. Bayesian and maximum-likelihood methods applied to mitochondrial proteins showed the earlier divergence of glaucophytes with rather weak support [88,89], whereas the maximum-likelihood phylogenetic tree based on aligned nucleotides indicated the earlier separation of red algae [89]. Alternative topologies positioning rhodophytes or green plants as the earliest branch within the Archaeplastida were not rejected by the approximately unbiased test in the protein tree.

Reasons behind inconsistencies in inferring relationships between Archaeplastida lineages

Disagreements over the phylogeny of Archaeplastida may result from different methodological limitations of phylogenetic inferences and complex gene histories, especially those related to plastids as discussed above. The influence of fast-evolving sites was tested by Rodriguez-Ezpeleta et al. [172] and Hackett et al. [84]. Their trees based on complete alignments of nuclear-encoded proteins placed red algae at the basal position of the clade containing glaucophytes and green plants. However, the removal of fast-evolving sites recovered glaucophytes as the basal clade to the sister-group of red algae and green plants with a stronger support.

An important process that may influence Archaeplastida phylogenies is heterotachy, or lineage-specific rate variations across sites within a protein through time. It results from structural and functional constraints acting on proteins [173]. Vogl et al. [174] showed that many plastid genes have significantly different phylogenies under standard substitution models and proposed that this incongruence could indeed result from heterotachy. In agreement with that, the application of heterotachy models found that many plastid genes of red and green plants are, in fact, subject to such evolution [175–177], and this process can cause errors in inference of phylogenetic trees leading to the LBA effect [178,179].

Deschamps and Moreira [46] studied conflicts in Archaeplastida relationships related to nuclear-encoded proteins in great detail. Following the findings of Leigh et al. [119], who noticed an attraction between rhodophytes and stramenopiles (containing red alga-derived plastid), Deschamps and Moreira assumed that the basal position of rhodophytes in relation to other Archaeplastida lineages [45,84,160,172] could result from endosymbiotic gene transfers linked to the secondary plastid endosymbiosis (Fig. 2). In agreement with that, the exclusion of eukaryotic representatives involved in such endosymbioses shifted the position of rhodophytes in phylogenetic trees from the base of the Archaeplastida clade to sister to the "green" lineage.

Comparable EGT-related effects could be responsible for the basal position of the Viridiplantae lineage, as its members were also engaged in secondary plastid endosymbioses. Such a position was obtained in trees using proteins of plastid origin encoded in nuclear genomes. However, the exclusion of fast-evolving proteins in the Viridiplantae did not change this result. Conserved plastid-encoded proteins also favored this topology, even under methods designed to decrease LBA effects and are more robust against compositional and evolutionary rate biases (i.e. amino acid recoding and a site-heterogeneous mixture CAT model). In contrast, a tree based on these proteins, but constructed using a simpler model that is more sensitive to systematic error, produced a topology with the basal placement of the Glaucophyta. This would seem to suggest that this topology could be artificial and result from (*i*) an attraction of glaucophyte sequences to a cyanobacterial outgroup due to a compositional bias, and (*ii*) the affiliation of "green" and "red" lineages because of a LBA artefact [46].

Criscuolo and Gribaldo [47] studying more plastid-encoded proteins at the amino acid, degenerate codon, and recoded amino acid levels obtained the earlier divergence of Rhodophyta. The same topology was favored using a joint set of cyanobacterial homologous plastid- and nuclear-encoded proteins for degenerated codons and recoded amino acids, but the Viridiplantae-first scenario was supported by the combined set and the nuclear-encoded proteins alone at the amino acid level. The early-branching Glaucophyta topology was in turn recovered from the nuclear-encoded proteins using degenerated codons and recoded amino acids. These results clearly indicate an influence of data types and methods on inferences of relationships within the Archaeplastida.

These discrepancies among Archaeplastida phylogenies may be caused by conflicts between protein-coding genes and their amino acid translations into proteins, which persists even when sophisticated and better-fitting models are used, i.e. the CAT model and the nonstationary (tree-heterogeneous) composition model [48]. Since convergent composition biases are induced at synonymous codon positions, analyses based on amino acids should be preferred. Such analyses using better-fitting models on amino acid alignments showed the early divergence of Glaucophyta, whereas the ancient (and probably artificial) separation of the Viridiplantae was recovered using standard homogeneous models at the amino acid level as well as standard and advanced models at the nucleotide level [48].

Another reason for the inconsistency in inferring relationships within Archaeplastida could be ancient gene paralogy [163]. Comparisons of the reduced genome of *Paulinella plastids* (acquired independently from those of the Archaeplastida via a primary cyanobacterial endosymbiosis) with free-living strains of the picocyanobacterium *Prochlorococcus* showed differential gene loss and concerted evolution among paralogous gene families. If similar events occurred after the primary endosymbiosis, they could mislead phylogenetic analyses based on plastid genes (Fig. 3). In fact, phylogenetic analyses of serially duplicated genes for photosystem II (*psbA*, *psbB*, *psbC* and *psbD*) indicated the basal emergence of Viridiplantae and a strong sisterhood of the Rhodophyta and Glaucophyta. In contrast, trees inferred from single-copy orthologs and anciently duplicated single-copy genes (*atpA*, *atpB*, *psaA* and *psaB*) favored the early-branching of glaucophytes, with significant support for a Rhodophyta-Viridiplantae clade. It should be noted that the tree built using the concatenated alignment of all these genes supported the Viridiplantae-first scenario, which suggests that phylogenetic signal from unrecognized paralogs dominated this data set [163].

A similar effect also could be caused by horizontal gene transfer into plastid genomes, which cannot be excluded. Although such transfers are less numerous compared to mitochondrial genomes [117], at least seven have been reported so far. The best known is the ribosomal protein *rpl36* gene, which was transferred from a proteobacterium or a planctomycete bacterium to plastid genomes of cryptophytes and haptophytes [180]. The other examples are genes encoding the large and small subunits of RuBisCO form I, acquired by the primary plastid genome of the common ancestor of red algae from a proteobacterium [152] (for additional cases, see [181,182] and references therein).

The difficulties in resolving relationships among the Archaeplastida lineages through phylogenetic methods suggest that early Archaeplastida members, just after the origin of primary plastids, underwent a rapid radiation including many molecular evolutionary processes, such as diversification (often increase) in substitution rate, and replacements, transfers, duplications and losses of genes. It seems that the much smaller subset of plastid genes is in general more influenced by these phenomena than nuclear genes, which quite unanimously indicate the earlier divergence of Glaucophyta from Rhodophyta and Viridiplantae lineages.

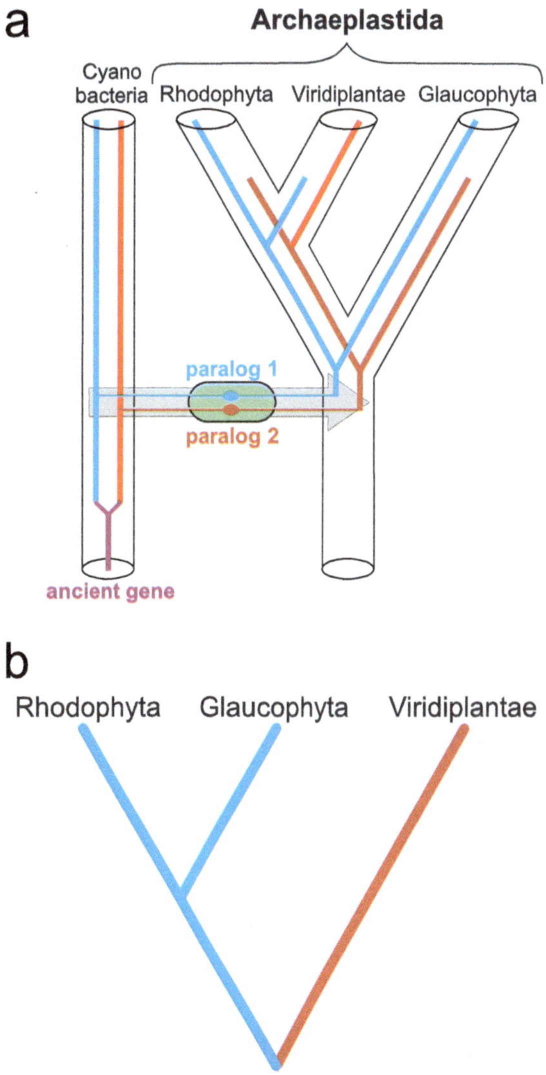

Fig. 3 Influence of differential loss of duplicated genes on the inference of phylogenetic relationships among Archaeplastida lineages. An ancient gene was duplicated in a cyanobacteria lineage into two paralogous genes, which were subsequently transferred via the primary endosymbiosis to the common ancestor of three Archaeplastida lineages (**a**). The genes were next subjected to differential losses. The paralog 1 stayed in rhodophytes and glaucophytes but was lost in viridiplants, which retained the paralog 2 that in turn decayed in rhodophytes and glaucophytes. This process misleads phylogenetic inference about the branching order of Archaeplastida lineages because the phylogenetic tree based on these genes indicates the earlier divergence of Viridiplantae (**b**) but, in fact, Glaucophyta separated as the first lineage (**a**). The Viridiplantae branch of the tree coalesces with Rhodophyta-Glaucophyta branch to the duplication event but not to the lineage separation.

Conclusions

Although it is commonly accepted that primary plastids of glaucophytes, red algae and green plants are of cyanobacterial origin, the monophyly of these groups is not unquestionably supported by all phylogenetic studies. Phylogenies based on plastid genes clearly recover monophyly of the three Archaeplastida plastid lineages; however, trees based on nuclear markers, rRNA and protein coding genes, vary substantially and have yet to resolve clear evolutionary positions of Archaeplastida host lineages. When non-monophyly is recovered, the primary plastid-containing eukaryotes often cluster with cryptophytes, haptophytes and related taxa. The new results based on mitochondrial genes suggest the monophyly of glaucophytes, red algae and green plants but do not exclude the other possibilities. The discrepancies among tree topologies obtained could result from many methodological limitations, for example stochastic and systematic errors involving horizontal/endosymbiotic gene transfers and hidden paralogy. However, even if these problems are taken into account, it is difficult to definitively exclude alternative scenarios for the origin of primary plastids. Nevertheless, trees based on nuclear genes that support the monophyly of Archaeplastida generally point to an earlier separation of glaucophytes in contrast to phylogenies based on mitochondrial and plastid genes. The smaller sets in the latter two cases appear more disturbed by complicated evolutionary phenomena. To conclusively solve the problem of the monophyly and the branching order of Archaeplastida plastids, more sequence data are required, especially those including glaucophytes and other poorly studied eukaryotic lineages. Giving the problems in phylogenetic reconstruction, other methods are desirable. EGT/HGT events can be identified by diverse composition-based methods. Other approaches alternative to phylogenetic analyses can include phylogenetic profiles, genomic contexts of studied genes, and presence or absence of genes in genomes. Analyses of rare mutation events called molecular signatures (e.g. conserved indels and gene fusion or splits) can also be helpful.

Acknowledgments

We are very grateful to the Reviewers for their excellent comments and insightful remarks that significantly improved the paper. This work was supported by Polish National Science Centre grant No. 2011/01/N/NZ8/00150 to P.G. and P.M.

Authors' contributions

The following declarations about authors' contributions to the research have been made: writing the manuscript and collecting data: PM, PG.

Competing interests

No competing interests have been declared.

References

1. Delwiche CF, Timme RE. Plants. Curr Biol. 2011;21(11):R417–R422.
2. Schenk HE. Glaucocystophytes. In: Encyclopedia of life sciences. Chichester: John Wiley & Sons; 2001.
3. Thomas DN. Seaweeds. London: Natural History Museum; 2002.
4. Lewis LA, McCourt RM. Green algae and the origin of land plants. Am J Bot. 2004;91(10):1535–1556.
5. Adl SM, Simpson AGB, Farmer MA, Andersen RA, Anderson OR, Barta JR, et al. The new higher level classification of eukaryotes with emphasis on the taxonomy of protists. J Eukaryot Microbiol. 2005;52(5):399–451.
6. Adl SM, Simpson AGB, Lane CE, Lukeš J, Bass D, Bowser SS, et al. The revised classification of eukaryotes. J Eukaryot Microbiol. 2012;59(5):429–514.
7. Keeling PJ, Burger G, Durnford DG, Lang BF, Lee RW, Pearlman RE, et al. The tree of eukaryotes. Trends Ecol Evol. 2005;20(12):670–676.
8. Gould SB, Waller RF, McFadden GI. Plastid evolution. Ann Rev Plant Biol. 2008;59(1):491–517.
9. Keeling PJ. The endosymbiotic origin, diversification and fate of plastids. Phil Trans R Soc Lond B. 2010;365(1541):729–748.
10. Reyes-Prieto A, Weber APM, Bhattacharya D. The origin and establishment of the plastid in algae and plants. Annu Rev Genet. 2007;41(1):147–168.
11. Stoebe B, Kowallik KV. Gene-cluster analysis in chloroplast genomics. Trends Genet. 1999;15(9):344–347.
12. Stoebe B, Martin W, Kowallik KV. Distribution and nomenclature of protein-coding genes in 12 sequenced chloroplast genomes. Plant Mol Biol Rep. 1998;16(3):243–255.
13. Besendahl A, Qiu YL, Lee J, Palmer JD, Bhattacharya D. The cyanobacterial origin and vertical transmission of the plastid tRNA(Leu) group-I intron. Curr Genet. 2000;37(1):12–23.
14. Pfanzagl B, Zenker A, Pittenauer E, Allmaier G, Martinez-Torrecuadrada J, Schmid ER, et al. Primary structure of cyanelle peptidoglycan of *Cyanophora paradoxa*: a prokaryotic cell wall as part of an organelle envelope. J Bacteriol. 1996;178(2):332–339.
15. Burey SC, Fathi-Nejad S, Poroyko V, Steiner JM, Löffelhardt W, Bohnert HJ. The central body of the cyanelles of *Cyanophora paradoxa*: a eukaryotic carboxysome? Can J Bot. 2005;83(7):758–764.
16. Kies L, Kremer BP. Phylum Glaucocystophyta. In: Margulis L, editor. Handbook of protoctista. Boston, MA: Jones and Bartlett Publishers; 1990. p. 152–166.
17. Archibald JM. The puzzle of plastid evolution. Curr Biol. 2009;19(2):R81–R88.
18. Bodył A, Stiller JW, Mackiewicz P. Chromalveolate plastids: direct descent or multiple endosymbioses? Trends Ecol Evol. 2009;24(3):119–121.
19. Yusa F, Steiner JM, Loffelhardt W. Evolutionary conservation of dual Sec translocases in the cyanelles of *Cyanophora paradoxa*. BMC Evol Biol. 2008;8(1):304.
20. Price DC, Chan CX, Yoon HS, Yang EC, Qiu H, Weber APM, et al. *Cyanophora paradoxa* genome elucidates origin of photosynthesis in algae and plants. Science. 2012;335(6070):843–847.
21. Chan CX, Gross J, Yoon HS, Bhattacharya D. Plastid origin and evolution: new models provide insights into old problems. Plant Physiol. 2011;155(4):1552–1560.
22. McFadden GI, van Dooren GG. Evolution: red algal genome affirms a common origin of all plastids. Curr Biol. 2004;14(13):R514–R516.
23. Reyes-Prieto A, Bhattacharya D. Phylogeny of Calvin cycle enzymes supports Plantae monophyly. Mol Phylogenet Evol. 2007;45(1):384–391.
24. Cavalier-Smith T, Lee JJ. Protozoa as hosts for endosymbioses and the conversion of symbionts into organelles. J Eukaryot Microbiol. 1985;32(3):376–379.
25. Cavalier-Smith T. Membrane heredity and early chloroplast evolution. Trends Plant Sci. 2000;5(4):174–182.

26. Cavalier-Smith T. The origins of plastids. Bot J Linn Soc. 1982;17(3):289–306.

27. Palmer JD. The symbiotic birth and spread of plastids: how many times and whodunit? J Phycol. 2003;39(1):4–12.

28. Howe C, Barbrook A, Nisbet RE, Lockhart P, Larkum AW. The origin of plastids. Philos Trans R Soc Lond B Biol Sci. 2008;363(1504):2675–2685.

29. Nozaki H. A new scenario of plastid evolution: plastid primary endosymbiosis before the divergence of the "Plantae", emended. J Plant Res. 2005;118(4):247–255.

30. Larkum AWD, Lockhart PJ, Howe CJ. Shopping for plastids. Trends Plant Sci. 2007;12(5):189–195.

31. Nozaki H, Maruyama S, Matsuzaki M, Nakada T, Kato S, Misawa K. Phylogenetic positions of Glaucophyta, green plants (Archaeplastida) and Haptophyta (Chromalveolata) as deduced from slowly evolving nuclear genes. Mol Phylogenet Evol. 2009;53(3):872–880.

32. Stiller JW, Reel DC, Johnson JC. A single origin of plastids revisited: convergent evolution in organellar genome content. J Phycol. 2003;39(1):95–105.

33. Stiller JW, Hall BD. The origin of red algae: implications for plastid evolution. Proc Natl Acad Sci USA. 1997;94(9):4520–4525.

34. Douglas SE, Turner S. Molecular evidence for the origin of plastids from a cyanobacterium-like ancestor. J Mol Evol. 1991;33(3):267–273.

35. Giovannoni SJ, Wood N, Huss V. Molecular phylogeny of oxygenic cells and organelles based on small-subunit ribosomal RNA sequences. In: Lewin RA, editor. Origins of plastids. New York, NY: Chapman and Hall; 1993. p. 159–170.

36. Olsen GJ, Woese CR, Overbeek R. The winds of (evolutionary) change: breathing new life into microbiology. J Bacteriol. 1994;176(1):1–6.

37. Marin B, Nowack EC, Melkonian M. A plastid in the making: evidence for a second primary endosymbiosis. Protist. 2005;156(4):425–432.

38. Ochoa de Alda JAG, Esteban R, Diago ML, Houmard J. The plastid ancestor originated among one of the major cyanobacterial lineages. Nat Commun. 2014;5:4937.

39. Delwiche C. Phylogenetic analysis of tufa sequences indicates a cyanobacterial origin of all plastids. Mol Phylogenet Evol. 1995;4(2):110–128.

40. Morden CW, Delwiche CF, Kuhsel M, Palmer JD. Gene phylogenies and the endosymbiotic origin of plastids. Biosystems. 1992;28(1–3):75–90.

41. Palmer JD, Delwiche CF. The origin and evolution of plastids and their genomes. In: Soltis DE, Soltis PS, Doyle JJ, editors. Molecular systematics of plants II. Boston, MA: Kluwer Academic Publishers; 1998. p. 375–409.

42. Ohta N, Sato N, Nozaki H, Kuroiwa T. Analysis of the cluster of ribosomal protein genes in the plastid genome of a unicellular red alga *Cyanidioschyzon merolae*: translocation of the str cluster as an early event in the rhodophyte-chromophyte lineage of plastid evolution. J Mol Evol. 1997;45(6):688–695.

43. Adachi J, Waddell PJ, Martin W, Hasegawa M. Plastid genome phylogeny and a model of amino acid substitution for proteins encoded by chloroplast DNA. J Mol Evol. 2000;50(4):348–358.

44. Yoon HS, Hackett JD, Ciniglia C, Pinto G, Bhattacharya D. A molecular timeline for the origin of photosynthetic eukaryotes. Mol Biol Evol. 2004;21(5):809–818.

45. Rodríguez-Ezpeleta N, Brinkmann H, Burey SC, Roure B, Burger G, Löffelhardt W, et al. Monophyly of primary photosynthetic eukaryotes: green plants, red algae, and glaucophytes. Curr Biol. 2005;15(14):1325–1330.

46. Deschamps P, Moreira D. Signal conflicts in the phylogeny of the primary photosynthetic eukaryotes. Mol Biol Evol. 2009;26(12):2745–2753.

47. Criscuolo A, Gribaldo S. Large-scale phylogenomic analyses indicate a deep origin of primary plastids within cyanobacteria. Mol Biol Evol. 2011;28(11):3019–3032.

48. Li B, Lopes JS, Foster PG, Embley TM, Cox CJ. Compositional biases among synonymous substitutions cause conflict between gene and protein trees for plastid origins. Mol Biol Evol. 2014;31(7):1697–1709.

49. Chu KH, Qi J, Yu ZG, Anh V. Origin and phylogeny of chloroplasts revealed by a simple correlation analysis of complete genomes. Mol Biol Evol. 2003;21(1):200–206.

50. Shih PM, Wu D, Latifi A, Axen SD, Fewer DP, Talla E, et al. Improving the coverage of the cyanobacterial phylum using diversity-driven genome sequencing. Proc Natl Acad Sci USA. 2013;110(3):1053–1058.

51. Marin B, Nowack EC, Glöckner G, Melkonian M. The ancestor of the *Paulinella chromatophore* obtained a carboxysomal operon by horizontal gene transfer from a *Nitrococcus*-like γ-proteobacterium. BMC Evol Biol. 2007;7(1):85.

52. Sogin ML. The phylogenetic significance of sequence diversity and length variations in eukaryotic small subunit ribosomal RNA coding regions. In: Warren L, Koprowski H, editors. New perspectives on evolution. New York, NY: Wiley-Liss; 1991. p. 175–188. (Wistar symposium series; vol 4).

53. Sogin ML, Elwood HJ, Gunderson JH. Evolutionary diversity of eukaryotic small-subunit rRNA genes. Proc Natl Acad Sci USA. 1986;83(5):1383–1387.

54. Bhattacharya D, Medlin L. The phylogeny of plastids: a review based on comparisons of small-subunit ribosomal RNA coding regions. J Phycol. 1995;31(4):489–498.

55. Bhattacharya D, Helmchen T, Bibeau C, Melkonian M. Comparisons of nuclear-encoded small-subunit ribosomal RNAs reveal the evolutionary position of the Glaucocystophyta. Mol Biol Evol. 1995;12(3):415–420.

56. van de Peer Y, Rensing SA, Maier UG, de Wachter R. Substitution rate calibration of small subunit ribosomal RNA identifies chlorarachniophyte endosymbionts as remnants of green algae. Proc Natl Acad Sci USA. 1996;93(15):7732–7736.

57. van de Peer Y, de Wachter R. Evolutionary relationships among the eukaryotic crown taxa taking into account site-to-site rate variation in 18S rRNA. J Mol Evol. 1997;45(6):619–630.

58. van de Peer Y, Baldauf SL, Doolittle WF, Meyer A. An updated and comprehensive rRNA phylogeny of (crown) eukaryotes based on rate-calibrated evolutionary distances. J Mol Evol. 2000;51(6):565–576.

59. Cavalier-Smith T. The phagotrophic origin of eukaryotes and phylogenetic classification of Protozoa. Int J Syst Evol Microbiol. 2002;52(2):297–354.

60. Cavalier-Smith T. Only six kingdoms of life. Proc Biol Sci. 2004;271(1545):1251–1262.

61. Okamoto N, Inouye I. The katablepharids are a distant sister group of the Cryptophyta: a proposal for Katablepharidophyta divisio nova/kathablepharida phylum novum based on SSU rDNA and beta-tubulin phylogeny. Protist. 2005;156(2):163–179.

62. Cuvelier ML, Ortiz A, Kim E, Moehlig H, Richardson DE, Heidelberg JF, et al. Widespread distribution of a unique marine protistan lineage. Environ Microbiol. 2008;10(6):1621–1634.

63. Ishida K, Inagaki Y, Sakaguchi M, Oiwa A, Kai A, Suzuki M, et al. Comprehensive SSU rRNA phylogeny of eukaryota. Endocytobiosis Cell Res. 2010;20:81–88.

64. Yoon HS, Price DC, Stepanauskas R, Rajah VD, Sieracki ME, Wilson WH, et al. Single-cell genomics reveals organismal interactions in uncultivated marine protists. Science. 2011;332(6030):714–717.

65. Seenivasan R, Sausen N, Medlin LK, Melkonian M. *Picomonas judraskeda* gen. et sp. nov.: the first identified member of the Picozoa

phylum nov., a widespread group of Picoeukaryotes, formerly known as "picobiliphytes". PLoS ONE. 2013;8(3):e59565.

66. Moreira D, von der Heyden S, Bass D, López-García P, Chao E, Cavalier-Smith T. Global eukaryote phylogeny: combined small- and large-subunit ribosomal DNA trees support monophyly of Rhizaria, Retaria and Excavata. Mol Phylogenet Evol. 2007;44(1):255–266.
67. Zhao S, Burki F, Brate J, Keeling PJ, Klaveness D, Shalchian-Tabrizi K. *Collodictyon* – an ancient lineage in the tree of eukaryotes. Mol Biol Evol. 2012;29(6):1557–1568.
68. Yabuki A, Inagaki Y, Ishida K. *Palpitomonas bilix* gen. et sp. nov.: a novel deep-branching heterotroph possibly related to Archaeplastida or Hacrobia. Protist. 2010;161(4):523–538.
69. Kim E, Simpson AGB, Graham LE. Evolutionary relationships of apusomonads inferred from taxon-rich analyses of 6 nuclear encoded genes. Mol Biol Evol. 2006;23(12):2455–2466.
70. Bhattacharya D, Weber K. The actin gene of the glaucocystophyte *Cyanophora paradoxa*: analysis of the coding region and introns, and an actin phylogeny of eukaryotes. Curr Genet. 1997;31(5):439–446.
71. Baldauf SL, Roger AJ, Wenk-Siefert I, Doolittle WF. A kingdom-level phylogeny of eukaryotes based on combined protein data. Science. 2000;290(5493):972–977.
72. Lartillot N. A bayesian mixture model for across-site heterogeneities in the amino-acid replacement process. Mol Biol Evol. 2004;21(6):1095–1109.
73. Rodríguez-Ezpeleta N, Brinkmann H, Roure B, Lartillot N, Lang BF, Philippe H. Detecting and overcoming systematic errors in genome-scale phylogenies. Syst Biol. 2007;56(3):389–399.
74. Reeb VC, Peglar MT, Yoon HS, Bai JR, Wu M, Shiu P, et al. Interrelationships of chromalveolates within a broadly sampled tree of photosynthetic protists. Mol Phylogenet Evol. 2009;53(1):202–211.
75. Nikolaev SI, Berney C, Fahrni JF, Bolivar I, Polet S, Mylnikov AP, et al. The twilight of Heliozoa and rise of Rhizaria, an emerging supergroup of amoeboid eukaryotes. Proc Natl Acad Sci USA. 2004;101(21):8066–8071.
76. Yoon HS, Grant J, Tekle YI, Wu M, Chaon BC, Cole JC, et al. Broadly sampled multigene trees of eukaryotes. BMC Evol Biol. 2008;8(1):14.
77. Parfrey LW, Grant J, Tekle YI, Lasek-Nesselquist E, Morrison HG, Sogin ML, et al. Broadly sampled multigene analyses yield a well-resolved eukaryotic tree of life. Syst Biol. 2010;59(5):518–533.
78. Kim E, Graham LE. EEF2 analysis challenges the monophyly of Archaeplastida and Chromalveolata. PLoS ONE. 2008;3(7):e2621.
79. Hampl V, Hug L, Leigh JW, Dacks JB, Lang BF, Simpson AGB, et al. Phylogenomic analyses support the monophyly of Excavata and resolve relationships among eukaryotic "supergroups". Proc Natl Acad Sci USA. 2009;106(10):3859–3864.
80. Cavalier-Smith T, Chao EE, Snell EA, Berney C, Fiore-Donno AM, Lewis R. Multigene eukaryote phylogeny reveals the likely protozoan ancestors of opisthokonts (animals, fungi, choanozoans) and Amoebozoa. Mol Phylogenet Evol. 2014;81:71–85.
81. Burki F, Okamoto N, Pombert JF, Keeling PJ. The evolutionary history of haptophytes and cryptophytes: phylogenomic evidence for separate origins. Proc Biol Sci. 2012;279(1736):2246–2254.
82. Tekle YI, Grant J, Anderson OR, Nerad TA, Cole JC, Patterson DJ, et al. Phylogenetic placement of diverse amoebae inferred from multigene analyses and assessment of clade stability within "Amoebozoa" upon removal of varying rate classes of SSU-rDNA. Mol Phylogenet Evol. 2008;47(1):339–352.
83. Nozaki H, Matsuzaki M, Takahara M, Misumi O, Kuroiwa H, Hasegawa M, et al. The phylogenetic position of red algae revealed by multiple nuclear genes from mitochondria-containing eukaryotes and an alternative hypothesis on the origin of plastids. J Mol Evol. 2003;56(4):485–497.
84. Hackett JD, Yoon HS, Li S, Reyes-Prieto A, Rummele SE, Bhattacharya D. Phylogenomic analysis supports the monophyly of cryptophytes and haptophytes and the association of rhizaria with chromalveolates. Mol Biol Evol. 2007;24(8):1702–1713.
85. Burki F, Shalchian-Tabrizi K, Pawlowski J. Phylogenomics reveals a new "megagroup" including most photosynthetic eukaryotes. Biol Lett. 2008;4(4):366–369.
86. Burki F, Inagaki Y, Brate J, Archibald JM, Keeling PJ, Cavalier-Smith T, et al. Large-scale phylogenomic analyses reveal that two enigmatic protist lineages, telonemia and centroheliozoa, are related to photosynthetic chromalveolates. Genome Biol Evol. 2009;1:231–238.
87. Yabuki A, Kamikawa R, Ishikawa SA, Kolisko M, Kim E, Tanabe AS, et al. *Palpitomonas bilix* represents a basal cryptist lineage: insight into the character evolution in Cryptista. Sci Rep. 2014;4:4641.
88. Zhao S, Shalchian-Tabrizi K, Klaveness D. Sulcozoa revealed as a paraphyletic group in mitochondrial phylogenomics. Mol Phylogenet Evol. 2013;69(3):462–468.
89. Jackson CJ, Reyes-Prieto A. The mitochondrial genomes of the Glaucophytes *Gloeochaete wittrockiana* and *Cyanoptyche gloeocystis*: multilocus phylogenetics suggests a monophyletic archaeplastida. Genome Biol Evol. 2014;6(10):2774–2785.
90. Stiller JW, Riley J, Hall BD. Are red algae plants? A critical evaluation of three key molecular data sets. J Mol Evol. 2001;52(6):527–539.
91. Roger AJ, Hug LA. The origin and diversification of eukaryotes: problems with molecular phylogenetics and molecular clock estimation. Philos Trans R Soc Lond B Biol Sci. 2006;361(1470):1039–1054.
92. Delsuc F, Brinkmann H, Philippe H. Phylogenomics and the reconstruction of the tree of life. Nat Rev Genet. 2005;6(5):361–375.
93. Brown MW, Sharpe SC, Silberman JD, Heiss AA, Lang BF, Simpson AGB, et al. Phylogenomics demonstrates that breviate flagellates are related to opisthokonts and apusomonads. Proc Biol Sci. 2013;280(1769):20131755.
94. Felsenstein J. Cases in which parsimony or compatibility methods will be positively misleading. Syst Zool. 1978;27(4):401.
95. Stiller JW, Harrell L. The largest subunit of RNA polymerase II from the Glaucocystophyta: functional constraint and short-branch exclusion in deep eukaryotic phylogeny. BMC Evol Biol. 2005;5(1):71.
96. Roure B, Baurain D, Philippe H. Impact of missing data on phylogenies inferred from empirical phylogenomic data sets. Mol Biol Evol. 2013;30(1):197–214.
97. Philippe H, Snell EA, Bapteste E, Lopez P, Holland PW, Casane D. Phylogenomics of eukaryotes: impact of missing data on large alignments. Mol Biol Evol. 2004;21(9):1740–1752.
98. Nozaki H, Iseki M, Hasegawa M, Misawa K, Nakada T, Sasaki N, et al. Phylogeny of primary photosynthetic eukaryotes as deduced from slowly evolving nuclear genes. Mol Biol Evol. 2007;24(8):1592–1595.
99. Inagaki Y, Nakajima Y, Sato M, Sakaguchi M, Hashimoto T. Gene sampling can bias multi-gene phylogenetic inferences: the relationship between red algae and green plants as a case study. Mol Biol Evol. 2009;26(5):1171–1178.
100. Brinkmann H, van der Giezen M, Zhou Y, de Raucourt GP, Philippe

H. An empirical assessment of long-branch attraction artefacts in deep eukaryotic phylogenomics. Syst Biol. 2005;54(5):743–757.

101. Susko E, Roger AJ. On reduced amino acid alphabets for phylogenetic inference. Mol Biol Evol. 2007;24(9):2139–2150.

102. Tuffley C, Steel M. Modeling the covarion hypothesis of nucleotide substitution. Math Biosci. 1998;147(1):63–91.

103. Chan CX, Yang EC, Banerjee T, Yoon HS, Martone PT, Estevez JM, et al. Red and green algal monophyly and extensive gene sharing found in a rich repertoire of red algal genes. Curr Biol. 2011;21(4):328–333.

104. Maddison WP. Gene trees in species trees. Syst Biol. 1997;46(3):523–536.

105. Page R. From gene to organismal phylogeny: reconciled trees and the gene tree/species tree problem. Mol Phylogenet Evol. 1997;7(2):231–240.

106. Chan CX, Bhattacharya D. Analysis of horizontal genetic transfer in red algae in the post-genomics age. Mob Genet Elem. 2013;3(6):e27669.

107. Chan CX, Bhattacharya D, Reyes-Prieto A. Endosymbiotic and horizontal gene transfer in microbial eukaryotes: impacts on cell evolution and the tree of life. Mob Genet Elem. 2012;2(2):101–105.

108. Huang J, Yue J. Horizontal gene transfer in the evolution of photosynthetic eukaryotes: HGT in plants. J Syst Evol. 2013;51(1):13–29.

109. Schönknecht G, Weber APM, Lercher MJ. Horizontal gene acquisitions by eukaryotes as drivers of adaptive evolution. BioEssays. 2014;36(1):9–20.

110. Keeling PJ, Palmer JD. Horizontal gene transfer in eukaryotic evolution. Nat Rev Genet. 2008;9(8):605–618.

111. Keeling PJ. Role of horizontal gene transfer in the evolution of photosynthetic eukaryotes and their plastids. In: Gogarten MB, Gogarten JP, Olendzenski LC, editors. Horizontal gene transfer. Totowa, NJ: Humana Press; 2009. p. 501–515. (vol 532).

112. Richardson AO, Palmer JD. Horizontal gene transfer in plants. J Exp Bot. 2006;58(1):1–9.

113. Bergthorsson U, Richardson AO, Young GJ, Goertzen LR, Palmer JD. Massive horizontal transfer of mitochondrial genes from diverse land plant donors to the basal angiosperm *Amborella*. Proc Natl Acad Sci USA. 2004;101(51):17747–17752.

114. Bergthorsson U, Adams KL, Thomason B, Palmer JD. Widespread horizontal transfer of mitochondrial genes in flowering plants. Nature. 2003;424(6945):197–201.

115. Woloszynska M, Bocer T, Mackiewicz P, Janska H. A fragment of chloroplast DNA was transferred horizontally, probably from non-eudicots, to mitochondrial genome of *Phaseolus*. Plant Mol Biol. 2004;56(5):811–820.

116. Timmis JN, Ayliffe MA, Huang CY, Martin W. Endosymbiotic gene transfer: organelle genomes forge eukaryotic chromosomes. Nat Rev Genet. 2004;5(2):123–135.

117. Kleine T, Maier UG, Leister D. DNA transfer from organelles to the nucleus: the idiosyncratic genetics of endosymbiosis. Annu Rev Plant Biol. 2009;60(1):115–138.

118. Lane CE, Archibald JM. The eukaryotic tree of life: endosymbiosis takes its TOL. Trends Ecol Evol. 2008;23(5):268–275.

119. Leigh JW, Susko E, Baumgartner M, Roger AJ. Testing congruence in phylogenomic analysis. Syst Biol. 2008;57(1):104–115.

120. Stiller JW. Experimental design and statistical rigor in phylogenomics of horizontal and endosymbiotic gene transfer. BMC Evol Biol. 2011;11(1):259.

121. Andersson JO, Roger AJ. A cyanobacterial gene in nonphotosynthetic protists – an early chloroplast acquisition in eukaryotes? Curr Biol. 2002;12(2):115–119.

122. Cavalier-Smith T. The origin, losses and gains of chloroplasts. In: Lewin RA, editor. Origins of plastids. New York, NY: Chapman and Hall; 1993. p. 291–348.

123. Dorrell RG, Smith AG. Do red and green make brown?: perspectives on plastid acquisitions within chromalveolates. Eukaryot Cell. 2011;10(7):856–868.

124. Vesteg M, Vacula R, Krajčovič J. On the origin of chloroplasts, import mechanisms of chloroplast-targeted proteins, and loss of photosynthetic ability – review. Folia Microbiol Praha. 2009;54(4):303–321.

125. Krause K. From chloroplasts to "cryptic" plastids: evolution of plastid genomes in parasitic plants. Curr Genet. 2008;54(3):111–121.

126. Borza T, Popescu CE, Lee RW. Multiple metabolic roles for the nonphotosynthetic plastid of the green alga *Prototheca wickerhamii*. Eukaryot Cell. 2005;4(2):253–261.

127. Mazumdar J, Wilson EH, Masek K, Hunter CA, Striepen B. Apicoplast fatty acid synthesis is essential for organelle biogenesis and parasite survival in *Toxoplasma gondii*. Proc Natl Acad Sci USA. 2006;103(35):13192–13197.

128. Bodył A, Mackiewicz P, Gagat P. Organelle evolution: *Paulinella* breaks a paradigm. Curr Biol. 2012;22(9):R304–R306.

129. Gagat P, Bodył A, Mackiewicz P, Stiller JW. Tertiary plastid endosymbioses in dinoflagellates. In: Löffelhardt W, editor. Endosymbiosis. Vienna: Springer; 2014. p. 233–290.

130. Kies L, Kremer BP. Function of cyanelles in the tecamoeba *Paulinella chromatophora*. Naturewissenschaften. 1979;66:578–579.

131. Bodył A, Mackiewicz P, Stiller JW. The intracellular cyanobacteria of *Paulinella chromatophora*: endosymbionts or organelles? Trends Microbiol. 2007;15(7):295–296.

132. Nowack ECM, Melkonian M, Glöckner G. Chromatophore genome sequence of *Paulinella* sheds light on acquisition of photosynthesis by eukaryotes. Curr Biol. 2008;18(6):410–418.

133. Reyes-Prieto A, Yoon HS, Moustafa A, Yang EC, Andersen RA, Boo SM, et al. Differential gene retention in plastids of common recent origin. Mol Biol Evol. 2010;27(7):1530–1537.

134. Nowack ECM, Vogel H, Groth M, Grossman AR, Melkonian M, Glockner G. Endosymbiotic gene transfer and transcriptional regulation of transferred genes in *Paulinella chromatophora*. Mol Biol Evol. 2011;28(1):407–422.

135. Nakayama T, Ishida K. Another acquisition of a primary photosynthetic organelle is underway in *Paulinella chromatophora*. Curr Biol. 2009;19(7):R284–R285.

136. Nowack ECM, Grossman AR. Trafficking of protein into the recently established photosynthetic organelles of *Paulinella chromatophora*. Proc Natl Acad Sci USA. 2012;109(14):5340–5345.

137. Mackiewicz P, Bodył A, Gagat P. Protein import into the photosynthetic organelles of *Paulinella chromatophora* and its implications for primary plastid endosymbiosis. Symbiosis. 2012;58(1–3):99–107.

138. Mackiewicz P, Bodył A, Gagat P. Possible import routes of proteins into the cyanobacterial endosymbionts/plastids of *Paulinella chromatophora*. Theory Biosci. 2012;131(1):1–18.

139. Bodył A, Mackiewicz P, Stiller JW. Comparative genomic studies suggest that the cyanobacterial endosymbionts of the amoeba *Paulinella chromatophora* possess an import apparatus for

nuclear-encoded proteins. Plant Biol. 2009;12:639–649.

140. Carpenter EJ, Foster RA. Marine cyanobacterial symbioses. In: Rai AN, Bergman B, Rasmussen U, editors. Cyanobacteria in symbiosis. Dordrecht: Kluwer Academic Publishers; 2002. p. 11–17.

141. Raven JA. Evolution of cyanobacterial symbioses. In: Rai AN, Bergman B, Rasmussen U, editors. Cyanobacteria in symbiosis. Dordrecht: Kluwer Academic Publishers; 2002. p. 329–346.

142. Kneip C, Lockhart P, Voß C, Maier UG. Nitrogen fixation in eukaryotes – new models for symbiosis. BMC Evol Biol. 2007;7(1):55.

143. Kneip C, Voβ C, Lockhart PJ, Maier UG. The cyanobacterial endosymbiont of the unicellular algae *Rhopalodia gibba* shows reductive genome evolution. BMC Evol Biol. 2008;8(1):30.

144. Rogers M, Keeling PJ. Lateral transfer and recompartmentalization of Calvin cycle enzymes of plants and algae. J Mol Evol. 2004;58(4):367–375.

145. Durnford DG, Deane JA, Tan S, McFadden GI, Gantt E, Green BR. A phylogenetic assessment of the eukaryotic light-harvesting antenna proteins, with implications for plastid evolution. J Mol Evol. 1999;48(1):59–68.

146. Rissler HM, Durnford DG. Isolation of a novel carotenoid-rich protein in *Cyanophora paradoxa* that is immunologically related to the light-harvesting complexes of photosynthetic eukaryotes. Plant Cell Physiol. 2005;46(3):416–424.

147. Plancke C, Colleoni C, Deschamps P, Dauvillee D, Nakamura Y, Haebel S, et al. Pathway of cytosolic starch synthesis in the model glaucophyte *Cyanophora paradoxa*. Eukaryot Cell. 2008;7(2):247–257.

148. Deschamps P, Haferkamp I, d'Hulst C, Neuhaus HE, Ball SG. The relocation of starch metabolism to chloroplasts: when, why and how. Trends Plant Sci. 2008;13(11):574–582.

149. Deschamps P, Colleoni C, Nakamura Y, Suzuki E, Putaux JL, Buleon A, et al. Metabolic symbiosis and the birth of the plant kingdom. Mol Biol Evol. 2008;25(3):536–548.

150. Ball S, Colleoni C, Cenci U, Raj JN, Tirtiaux C. The evolution of glycogen and starch metabolism in eukaryotes gives molecular clues to understand the establishment of plastid endosymbiosis. J Exp Bot. 2011;62(6):1775–1801.

151. Cavalier-Smith T. Eukaryote kingdoms: seven or nine? Biosystems. 1981;14(3–4):461–481.

152. Delwiche CF, Palmer JD. Rampant horizontal transfer and duplication of rubisco genes in eubacteria and plastids. Mol Biol Evol. 1996;13(6):873–882.

153. Shibata M, Kashino Y, Satoh K, Koike H. Isolation and characterization of oxygen-evolving thylakoid membranes and photosystem II particles from a glaucocystophyte, *Cyanophora paradoxa*. Plant Cell Physiol. 2001;42(7):733–741.

154. Koike H, Shibata M, Yasutomi K, Kashino Y, Satoh K. Identification of photosystem I components from a glaucocystophyte, *Cyanophora paradoxa*: the PsaD protein has an N-terminal stretch homologous to higher plants. Photosynth Res. 2000;65(3):207–217.

155. Machida M, Takechi K, Sato H, Chung SJ, Kuroiwa H, Takio S, et al. Genes for the peptidoglycan synthesis pathway are essential for chloroplast division in moss. Proc Natl Acad Sci USA. 2006;103(17):6753–6758.

156. Takano H, Takechi K. Plastid peptidoglycan. Biochim Biophys Acta. 2010;1800(2):144–151.

157. Lockhart PJ, Howe CJ, Bryant DA, Beanland TJ, Larkum AWD. Substitutional bias confounds inference of cyanelle origins from sequence data. J Mol Evol. 1992;34(2):153–162.

158. Lockhart PJ, Penny D, Hendy MD, Howe CJ, Beanland TJ, Larkum AWD. Controversy on chloroplast origins. FEBS Lett. 1992;301(2):127–131.

159. Okamoto N, Chantangsi C, Horák A, Leander BS, Keeling PJ. Molecular phylogeny and description of the novel katablepharid *Roombia truncata* gen. et sp. nov., and establishment of the hacrobia taxon nov. PLoS ONE. 2009;4(9):e7080.

160. Burki F, Shalchian-Tabrizi K, Minge M, Skjæveland Å, Nikolaev SI, Jakobsen KS, et al. Phylogenomics reshuffles the eukaryotic supergroups. PLoS ONE. 2007;2(8):e790.

161. Turner S, Pryer KM, Miao VP, Palmer JD. Investigating deep phylogenetic relationships among cyanobacteria and plastids by small subunit rRNA sequence analysis. J Eukaryot Microbiol. 1999;46(4):327–338.

162. Reyes-Prieto A, Bhattacharya D. Phylogeny of nuclear-encoded plastid-targeted proteins supports an early divergence of glaucophytes within Plantae. Mol Biol Evol. 2007;24(11):2358–2361.

163. Qiu H, Yang EC, Bhattacharya D, Yoon HS. Ancient gene paralogy may mislead inference of plastid phylogeny. Mol Biol Evol. 2012;29(11):3333–3343.

164. Helmchen TA, Bhattacharya D, Melkonian M. Analyses of ribosomal RNA sequences from glaucocystophyte cyanelles provide new insights into the evolutionary relationships of plastids. J Mol Evol. 1995;41(2):203–210.

165. Yoon HS, Hackett JD, van Dolah FM, Nosenko T, Lidie KL, Bhattacharya D. Tertiary endosymbiosis driven genome evolution in dinoflagellate algae. Mol Biol Evol. 2005;22(5):1299–1308.

166. Yoon HS, Nakayama T, Reyes-Prieto A, Andersen RA, Boo SM, Ishida K, et al. A single origin of the photosynthetic organelle in different *Paulinella* lineages. BMC Evol Biol. 2009;9(1):98.

167. Falcón LI, Magallón S, Castillo A. Dating the cyanobacterial ancestor of the chloroplast. ISME J. 2010;4(6):777–783.

168. Auch AF, Henz SR, Holland BR, Göker M. Genome BLAST distance phylogenies inferred from whole plastid and whole mitochondrion genome sequences. BMC Bioinformatics. 2006;7(1):350.

169. Nozaki H, Ohta N, Matsuzaki M, Misumi O, Kuroiwa T. Phylogeny of plastids based on cladistic analysis of gene loss inferred from complete plastid genome sequences. J Mol Evol. 2003;57(4):377–382.

170. Sanchez-Puerta MV, Bachvaroff TR, Delwiche CF. Sorting wheat from chaff in multi-gene analyses of chlorophyll *c*-containing plastids. Mol Phylogenet Evol. 2007;44(2):885–897.

171. de Las Rivas J. Comparative analysis of chloroplast genomes: functional annotation, genome-based phylogeny, and deduced evolutionary patterns. Genome Res. 2002;12(4):567–583.

172. Rodríguez-Ezpeleta N, Brinkmann H, Burger G, Roger AJ, Gray MW, Philippe H, et al. Toward resolving the eukaryotic tree: the phylogenetic positions of jakobids and cercozoans. Curr Biol. 2007;17(16):1420–1425.

173. Lopez P, Casane D, Philippe H. Heterotachy, an important process of protein evolution. Mol Biol Evol. 2002;19(1):1–7.

174. Vogl C, Badger J, Kearney P, Li M, Clegg M, Jiang T. Probabilistic analysis indicates discordant gene trees in chloroplast evolution. J Mol Evol. 2003;56(3):330–340.

175. Ane C. Covarion structure in plastid genome evolution: a new statistical test. Mol Biol Evol. 2005;22(4):914–924.

176. Whelan S, Blackburne BP, Spencer M. Phylogenetic substitution models for detecting heterotachy during plastid evolution. Mol Biol Evol. 2011;28(1):449–458.

177. Lockhart PJ, Steel MA, Barbrook AC, Huson DH, Charleston MA, Howe CJ. A covariotide model explains apparent phylogenetic

structure of oxygenic photosynthetic lineages. Mol Biol Evol. 1998;15(9):1183–1188.

178. Lockhart P, Steel M. A tale of two processes. Syst Biol. 2005;54(6):948–951.

179. Lockhart P. Heterotachy and tree building: a case study with plastids and eubacteria. Mol Biol Evol. 2005;23(1):40–45.

180. Rice DW, Palmer JD. An exceptional horizontal gene transfer in plastids: gene replacement by a distant bacterial paralog and evidence that haptophyte and cryptophyte plastids are sisters. BMC Biol. 2006;4(1):31.

181. Mackiewicz P, Bodył A, Moszczyński K. The case of horizontal gene transfer from bacteria to the peculiar dinoflagellate plastid genome. Mob Genet Elem. 2013;3(4):e25845.

182. Moszczynski K, Mackiewicz P, Bodyl A. Evidence for horizontal gene transfer from bacteroidetes bacteria to dinoflagellate minicircles. Mol Biol Evol. 2012;29(3):887–892.

183. Bachvaroff TR, Sanchez-Puerta MV, Delwiche CF. Chlorophyll *c*-containing plastid relationships based on analyses of a multigene data set with all four chromalveolate lineages. Mol Biol Evol. 2005;22(9):1772–1782.

184. Cuvelier ML, Allen AE, Monier A, McCrow JP, Messie M, Tringe SG, et al. Targeted metagenomics and ecology of globally important uncultured eukaryotic phytoplankton. Proc Natl Acad Sci USA. 2010;107(33):14679–14684.

185. Hagopian JC, Reis M, Kitajima JP, Bhattacharya D, de Oliveira MC. Comparative analysis of the complete plastid genome sequence of the red alga *Gracilaria tenuistipitata* var. *liui* provides insights into the evolution of rhodoplasts and their relationship to other plastids. J Mol Evol. 2004;59(4):464–477.

186. Janouskovec J, Horak A, Obornik M, Lukes J, Keeling PJ. A common red algal origin of the apicomplexan, dinoflagellate, and heterokont plastids. Proc Natl Acad Sci USA. 2010;107(24):10949–10954.

187. Janouškovec J, Horák A, Barott KL, Rohwer FL, Keeling PJ. Global analysis of plastid diversity reveals apicomplexan-related lineages in coral reefs. Curr Biol. 2012;22(13):R518–R519.

188. Khan H, Parks N, Kozera C, Curtis BA, Parsons BJ, Bowman S, et al. Plastid genome sequence of the cryptophyte alga *Rhodomonas salina* CCMP1319: lateral transfer of putative DNA replication machinery and a test of chromist plastid phylogeny. Mol Biol Evol. 2007;24(8):1832–1842.

189. Kim E, Harrison JW, Sudek S, Jones MDM, Wilcox HM, Richards TA, et al. Newly identified and diverse plastid-bearing branch on the eukaryotic tree of life. Proc Natl Acad Sci USA. 2011;108(4):1496–1500.

190. Le Corguillé G, Pearson G, Valente M, Viegas C, Gschloessl B, Corre E, et al. Plastid genomes of two brown algae, *Ectocarpus siliculosus* and *Fucus vesiculosus*: further insights on the evolution of red-algal derived plastids. BMC Evol Biol. 2009;9(1):253.

191. Martin W, Stoebe B, Goremykin V, Hansmann S, Hasegawa M, Kowallik KV. Gene transfer to the nucleus and the evolution of chloroplasts. Nature. 1998;393(6681):162–165.

192. Martin W, Rujan T, Richly E, Hansen A, Cornelsen S, Lins T, et al. Evolutionary analysis of *Arabidopsis*, cyanobacterial, and chloroplast genomes reveals plastid phylogeny and thousands of cyanobacterial genes in the nucleus. Proc Natl Acad Sci USA. 2002;99(19):12246–12251.

193. Nelissen B, van de Peer Y, Wilmotte A, de Wachter R. An early origin of plastids within the cyanobacterial divergence is suggested by evolutionary trees based on complete 16S rRNA sequences. Mol Biol Evol. 1995;12(6):1166–1173.

194. Ohta N, Matsuzaki M, Misumi O, Miyagishima SY, Nozaki H, Tanaka K, et al. Complete sequence and analysis of the plastid genome of the unicellular red alga *Cyanidioschyzon merolae*. DNA Res. 2003;10(2):67–77.

195. Rogers MB, Gilson PR, Su V, McFadden GI, Keeling PJ. The complete chloroplast genome of the chlorarachniophyte *Bigelowiella natans*: evidence for independent origins of chlorarachniophyte and euglenid secondary endosymbionts. Mol Biol Evol. 2007;24(1):54–62.

196. Sato N. Origin and evolution of plastids: genomic view on the unification and diversity of plastids. In: Wise RR, Hoober JK, editors. The structure and function of plastids. Dordrecht: Springer; 2006. p. 75–102. (Advances in photosynthesis and respiration).

197. Tengs T, Dahlberg OJ, Shalchian-Tabrizi K, Klaveness D, Rudi K, Delwiche CF, et al. Phylogenetic analyses indicate that the 19'Hexanoyloxy-fucoxanthin-containing dinoflagellates have tertiary plastids of haptophyte origin. Mol Biol Evol. 2000;17(5):718–729.

198. Wang Y, Joly S, Morse D. Phylogeny of dinoflagellate plastid genes recently transferred to the nucleus supports a common ancestry with red algal plastid genes. J Mol Evol. 2008;66(2):175–184.

199. Baurain D, Brinkmann H, Petersen J, Rodriguez-Ezpeleta N, Stechmann A, Demoulin V, et al. Phylogenomic evidence for separate acquisition of plastids in cryptophytes, haptophytes, and stramenopiles. Mol Biol Evol. 2010;27(7):1698–1709.

200. Minge MA, Silberman JD, Orr RJ, Cavalier-Smith T, Shalchian-Tabrizi K, Burki F, et al. Evolutionary position of breviate amoebae and the primary eukaryote divergence. Proc Biol Sci. 2009;276(1657):597–604.

201. Moreira D, Le Guyader H, Philippe H. The origin of red algae and the evolution of chloroplasts. Nature. 2000;405(6782):69–72.

202. Patron NJ, Inagaki Y, Keeling PJ. Multiple gene phylogenies support the monophyly of cryptomonad and haptophyte host lineages. Curr Biol. 2007;17(10):887–891.

203. Tekle YI, Grant J, Cole JC, Nerad TA, Anderson OR, Patterson DJ, et al. A multigene analysis of *Corallomyxa tenera* sp. nov. suggests its membership in a clade that includes Gromia, Haplosporidia and Foraminifera. Protist. 2007;158(4):457–472.

204. Tekle YI, Parfrey LW, Katz LA. Molecular data are transforming hypotheses on the origin and diversification of eukaryotes. Bioscience. 2009;59(6):471–481.

Application and comparative performance of network modularity algorithms to ecological communities classification

Thorsten Thiergart[1], Ulf Schmitz[1], Giddy Landan[2], William F. Martin[1], Tal Dagan[2]*

[1] Institute of Molecular Evolution, Heinrich-Heine University Düsseldorf, Universitätsstrasse 1, 40225 Düsseldorf, Germany

[2] Genomic Microbiology Group, Institute of Microbiology, Christian-Albrechts-University of Kiel, Christian-Albrechts-Platz 4, 24118 Kiel, Germany

Abstract

Network modularity is a well-studied large-scale connectivity pattern in networks. The detection of modules in real networks constitutes a crucial step towards a description of the network building blocks and their evolutionary dynamics. The performance of modularity detection algorithms is commonly quantified using simulated networks data. However, a comparison of the modularity algorithms utility for real biological data is scarce. Here we investigate the utility of network modularity algorithms for the classification of ecological plant communities. Plant community classification by the traditional approaches requires prior knowledge about the characteristic and differential species, which are derived from a manual inspection of vegetation tables. Using the raw species abundance data we constructed six different networks that vary in their edge definitions. Four network modularity algorithms were examined for their ability to detect the traditionally recognized plant communities. The use of more restrictive edge definitions significantly increased the accuracy of community detection, that is, the correspondence between network-based and traditional community classification. Random-walk based modularity methods yielded slightly better results than approaches based on the modularity function. For the whole network, the average agreement between the manual classification and the network-based modules is 76% with varying congruence levels for different communities ranging between 11% and 100%. The network-based approach recovered the known ecological gradient from riverside – sand and gravel bank vegetation – to dryer habitats like semidry grassland on dykes. Our results show that networks modularity algorithms offer new avenues of pursuit for the computational analysis of species communities.

Keywords: ecoinformatics; information visualization; vegetation unit; phytosociology; networks; classification

Introduction

Network modularity (community structure) is a well-studied large-scale connectivity pattern in networks [1,2], with several detection algorithms described in the literature (for a review, see [2]). A comparative study of module assignment accuracy by the different algorithms using benchmark networks showed that algorithm performance varies according to network size and the level of inter-module mixing [3]. A comparative evaluation of different community detection algorithms using real ecological data is however lacking so far. Here we test the applicability of network modularity algorithms as a method to classify plant species communities.

Community ecologists seek to understand the processes underlying organism and environment interaction dynamics of diversity, abundance, and composition of species in communities [4]. Vegetation science focuses on the ecology and composition of plant communities [5]. A basic task of the vegetation ecologist is to characterize, identify and distinguish different vegetation units that comprise plant species with similar habitat preferences. A common traditional approach is the making of relevés [6], which comprise a catalog of all plant species that occur in a vegetation plot together with their respective degree of coverage (i.e., frequency). Plant communities are ascertained by sorting the relevés in vegetation tables according to the occurrence of diagnostic species. Large and complex vegetation tables can however become error-prone and do not provide a concise overview of the whole data. A potentially more critical limitation is that the method demands an a priori knowledge about the respective diagnostic species whose identity can be a matter of debate. Diagnostic species include those particular species whose occurrence in the relevés may serve as an important telltale for the plant community classification. These include the character species, whose occurrence is typical to specific plant communities, and the differential species, whose occurrence can be used to distinguish related plant communities, but are not limited to a single community. Computer based methods including network applications have been used in

* Corresponding author. Email: tdagan@ifam.uni-kiel.de

Handling Editor: Andrzej Bodył

the classification of plant communities [7–11]. De Cáceres et al. [7,8] tested the performance of fuzzy clustering for the classification of communities in mesophytic and xerophytic pastures of Spanish highlands. Oliver et al. [9] employed a different approach including construction of dendrogram groups using hierarchical clustering for the classification of a broad diversity of communities from New South Wales. Classification success rates using those methods range between 75 and 80% [7,9].

For the present study, we investigated four different modularity functions that do not require a pre-determined number of expected modules. These four algorithms are based on two main approaches. The modularity maximization (ModMax) function by Girvan and Newman [1] is an application of a binary recursion algorithm that iteratively splits the network into modules. The algorithm seeks to maximize the modularity function defined as the ratio of edge frequency within modules and edge frequency outside modules. In the simulated annealing (SimAnn) algorithm the modularity function is calculated similarly to the ModMax algorithm and optimized by a simulated annealing approach [12]. The Markov cluster algorithm (MCL) [13] applies a flow simulation approach that is equivalent to a random walk along the edges in the network and measuring the probability to pass between different nodes. Edges that link highly connected nodes comprising a module are assumed to be more frequently travelled than edges that connect nodes from different modules. The less travelled edges are gradually omitted and the remaining connected nodes are the resulting modules. The information flow map (InfoMap) algorithm [14] applies a similar strategy to that of MCL, namely that the path of a random walk along the network edges will pass more frequently between nodes in the same module. The protocol of InfoMap incorporates principles from the field of information theory that are used to eliminate uninformative edges from the network. All algorithms except SimAnn include an implementation of weighted networks. All selected algorithms assign the nodes into a single module only.

The utility of the four modularity algorithms for plant communities classification was tested using data surveyed in the Lower Rhine floodplain vegetation plots. We further investigated the effects of several parameters affecting network based classification and determine to what extent this approach, entailing minimal a priori information, can approximate the results of manual classification for the same data.

Material and methods

Data collection and syntaxonomy

The ecological data consisted of 282 vegetation plots (relevés) that were recorded between 1996 and 2006 in the floodplains of the lower Rhine area, Germany [15,16]. The sampled plot size ranges between 4 m^2 and 100 m^2. Species communities in the plots were classified according to the Braun-Blanquet method considering characteristic and differential species yielding 13 different communities [15]. The communities span a geographical gradient including sand and gravel bank vegetation, communities of moist meadows, flood swards, reeds, nitrophyte vegetation, riparian forests and communities of semidry to fresh meadows (Fig. 1a). Plant species in each plot where identified and ranked by their coverage following the scale of Braun-Blanquet [6] that ranges between 1 (1–5%) and 5 (75–100%). A total of 232 different species were identified during the survey, species with coverage <1% were excluded from network calculations. The frequency of different species per plot with coverage >1% ranges between 1 and 27 with a median of 6. Syntaxa were named according to Pott [17] and LANUV [18], complemented by Schmitz and Lösch [19].

Vegetation network structure and properties

In the vegetation network, vertices correspond to plots in the vegetation table while the edges designate species composition similarity between the plots that they connect. Edge weights in the network were calculated by two different species similarity measures: the Sørensen similarity index [20] and the weighted similarity index [21].

The Sørensen similarity index (*SSI*) is calculated as:

$$SSI = \frac{2C}{A + B} \tag{1}$$

where A and B are the number of species in plots A and B respectively, and C is the number of species shared by the two plots.

The weighted similarity index (*wSI*) is calculated from the species coverage rankings as:

$$wSi = \frac{{}^{1}/_{2}\, M_c}{M_A + M_B + {}^{1}/_{2}\, M_C} \tag{2}$$

where M_A and M_B are the total coverage rankings of species that are present only in plot A and B respectively, and M_C is calculated by the total rankings of all species present in both plots.

To estimate the level of connectivity among plots of different communities we used the mixing parameter μ_t that quantifies the proportion of links connecting a certain node with nodes outside the community [3]:

$$\mu_t = \frac{k_i^{out}}{k_i^{in} + k_i^{out}} \tag{3}$$

The variables k_i^{out} and k_i^{in} designate the number of edges connecting node i with nodes outside and within the community respectively. The weighted mixing parameter μ_w quantifies the inter-community connectivity strength [3]:

$$\mu_w = \frac{w_i^{out}}{w_i^{in} + w_i^{out}} \tag{4}$$

The variables w_i^{out} and w_i^{in} designate the sum of edge weights for edges connecting node i with nodes from different or the same community respectively. It is commonly agreed that in a network where the average proportion of edges connecting a given node with nodes outside its community is >50% (i.e., $\mu > 0.5$) the modularity structure (if it exists) would be difficult to detect using computational methods [3]. Network graphs were generated with Cytoscape [22] version 2.8.0 using the "force directed" layout with default parameters [23].

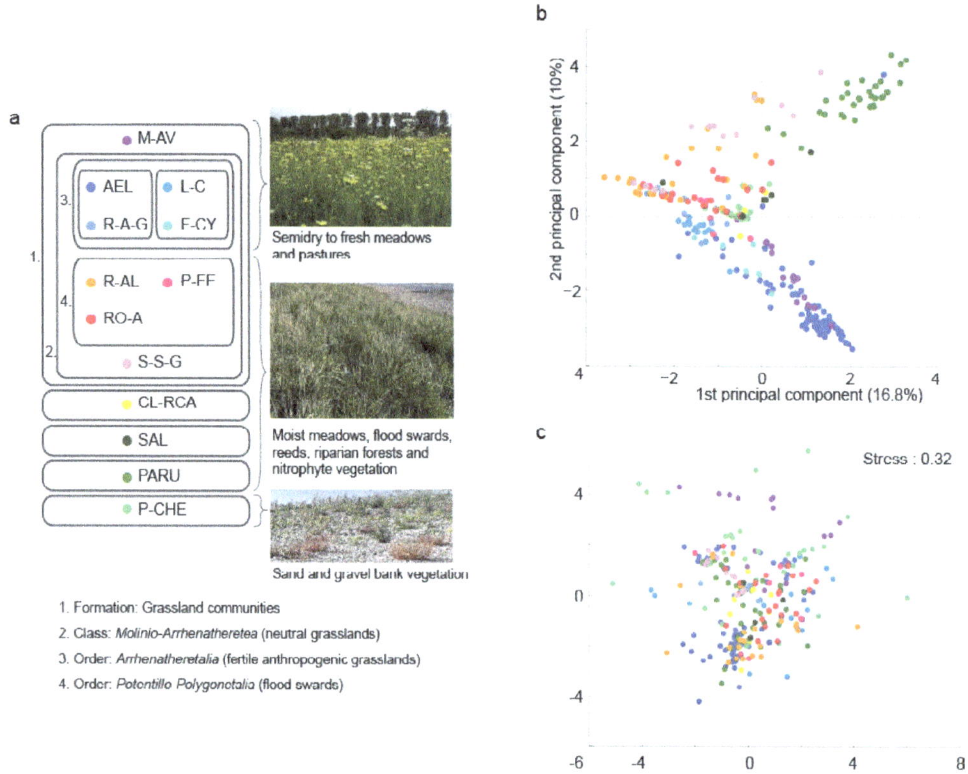

Fig. 1 A hierarchical representation of the ecological communities, results of PCA and MDS. **a** Each colored dot represents one of the 13 plant communities that have been classified in the vegetation data. The boxes describe relationships between plant communities with groups sharing the same syntaxonomy marked by a box. Descriptions next to the boxes provide the threefold gradient present in the data, which is independent from the syntaxonomic community classification. **b** Principal component analysis result. Each dot signifies a plot with colors according to the legend above. **c** Nonmetric multidimensional scaling analysis of the vegetation units based on euclidian distances. AEL – *Arrhenatheretum elatioris*; CL-RCA – *Cuscuto lupuliformis-Rubetum caesii*; F-CY – *Festuco-Cynosuretum*/*Luzulo-Cynosuretum*; L-C – *Lolio-Cynosuretum*; M-AV – *Medicagini-Avenetum*/*Mesobrometum alluviale*; P-CHE – *Polygono-Chenopodietum*; P-FE – *Potentillo-Festucetum arundinaceae*; PARU – *Phalaridetum arundinaceae*; R-A-G – *Ranunculus repens-Alopecurus pratensis* community; R-AL – *Ranunculo repentis-Alopecuretum geniculati*; RO-A – *Rorippo-Agrostietum stoloniferae*; SAL – *Salicetum albae*; S-S-G – *Sanguisorba officinalis-Silaum silaus* community.

Detection of modules within the network

Modularity applications for SimAn, MCL and InfoMap were downloaded from their dedicated websites. Module detection using ModMax was implemented with a MatLab™ script for the modularity maximization function as described by Newman [24]. The modularity detection accuracy was quantified by the quality of overlap between the modules obtained and the species communities as defined by the expert ecologists. The quality measures were calculated as described by Brohée and van Helden [25]. Standard data analysis and statistical tests were performed using MatLab™ version R2012a (7.14.0.739).

Results

Results of principal component analysis and multidimensional scaling

To test the level of similarity among plots in the different plant communities using traditional methods, we analyzed the data using principal component analysis (PCA) where plant species comprise the variables and the plots are defined as observations. The result does not reveal a clear distinction between plots in the different communities, yet their distribution in the PCA largely conforms to the plots ecological distribution. Plots of the *Phalaridetum arundinaceae* (PARU) and the *Arrhenatheretum elatioris* (AEL) tend to group together but the rest of the communities appear as intermixed (Fig. 1b). The first two components explain in total 26.8% of variability in the data. A further analysis of the plot composition dissimilarity using a multidimensional scaling (MDS) approach results in mixed distribution of the plots and no clear distinction between the communities (Fig. 1c). The results of the multivariate methods reveal some order in the data but they cannot be used in order to classify the plant communities.

Edge definition and species community mixing

The definition of edge weight is expected to have a strong influence on the network structure and the modularity algorithms performance. To investigate the effect of plot similarity estimation on the network modularity, we constructed networks using six different combinations of edge weight assignment and connectivity rules. Those include networks in which the edge weight is calculated by *SSI* or *wSI* measures as well as different edge exclusion regimes according to the shared dominant species (the species having the highest coverage within the plot; Tab. 1). Network construction based on plant composition similarity among the plots (Tab. 1: network A, Fig. 2a) results in a highly connected network where each plot is, on average, connected to half of the plots in the network (Tab. 1). The average weighted mixing parameter of nodes in network A (Tab. 1) is >0.5, indicating that species communities in this network will be difficult to detect using any modularity function.

Plots of the *Arrhenatherum elatioris* (AEL) and *Phalaridetum arundinaceae* (PARU) species communities (for abbreviations see legend of Fig. 1) are frequently connected with plots in the same community. Consequently the mixing parameter of these communities is low (PARU $\mu_w = 0.52$; AEL: $\mu_w = 0.42$; Fig. 2f), suggesting that these species communities are relatively well defined. This result is in agreement with the distinction level of these two communities in the PCA result.

In network B the edge weight was calculated by a weighted similarity index to include information about the plot coverage of each species (Tab. 1). This results in a slight decrease of the average μ_w in the total network. In addition, the average mixing parameter of nodes in the *Cuscuto lupuliformis-Rubetum caesii* (CL-RCA) species community drops below 0.5 (Fig. 2f). Furthermore, plots in this community are largely indistinguishable in any visible way (Fig. 1b, Fig. 2b).

To reinforce the species community structure within the network we adopted a procedure from the ecological definition of species communities relying on the dominant species for the characterization of plant communities [26]. In network C the plots are connected only if the dominant species is identical. The resulting network is sparser, with 73% fewer edges than networks A or B, and the node connectivity is reduced to 36 nearest neighbors per node on average (Tab. 1). The grouping of plots from several species communities is visible in the network, including M-AV, AEL, L-C, PARU, and CL-RCA (Fig. 2c). Accordingly, the mean weighted mixing parameter of these five communities is <0.5 (Fig. 2f). The overall mean mixing parameter in the network is significantly lower in comparison to networks A and B with an average of 38% inter-community edges per node (Tab. 1). Seven plots do not share their dominant species with any of the plots and are disconnected from the network main component (Fig. 2c).

In network D the connectivity rule makes use of the dominant species coverage information so that only plots sharing their dominant species in similar levels of coverage are connected. This restriction results in a decrease of 22% of the edges in the network (Tab. 1). The mixing parameter is reduced as well with an average of 30% inter-community links per plot. Nodes in plant community CL-RCA that are characterized by *Cuscuta lupuliformis* (willow dodder) and *Rubus caesius* (European dewberry), form a homogeneous cluster that is disconnected from the main network component (Fig. 2d). Plots of moist meadow and flood sward communities (R-A-G, R-AL, RO-A, P-FE and S-S-G) are frequently inter-connected (Fig. 2d). The high mixing parameter of plots in these communities (Fig. 2e) indicates that their species composition and dominant species are similar, hence the success of the computational approach to distinguish between species communities in this class is low. Plots in the F-CY community have an average $\mu_w = 0.77$ and are mixed accordingly with other communities in the order *Arrhenatheretalia* (fertile anthropogenic meadows and pastures; Fig. 2d). Indeed, network D does not present a complete division between all species communities, yet there is a certain distinction between rarely and often flooded habitats in which the plots were sampled. The top part of the network comprises species communities sampled in semidry to fresh meadows and pastures, the waistline presents plots sampled in moist habitats, while the bottom portraits plots sampled near the Rhine waterside on sand and gravel banks (Fig. 2d).

Using the dominant species for the connectivity rule resulted in a significant overall decrease in the plot inter-community mixing level (Fig. 2g). Further increase of the network modularity may be achieved by using a different edge weight calculation leading to stronger connectivity among plots classified into the same species community. Networks C_w and D_w are constructed using the same connectivity rules as in networks C and D respectively, but with edge weights calculated by the weighted similarity index (*wSI*; Tab. 1). The mixing parameter distribution of nodes in the resulting networks does not differ significantly from the networks where the *SSI* was used for the edge weight calculation ($P_{Cw} = 0.58$, $n_{Cw} = 275$ and $P_{Dw} = 0.72$, $n_{Dw} = 272$, using Kruskal–Wallis test and Tukey posthoc comparisons with $\alpha = 0.05$; Fig. 2g). This counter intuitive result may be explained by the similar distributions of *SSI* and *wSI* in our data. The two similarity measures are significantly linearly correlated ($SSI = 0.89 \times wSI + 0.05$; $P = 0$, $n = 3888$) where changes in *SSI* explain 84% of the variability in *wSI* ($R^2 = 0.84$).

Modules within the network of plant communities

To test the applicability of network modularity algorithms for species community classification we constructed modules in the network using four different algorithms. The results reveal that indeed module detection accuracy improves when the network is constructed using connectivity rules yielding lower average mixing (Fig. 3). The modularity detection in networks A and B using ModMax, MCL, and InfoMap algorithms yielded overall similar results while SimAnn was an exception with 94 modules. The 94 modules included 3 large modules comprising 67% of the plots and 91 single-plot modules (Tab. 2). In what follows, we discuss the resulting modules in network D, which is characterized by the lowest mixing parameter distribution and overall higher module detection accuracy. The modularity maximization based methods ModMax and SimAnn (with default starting resolution) resulted in nine and seven modules accordingly, while the information flow algorithms MCL (with default Inflation parameter) and InfoMap resulted in fourteen modules each. The latter two methods resulted in a higher positive predictive value (*PPV*) in comparison to the

Tab. 1 Vegetation plot network definitions and properties.

Network	Edge weight	Connectivity rule – plots *i* and *j* are connected if:	No. of edges	Mean connectivity[2]	Mean μ_t	Mean μ_w
A	$a_{ij} = SSI$	$a_{ij} > 0$	18.011	127	0.7	0.61
B	$a_{ij} = wSI$	$a_{ij} > 0$	18.011	127	0.7	0.55
C	$a_{ij} = SSI$	One or more of the most frequent species[1] are identical.	4.988	36	0.38	0.34
D	$a_{ij} = SSI$	One or more of the most frequent species[1] are identical, and the difference in their plot coverage is ±1.	3.888	28	0.31	0.3
C_w	$a_{ij} = wSI$	As in C.	4.988	36	0.38	0.34
D_w	$a_{ij} = wSI$	As in D.	3.888	28	0.31	0.29

[1] The most frequent species are species having the maximum coverage ranking in the plot. [2] Connectivity (C_i) is calculated as the number of nodes connected to node *i* by a single edge.

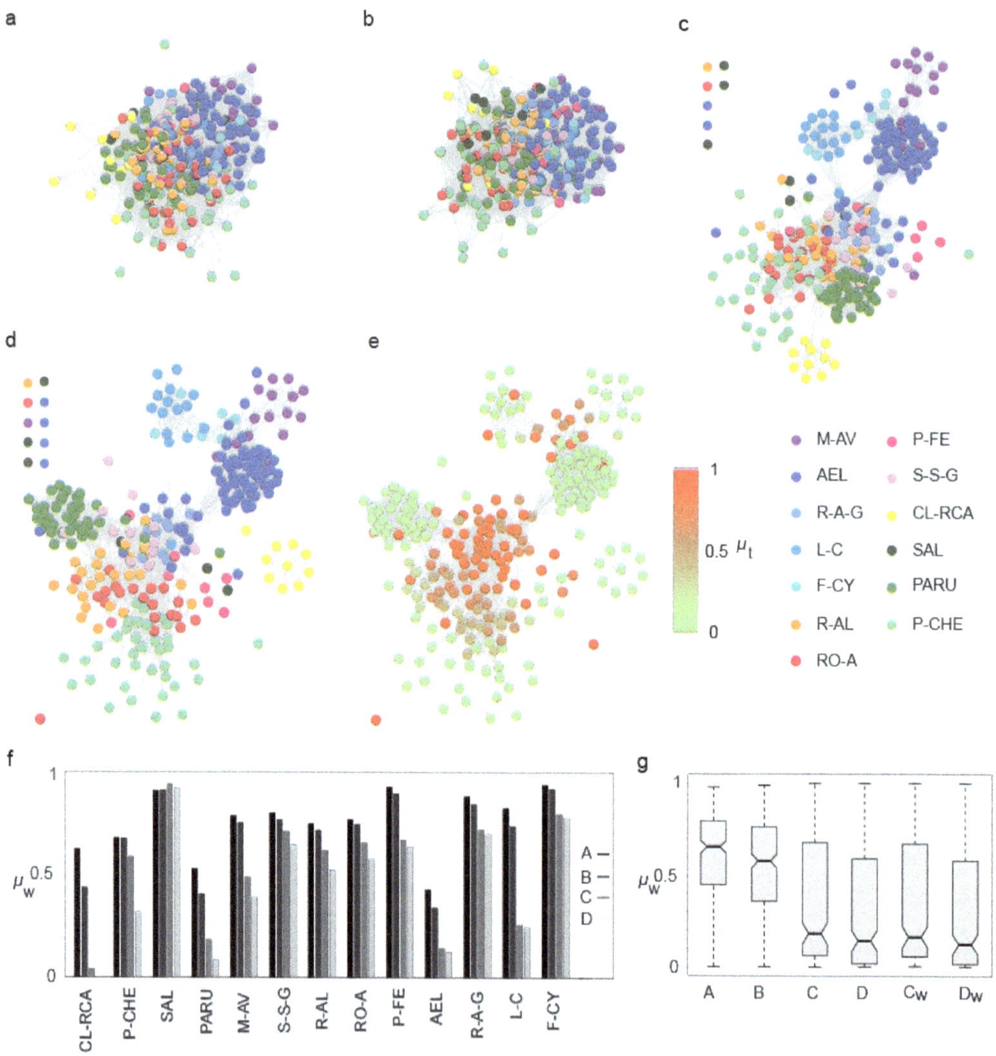

Fig. 2 Plant community networks. **a–d** Networks A–D calculated according to their definitions in Tab. 1. **e** Distribution of the node mixing parameter, μ_t, in network D. The nodes are painted by their μ_t, ranging between 0 and 1 according to the color scale-bar on the right. **f** The distribution of community μ_w in networks A–D. **g** Distribution of node μ_w in the six tested networks (Tab. 1).

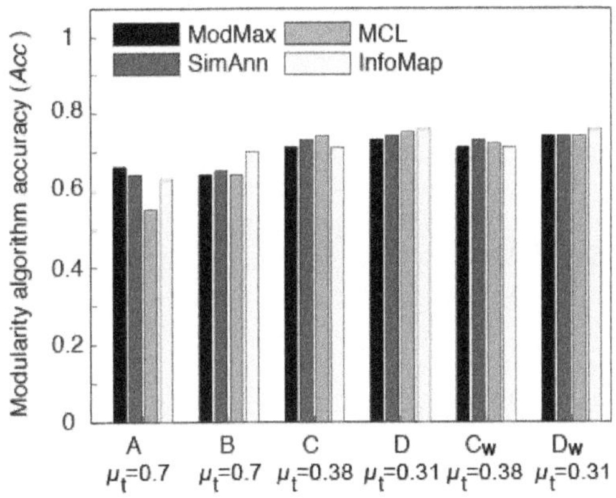

Fig. 3 Module detection accuracy.

modularity maximization methods, indicating that modules by the MCL and InfoMap result in a better prediction of the species communities (Tab. 2).

The sensitivity (*Sn*) measure used to quantify the success rate of assigning plots from the same plant community into the same module is higher using the ModMax and SimAnn functions (Tab. 2). These algorithms yield fewer and larger modules that contain whole communities, leading to higher sensitivity values. Weighting the *PPV* and *Sn* values to assess the algorithm accuracy (*Acc*) reveals somewhat better precision using the flow information methods with a slight advantage to InfoMap (Tab. 2). The community separation (Sep_{com}) reveals that MCL yields the best overlap between species communities and modules, while the module separation (Sep_{mod}) shows that modules in the SimAnn result in higher community mixing and somewhat better overall separation in comparison to the other methods (Tab. 2). Testing the impact of the SimAnn starting resolution parameter and the MCL inflation parameter on the performance of these two algorithms showed that the best accuracy was achieved using their default parameters (Tab. 3).

All modularity methods except SimAnn yield modules that overlap well with species communities of the fresh meadows and pastures (Fig. 4). From among the modules assigned to these communities, the module that specifies the A-EL community has the best overlap, followed closely by plots of the M-AV community (Tab. 4). Plots in the L-C and F-CY species communities are grouped together in all methods (Fig. 4). This result supports the view expressed by several authors that F-CY should be included in the L-C community as a nutrient-poor variant of the latter [27,28]. The ModMax algorithm resulted in a slightly higher separation value of L-C community than the other algorithms because it succeeds in identifying a plot that is densely connected with 39 plots from other communities (the plot is marked by an arrow in Fig. 3a; Tab. 4). The multitude of inter-community connections of this plot is due to the presence of six equally coverage-ranked species in the plot that are all defined as dominant using our connectivity rule. These species are dominant in multiple species communities and hence the high connectivity of this plot.

Tab. 2 Comparison of community detection algorithms.

	ModMax	SimAnn	MCL	InfoMap
Network A				
No. modules	3	94	3	3
PPV	0.51	0.73	0.31	0.43
Sn	0.86	0.57	0.96	0.92
Acc	0.66	0.64	0.55	0.63
Sep_{com}	0.18	0.57	0.15	0.20
Sep_{mod}	0.80	0.08	0.66	0.86
Sep	0.38	0.21	0.32	0.41
Network B				
No. modules	5	94	5	5
PPV	0.51	0.73	0.45	0.56
Sn	0.79	0.57	0.90	0.89
Acc	0.64	0.65	0.64	0.70
Sep_{com}	0.23	0.57	0.22	0.26
Sep_{mod}	0.59	0.08	0.57	0.68
Sep	0.37	0.21	0.36	0.42
Network C				
No. modules	8	7	11	13
PPV	0.65	0.65	0.67	0.71
Sn	0.76	0.82	0.82	0.70
Acc	0.71	0.73	0.74	0.71
Sep_{com}	0.42	0.42	0.50	0.51
Sep_{mod}	0.68	0.78	0.59	0.51
Sep	0.54	0.57	0.54	0.51
Network C_w				
No. modules	8	7	14	14
PPV	0.62	0.65	0.68	0.72
Sn	0.80	0.83	0.77	0.69
Acc	0.71	0.73	0.72	0.71
Sep_{com}	0.39	0.43	0.51	0.52
Sep_{mod}	0.63	0.80	0.47	0.48
Sep	0.49	0.58	0.49	0.50
Network D				
No. modules	9	7	14	14
PPV	0.65	0.67	0.75	0.75
Sn	0.82	0.82	0.75	0.76
Acc	0.73	0.74	0.75	0.76
Sep_{com}	0.43	0.42	0.57	0.55
Sep_{mod}	0.61	0.78	0.53	0.51
Sep	0.51	0.57	0.55	0.53
Network D_w				
No. modules	9	7	14	14
PPV	0.65	0.67	0.71	0.75
Sn	0.83	0.82	0.76	0.76
Acc	0.74	0.74	0.74	0.76
Sep_{com}	0.43	0.42	0.55	0.55
Sep_{mod}	0.63	0.78	0.51	0.51
Sep	0.52	0.57	0.53	0.53

Abbreviations are explained in the text.

Tab. 3 Comparison of community detection accuracy in network D by MCL and SimAnn algorithms using different parameters.

MCL	Inflation parameter	Accuracy
	5.00	0.693
	4.00	0.689
	3.00	0.749
	2.50	0.745
	2.20	0.741
	2.00	0.752
	1.80	0.741
	1.50	0.738
	1.20	0.660
SimAnn	**Resolution**	**Accuracy**
	0.30	0.680
	0.35	0.700
	0.40	0.717
	0.45	0.738
	0.50	0.739
	0.55	0.739
	0.60	0.737
	0.65	0.735

All modularity functions have 100% success rate in specifying the CL-RCA community. Using the dominant species and coverage for the connectivity rule disconnected the CL-RCA plots from the network and enabled the accurate detection of this plant community (Tab. 4, Fig. 4). The next best-identified community is PARU with 94% of its plots classified into one module containing a marginal frequency (5%) of plots from other communities (Tab. 4). Plots of the orders *Arrhenatheretalia* (fertile anthropogenic grasslands) and *Potentillo-Polygonetalia* (flood swards) are classified into two modules or more, but better distinction between the communities was unsuccessful. The P-CHE community comprising sand and gravel bank vegetation is well detected by all algorithms except ModMax (Tab. 4). ModMax yielded a low-resolution modularity structure where the P-CHE plots are grouped together with R-AL and RO-A plots into one module.

Discussion

Various statistical methods such as PCA and MDS have been used in the past to analyze and identify species communities. They provide a general overview of the similarity distribution across the sampled plots, yet these methods are not suited for de novo classification of species communities.

In the present work, we found that networks can attain an accuracy of 76% relative to manual classification using existing modularity algorithms. The network modularity structure, even in cases where it is not highly accurate relative to manual methods, is helpful in providing a general overview over the global distribution of shared plot composition in a given dataset. This includes gradients in ecological properties of different habitats, such as moisture or temperature, having an impact on observed species composition.

None of modularity functions were able to distinguish between plots in the L-C (*Lolio-Cynosuretum*) and F-CY (*Festuco-Cynosuretum*) species communities. These two communities were distinguished based on species composition by many authors [29–33]. However, smooth transitions in plant community assembly between both communities occur frequently [34]. Consequently, several studies have considered the F-CY community as a nutrient-poor form of L-C community and included it into the latter [27,28]. The modularity structure of our network supports that view. Such community continuums pose a challenge to the network-based community detection methods and would require the addition of either additional ecological/biological information or further analysis of network higher-order structure. On the other hand, the distinctiveness of the CL-RCA (*Cuscuto lupuliformis-Rubetum caesii*), which was first described by Schmitz and Lösch [19], is strongly recovered in the networks.

Using the a priori information of species communities in this study enabled the test of different connectivity rules while studying their impact on the network modularity. For example, the use of dominant species as the hallmark of species communities in the connectivity rule increased the network modularity and improved the performance of the modularity functions considerably. However, the rigorous use of the dominant species may also bias the modularity results, especially in plots with more than one dominant species of the same coverage. Furthermore, the dominant species is not necessarily the character species, which was used for the manual ecological classification of species communities in our data. Our computational approach is hence more similar to the "Uppsala school" (dominant species) rather than the "Zurich-Montpelier school" (character species) in vegetation studies. This difference between the manual and computational classification approaches here stems from the difficulty in finding a rigorous definition of character species based on their coverage ranking alone. From the present study, a critical parameter for the utility of networks for classifying species communities are the connectivity rule and the function used to identify characteristic species.

The ranking of modularity function performance for classification of vegetation plots in our study differs from their ranking based on the analysis of benchmark networks. A recent comparison of community detection algorithms using a simulated (artificial) weighted network in which the modules are known revealed that the InfoMap function performs much better than MCL or SimAnn algorithms [3]. In our networks comprised of real data, MCL and InfoMap performed similarly well. Indeed, the present analyses embrace the notion that lower mixing parameter leads to

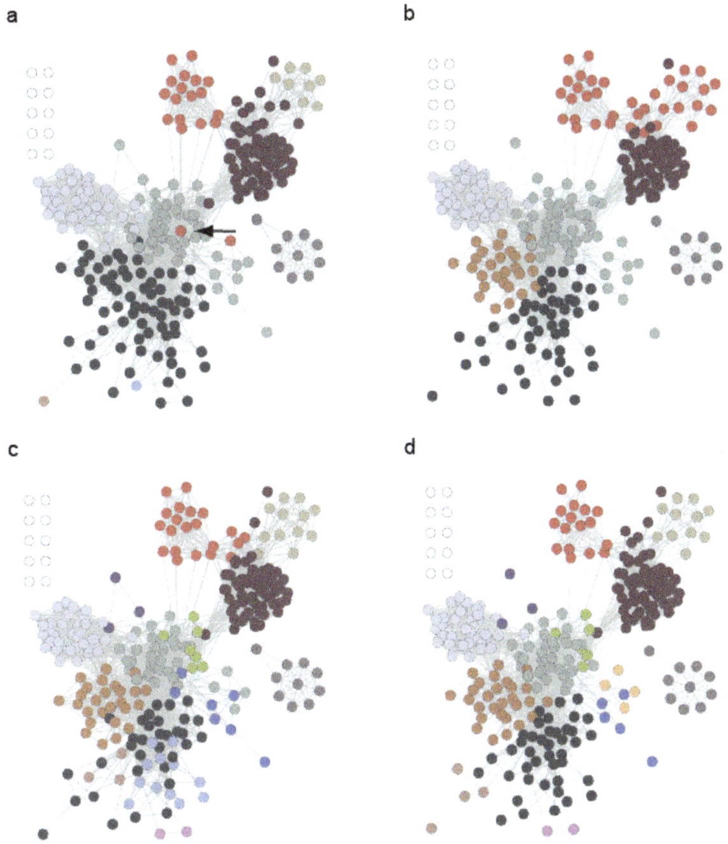

Fig. 4 Modules detected in network D by the four modularity functions. Nodes having the same color were grouped into the same module. **a** ModMax. **b** MCL. **c** Infomap. **d** SimAnn.

Tab. 4 Quality of community separation using the different algorithms for network D.

Species community	No. of plots	No. of classified plots	μ_w (mean ±std)	Sepcom			
				ModMax	SimAnn	MCL	Infomap
M-AV	16	15	0.26 ±0.35	0.58	0.32	0.8	0.74
AEL	76	72	0.19 ±0.28	0.74	0.79	0.82	0.79
R-A-G	10	10	0.67 ±0.17	0.2	0.18	0.23	0.24
L-C	15	15	0.21 ±0.2	0.78	0.37	0.62	0.77
F-CY	6	6	0.76 ±0.06	0.1	0.17	0.28	0.11
R-AL	26	25	0.48 ±0.24	0.19	0.45	0.53	0.45
RO-A	21	20	0.54 ±0.19	0.3	0.26	0.35	0.27
P-fe	8	8	0.45 ±0.47	0.12	0.11	0.65	0.43
S-S-G	13	13	0.54 ±0.26	0.15	0.18	0.38	0.4
Cl-RCA	10	10	0	1	1	1	1
SAL	6	3	0.82 ±0.15	0.06	0.05	0.08	0.34
PARU	39	39	0.12 ±0.21	0.78	0.87	0.9	0.9
P-Che	36	36	0.18 ±0.25	0.48	0.67	0.74	0.69

better community detection performance [3]. Variation in connectivity rules had a strong impact on community detection accuracy. This result stresses the importance of edge definition in networks constructed from real data. Furthermore, our findings reveal a more heterogeneous view of mixing parameter distribution over the network nodes in comparison to benchmark networks. Consequently, different communities may vary in their detection accuracy, in that several communities detected in high accuracy (e.g., PARU) and some that are not identified at all (e.g., SAL). Indeed, an estimation of network modularity (community mixing) for real data is impossible. Yet, our results show that for a network based species communities' analysis it is recommended to eliminate edges connecting plots that do not share any species, and in addition to employ information regarding diagnostic species in the connectivity rule.

Our results show that network methods can readily be used to visualize and analyze vegetation tables for the identification and study of plant communities. Modularity detection algorithms could also be applied to other biological systems, for example the study of microbial species community structure, which entails even larger datasets [35].

Acknowledgments

Work in the authors' laboratories is supported by the European Research Council (grant No. 232975 to WFM; grant No. 281357 to TD).

Authors' contributions

The following declarations about authors' contributions to the research have been made: carried out the computational analysis and drafted the manuscript: TT; collected the underlying ecological data and drafted parts of the manuscript: US; contributed to the computational analysis: GD; conceived the study and drafted the manuscript: WFM, TD.

References

1. Girvan M, Newman MEJ. Community structure in social and biological networks. Proc Natl Acad Sci USA. 2002;99(12):7821–7826.
2. Fortunato S. Community detection in graphs. Phys Rep. 2010;486(3–5):75–174.
3. Lancichinetti A, Fortunato S. Community detection algorithms: a comparative analysis. Phys Rev E Stat Nonlin Soft Matter Phys. 2009;80(5).
4. Vellend M. Conceptual synthesis in community ecology. Q Rev Biol. 2010;85(2):183–206.
5. Mucina L. Classification of vegetation: past, present and future. J Veg Sci. 1997;8(6):751–760.
6. Braun-Blanquet J, Fuller GD, Conrad HS. Plant sociology: the study of plant communities. New York, NY: Hafner Press; 1965.
7. de Cáceres M, Font X, Vicente P, Oliva F. Numerical reproduction of traditional classifications and automatic vegetation identification. J Veg Sci. 2009;20(4):620–628.
8. de Cáceres M, Font X, Oliva F. The management of vegetation classifications with fuzzy clustering: fuzzy clustering in vegetation classifications. J Veg Sci. 2010;21(6):1138–1151.
9. Oliver I, Broese EA, Dillon ML, Sivertsen D, McNellie MJ. Semi-automated assignment of vegetation survey plots within an classification of vegetation types. Methods Ecol Evol. 2013;4(1):73–81.
10. Roleček J, Tichý L, Zelený D, Chytrý M. Modified TWINSPAN classification in which the hierarchy respects cluster heterogeneity. J Veg Sci. 2009;20(4):596–602.
11. Tichý L, Chytrý M, Hájek M, Talbot SS, Botta-Dukát Z. OptimClass: using species-to-cluster fidelity to determine the optimal partition in classification of ecological communities. J Veg Sci. 2010;21(2):287–299.
12. Guimerà R, Sales-Pardo M, Amaral L. Modularity from fluctuations in random graphs and complex networks. Phys Rev E Stat Nonlin Soft Matter Phys. 2004;70(2).
13. van Dongen S. Graph clustering by flow simulations [PhD thesis]. Utrecht: University of Utrecht; 2000.
14. Rosvall M, Bergstrom CT. Maps of random walks on complex networks reveal community structure. Proc Natl Acad Sci USA. 2008;105(4):1118–1123.
15. Schmitz U, Lösch R. Neophyten und C4-Pflanzen in der Auenvegetation des Niederrheins. Decheniana. 2005;158:55–77.
16. Schmitz U. Increase of alien and C4 plant species in annual river bank vegetation of the River Rhine. Phytocoenologia. 2006;36(3):393–402.
17. Pott R. Die Pflanzengesellschaften Deutschlands. 2nd ed. Stuttgart: E. Ulmer Verlag; 1995.
18. LANUV. Vegetationstypenliste (list of vegetation types and their abbreviation) [Internet]. 2014.
19. Schmitz U, Lösch R. Vorkommen und Soziologie der Cuscuta-Arten in der Ufervegetation des Niederrheins. Tuexenia. 1995;15:373–385.
20. Sørensen TJ. A method of establishing groups of equal amplitude in plant sociology based on similarity of species content and its application to analyses of the vegetation on Danish commons. Copenhagen: I kommission hos E. Munksgaard; 1948. (Biologiske Skrifter; vol 5).
21. Ellenberg H, Walter H. Einführung in die Phytologie. Stuttgart: Ulmer; 1956.
22. Shannon P, Markiel A, Ozier O, Baliga NS, Wang JT, Ramage D, et al. Cytoscape: a software environment for integrated models of biomolecular interaction networks. Genome Res. 2003;13(11):2498–2504.
23. Fruchterman TMJ, Reingold EM. Graph drawing by force-directed placement. Softw Pr Exp. 1991;21(11):1129–1164.
24. Newman MEJ. Modularity and community structure in networks. Proc Natl Acad Sci USA. 2006;103(23):8577–8582.
25. Brohée S, van Helden J. Evaluation of clustering algorithms for protein-protein interaction networks. BMC Bioinformatics. 2006;7(1):488.
26. Whittaker RH. Ordination and classification of communities. The Hague: Junk; 1973.
27. Dierschke H. *Molinio-Arrhenatheretea* (E1). Kulturgrasland und verwandte Vegetationstypen. Teil 1: *Arrhenatheretalia* Wiesen und Weiden frischer standorte. Synopsis der Pflanzengesellschaften Deutschlands. Göttingen: Selbstverlag der Floristisch-soziologischen Arbeitsgemeinschaft; 1997.
28. Rennwald E. Rote Liste der Pflanzengesellschaften Deutschlands mit Anmerkungen zur Gefährdung. Schriftenreihe Für Veg. 2000;35:393–592.
29. Runge F. Die Pflanzengesellschaften Mitteleuropas. Münster: Aschendorff; 1990.
30. Wilmanns O. Ökologische Pflanzensoziologie. Heidelberg: Quelle & Meyer; 1998.
31. Schubert R, Hilbig W, Klotz S. Bestimmungsbuch der Pflanzengesellschaften Mittel- und Nordostdeutschlands. Wiesbaden: Spektrum Akademischer Verlag; 2001.

32. Foerster E. Pflanzengesellschaften des Grünlandes in Nordrhein-Westfalen. Münster: Landwirtschaftsverlag; 1983.

33. Verbücheln G, Hinterlang D, Pardey A, Pott R, Raabe U, van de Weyer K. Rote Liste der Pflanzengesellschaften in Nordrhein-Westfalen. Recklinghausen: LÖBF-Schriftenreihe; 1995.

34. Ellenberg H. Vegetation Mitteleuropas mit den Alpen in ökologischer, dynamischer und historischer Sicht. Stuttgart: Eugen Ulmer; 1996.

35. Gonzalez A, Clemente JC, Shade A, Metcalf JL, Song S, Prithiviraj B, et al. Our microbial selves: what ecology can teach us. EMBO Rep. 2011;12(8):775–784.

Permissions

All chapters in this book were first published in ASBP, by Polish Botanical Society; hereby published with permission under the Creative Commons Attribution License or equivalent. Every chapter published in this book has been scrutinized by our experts. Their significance has been extensively debated. The topics covered herein carry significant findings which will fuel the growth of the discipline. They may even be implemented as practical applications or may be referred to as a beginning point for another development.

The contributors of this book come from diverse backgrounds, making this book a truly international effort. This book will bring forth new frontiers with its revolutionizing research information and detailed analysis of the nascent developments around the world.

We would like to thank all the contributing authors for lending their expertise to make the book truly unique. They have played a crucial role in the development of this book. Without their invaluable contributions this book wouldn't have been possible. They have made vital efforts to compile up to date information on the varied aspects of this subject to make this book a valuable addition to the collection of many professionals and students.

This book was conceptualized with the vision of imparting up-to-date information and advanced data in this field. To ensure the same, a matchless editorial board was set up. Every individual on the board went through rigorous rounds of assessment to prove their worth. After which they invested a large part of their time researching and compiling the most relevant data for our readers.

The editorial board has been involved in producing this book since its inception. They have spent rigorous hours researching and exploring the diverse topics which have resulted in the successful publishing of this book. They have passed on their knowledge of decades through this book. To expedite this challenging task, the publisher supported the team at every step. A small team of assistant editors was also appointed to further simplify the editing procedure and attain best results for the readers.

Apart from the editorial board, the designing team has also invested a significant amount of their time in understanding the subject and creating the most relevant covers. They scrutinized every image to scout for the most suitable representation of the subject and create an appropriate cover for the book.

The publishing team has been an ardent support to the editorial, designing and production team. Their endless efforts to recruit the best for this project, has resulted in the accomplishment of this book. They are a veteran in the field of academics and their pool of knowledge is as vast as their experience in printing. Their expertise and guidance has proved useful at every step. Their uncompromising quality standards have made this book an exceptional effort. Their encouragement from time to time has been an inspiration for everyone.

The publisher and the editorial board hope that this book will prove to be a valuable piece of knowledge for researchers, students, practitioners and scholars across the globe.

List of Contributors

Krystyna Towpasz
Institute of Botany, Jagiellonian University, Kopernika 27, 31-501 Kraków, Poland

Alina Stachurska-Swakoń
Institute of Botany, Jagiellonian University, Kopernika 27, 31-501 Kraków, Poland

Agnieszka K. Kowalkowska
Department of Plant Cytology and Embryology, University of Gdańsk, Wita Stwosza 59, 80-308 Gdańsk, Poland

Hanna B. Margońska
Department of Plant Taxonomy and Nature Conservation, University of Gdańsk, Wita Stwosza 59, 80-308 Gdańsk, Poland

Gorana Jelic Mrcelic
Maritime Faculty, University of Split, Zrinsko Frankopanska 38, 21 000 Split, Croatia

Merica Sliskovic
Maritime Faculty, University of Split, Zrinsko Frankopanska 38, 21 000 Split, Croatia

Boris Antolic
Institute of Oceanography and Fisheries, Setaliste Ivana Mestrovica 63, 21 000 Split, Croatia

Janetta Niemann
Department of Genetics and Plant Breeding, Poznań University of Life Sciences, Dojazd 11, 60-632 Poznań, Poland

Andrzej Wojciechowski
Department of Genetics and Plant Breeding, Poznań University of Life Sciences, Dojazd 11, 60-632 Poznań, Poland

Jowita Janowicz
Department of Genetics and Plant Breeding, Poznań University of Life Sciences, Dojazd 11, 60-632 Poznań, Poland

Marcin Kiedrzyński
Department of Geobotany and Plant Ecology, Faculty of Biology and Environmental Protection, University of Łódź, Banacha 2/16, 90-237 Łódź, Poland

Janina Jakubowska-Gabara
Department of Geobotany and Plant Ecology, Faculty of Biology and Environmental Protection, University of Łódź, Banacha 2/16, 90-237 Łódź, Poland

Bożena Denisow
Department of Botany, University of Life Sciences in Lublin, Akademicka 15, 20-950 Lublin, Poland

Małgorzata Wrzesień
Institute of Biology and Biochemistry, Maria Curie-Skłodowska University, Akademicka 19, 20-033 Lublin, Poland

Anna Cwener
Institute of Biology and Biochemistry, Maria Curie-Skłodowska University, Akademicka 19, 20-033 Lublin, Poland

Katarzyna Marcysiak
Department of Botany, Kazimierz Wielki University, Ossolińskich 12, 85-093 Bydgoszcz, Poland

Ingvar Svanberg
Uppsala Centre for Russian and Eurasian Studies, Uppsala University, Box 514, 751 20 Uppsala, Sweden

Renata Soukand
Estonian Literary Museum, Vanemuise 42, 51003 Tartu, Estonia

Łukasz Łuczaj
Department of Botany and Biotechnology of Economic Plants, University of Rzeszów, Werynia 502, 36-100 Kolbuszowa, Poland

Raivo Kalle
Institute of Veterinary Medicine and Animal Sciences, Estonian University of Life Sciences, Kreutzwaldi 62, 51014 Tartu, Estonia

Olga Zyryanova
V. N. Sukachev Institute of Forest, Russian Academy of Sciences, Academgorodok 50, 660036 Krasnoyarsk, Russia

Andrea Dénes
Natural History Department, Janus Pannonius Museum, Box 158, 7601 Pécs, Hungary

Nóra Papp
Department of Pharmacognosy, University of Pécs, Rókus 2, 7624 Pécs, Hungary

Aneli Nedelcheva
Department of Botany, Sofia University "St. Kliment Ohridski", Dragan Tzankov 8, 1164 Sofia, Bulgaria

Daiva Šeškauskaitė
Kaunas Forestry and Environmental Engineering University of Applied Sciences, Liepu 1, Girionys, LT 53101 Kaunas, Lithuania

Iwona Kołodziejska-Degórska
Warsaw University Botanic Garden, Aleje Ujazdowskie 4, 00-478 Warsaw, Poland
Institute for Interdisciplinary Research "Artes Liberales", University of Warsaw, Nowy Świat 69, 00-046 Warsaw, Poland

Valeria Kolosova
Institute for Linguistic Studies, Russian Academy of Sciences, Tuchkov pereulok 9, 199053 Saint-Petersburg, Russia

Arkadiusz Sebastian Nowak
Department of Biosystematics, Opole University, Oleska 22, 45-052 Opole, Poland

Marcin Nobis
Institute of Botany, Jagiellonian University, Kopernika 27, 31-501 Kraków, Poland
Laboratory of Biodiversity and Ecology, Tomsk State University, Lenin Prospekt 36, Tomsk, 634050, Russian Federation

Sylwia Nowak
Department of Biosystematics, Opole University, Oleska 22, 45-052 Opole, Poland

Agnieszka Nobis
Institute of Botany, Jagiellonian University, Kopernika 27, 31-501 Kraków, Poland

Beata Woziwoda
Department of Geobotany and Plant Ecology, University of Łódź, Banacha 12/16, 90-237 Łódź, Poland

Dominik Kopeć
Department of Geobotany and Plant Ecology, University of Łódź, Banacha 12/16, 90-237 Łódź, Poland

Janusz Witkowski
Regional Directorate of the State Forests in Łódź, Jana Matejki 16, 91-402 Łódź, Poland

Wioletta Brosowska-Arendt
Department of Biochemistry, University of Warmia and Mazury in Olsztyn, Oczapowskiego 2, 10-719 Olsztyn, Poland

Stanisław Weidner
Department of Biochemistry, University of Warmia and Mazury in Olsztyn, Oczapowskiego 2, 10-719 Olsztyn, Poland

Grzegorz Migdałek
Institute of Biology, Pedagogical University of Cracow, Podchorążych 2, 30-084 Cracow, Poland

Jolanta Kolczyk
Department of Plant Cytology and Embryology, Institute of Botany, Jagiellonian University, Gronostajowa 9, 30-387 Cracow, Poland

Artur Pliszko
Department of Plant Taxonomy, Phytogeography and Herbarium, Institute of Botany, Jagiellonian University, Kopernika 27, 31 501 Cracow, Poland

Maria Kościńska-Pająk
Department of Plant Cytology and Embryology, Institute of Botany, Jagiellonian University, Gronostajowa 9, 30-387 Cracow, Poland

Aneta Słomka
Department of Plant Cytology and Embryology, Institute of Botany, Jagiellonian University, Gronostajowa 9, 30-387 Cracow, Poland

Nóra Papp
Department of Pharmacognosy, University of Pécs, Rókus 2, 7624 Pécs, Hungary

Dóra Czégényi
Department of Hungarian Ethnography and Anthropology, University of Babeş-Bolyai of Cluj-Napoca, Horea 31, 400202 Cluj-Napoca, Romania

Anita Hegedűs
Department of Languages for Specific Purposes, University of Pécs, Rókus 2, 7624 Pécs, Hungary

Tamás Morschhauser
Department of Plant Taxonomy and Geobotany, University of Pécs, Ifjúság 6, 7624 Pécs, Hungary

Cassandra L. Quave
Center for the Study of Human Health, Emory University, 550 Asbury Circle, Candler Library 107, Atlanta, GA 30322, USA

Kevin Cianfaglione
School of Biosciences and Veterinary Medicine, University of Camerino, Via Pontoni 5, 62032 Camerino (MC), Italy

Andrea Pieroni
University of Gastronomic Sciences, Piazza Vittorio Emanuele 9, I-12060 Fr. Pollenzo, Bra (CN), Italy

Maria Pilarska
Department of Plant Cytology and Embryology, Jagiellonian University, Gronostajowa 9, 30-387 Cracow, Poland

Marzena Popielarska-Konieczna
Department of Plant Cytology and Embryology, Jagiellonian University, Gronostajowa 9, 30-387 Cracow, Poland

Halina Ślesak
Department of Plant Cytology and Embryology, Jagiellonian University, Gronostajowa 9, 30-387 Cracow, Poland

Małgorzata Kozieradzka-Kiszkurno
Department of Plant Cytology and Embryology, University of Gdańsk, Wita Stwosza 59, 80-308 Gdańsk, Poland

Grzegorz Góralski
Department of Plant Cytology and Embryology, Jagiellonian University, Gronostajowa 9, 30-387 Cracow, Poland

Robert Konieczny
Department of Plant Cytology and Embryology, Jagiellonian University, Gronostajowa 9, 30-387 Cracow, Poland

Jerzy Bohdanowicz
Department of Plant Cytology and Embryology, University of Gdańsk, Wita Stwosza 59, 80-308 Gdańsk, Poland

Elżbieta Kuta
Department of Plant Cytology and Embryology, Jagiellonian University, Gronostajowa 9, 30-387 Cracow, Poland

Krystyna Winiarczyk
Department of Plant Anatomy and Cytology, Maria Curie-Skłodowska University, Akademicka 19, 20-033 Lublin, Poland

Katarzyna Skrzypczak
Faculty of Food Science and Biotechnology, University of Life Sciences in Lublin, Skromna 8, 20-704 Lublin, Poland

Jolanta Jaroszuk-Ścise
Department of Environmental Microbiology, Maria Curie-Skłodowska University, Akademicka 19, 20-033 Lublin, Poland

Jan Bocianowski
Department of Mathematical and Statistical Methods, Poznań University of Life Sciences, Wojska Polskiego 28, 60-637 Poznań, Poland

Anna Geraci
Department of Environmental Biology and Biodiversity, University of Palermo, Via Archirafi, 38-90123, Italy

Vincenza Polizzano
Department of Environmental Biology and Biodiversity, University of Palermo, Via Archirafi, 38-90123, Italy

Pasquale Marino
Department of Environmental Biology and Biodiversity, University of Palermo, Via Archirafi, 38-90123, Italy

Rosario Schicchi
Department of Environmental Biology and Biodiversity, University of Palermo, Via Archirafi, 38-90123, Italy

Agnieszka Niedziela
Department of Plant Physiology and Biochemistry, Plant Breeding and Acclimatization Institute - National Research Institute, Radzików, 05-870 Błonie, Poland

Jan Jarosław Rybczyński
Department of Experimental Plant Biology, Botanical Garden - Center for Biological Diversity Conservation of the Polish Academy of Sciences in Powsin, Prawdziwka 2, 02-973 Warsaw, Poland

Marta Kolanowska
Department of Plant Taxonomy and Nature Conservation, University of Gdańsk, Wita Stwosza 59, 80-308 Gdańsk-Oliwa, Poland

Magdalena Marta Wróblewska
Department of Developmental Plant Biology, Institute of Experimental Biology, University of Wrocław, Kanonia 6/8, 50-328 Wrocław, Poland

Romana Prausová
Department of Biology, Faculty of Science, University of Hradec Králové, 500 02 Hradec Králové, Czech Republic

Zuzana Kozelková
Department of Biology, Faculty of Science, University of Hradec Králové, 500 02 Hradec Králové, Czech Republic

Lenka Šafářová
East Bohemian Museum in Pardubice, Zámek 2, 530 02 Pardubice, Czech Republic

Qia Wang
Key Laboratory for Plant Diversity and Biogeography of East Asia, Kunming Institute of Botany, Chinese Academy of Sciences, Kunming 650201, China
University of Chinese Academy of Sciences, Beijing 10049, China

Hang Sun
Key Laboratory for Plant Diversity and Biogeography of East Asia, Kunming Institute of Botany, Chinese Academy of Sciences, Kunming 650201, China

Jinling Huang
Department of Biology, East Carolina University, Greenville, NC 27858, USA

Yukun Sun
Department of Biological and Chemical Sciences, Illinois Institute of Technology, Chicago, IL 60616, USA

Jean-Francois Pombert
Department of Biological and Chemical Sciences, Illinois Institute of Technology, Chicago, IL 60616, USA

Paulina Brzeska
Department of Aquatic Ecology, Maritime Institute in Gdańsk, Długi Targ 41/42, 80-830 Gdańsk, Poland

Adam Woźniczka
Research Station in Świnoujście, National Marine Fisheries Research Institute, Plac Słowiański 11, 72-600 Świnoujście, Poland

Mariusz Pełechaty
Department of Hydrobiology, Faculty of Biology, Adam Mickiewicz University in Poznań, Umultowska 89, 61-614 Poznań, Poland

Irmgard Blindow
Biological Station of Hiddensee, University of Greifswald, Biologenweg 15, 18565 Kloster, Germany

Karl J. Niklas
Section of Plant Biology, School of Integrative Plant Science, Cornell University, Ithaca, NY 14853, USA

Edward D. Cobb
Section of Plant Biology, School of Integrative Plant Science, Cornell University, Ithaca, NY 14853, USA

A. Keith Dunker
Center for Computational Biology and Bioinformatics, School of Medicine, Indiana University, Indianapolis, IN 46202, USA

Maria Virginia Sanchez-Puerta
IBAM-CONICET & FCA, FCEN, Universidad Nacional de Cuyo, Chacras de Coria, Mendoza 5505, Argentina

Zofia Szweykowska-Kulińska
Institute of Molecular Biology and Biotechnology, Faculty of Biology, Adam Mickiewicz University, Umultowska 89, 61-614 Poznań, Poland

Zbigniew Mirek
W. Szafer Institute of Botany, Polish Academy of Sciences, Lubicz 46, 31-512 Kraków, Poland

Agnieszka Nikel
W. Szafer Institute of Botany, Polish Academy of Sciences, Lubicz 46, 31-512 Kraków, Poland

Łukasz Wilk
W. Szafer Institute of Botany, Polish Academy of Sciences, Lubicz 46, 31-512 Kraków, Poland

Małgorzata Baranek
Department of Plant Physiology, Institute of Experimental Biology, Faculty of Biology, Adam Mickiewicz University, Umultowska 89, 61-614 Poznań, Poland

Tomasz P. Wyka
Department of General Botany, Institute of Experimental Biology, Faculty of Biology, Adam Mickiewicz University, Umultowska 89, 61-614 Poznań, Poland

Grzegorz Jackowski
Department of Plant Physiology, Institute of Experimental Biology, Faculty of Biology, Adam Mickiewicz University, Umultowska 89, 61-614 Poznań, Poland

Eric D. Salomaki
Department of Biological Sciences, University of Rhode Island, Kingston, RI 02881, USA

Christopher E. Lane
Department of Biological Sciences, University of Rhode Island, Kingston, RI 02881, USA

Jan de Vries
Institute for Molecular Evolution, Heinrich Heine-University, Düsseldorf 40225, Germany

Cessa Rauch
Institute for Molecular Evolution, Heinrich Heine-University, Düsseldorf 40225, Germany

Gregor Christa
Institute for Molecular Evolution, Heinrich Heine-University, Düsseldorf 40225, Germany
CESAM, University of Aveiro, 3810-193 Aveiro, Portugal

Sven B. Gould
Institute for Molecular Evolution, Heinrich Heine-University, Düsseldorf 40225, Germany

Paweł Mackiewicz
Department of Genomics, Faculty of Biotechnology, University of Wrocław, Fryderyka Joliot-Curie 14a, 50-383 Wrocław, Poland

Przemysław Gagat
Department of Genomics, Faculty of Biotechnology, University of Wrocław, Fryderyka Joliot-Curie 14a, 50-383 Wrocław, Poland

Thorsten Thiergart
Institute of Molecular Evolution, Heinrich-Heine University Düsseldorf, Universitätsstrasse 1, 40225 Düsseldorf, Germany

Ulf Schmitz
Institute of Molecular Evolution, Heinrich-Heine University Düsseldorf, Universitätsstrasse 1, 40225 Düsseldorf, Germany

Giddy Landan
Genomic Microbiology Group, Institute of Microbiology, Christian-Albrechts-University of Kiel, Christian-Albrechts-Platz 4, 24118 Kiel, Germany

William F. Martin
Institute of Molecular Evolution, Heinrich-Heine University Düsseldorf, Universitätsstrasse 1, 40225 Düsseldorf, Germany

Tal Dagan
Genomic Microbiology Group, Institute of Microbiology, Christian-Albrechts-University of Kiel, Christian-Albrechts-Platz 4, 24118 Kiel, Germany